The Science of The Big Bang Theory

An Overview of our Universe

Volume 1 of 2

Contents

Chapter 1

The Big Bang Theory

The **Big Bang** theory is the prevailing cosmological model for the universe from the earliest known periods through its subsequent large-scale evolution.[1][2][3] It states that the universe expanded from a very high density state,[4][5] and offers a comprehensive explanation for a broad range of observed phenomena, including the abundance of light elements, the cosmic microwave background, large scale structure, and Hubble's Law.[6] If the known laws of physics are extrapolated beyond where they are valid, there is a singularity. Modern measurements place this moment at approximately 13.8 billion years ago, which is thus considered the age of the universe.[7] After the initial expansion, the universe cooled sufficiently to allow the formation of subatomic particles, and later simple atoms. Giant clouds of these primordial elements later coalesced through gravity to form stars and galaxies.

In the mid-20th century, three British astrophysicists, Stephen Hawking, George F. R. Ellis, and Roger Penrose turned their attention to the theory of relativity and its implications regarding our notions of time. In 1968 and 1970, they published papers in which they extended Einstein's theory of general relativity to include measurements of time and space.[8][9] According to their calculations, time and space had a finite beginning that corresponded to the origin of matter and energy.

Since Georges Lemaître first noted, in 1927, that an expanding universe might be traced back in time to an originating single point, scientists have built on his idea of cosmic expansion. While the scientific community was once divided between supporters of two different expanding universe theories, the Big Bang and the Steady State theory, accumulated empirical evidence provides strong support for the former.[10] In 1929, from analysis of galactic redshifts, Edwin Hubble concluded that galaxies are drifting apart, important observational evidence consistent with the hypothesis of an expanding universe. In 1964, the cosmic microwave background radiation was discovered, which was crucial evidence in favor of the Big Bang model, since that theory predicted the existence of background radiation throughout the universe before it was discovered. More recently, measurements of the redshifts of supernovae indicate that the expansion of the universe is accelerating, an observation attributed to dark energy.[11] The known physical laws of nature can be used to calculate the characteristics of the universe in detail back in time to an initial state of extreme density and temperature.[12][13][14]

1.1   Overview

Hubble observed that the distances to faraway galaxies were strongly correlated with their redshifts. This was interpreted to mean that all distant galaxies and clusters are receding away from our vantage point with an apparent velocity proportional to their distance: that is, the farther they are, the faster they move away from us, regardless of direction.[19] Assuming the Copernican principle (that the Earth is not the center of the universe), the only remaining interpretation is that all observable regions of the universe are receding from all others. Since we know that the distance between galaxies increases today, it must mean that in the past galaxies were closer together. The continuous expansion of the universe implies that

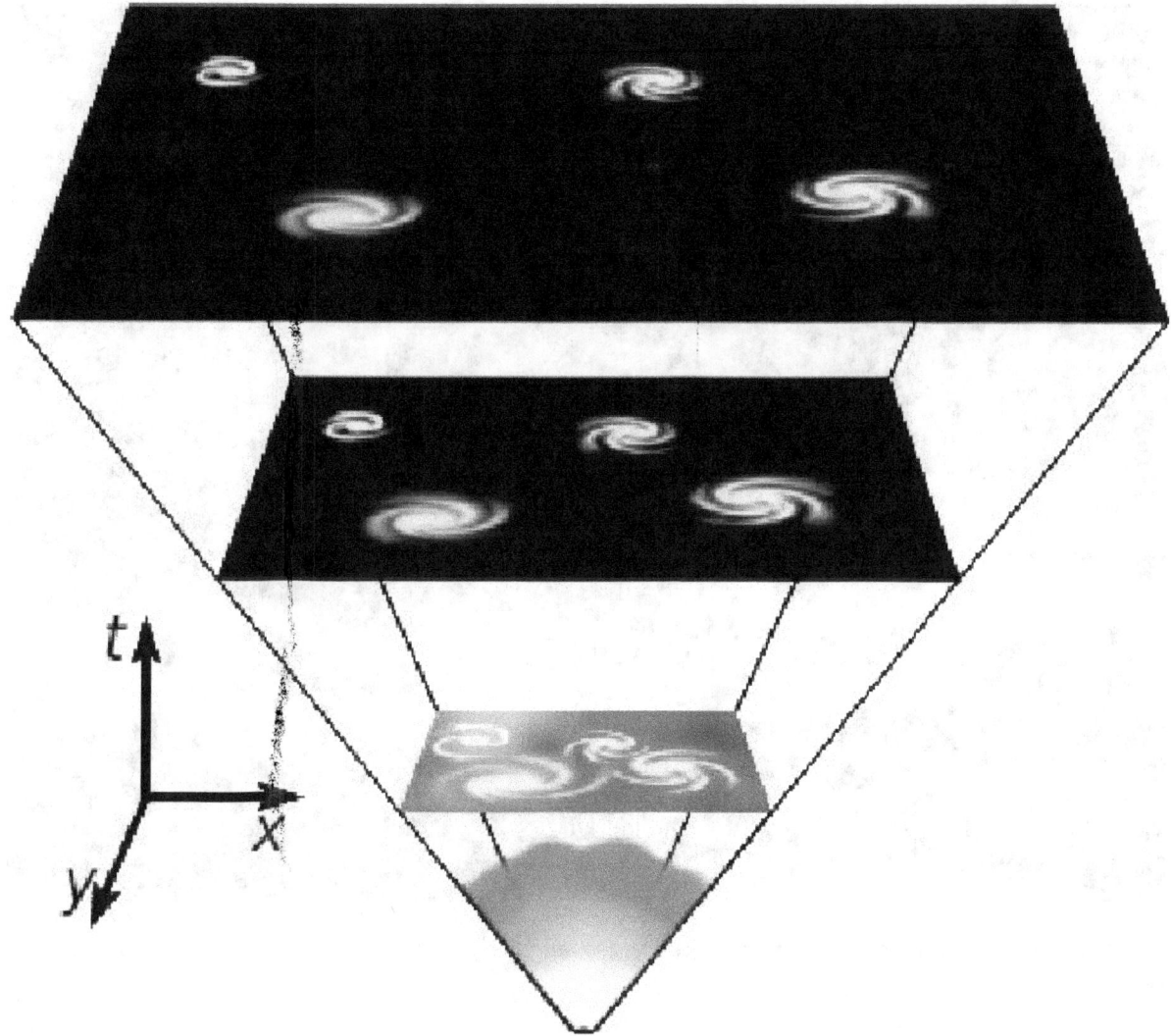

According to the Big Bang model, the universe expanded from an extremely dense and hot state and continues to expand today.

the universe was denser and hotter in the past.

Large particle accelerators can replicate the conditions that prevailed after the early moments of the universe, resulting in confirmation and refinement of the details of the Big Bang model. However, these accelerators can only probe so far into high energy regimes. Consequently, the state of the universe in the earliest instants of the Big Bang expansion is still poorly understood and an area of open investigation and indeed, speculation.

The first subatomic particles included protons, neutrons, and electrons. Though simple atomic nuclei formed within the first three minutes after the Big Bang, thousands of years passed before the first electrically neutral atoms formed. The majority of atoms produced by the Big Bang were hydrogen, along with helium and traces of lithium. Giant clouds of these primordial elements later coalesced through gravity to form stars and galaxies, and the heavier elements were synthesized either within stars or during supernovae.

The Big Bang theory offers a comprehensive explanation for a broad range of observed phenomena, including the abundance of light elements, the cosmic microwave background, large scale structure, and Hubble's Law.[6] The framework for the Big Bang model relies on Albert Einstein's theory of general relativity and on simplifying assumptions such as homogeneity and isotropy of space. The governing equations were formulated by Alexander Friedmann, and similar solutions were worked on by Willem de Sitter. Since then, astrophysicists have incorporated observational and theoretical additions into the Big Bang model, and its parametrization as the Lambda-CDM model serves as the framework for cur-

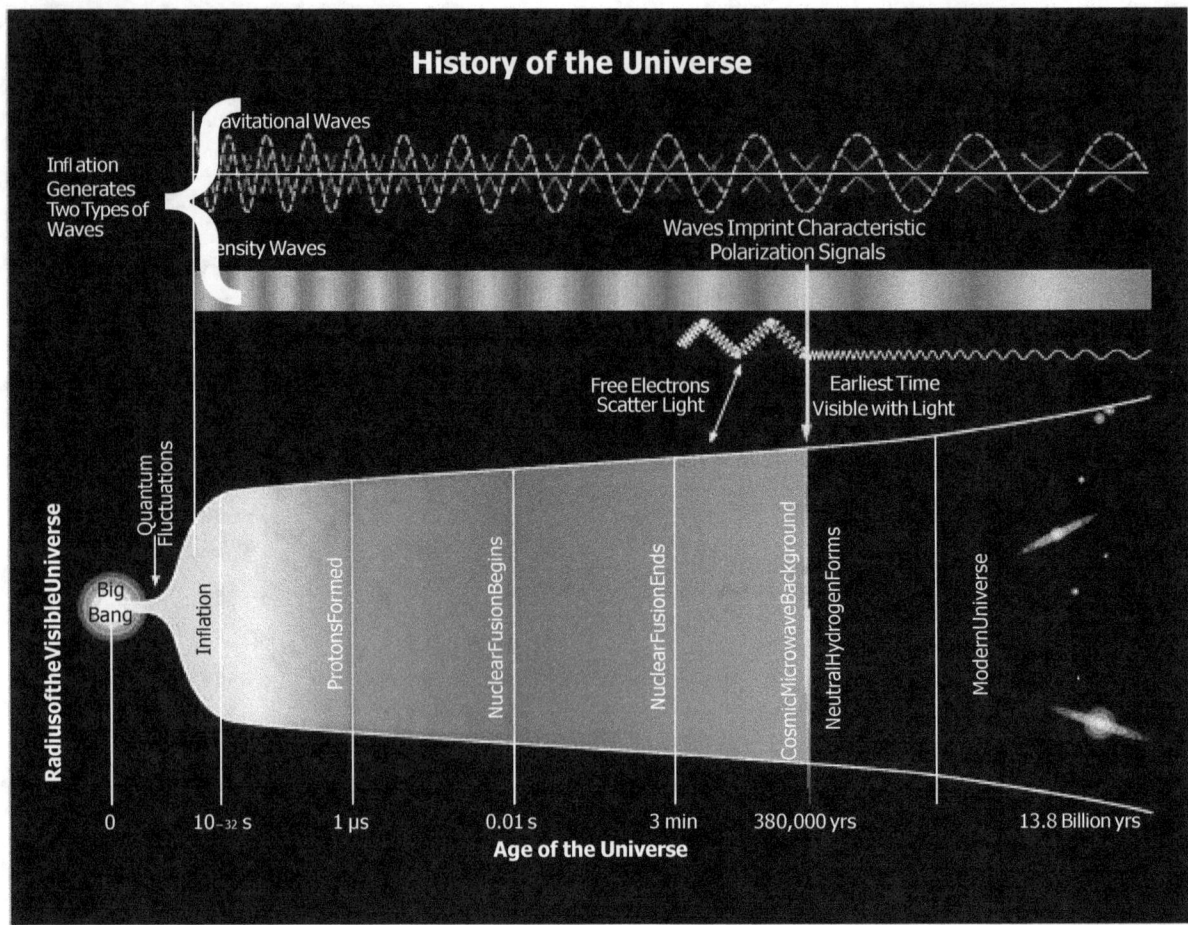

History of the Universe - gravitational waves are hypothesized to arise from cosmic inflation, an expansion just after the Big Bang.[15][16][17][18]

rent investigations of theoretical cosmology. The Lambda-CDM model is the standard model of Big Bang cosmology, the simplest model that provides a reasonably good account of various observations about the universe.

1.2 Timeline of the Big Bang

Main article: Timeline of the Big Bang

1.2.1 Singularity

See also: Gravitational singularity and Planck epoch

Extrapolation of the expansion of the universe backwards in time using general relativity yields an infinite density and temperature at a finite time in the past.[20] This singularity signals the breakdown of general relativity and thus, all the laws of physics. How closely we can extrapolate towards the singularity is debated—certainly no closer than the end of the Planck epoch. This singularity is sometimes called "the Big Bang",[21] but the term can also refer to the early hot, dense phase itself,[22][notes 1] which can be considered the "birth" of our universe. Based on measurements of the expansion using Type Ia supernovae, measurements of temperature fluctuations in the cosmic microwave background, and measurements of the correlation function of galaxies, the universe has an estimated age of 13.798 ± 0.037 billion years.[23] The agreement of these three independent measurements strongly supports the ΛCDM model that describes in detail the contents of the universe.

1.2.2   Inflation and baryogenesis

Main articles: Cosmic inflation and baryogenesis

The earliest phases of the Big Bang are subject to much speculation. In the most common models the universe was filled homogeneously and isotropically with an incredibly high energy density and huge temperatures and pressures and was very rapidly expanding and cooling. Approximately 10^{-37} seconds into the expansion, a phase transition caused a cosmic inflation, during which the universe grew exponentially.[24] After inflation stopped, the universe consisted of a quark–gluon plasma, as well as all other elementary particles.[25] Temperatures were so high that the random motions of particles were at relativistic speeds, and particle–antiparticle pairs of all kinds were being continuously created and destroyed in collisions.[4] At some point an unknown reaction called baryogenesis violated the conservation of baryon number, leading to a very small excess of quarks and leptons over antiquarks and antileptons—of the order of one part in 30 million. This resulted in the predominance of matter over antimatter in the present universe.[26]

1.2.3   Cooling

Main articles: Big Bang nucleosynthesis and cosmic microwave background radiation
 The universe continued to decrease in density and fall in temperature, hence the typical energy of each particle was

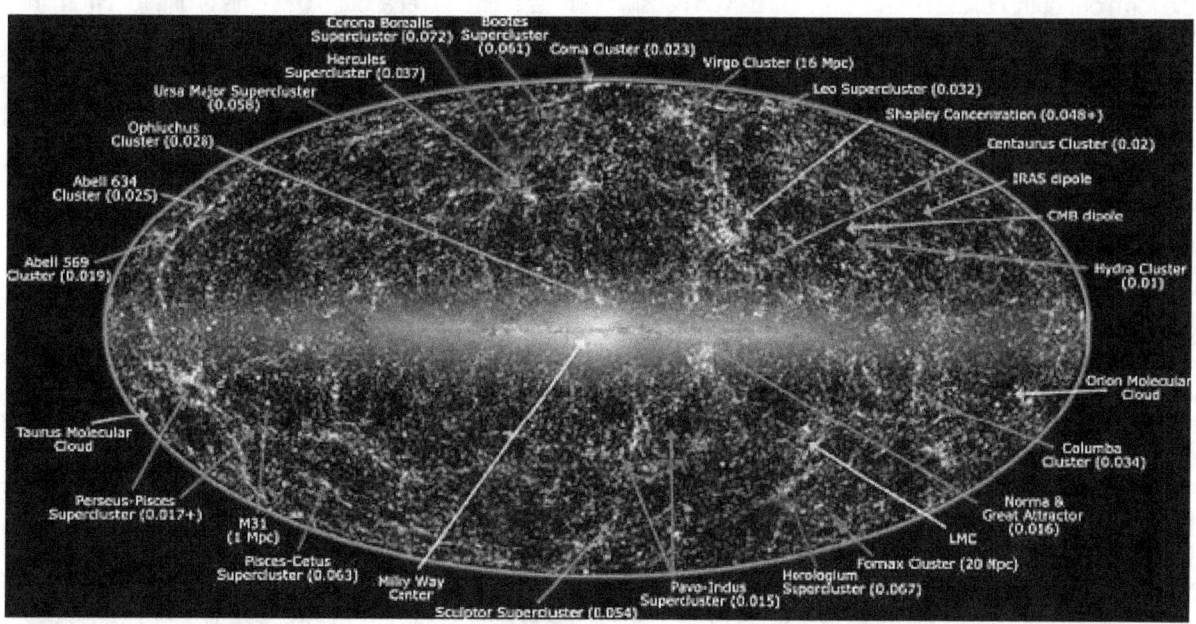

Panoramic view of the entire near-infrared sky reveals the distribution of galaxies beyond the Milky Way. Galaxies are color-coded by redshift.

decreasing. Symmetry breaking phase transitions put the fundamental forces of physics and the parameters of elementary particles into their present form.[27] After about 10^{-11} seconds, the picture becomes less speculative, since particle energies drop to values that can be attained in particle physics experiments. At about 10^{-6} seconds, quarks and gluons combined to form baryons such as protons and neutrons. The small excess of quarks over antiquarks led to a small excess of baryons over antibaryons. The temperature was now no longer high enough to create new proton–antiproton pairs (similarly for neutrons–antineutrons), so a mass annihilation immediately followed, leaving just one in 10^{10} of the original protons and neutrons, and none of their antiparticles. A similar process happened at about 1 second for electrons and positrons. After these annihilations, the remaining protons, neutrons and electrons were no longer moving relativistically and the energy density of the universe was dominated by photons (with a minor contribution from neutrinos).

A few minutes into the expansion, when the temperature was about a billion (one thousand million; 10^9; SI prefix giga-) kelvin and the density was about that of air, neutrons combined with protons to form the universe's deuterium and helium nuclei in a process called Big Bang nucleosynthesis.[28] Most protons remained uncombined as hydrogen nuclei. As the universe cooled, the rest mass energy density of matter came to gravitationally dominate that of the photon radiation. After about 379,000 years the electrons and nuclei combined into atoms (mostly hydrogen); hence the radiation decoupled from matter and continued through space largely unimpeded. This relic radiation is known as the cosmic microwave background radiation.[29] The chemistry of life may have begun shortly after the Big Bang, 13.8 billion years ago, during a habitable epoch when the universe was only 10–17 million years old.[30][31][32]

1.2.4 Structure formation

Main article: Structure formation

Over a long period of time, the slightly denser regions of the nearly uniformly distributed matter gravitationally attracted nearby matter and thus grew even denser, forming gas clouds, stars, galaxies, and the other astronomical structures observable today.[4] The details of this process depend on the amount and type of matter in the universe. The four possible types of matter are known as cold dark matter, warm dark matter, hot dark matter, and baryonic matter. The best measurements available (from WMAP) show that the data is well-fit by a Lambda-CDM model in which dark matter is assumed to be cold (warm dark matter is ruled out by early reionization[33]), and is estimated to make up about 23% of the matter/energy of the universe, while baryonic matter makes up about 4.6%.[34] In an "extended model" which includes hot dark matter in the form of neutrinos, then if the "physical baryon density" Ω h^2 is estimated at about 0.023 (this is different from the 'baryon density' Ω expressed as a fraction of the total matter/energy density, which as noted above is about 0.046), and the corresponding cold dark matter density Ω h^2 is about 0.11, the corresponding neutrino density $\Omega_v h^2$ is estimated to be less than 0.0062.[34]

1.2.5 Cosmic acceleration

Main article: Accelerating universe

Independent lines of evidence from Type Ia supernovae and the CMB imply that the universe today is dominated by a mysterious form of energy known as dark energy, which apparently permeates all of space. The observations suggest 73% of the total energy density of today's universe is in this form. When the universe was very young, it was likely infused with dark energy, but with less space and everything closer together, gravity predominated, and it was slowly braking the expansion. But eventually, after numerous billion years of expansion, the growing abundance of dark energy caused the expansion of the universe to slowly begin to accelerate. Dark energy in its simplest formulation takes the form of the cosmological constant term in Einstein's field equations of general relativity, but its composition and mechanism are unknown and, more generally, the details of its equation of state and relationship with the Standard Model of particle physics continue to be investigated both observationally and theoretically.[11]

All of this cosmic evolution after the inflationary epoch can be rigorously described and modelled by the ΛCDM model of cosmology, which uses the independent frameworks of quantum mechanics and Einstein's General Relativity. There is no well-supported model describing the action prior to 10^{-15} seconds or so. Apparently a new unified theory of quantum gravitation is needed to break this barrier. Understanding this earliest of eras in the history of the universe is currently one of the greatest unsolved problems in physics.

1.3 Underlying assumptions

The Big Bang theory depends on two major assumptions: the universality of physical laws and the cosmological principle. The cosmological principle states that on large scales the universe is homogeneous and isotropic.

Abell 2744 galaxy cluster - Hubble Frontier Fields view.[35]

These ideas were initially taken as postulates, but today there are efforts to test each of them. For example, the first assumption has been tested by observations showing that largest possible deviation of the fine structure constant over much of the age of the universe is of order 10^{-5}.[36] Also, general relativity has passed stringent tests on the scale of the Solar System and binary stars.[notes 2]

If the large-scale universe appears isotropic as viewed from Earth, the cosmological principle can be derived from the simpler Copernican principle, which states that there is no preferred (or special) observer or vantage point. To this end, the cosmological principle has been confirmed to a level of 10^{-5} via observations of the CMB. The universe has been measured to be homogeneous on the largest scales at the 10% level.[37]

1.3.1 Expansion of space

Main articles: Friedmann–Lemaître–Robertson–Walker metric and Metric expansion of space

General relativity describes spacetime by a metric, which determines the distances that separate nearby points. The points, which can be galaxies, stars, or other objects, themselves are specified using a coordinate chart or "grid" that is laid down over all spacetime. The cosmological principle implies that the metric should be homogeneous and isotropic on large scales, which uniquely singles out the Friedmann–Lemaître–Robertson–Walker metric (FLRW metric). This metric contains a scale factor, which describes how the size of the universe changes with time. This enables a convenient choice of a coordinate system to be made, called comoving coordinates. In this coordinate system the grid expands along with the universe, and objects that are moving only due to the expansion of the universe remain at fixed points on the grid. While their *coordinate* distance (comoving distance) remains constant, the *physical* distance between two such comoving points expands proportionally with the scale factor of the universe.[38]

The Big Bang is not an explosion of matter moving outward to fill an empty universe. Instead, space itself expands with time everywhere and increases the physical distance between two comoving points. In other words, the Big Bang is not an explosion *in space*, but rather an expansion *of space*.[4] Because the FLRW metric assumes a uniform distribution of mass and energy, it applies to our universe only on large scales—local concentrations of matter such as our galaxy are gravitationally bound and as such do not experience the large-scale expansion of space.[39]

1.3.2 Horizons

Main article: Cosmological horizon

An important feature of the Big Bang spacetime is the presence of horizons. Since the universe has a finite age, and light travels at a finite speed, there may be events in the past whose light has not had time to reach us. This places a limit or a *past horizon* on the most distant objects that can be observed. Conversely, because space is expanding, and more distant objects are receding ever more quickly, light emitted by us today may never "catch up" to very distant objects. This defines a *future horizon*, which limits the events in the future that we will be able to influence. The presence of either type of horizon depends on the details of the FLRW model that describes our universe. Our understanding of the universe back to very early times suggests that there is a past horizon, though in practice our view is also limited by the opacity of the universe at early times. So our view cannot extend further backward in time, though the horizon recedes in space. If the expansion of the universe continues to accelerate, there is a future horizon as well.[40]

1.4 History

Main article: History of the Big Bang theory
See also: Timeline of cosmology

1.4.1 Etymology

English astronomer Fred Hoyle is credited with coining the term "Big Bang" during a 1949 BBC radio broadcast. It is popularly reported that Hoyle, who favored an alternative "steady state" cosmological model, intended this to be pejorative, but Hoyle explicitly denied this and said it was just a striking image meant to highlight the difference between the two models.[41][42][43]:129

1.4.2   Development

Hubble eXtreme Deep Field (XDF)

XDF size compared to the size of the moon - several thousand galaxies, each consisting of billions of stars, are in this small view.

XDF (2012) view - each light speck is a galaxy - some of these are as old as 13.2 billion years[44] - the universe is estimated to contain 200 billion galaxies.

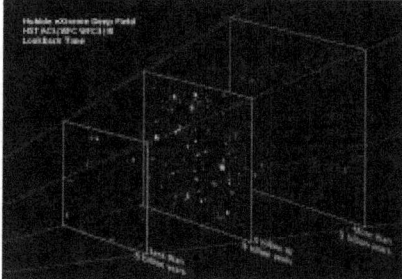

XDF image shows fully mature galaxies in the foreground plane - nearly mature galaxies from 5 to 9 billion years ago - protogalaxies, blazing with young stars, beyond 9 billion years.

The Big Bang theory developed from observations of the structure of the universe and from theoretical considerations. In 1912 Vesto Slipher measured the first Doppler shift of a "spiral nebula" (spiral nebula is the obsolete term for spiral galaxies), and soon discovered that almost all such nebulae were receding from Earth. He did not grasp the cosmological implications of this fact, and indeed at the time it was highly controversial whether or not these nebulae were "island universes" outside our Milky Way.[45][46] Ten years later, Alexander Friedmann, a Russian cosmologist and mathematician, derived the Friedmann equations from Albert Einstein's equations of general relativity, showing that the universe might be expanding in contrast to the static universe model advocated by Einstein at that time.[47] In 1924 Edwin Hubble's measurement of the great distance to the nearest spiral nebulae showed that these systems were indeed other galaxies. Independently deriving Friedmann's equations in 1927, Georges Lemaître, a Belgian physicist and Roman Catholic priest, proposed that the inferred recession of the nebulae was due to the expansion of the universe.[48]

In 1931 Lemaître went further and suggested that the evident expansion of the universe, if projected back in time, meant that the further in the past the smaller the universe was, until at some finite time in the past all the mass of the universe was concentrated into a single point, a "primeval atom" where and when the fabric of time and space came into existence.[49]

Starting in 1924, Hubble painstakingly developed a series of distance indicators, the forerunner of the cosmic distance ladder, using the 100-inch (2,500 mm) Hooker telescope at Mount Wilson Observatory. This allowed him to estimate distances to galaxies whose redshifts had already been measured, mostly by Slipher. In 1929 Hubble discovered a correlation between distance and recession velocity—now known as Hubble's law.[19][50] Lemaître had already shown that this was expected, given the cosmological principle.[11]

In the 1920s and 1930s almost every major cosmologist preferred an eternal steady state universe, and several complained that the beginning of time implied by the Big Bang imported religious concepts into physics; this objection was later repeated by supporters of the steady state theory.[51] This perception was enhanced by the fact that the originator of the Big Bang theory, Monsignor Georges Lemaître, was a Roman Catholic priest.[52] Arthur Eddington agreed with Aristotle that the universe did not have a beginning in time, viz., that matter is eternal. A beginning in time was "repugnant" to him.[53][54] Lemaître, however, thought that

> If the world has begun with a single quantum, the notions of space and time would altogether fail to have any meaning at the beginning; they would only begin to have a sensible meaning when the original quantum had been divided into a sufficient number of quanta. If this suggestion is correct, the beginning of the world happened a little before the beginning of space and time.[55]

During the 1930s other ideas were proposed as non-standard cosmologies to explain Hubble's observations, including the Milne model,[56] the oscillatory universe (originally suggested by Friedmann, but advocated by Albert Einstein and Richard Tolman)[57] and Fritz Zwicky's tired light hypothesis.[58]

After World War II, two distinct possibilities emerged. One was Fred Hoyle's steady state model, whereby new matter would be created as the universe seemed to expand. In this model the universe is roughly the same at any point in time.[59] The other was Lemaître's Big Bang theory, advocated and developed by George Gamow, who introduced big bang nucleosynthesis (BBN)[60] and whose associates, Ralph Alpher and Robert Herman, predicted the cosmic microwave background radiation (CMB).[61] Ironically, it was Hoyle who coined the phrase that came to be applied to Lemaître's theory, referring to it as "this *big bang* idea" during a BBC Radio broadcast in March 1949.[43]:129[notes 3] For a while, support was split between these two theories. Eventually, the observational evidence, most notably from radio source counts, began to favor Big Bang over Steady State. The discovery and confirmation of the cosmic microwave background radiation in 1964[63] secured the Big Bang as the best theory of the origin and evolution of the universe. Much of the current work in cosmology includes understanding how galaxies form in the context of the Big Bang, understanding the physics of the universe at earlier and earlier times, and reconciling observations with the basic theory.

Significant progress in Big Bang cosmology have been made since the late 1990s as a result of advances in telescope technology as well as the analysis of data from satellites such as COBE,[64] the Hubble Space Telescope and WMAP.[65] Cosmologists now have fairly precise and accurate measurements of many of the parameters of the Big Bang model, and have made the unexpected discovery that the expansion of the universe appears to be accelerating.

1.5 Observational evidence

"[The] big bang picture is too firmly grounded in data from every area to be proved invalid in its general features."
Lawrence Krauss[66]

The earliest and most direct observational evidence of the validity of the theory are the expansion of the universe according to Hubble's law (as indicated by the redshifts of galaxies), discovery and measurement of the cosmic microwave background and the relative abundances of light elements produced by Big Bang nucleosynthesis. More recent evidence includes observations of galaxy formation and evolution, and the distribution of large-scale cosmic structures,[67] These are sometimes called the "four pillars" of the Big Bang theory.[68]

Artist's depiction of the WMAP satellite gathering data to help scientists understand the Big Bang

Precise modern models of the Big Bang appeal to various exotic physical phenomena that have not been observed in terrestrial laboratory experiments or incorporated into the Standard Model of particle physics. Of these features, dark matter is currently subjected to the most active laboratory investigations.[69] Remaining issues include the cuspy halo problem and the dwarf galaxy problem of cold dark matter. Dark energy is also an area of intense interest for scientists, but it is not clear whether direct detection of dark energy will be possible.[70] Inflation and baryogenesis remain more speculative features of current Big Bang models. Viable, quantitative explanations for such phenomena are still being sought. These are currently unsolved problems in physics.

1.5.1   Hubble's law and the expansion of space

Main articles: Hubble's law and Metric expansion of space
See also: Distance measures (cosmology) and Scale factor (universe)

Observations of distant galaxies and quasars show that these objects are redshifted—the light emitted from them has been shifted to longer wavelengths. This can be seen by taking a frequency spectrum of an object and matching the spectroscopic pattern of emission lines or absorption lines corresponding to atoms of the chemical elements interacting with the light. These redshifts are uniformly isotropic, distributed evenly among the observed objects in all directions. If the redshift is interpreted as a Doppler shift, the recessional velocity of the object can be calculated. For some galaxies, it is possible to estimate distances via the cosmic distance ladder. When the recessional velocities are plotted against these distances, a linear relationship known as Hubble's law is observed:[19]

$$v = H_0 D,$$

where

- v is the recessional velocity of the galaxy or other distant object,

- D is the comoving distance to the object, and

- H_0 is Hubble's constant, measured to be 70.4+1.3
  −1.4 km/s/Mpc by the WMAP probe.[34]

Hubble's law has two possible explanations. Either we are at the center of an explosion of galaxies—which is untenable given the Copernican principle—or the universe is uniformly expanding everywhere. This universal expansion was predicted from general relativity by Alexander Friedmann in 1922[47] and Georges Lemaître in 1927,[48] well before Hubble made his 1929 analysis and observations, and it remains the cornerstone of the Big Bang theory as developed by Friedmann, Lemaître, Robertson, and Walker.

The theory requires the relation $v = HD$ to hold at all times, where D is the comoving distance, v is the recessional velocity, and v, H, and D vary as the universe expands (hence we write H_0 to denote the present-day Hubble "constant"). For distances much smaller than the size of the observable universe, the Hubble redshift can be thought of as the Doppler shift corresponding to the recession velocity v. However, the redshift is not a true Doppler shift, but rather the result of the expansion of the universe between the time the light was emitted and the time that it was detected.[71]

That space is undergoing metric expansion is shown by direct observational evidence of the Cosmological principle and the Copernican principle, which together with Hubble's law have no other explanation. Astronomical redshifts are extremely isotropic and homogeneous,[19] supporting the Cosmological principle that the universe looks the same in all directions, along with much other evidence. If the redshifts were the result of an explosion from a center distant from us, they would not be so similar in different directions.

Measurements of the effects of the cosmic microwave background radiation on the dynamics of distant astrophysical systems in 2000 proved the Copernican principle, that, on a cosmological scale, the Earth is not in a central position.[72] Radiation from the Big Bang was demonstrably warmer at earlier times throughout the universe. Uniform cooling of the cosmic microwave background over billions of years is explainable only if the universe is experiencing a metric expansion, and excludes the possibility that we are near the unique center of an explosion.

1.5.2   Cosmic microwave background radiation

Main article: Cosmic microwave background radiation
In 1964 Arno Penzias and Robert Wilson serendipitously discovered the cosmic background radiation, an omnidirectional signal in the microwave band.[63] Their discovery provided substantial confirmation of the general CMB predictions: the radiation was found to be consistent with an almost perfect black body spectrum in all directions; this spectrum has been redshifted by the expansion of the universe, and today corresponds to approximately 2.725 K. This tipped the balance of evidence in favor of the Big Bang model, and Penzias and Wilson were awarded a Nobel Prize in 1978.

The *surface of last scattering* corresponding to emission of the CMB occurs shortly after *recombination*, the epoch when neutral hydrogen becomes stable. Prior to this, the universe comprised a hot dense photon-baryon plasma sea where photons were quickly scattered from free charged particles. Peaking at around 372±14 kyr,[33] the mean free path for a photon becomes long enough to reach the present day and the universe becomes transparent.

In 1989 NASA launched the Cosmic Background Explorer satellite (COBE). Its findings were consistent with predictions regarding the CMB, finding a residual temperature of 2.726 K (more recent measurements have revised this figure down slightly to 2.725 K) and providing the first evidence for fluctuations (anisotropies) in the CMB, at a level of about one part in 10^5.[64] John C. Mather and George Smoot were awarded the Nobel Prize for their leadership in this work. During the following decade, CMB anisotropies were further investigated by a large number of ground-based and balloon experiments. In 2000–2001 several experiments, most notably BOOMERanG, found the shape of the universe to be spatially almost flat by measuring the typical angular size (the size on the sky) of the anisotropies.[77][78][79]

In early 2003 the first results of the Wilkinson Microwave Anisotropy Probe (WMAP) were released, yielding what were at the time the most accurate values for some of the cosmological parameters. The results disproved several specific cosmic inflation models, but are consistent with the inflation theory in general.[65] The Planck space probe was launched in May 2009. Other ground and balloon based cosmic microwave background experiments are ongoing.

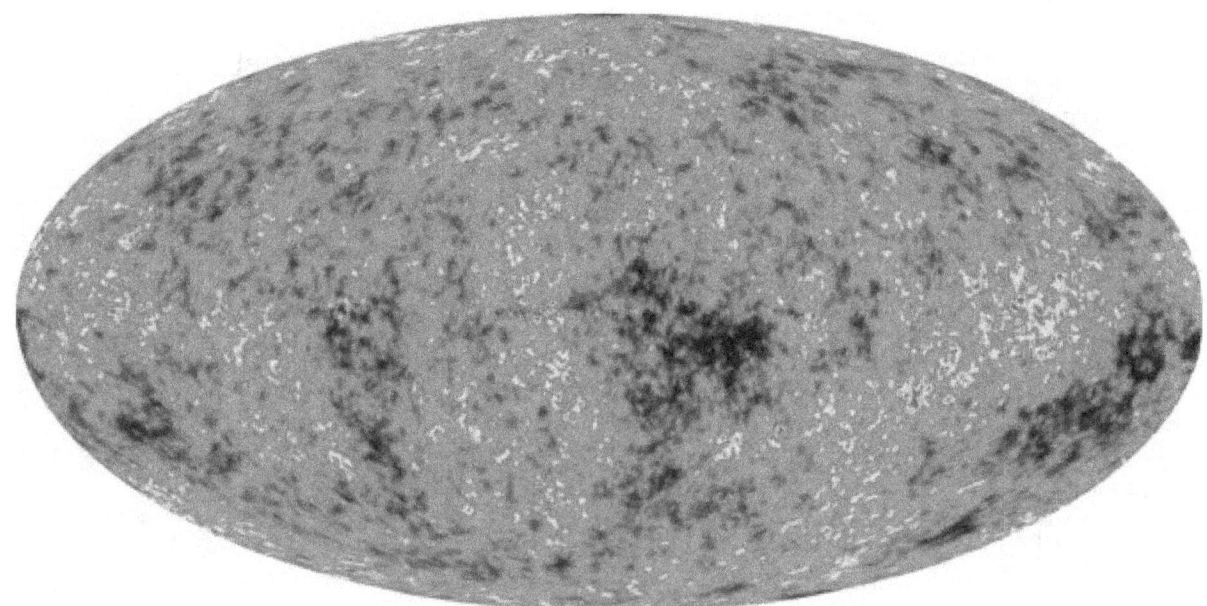

9 year WMAP image of the cosmic microwave background radiation (2012).[73][74] The radiation is isotropic to roughly one part in 100,000.[75]

1.5.3   Abundance of primordial elements

Main article: Big Bang nucleosynthesis

Using the Big Bang model it is possible to calculate the concentration of helium-4, helium-3, deuterium, and lithium-7 in the universe as ratios to the amount of ordinary hydrogen.[28] The relative abundances depend on a single parameter, the ratio of photons to baryons. This value can be calculated independently from the detailed structure of CMB fluctuations. The ratios predicted (by mass, not by number) are about 0.25 for 4He/H, about 10^{-3} for 2H/H, about 10^{-4} for 3He/H and about 10^{-9} for 7Li/H.[28]

The measured abundances all agree at least roughly with those predicted from a single value of the baryon-to-photon ratio. The agreement is excellent for deuterium, close but formally discrepant for 4He, and off by a factor of two for 7Li; in the latter two cases there are substantial systematic uncertainties. Nonetheless, the general consistency with abundances predicted by Big Bang nucleosynthesis is strong evidence for the Big Bang, as the theory is the only known explanation for the relative abundances of light elements, and it is virtually impossible to "tune" the Big Bang to produce much more or less than 20–30% helium.[80] Indeed, there is no obvious reason outside of the Big Bang that, for example, the young universe (i.e., before star formation, as determined by studying matter supposedly free of stellar nucleosynthesis products) should have more helium than deuterium or more deuterium than 3He, and in constant ratios, too.[81]:182–185

1.5.4   Galactic evolution and distribution

Main articles: Galaxy formation and evolution and Structure formation

Detailed observations of the morphology and distribution of galaxies and quasars are in agreement with the current state of the Big Bang theory. A combination of observations and theory suggest that the first quasars and galaxies formed about a billion years after the Big Bang, and since then larger structures have been forming, such as galaxy clusters and superclusters. Populations of stars have been aging and evolving, so that distant galaxies (which are observed as they were in the early universe) appear very different from nearby galaxies (observed in a more recent state). Moreover, galaxies that formed relatively recently appear markedly different from galaxies formed at similar distances but shortly after the Big Bang. These observations are strong arguments against the steady-state model. Observations of star formation, galaxy

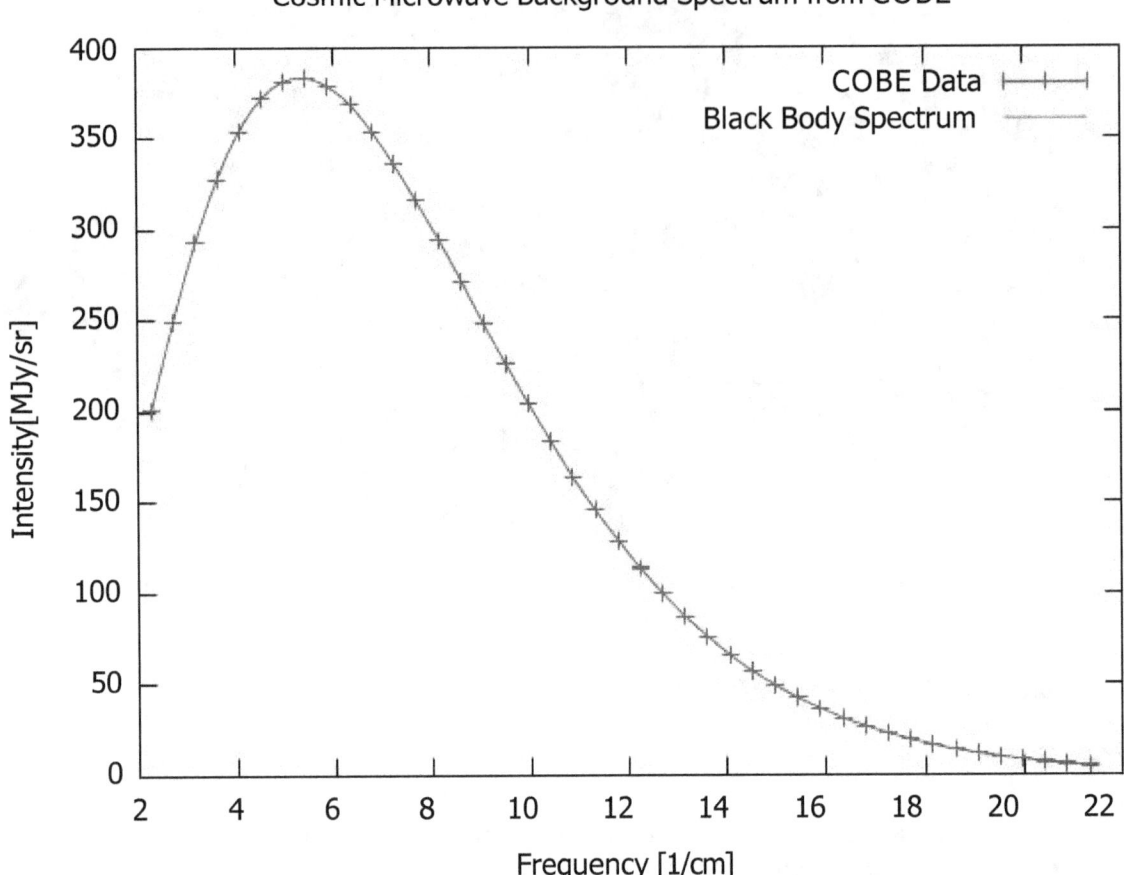

The cosmic microwave background spectrum measured by the FIRAS instrument on the COBE satellite is the most-precisely measured black body spectrum in nature.[76] The data points and error bars on this graph are obscured by the theoretical curve.

and quasar distributions and larger structures agree well with Big Bang simulations of the formation of structure in the universe and are helping to complete details of the theory.[82][83]

1.5.5 Primordial gas clouds

In 2011 astronomers found what they believe to be pristine clouds of primordial gas, by analyzing absorption lines in the spectra of distant quasars. Before this discovery, all other astronomical objects have been observed to contain heavy elements that are formed in stars. These two clouds of gas contain no elements heavier than hydrogen and deuterium.[84][85] Since the clouds of gas have no heavy elements, they likely formed in the first few minutes after the Big Bang, during Big Bang nucleosynthesis. Their composition matches the composition predicted from Big Bang nucleosynthesis. This provides direct evidence that there was a period in the history of the universe before the formation of the first stars, when most ordinary matter existed in the form of clouds of neutral hydrogen.

1.5.6 Other lines of evidence

The age of the universe as estimated from the Hubble expansion and the CMB is now in good agreement with other estimates using the ages of the oldest stars, both as measured by applying the theory of stellar evolution to globular clusters and through radiometric dating of individual Population II stars.[86]

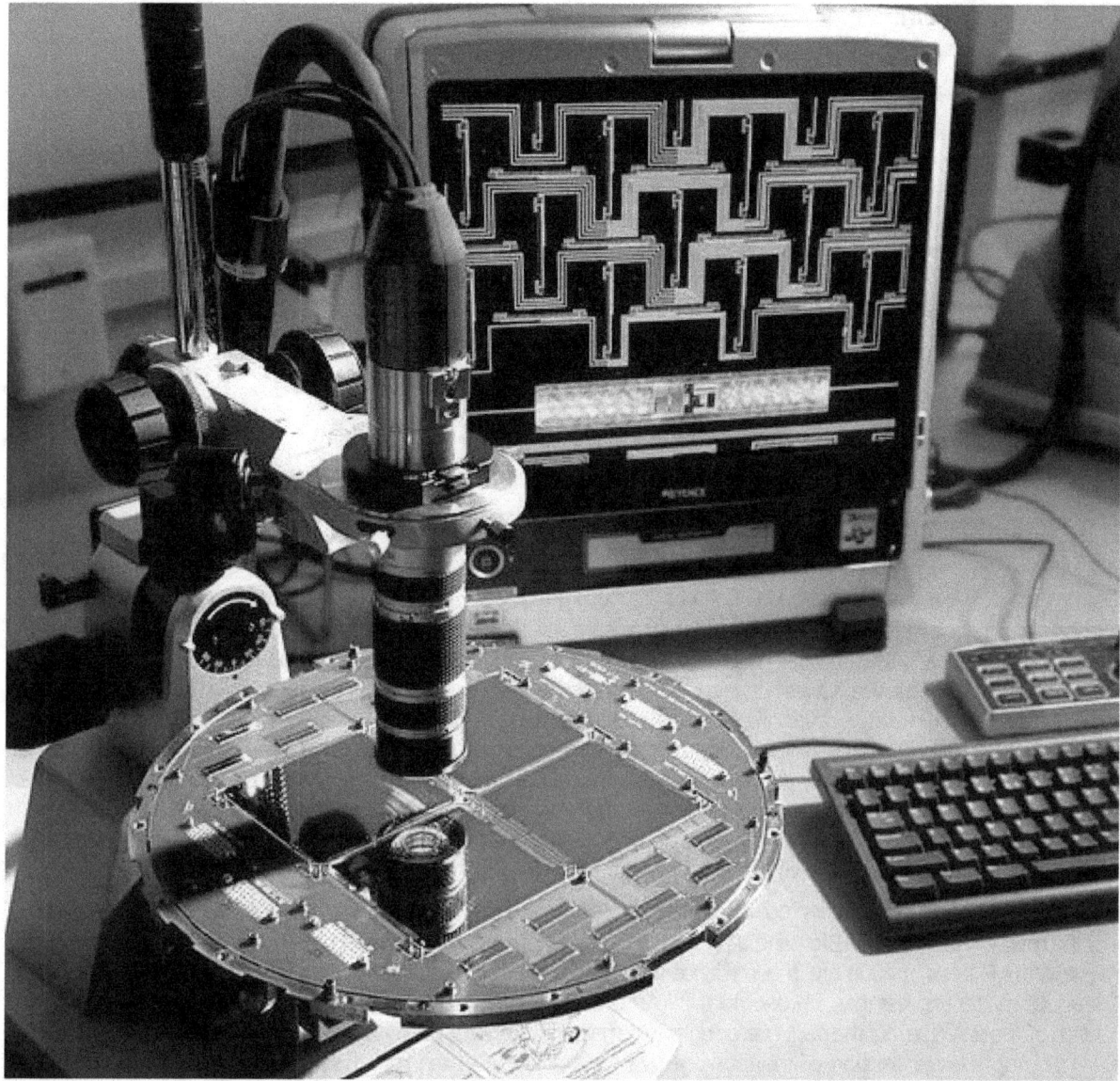

Focal plane of BICEP2 telescope under a microscope - may have detected gravitational waves from the infant universe.[15][16][17][18]

The prediction that the CMB temperature was higher in the past has been experimentally supported by observations of very low temperature absorption lines in gas clouds at high redshift.[87] This prediction also implies that the amplitude of the Sunyaev–Zel'dovich effect in clusters of galaxies does not depend directly on redshift. Observations have found this to be roughly true, but this effect depends on cluster properties that do change with cosmic time, making precise measurements difficult.[88][89]

On 17 March 2014, astronomers at the Harvard-Smithsonian Center for Astrophysics announced the apparent detection of primordial gravitational waves, which, if confirmed, may provide strong evidence for inflation and the Big Bang.[15][16][17][18] However, on 19 June 2014, lowered confidence in confirming the findings was reported;[90][91][92] and on 19 September 2014, even more lowered confidence.[93][94]

1.6   Related issues in physics

1.6.1 Baryon asymmetry

Main article: Baryon asymmetry

It is not yet understood why the universe has more matter than antimatter.[95] It is generally assumed that when the universe was young and very hot, it was in statistical equilibrium and contained equal numbers of baryons and antibaryons. However, observations suggest that the universe, including its most distant parts, is made almost entirely of matter. A process called baryogenesis was hypothesized to account for the asymmetry. For baryogenesis to occur, the Sakharov conditions must be satisfied. These require that baryon number is not conserved, that C-symmetry and CP-symmetry are violated and that the universe depart from thermodynamic equilibrium.[96] All these conditions occur in the Standard Model, but the effect is not strong enough to explain the present baryon asymmetry.

1.6.2 Dark energy

Main article: Dark energy

Measurements of the redshift–magnitude relation for type Ia supernovae indicate that the expansion of the universe has been accelerating since the universe was about half its present age. To explain this acceleration, general relativity requires that much of the energy in the universe consists of a component with large negative pressure, dubbed "dark energy".[11] Dark energy, though speculative, solves numerous problems. Measurements of the cosmic microwave background indicate that the universe is very nearly spatially flat, and therefore according to general relativity the universe must have almost exactly the critical density of mass/energy. But the mass density of the universe can be measured from its gravitational clustering, and is found to have only about 30% of the critical density.[11] Since theory suggests that dark energy does not cluster in the usual way it is the best explanation for the "missing" energy density. Dark energy also helps to explain two geometrical measures of the overall curvature of the universe, one using the frequency of gravitational lenses, and the other using the characteristic pattern of the large-scale structure as a cosmic ruler.

Negative pressure is believed to be a property of vacuum energy, but the exact nature and existence of dark energy remains one of the great mysteries of the Big Bang. Possible candidates include a cosmological constant and quintessence. Results from the WMAP team in 2008 are in accordance with a universe that consists of 73% dark energy, 23% dark matter, 4.6% regular matter and less than 1% neutrinos.[34] According to theory, the energy density in matter decreases with the expansion of the universe, but the dark energy density remains constant (or nearly so) as the universe expands. Therefore, matter made up a larger fraction of the total energy of the universe in the past than it does today, but its fractional contribution will fall in the far future as dark energy becomes even more dominant.

1.6.3 Dark matter

Main article: Dark matter

During the 1970s and 80s, various observations showed that there is not sufficient visible matter in the universe to account for the apparent strength of gravitational forces within and between galaxies. This led to the idea that up to 90% of the matter in the universe is dark matter that does not emit light or interact with normal baryonic matter. In addition, the assumption that the universe is mostly normal matter led to predictions that were strongly inconsistent with observations. In particular, the universe today is far more lumpy and contains far less deuterium than can be accounted for without dark matter. While dark matter has always been controversial, it is inferred by various observations: the anisotropies in the CMB, galaxy cluster velocity dispersions, large-scale structure distributions, gravitational lensing studies, and X-ray measurements of galaxy clusters.[97]

Indirect evidence for dark matter comes from its gravitational influence on other matter, as no dark matter particles have been observed in laboratories. Many particle physics candidates for dark matter have been proposed, and several projects to detect them directly are underway.[98]

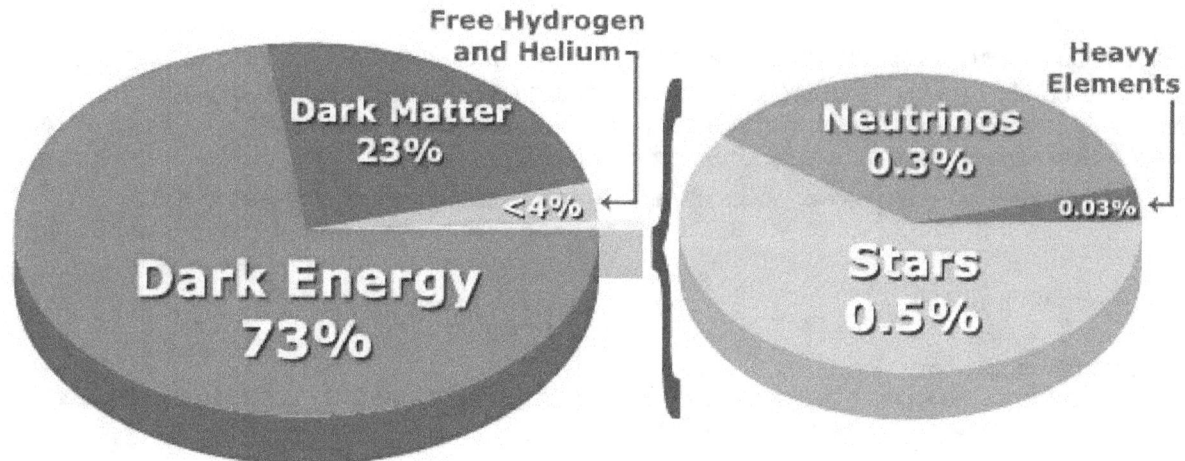

Chart shows the proportion of different components of the universe – about 95% is dark matter and dark energy.

1.6.4 Globular cluster age

Main article: Cosmic age problem

In the mid-1990s observations of globular clusters appeared to be inconsistent with the Big Bang theory. Computer simulations that matched the observations of the stellar populations of globular clusters suggested that they were about 15 billion years old, which conflicted with the 13.8 billion year age of the universe. This issue was partially resolved in the late 1990s when new computer simulations, which included the effects of mass loss due to stellar winds, indicated a much younger age for globular clusters.[99] There remain some questions as to how accurately the ages of the clusters are measured, but it is clear that observations of globular clusters no longer appear inconsistent with the Big Bang theory.

1.7 Problems

There are generally considered to be three outstanding problems with the Big Bang theory: the horizon problem, the flatness problem, and the magnetic monopole problem. The most common answer to these problems is inflationary theory; however, since this creates new problems, other options have been proposed, such as the Weyl curvature hypothesis.[100][101]

1.7.1 Horizon problem

Main article: Horizon problem

The horizon problem results from the premise that information cannot travel faster than light. In a universe of finite age this sets a limit—the particle horizon—on the separation of any two regions of space that are in causal contact.[102] The observed isotropy of the CMB is problematic in this regard: if the universe had been dominated by radiation or matter at all times up to the epoch of last scattering, the particle horizon at that time would correspond to about 2 degrees on the sky. There would then be no mechanism to cause wider regions to have the same temperature.[81]:191–202

A resolution to this apparent inconsistency is offered by inflationary theory in which a homogeneous and isotropic scalar energy field dominates the universe at some very early period (before baryogenesis). During inflation, the universe undergoes exponential expansion, and the particle horizon expands much more rapidly than previously assumed, so that regions presently on opposite sides of the observable universe are well inside each other's particle horizon. The observed isotropy of the CMB then follows from the fact that this larger region was in causal contact before the beginning of inflation.[24]:180–186

Heisenberg's uncertainty principle predicts that during the inflationary phase there would be quantum thermal fluctuations, which would be magnified to cosmic scale. These fluctuations serve as the seeds of all current structure in the universe.[81]:207 Inflation predicts that the primordial fluctuations are nearly scale invariant and Gaussian, which has been accurately confirmed by measurements of the CMB.[103]:sec 6

If inflation occurred, exponential expansion would push large regions of space well beyond our observable horizon.[24]:180–186

1.7.2 Flatness problem

Main article: Flatness problem

The flatness problem (also known as the oldness problem) is an observational problem associated with a Friedmann–

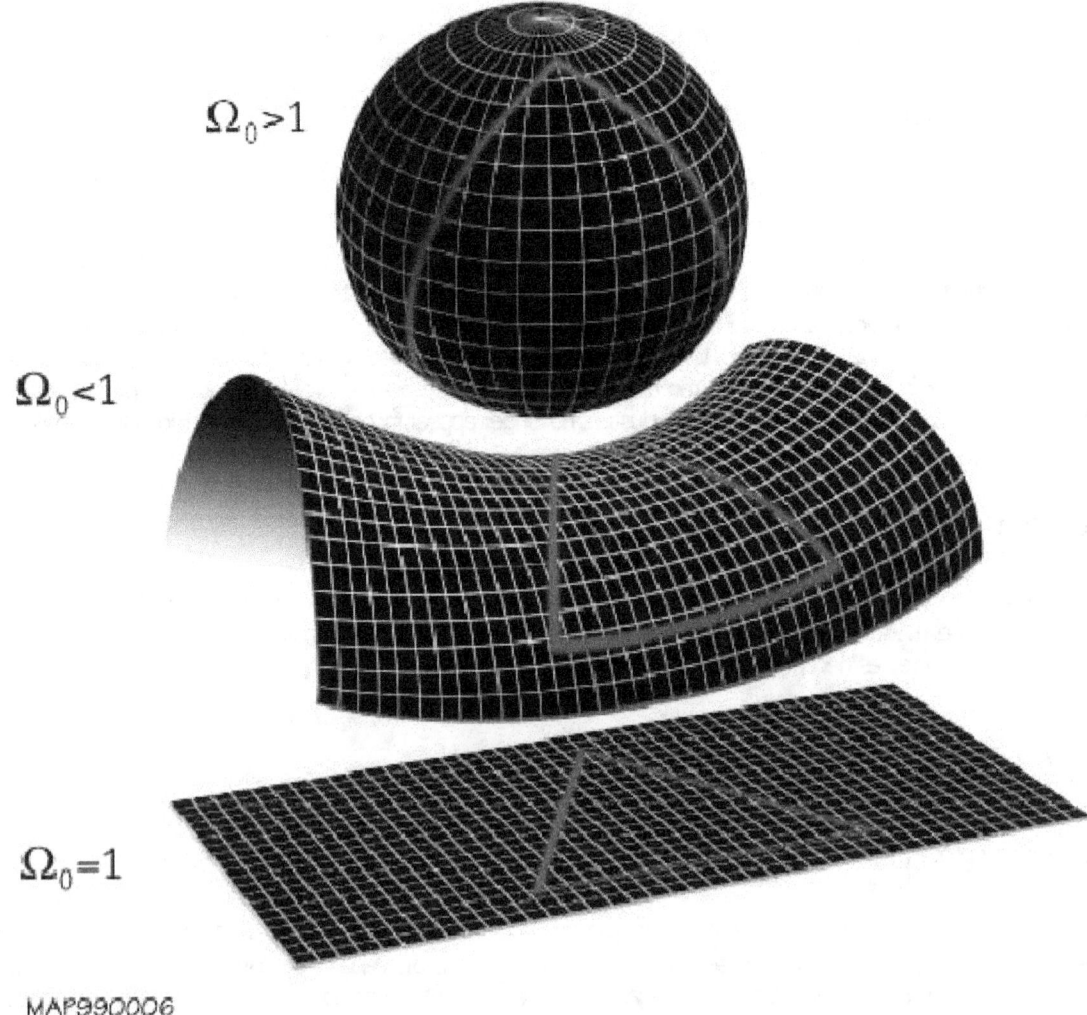

The overall geometry of the universe is determined by whether the Omega cosmological parameter is less than, equal to or greater than 1. Shown from top to bottom are a closed universe with positive curvature, a hyperbolic universe with negative curvature and a flat universe with zero curvature.

Lemaître–Robertson–Walker metric.[102] The universe may have positive, negative, or zero spatial curvature depending on its total energy density. Curvature is negative if its density is less than the critical density, positive if greater; and zero at the critical density, in which case space is said to be *flat*. The problem is that any small departure from the critical density grows with time, and yet the universe today remains very close to flat.[notes 4] Given that a natural timescale for departure from flatness might be the Planck time, 10^{-43} seconds,[4] the fact that the universe has reached neither a heat death nor a Big Crunch after billions of years requires an explanation. For instance, even at the relatively late age of a few minutes (the time of nucleosynthesis), the universe density must have been within one part in 10^{14} of its critical value, or it would not exist as it does today.[104]

1.7.3 Magnetic monopoles

Main article: Magnetic monopole

The magnetic monopole objection was raised in the late 1970s. Grand unification theories predicted topological defects in space that would manifest as magnetic monopoles. These objects would be produced efficiently in the hot early universe, resulting in a density much higher than is consistent with observations, given that no monopoles have been found. This problem is also resolved by cosmic inflation, which removes all point defects from the observable universe, in the same way that it drives the geometry to flatness.[102]

1.8 The future according to the Big Bang theory

Main article: Ultimate fate of the universe

Before observations of dark energy, cosmologists considered two scenarios for the future of the universe. If the mass density of the universe were greater than the critical density, then the universe would reach a maximum size and then begin to collapse. It would become denser and hotter again, ending with a state similar to that in which it started—a Big Crunch.[40] Alternatively, if the density in the universe were equal to or below the critical density, the expansion would slow down but never stop. Star formation would cease with the consumption of interstellar gas in each galaxy; stars would burn out leaving white dwarfs, neutron stars, and black holes. Very gradually, collisions between these would result in mass accumulating into larger and larger black holes. The average temperature of the universe would asymptotically approach absolute zero—a Big Freeze.[105] Moreover, if the proton were unstable, then baryonic matter would disappear, leaving only radiation and black holes. Eventually, black holes would evaporate by emitting Hawking radiation. The entropy of the universe would increase to the point where no organized form of energy could be extracted from it, a scenario known as heat death.[106]:sec VI.D

Modern observations of accelerating expansion imply that more and more of the currently visible universe will pass beyond our event horizon and out of contact with us. The eventual result is not known. The ΛCDM model of the universe contains dark energy in the form of a cosmological constant. This theory suggests that only gravitationally bound systems, such as galaxies, will remain together, and they too will be subject to heat death as the universe expands and cools. Other explanations of dark energy, called phantom energy theories, suggest that ultimately galaxy clusters, stars, planets, atoms, nuclei, and matter itself will be torn apart by the ever-increasing expansion in a so-called Big Rip.[107]

1.9 Before: Speculative physics beyond the Big Bang theory

Main article: cosmogony
See also: Planck epoch

While the Big Bang model is well established in cosmology, it is likely to be refined. The Big Bang Theory, built upon the equations of classical general relativity, indicates a singularity at the origin of cosmic time; this infinite energy density is regarded as impossible in physics. Still, it is known that the equations are not applicable before the time when the universe cooled down to the Planck temperature, and this conclusion depends on various assumptions, of which some could never be experimentally verified.

One proposed refinement to avoid this would-be singularity is to develop a correct treatment of quantum gravity.[108]

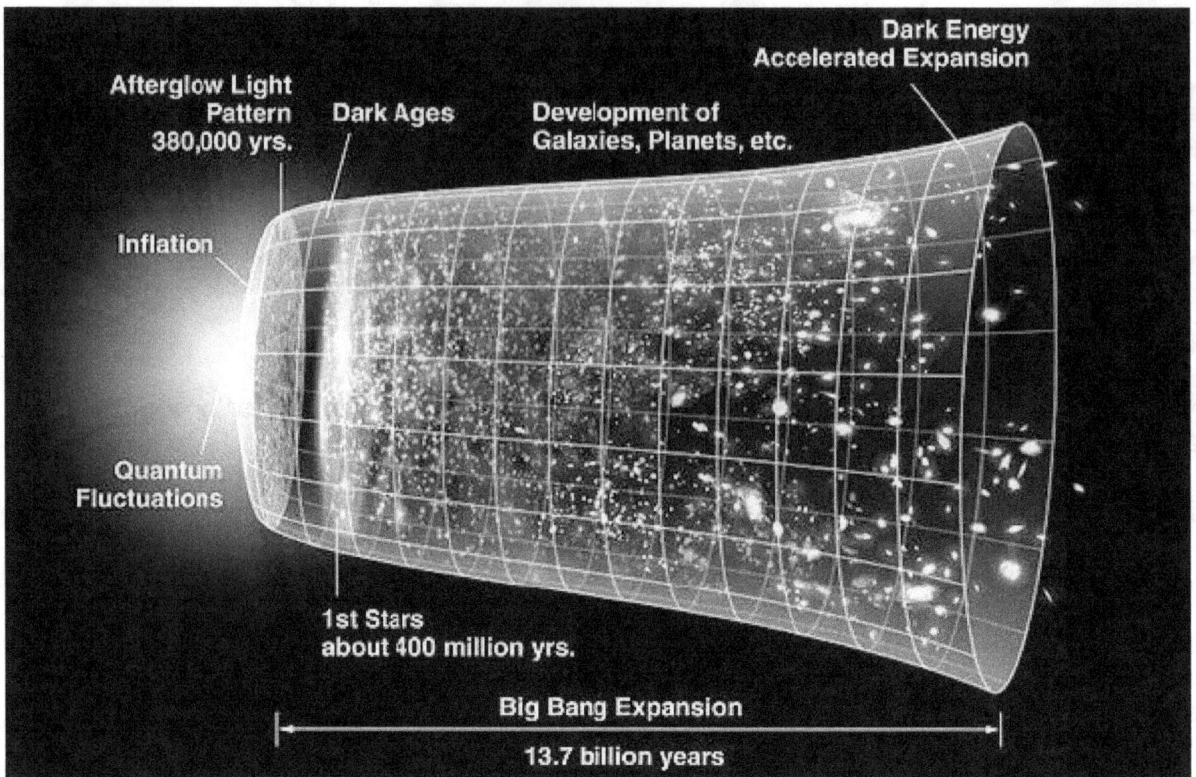

This is an artist's concept of the metric expansion of space, where space (including hypothetical non-observable portions of the universe) is represented at each time by the circular sections. Note on the left the dramatic expansion (not to scale) occurring in the inflationary epoch, and at the center the expansion acceleration. The scheme is decorated with WMAP images on the left and with the representation of stars at the appropriate level of development.

It is not known what could have preceded the hot dense state of the early universe or how and why it originated, though speculation abounds in the field of cosmogony.

Some proposals, each of which entails untested hypotheses, are:

- Models including the Hartle–Hawking no-boundary condition, in which the whole of space-time is finite; the Big Bang does represent the limit of time but without any singularity.[109]

- Big Bang lattice model, states that the universe at the moment of the Big Bang consists of an infinite lattice of fermions, which is smeared over the fundamental domain so it has rotational, translational and gauge symmetry. The symmetry is the largest symmetry possible and hence the lowest entropy of any state.[110]

- Brane cosmology models, in which inflation is due to the movement of branes in string theory; the pre-Big Bang model; the ekpyrotic model, in which the Big Bang is the result of a collision between branes and the cyclic model, a variant of the ekpyrotic model in which collisions occur periodically. In the latter model the Big Bang was preceded by a Big Crunch and the universe cycles from one process to the other.[111][112][113][114]

- Eternal inflation, in which universal inflation ends locally here and there in a random fashion, each end-point leading to a *bubble universe*, expanding from its own big bang.[115][116]

Proposals in the last two categories, see the Big Bang as an event in either a much larger and older universe or in a multiverse.

1.10 Religious and philosophical interpretations

Main article: Religious interpretations of the Big Bang theory

As a description of the origin of the universe, the Big Bang has significant bearing on religion and philosophy.[117][118] As a result, it has become one of the liveliest areas in the discourse between science and religion.[119] Some believe the Big Bang implies a creator,[120][121] and some see its mention in their holy books,[122] while others argue that Big Bang cosmology makes the notion of a creator superfluous.[118][123]

1.11 See also

- Big Crunch
- Cosmic Calendar
- Shape of the universe

1.12 Notes

[1] There is no consensus about how long the Big Bang phase lasted. For some writers this denotes only the initial singularity, for others the whole history of the universe. Usually, at least the first few minutes (during which helium is synthesized) are said to occur "during the Big Bang".

[2] Detailed information of and references for tests of general relativity are given in the article tests of general relativity.

[3] It is commonly reported that Hoyle intended this to be pejorative. However, Hoyle later denied that, saying that it was just a striking image meant to emphasize the difference between the two theories for radio listeners.[62]

[4] Strictly, dark energy in the form of a cosmological constant drives the universe towards a flat state; however, our universe remained close to flat for several billion years, before the dark energy density became significant.

1.13 References

[1] Joseph Silk (2009). *Horizons of Cosmology*. Templeton Press. p. 208.

[2] Simon Singh (2005). *Big Bang: The Origin of the Universe*. Harper Perennial. p. 560.

[3] Wollack, E. J. (10 December 2010). "Cosmology: The Study of the Universe". *Universe 101: Big Bang Theory*. NASA. Archived from the original on 14 May 2011. Retrieved 27 April 2011. The second section discusses the classic tests of the Big Bang theory that make it so compelling as the likely valid description of our universe.

[4] "First Second of the Big Bang". *How The Universe Works 3*. 2014. Discovery Science.

[5] "Big-bang model". *Encyclopedia Britannica*. Retrieved 11 February 2015.

[6] Wright, E. L. (9 May 2009). "What is the evidence for the Big Bang?". *Frequently Asked Questions in Cosmology*. UCLA, Division of Astronomy and Astrophysics. Retrieved 16 October 2009.

[7] "Planck reveals an almost perfect universe". *Planck*. ESA. 2013-03-21. Retrieved 2013-03-21.

[8] Hawking, S.; Ellis, G. F. (1968). "The Cosmic Black-Body Radiation and the Existence of Singularities in our Universe". *Astrophysical Journal, vol. 152, p.25*.

[9] Hawking, S.; Penrose, R. (27 January 1970). "The Singularities of Gravitational Collapse and Cosmology". *Proceedings of the Royal Society A: Mathematical, Physical & Engineering Sciences* (The Royal Society) **314** (1519): 529–548. Bibcode:1970RSPSA.314..529H. doi:10.1098/rspa.1970.0021. Retrieved 27 March 2015.

[10] Kragh, H. (1996). *Cosmology and Controversy*. Princeton University Press. p. 318. ISBN 0-691-02623-8.

[11] Peebles, P. J. E. and Ratra, Bharat (2003). "The cosmological constant and dark energy". *Reviews of Modern Physics* 75 (2): 559–606. arXiv:astro-ph/0207347. Bibcode:2003RvMP...75..559P. doi:10.1103/RevModPhys.75.559.

[12] Gibson, C. H. (2001). "The First Turbulent Mixing and Combustion" (PDF). *IUTAM Turbulent Mixing and Combustion*.

[13] Gibson, C. H. (2001). "Turbulence And Mixing In The Early Universe". arXiv:astro-ph/0110012 [astro-ph].

[14] Gibson, C. H. (2005). "The First Turbulent Combustion". arXiv:astro-ph/0501416 [astro-ph].

[15] Staff (17 March 2014). "BICEP2 2014 Results Release". *National Science Foundation*. Retrieved 18 March 2014.

[16] Clavin, Whitney (17 March 2014). "NASA Technology Views Birth of the Universe". NASA. Retrieved 17 March 2014.

[17] Overbye, Dennis (17 March 2014). "Detection of Waves in Space Buttresses Landmark Theory of Big Bang". *The New York Times*. Retrieved 17 March 2014.

[18] Overbye, Dennis (24 March 2014). "Ripples From the Big Bang". *New York Times*. Retrieved 24 March 2014.

[19] Hubble, E. (1929). "A Relation Between Distance and Radial Velocity Among Extra-Galactic Nebulae". *Proceedings of the National Academy of Sciences* 15 (3): 168–73. Bibcode:1929PNAS...15..168H. doi:10.1073/pnas.15.3.168. PMC 522427. PMID 16577160.

[20] Hawking, S. W.; Ellis, G. F. R. (1973). *The Large-Scale Structure of Space-Time*. Cambridge University Press. ISBN 0-521-20016-4.

[21] Roos, M. (2008). "Expansion of the Universe – Standard Big Bang Model". In Engvold, O.; Stabell, R.; Czerny, B.; Lattanzio, J. *Astronomy and Astrophysics*. Encyclopedia of Life Support Systems. UNESCO. arXiv:0802.2005. This singularity is termed the *Big Bang*.

[22] Drees, W. B. (1990). *Beyond the big bang: quantum cosmologies and God*. Open Court Publishing. pp. 223–224. ISBN 978-0-8126-9118-4.

[23] Planck Collaboration (2013). "Planck 2013 results. XVI. Cosmological parameters". *Astronomy & Astrophysics* ?. pp. ?. arXiv:1303.5076. doi:10.1051/0004-6361/201321591.

[24] Guth, A. H. (1998). *The Inflationary Universe: Quest for a New Theory of Cosmic Origins*. Vintage Books. ISBN 978-0-09-995950-2.

[25] Schewe, P. (2005). "An Ocean of Quarks". *Physics News Update* (American Institute of Physics) 728 (1).

[26] Kolb and Turner (1988), chapter 6

[27] Kolb and Turner (1988), chapter 7

[28] Kolb and Turner (1988), chapter 4

[29] Peacock (1999), chapter 9

[30] Loeb, Abraham (October 2014). "The Habitable Epoch of the Early Universe". *International Journal of Astrobiology* 13 (04): 337–339. arXiv:1312.0613. Bibcode:2014IJAsB..13..337L. doi:10.1017/S1473550414000196. Retrieved 15 December 2014.

[31] Loeb, Abraham (2 December 2013). "The Habitable Epoch of the Early Universe" (PDF). *Arxiv*. arXiv:1312.0613v3. Retrieved 15 December 2014.

[32] Dreifus, Claudia (2 December 2014). "Much-Discussed Views That Go Way Back - Avi Loeb Ponders the Early Universe, Nature and Life". *New York Times*. Retrieved 3 December 2014.

[33] Spergel, D. N. et al. (2003). "First year Wilkinson Microwave Anisotropy Probe (WMAP) observations: determination of cosmological parameters". *Astrophysical Journal Supplement* 148 (1): 175–194. arXiv:astro-ph/0302209. Bibcode:2003ApJS..148..175S. doi:10.1086/377226.

[34] Jarosik, N.; et al. (WMAP Collaboration) (2011). "Seven-Year Wilkinson Microwave Anisotropy Probe (WMAP) Observations: Sky Maps, Systematic Errors, and Basic Results" (PDF). NASA/GSFC. p. 39, Table 8. Retrieved 4 December 2010.

[35] Clavin, Whitney; Jenkins, Ann; Villard, Ray (7 January 2014). "NASA's Hubble and Spitzer Team up to Probe Faraway Galaxies". NASA. Retrieved 8 January 2014.

[36] Ivanchik, A. V.; Potekhin, A. Y.; Varshalovich, D. A. (1999). "The Fine-Structure Constant: A New Observational Limit on Its Cosmological Variation and Some Theoretical Consequences". *Astronomy and Astrophysics* 343: 459. arXiv:astro-ph/9810166. Bibcode:1999A&A...343..439I.

[37] Goodman, J. (1995). "Geocentrism Reexamined". *Physical Review D* 52 (4): 1821–1827. arXiv:astro-ph/9506068. Bibcode:1995PhRvD..52.1821G doi:10.1103/PhysRevD.52.1821.

[38] d'Inverno, R. (1992). "Chapter 23". *Introducing Einstein's Relativity.* Oxford University Press. ISBN 0-19-859686-3.

[39] Tamara M. Davis and Charles H. Lineweaver, *Expanding Confusion: common misconceptions of cosmological horizons and the superluminal expansion of the Universe.* astro-ph/0310808

[40] Kolb and Turner (1988), chapter 3

[41] "'Big bang' astronomer dies". BBC News. 22 August 2001. Archived from the original on 8 December 2008. Retrieved 7 December 2008.

[42] Croswell, K. (1995). "Chapter 9". *The Alchemy of the Heavens.* Anchor Books.

[43] Mitton. *Fred Hoyle: A Life in Science.* Cambridge University Press. ISBN 978-1-139-49595-0."To create a picture in the mind of the listener, Hoyle had likened the explosive theory of the universe's origin to a 'big bang'"

[44] Moskowitz, C. (25 September 2012). "Hubble Telescope Reveals Farthest View Into Universe Ever". Space.com. Retrieved 26 September 2012.

[45] Slipher, V. M. (1913). "The Radial Velocity of the Andromeda Nebula". *Lowell Observatory Bulletin* 1: 56–57. Bibcode:1913LowOB...2...56S.

[46] Slipher, V. M. (1915). "Spectrographic Observations of Nebulae". *Popular Astronomy* 23: 21–24. Bibcode:1915PA.....23Q..21S.

[47] Friedman, A. A. (1922). "Über die Krümmung des Raumes". *Zeitschrift für Physik* (in German) 10 (1): 377–386. Bibcode:1922ZPhy...10..377F. doi:10.1007/BF01332580.

   (English translation in: Friedman, A. (1999). "On the Curvature of Space". *General Relativity and Gravitation* 31 (12): 1991–2000. Bibcode:1999GReGr..31.1991F. doi:10.1023/A:1026751225741.)

[48] Lemaître, G. (1927). "Un univers homogène de masse constante et de rayon croissant rendant compte de la vitesse radiale des nébuleuses extragalactiques". *Annals of the Scientific Society of Brussels* (in French) 47A: 41.

   (Translated in: Lemaître, G. (1931). "A Homogeneous universe of Constant Mass and Growing Radius Accounting for the Radial Velocity of Extragalactic Nebulae". *Monthly Notices of the Royal Astronomical Society* 91: 483–490. Bibcode:1931MNRAS..91..483L. doi:10.1093/mnras/91.5.483.)

[49] Lemaître, G. (1931). "The Evolution of the universe: Discussion". *Nature* 128 (3234): 699–701. Bibcode:1931Natur.128..704L. doi:10.1038/128704a0.

[50] Christianson, E. (1995). *Edwin Hubble: Mariner of the Nebulae.* Farrar, Straus and Giroux. ISBN 0-374-14660-8.

[51] Kragh, H. (1996). *Cosmology and Controversy.* Princeton University Press. ISBN 0-691-02623-8.

[52] "People and Discoveries: Big Bang Theory". *A Science Odyssey.* PBS. Retrieved 9 March 2012.

[53] Eddington, A. (1931). "The End of the World: from the Standpoint of Mathematical Physics". *Nature* 127 (3203): 447–453. Bibcode:1931Natur.127..447E. doi:10.1038/127447a0.

[54] Appolloni, S. (2011). ""Repugnant", "Not Repugnant at All": How the Respective Epistemic Attitudes of Georges Lemaitre and Sir Arthur Eddington Influenced How Each Approached the Idea of a Beginning of the universe". *IBSU Scientific Journal* 5 (1): 19–44.

[55] Lemaître, G. (1931). "The Beginning of the World from the Point of View of Quantum Theory". *Nature* 127 (3210): 706. Bibcode:1931Natur.127..706L. doi:10.1038/127706b0.

[56] Milne, E. A. (1935). *Relativity, Gravitation and World Structure.* Oxford University Press. LCCN 35019093.

[57] Tolman, R. C. (1934). *Relativity, Thermodynamics, and Cosmology*. Clarendon Press. ISBN 0-486-65383-8. LCCN 34032023.

[58] Zwicky, F. (1929). "On the Red Shift of Spectral Lines through Interstellar Space". *Proceedings of the National Academy of Sciences* 15 (10): 773–779. Bibcode:1929PNAS...15..773Z. doi:10.1073/pnas.15.10.773. PMC 522555. PMID 16577237.

[59] Hoyle, F. (1948). "A New Model for the Expanding Universe". *Monthly Notices of the Royal Astronomical Society* 108: 372–382. Bibcode:1948MNRAS.108..372H. doi:10.1093/mnras/108.5.372.

[60] Alpher, R. A.; Bethe, H.; Gamow, G. (1948). "The Origin of Chemical Elements". *Physical Review* 73 (7): 803–804. Bibcode:1948PhRv...73..803A. doi:10.1103/PhysRev.73.803.

[61] Alpher, R. A.; Herman, R. (1948). "Evolution of the Universe". *Nature* 162 (4124): 774–775. Bibcode:1948Natur.162..774A. doi:10.1038/162774b0.

[62] Croswell, K. (1995). *The Alchemy of the Heavens*. Anchor Books. chapter 9. ISBN 978-0-385-47213-5.

[63] Penzias, A. A.; Wilson, R. W. (1965). "A Measurement of Excess Antenna Temperature at 4080 Mc/s". *Astrophysical Journal* 142: 419. Bibcode:1965ApJ...142..419P. doi:10.1086/148307.

[64] Boggess, N. W. et al. (1992). "The COBE Mission: Its Design and Performance Two Years after the launch". *Astrophysical Journal* 397: 420. Bibcode:1992ApJ...397..420B. doi:10.1086/171797.

[65] Spergel, D. N. et al. (2006). "Wilkinson Microwave Anisotropy Probe (WMAP) Three Year Results: Implications for Cosmology". *Astrophysical Journal Supplement* 170 (2): 377–408. arXiv:astro-ph/0603449. Bibcode:2007ApJS..170..377S. doi:10.1086/513700.

[66] Krauss, L. (2012). *A Universe From Nothing: Why there is Something Rather than Nothing*. Free Press. p. 118. ISBN 978-1-4516-2445-8.

[67] Gladders, M. D. et al. (2007). "Cosmological Constraints from the Red-Sequence Cluster Survey". *The Astrophysical Journal* 655 (1): 128–134. arXiv:astro-ph/0603588. Bibcode:2007ApJ...655..128G. doi:10.1086/509909.

[68] The Four Pillars of the Standard Cosmology

[69] Sadoulet, B. (2010). "Direct Searches for Dark Matter". *Astro2010: The Astronomy and Astrophysics Decadal Survey*. National Academies Press. Retrieved 12 March 2012.

[70] Cahn, R. (2010). "For a Comprehensive Space-Based Dark Energy Mission". *Astro2010: The Astronomy and Astrophysics Decadal Survey*. National Academies Press. Retrieved 12 March 2012.

[71] Peacock (1999), chapter 3

[72] Srianand, R.; Petitjean, P.; Ledoux, C. (2000). "The microwave background temperature at the redshift of 2.33771". *Nature* 408 (6815): 931–935. arXiv:astro-ph/0012222. Bibcode:2000Natur.408..931S. doi:10.1038/35050020. Lay summary – *European Southern Observatory* (December 2000).

[73] Bennett, C. L. et al. (2013). "Nine-Year Wilkinson Microwave Anisotropy Probe (WMAP) Observations: Final Maps and Results". arXiv:1212.5225.

[74] Gannon, M. (21 December 2012). "New 'Baby Picture' of Universe Unveiled". Space.com. Retrieved 21 December 2012.

[75] Wright, E. L. (2004). "Theoretical Overview of Cosmic Microwave Background Anisotropy". In W. L. Freedman. *Measuring and Modeling the Universe*. Carnegie Observatories Astrophysics Series. Cambridge University Press. p. 291. arXiv:astro-ph/0305591. ISBN 0-521-75576-X.

[76] White, M. (1999). *Anisotropies in the CMB. Proceedings of the Los Angeles Meeting, DPF 99* (UCLA). arXiv:astro-ph/9903232. Bibcode:1999dpf..conf.... W.

[77] A. Melchiorri et. al. (1999). "A measurement of Omega from the North American test flight of BOOMERANG". *Astrophys Journal* (Institute of Physics) (536). Retrieved 2015-05-15.

[78] P. de Bernardis et al. (2000). "A Flat Universe from High-Resolution Maps of the Cosmic Microwave Background Radiation". *Nature* (Nature Publishing Group) 404: 955–959. arXiv:astro-ph/0004404. doi:10.1038/35010035.

[79] A. D. Miller et al. (1999). "A Measurement of the Angular Power Spectrum of the Cosmic Microwave Background from l = 100 to 400". *The Astrophysical Journal Letters* 524 (1). arXiv:astro-ph/9906421. Bibcode:1999ApJ...524L...1M. doi:10.1086/312293.

[80] Steigman, G. (2005). "Primordial Nucleosynthesis: Successes And Challenges". *International Journal of Modern Physics E* **15**: 1–36. arXiv:astro-ph/0511534. Bibcode:2006IJMPE..15....1S. doi:10.1142/S0218301306004028.

[81] Barbara Sue Ryden (2003). *Introduction to cosmology*. Addison-Wesley. ISBN 978-0-8053-8912-8.

[82] Bertschinger, E. (2001). "Cosmological Perturbation Theory and Structure Formation". arXiv:astro-ph/0101009 [astro-ph].

[83] Bertschinger, E. (1998). "Simulations of Structure Formation in the Universe". *Annual Review of Astronomy and Astrophysics* **36** (1): 599–654. Bibcode:1998ARA&A..36..599B. doi:10.1146/annurev.astro.36.1.599.

[84] Fumagalli, M.; O'Meara, J. M.; Prochaska, J. X. (2011). "Detection of Pristine Gas Two Billion Years After the Big Bang". *Science* **334** (6060): 1245–9. arXiv:1111.2334. Bibcode:2011Sci...334.1245F. doi:10.1126/science.1213581. PMID 22075722.

[85] "Astronomers Find Clouds of Primordial Gas from the Early Universe, Just Moments After Big Bang". Science Daily. 10 November 2011. Retrieved 13 November 2011.

[86] Perley, D. (21 February 2005). "Determination of the Universe's Age, t_o". University of California Berkeley, Astronomy Department. Retrieved 27 January 2012.

[87] Srianand, R.; Noterdaeme, P.; Ledoux, C.; Petitjean, P. (2008). "First detection of CO in a high-redshift damped Lyman-α system". *Astronomy and Astrophysics* **482** (3): L39. Bibcode:2008A&A...482L..39S. doi:10.1051/0004-6361:200809727.

[88] Avgoustidis, A.; Luzzi, G.; Martins, C. J. A. P.; Monteiro, A. M. R. V. L. (2011). "Constraints on the CMB temperature-redshift dependence from SZ and distance measurements". arXiv:1112.1862v1 [astro-ph.CO].

[89] Belusevic, R. (2008). *Relativity, Astrophysics and Cosmology*. Wiley-VCH. p. 16. ISBN 3-527-40764-2.

[90] Overbye, Dennis (19 June 2014). "Astronomers Hedge on Big Bang Detection Claim". *New York Times*. Retrieved 20 June 2014.

[91] Amos, Jonathan (19 June 2014). "Cosmic inflation: Confidence lowered for Big Bang signal". *BBC News*. Retrieved 20 June 2014.

[92] Ade, P.A.R. (BICEP2 Collaboration) et al. (19 June 2014). "Detection of B-Mode Polarization at Degree Angular Scales by BI-CEP2". *Physical Review Letters* **112**: 241101. arXiv:1403.3985. Bibcode:2014PhRvL.112x1101A. doi:10.1103/PhysRevLett.112.241101. PMID 24996078.

[93] Planck Collaboration Team (19 September 2014). "Planck intermediate results. XXX. The angular power spectrum of polarized dust emission at intermediate and high Galactic latitudes". *ArXiv*. arXiv:1409.5738. Bibcode:2014arXiv1409.5738P. Retrieved 22 September 2014.

[94] Overbye, Dennis (22 September 2014). "Study Confirms Criticism of Big Bang Finding". *New York Times*. Retrieved 22 September 2014.

[95] Kolb and Turner, chapter 6

[96] Sakharov, A. D. (1967). "Violation of CP Invariance, C Asymmetry and Baryon Asymmetry of the Universe". *Zhurnal Eksperimental'noi i Teoreticheskoi Fiziki, Pisma* (in Russian) **5**: 32.

   (Translated in *Journal of Experimental and Theoretical Physics Letters* **5**, 24 (1967).)

[97] Keel, B. (October 2009). "Dark Matter". Retrieved 24 July 2013.

[98] Yao, W. M. et al. (2006). "Review of Particle Physics: Dark Matter" (PDF). *Journal of Physics G* **33** (1): 1–1232. arXiv:astro-ph/0601168. Bibcode:2006JPhG...33....1Y. doi:10.1088/0954-3899/33/1/001.

[99] Navabi, A. A.; Riazi, N. (2003). "Is the Age Problem Resolved?". *Journal of Astrophysics and Astronomy* **24** (1–2): 3–10. Bibcode:2003JApA...24....3N. doi:10.1007/BF03012187.

[100] Penrose, R. (1979). Hawking, S. W. (ed); Israel, W. (ed), ed. *Singularities and Time-Asymmetry. General Relativity: An Einstein Centenary Survey* (Cambridge University Press): 581–638.

[101] Penrose, R. (1989). Fergus, E. J. (ed), ed. *Difficulties with Inflationary Cosmology. Proceedings of the 14th Texas Symposium on Relativistic Astrophysics* (New York Academy of Sciences): 249–264. doi:10.1111/j.1749-6632.1989.tb50513.x.

[102] Kolb and Turner (1988), chapter 8

[103] D. N. Spergel et al. (2007). "Three-Year Wilkinson Microwave Anisotropy Probe (WMAP) Observations: Implications for Cosmology" (PDF). *The Astrophysical Journal Supplement Series* **170**: 377–408. arXiv:astro-ph/0603449. Bibcode:2007ApJS..170..377S. doi:10.1086/513700.

[104] Dicke, R. H.; Peebles, P. J. E. Hawking, S. W. (ed); Israel, W. (ed), ed. *The big bang cosmology—enigmas and nostrums. General Relativity: an Einstein centenary survey* (Cambridge University Press): 504–517.

[105] Griswold, Britt (2012). "What is the Ultimate Fate of the Universe?". *Universe 101 Big Bang Theory*. NASA.

[106] Fred C. Adams and Gregory Laughlin (1997). "A dying universe: the long-term fate and evolution of astrophysical objects". *Reviews of Modern Physics* **69** (2): 337–372. arXiv:astro-ph/9701131. Bibcode:1997RvMP...69..337A. doi:10.1103/RevModPhys.69.337..

[107] Caldwell, R. R; Kamionkowski, M.; Weinberg, N. N. (2003). "Phantom Energy and Cosmic Doomsday". *Physical Review Letters* **91** (7): 071301. arXiv:astro-ph/0302506. Bibcode:2003PhRvL..91g1301C. doi:10.1103/PhysRevLett.91.071301. PMID 12935004.

[108] Hawking, S. W.; Ellis, G. F. R. (1973). *The Large Scale Structure of Space-Time*. Cambridge (UK): Cambridge University Press. ISBN 0-521-09906-4.

[109] Hartle, J. H.; Hawking, S. (1983). "Wave Function of the Universe". *Physical Review D* **28** (12): 2960–2975. Bibcode:1983PhRvD..28.2960H. doi:10.1103/PhysRevD.28.2960.

[110] Bird, P. (2011). "Determining the Big Bang State Vector" (PDF).

[111] Langlois, D. (2002). "Brane Cosmology: An Introduction". *Progress of Theoretical Physics Supplement* **148**: 181–212. arXiv:hep-th/0209261. Bibcode:2002PThPS.148..181L. doi:10.1143/PTPS.148.181.

[112] Linde, A. (2002). "Inflationary Theory versus Ekpyrotic/Cyclic Scenario". arXiv:hep-th/0205259 [hep-th].

[113] Than, K. (2006). "Recycled Universe: Theory Could Solve Cosmic Mystery". Space.com. Retrieved 3 July 2007.

[114] Kennedy, B. K. (2007). "What Happened Before the Big Bang?". Archived from the original on 4 July 2007. Retrieved 3 July 2007.

[115] Linde, A. (1986). "Eternal Chaotic Inflation". *Modern Physics Letters A* **1** (2): 81–85. Bibcode:1986MPLA....1...81L. doi:10.1142/S0217732386000129.

[116] Linde, A. (1986). "Eternally Existing Self-Reproducing Chaotic Inflationary Universe". *Physics Letters B* **175** (4): 395–400. Bibcode:1986PhLB..175..395L. doi:10.1016/0370-2693(86)90611-8.

[117] Harris, J. F. (2002). *Analytic philosophy of religion*. Springer. p. 128. ISBN 978-1-4020-0530-5.

[118] Frame, T. (2009). *Losing my religion*. UNSW Press. pp. 137–141. ISBN 978-1-921410-19-2.

[119] Harrison, P. (2010). *The Cambridge Companion to Science and Religion*. Cambridge University Press. p. 9. ISBN 978-0-521-71251-4.

[120] Harris 2002, p. 129

[121] Craig, William Lane (1999). "The ultimate question of origins: God and the beginning of the Universe.". *Astrophysics and Space Science*. 269 - 270, No. 1 - 4: 723–740. doi:10.1007/978-94-011-4114-7_85.

[122] Asad, Muhammad (1984). *The Message of the Qu'rán*. Gibraltar, Spain: Dar al-Andalus Limited. ISBN 1904510000.

[123] Sagan, C. (1988). *introduction to A Brief History of Time by Stephen Hawking*. Bantam Books. pp. X. ISBN 0-553-34614-8. ... a universe with no edge in space, no beginning or end in time, and nothing for a Creator to do.

1.13.1   Books

- Farrell, John (2005). *The Day Without Yesterday: Lemaitre, Einstein, and the Birth of Modern Cosmology*. New York, NY: Thunder's Mouth Press. ISBN 1-56025-660-5.

- Kolb, E.; Turner, M. (1988). *The Early Universe*. Addison–Wesley. ISBN 0-201-11604-9.

- Masters, Ken (Ed.) (2015). *Origins: Before the Big Bang*. Lulu.com. ISBN 978-1-312-75326-6.

- Ostriker, Jeremiah P.; Mitton, Simon (2013). *Heart of Darkness: Unraveling the mysteries of the invisible Universe*. Princeton, NJ: Princeton University Press. ISBN 978-0-691-13430-7.

- Peacock, J. (1999). *Cosmological Physics*. Cambridge University Press. ISBN 0-521-42270-1.

- Woolfson, M. (2013). *Time, Space, Stars and Man: The Story of Big Bang (2nd edition)*. World Scientific Publishing. ISBN 978-1-84816-933-3.

1.14   Further reading

For an annotated list of textbooks and monographs, see physical cosmology.

- Alpher, R. A.; Herman, R. (1988). "Reflections on early work on 'big bang' cosmology". *Physics Today* 8 (8): 24–34. Bibcode:1988PhT....41h..24A. doi:10.1063/1.881126.

- "Cosmic Journey: A History of Scientific Cosmology". American Institute of Physics.

- Barrow, J. D. (1994). *The Origin of the Universe*. Weidenfeld & Nicolson. ISBN 0-297-81497-4.

- Davies, P. C. W. (1992). *The Mind of God: The scientific basis for a rational world*. Simon & Schuster. ISBN 0-671-71069-9.

- Feuerbacher, B.; Scranton, R. (2006). "Evidence for the Big Bang". TalkOrigins.

- Mather, J. C.; Boslough, J. (1996). *The very first light: the true inside story of the scientific journey back to the dawn of the Universe*. Basic Books. p. 300. ISBN 0-465-01575-1.

- Riordan, Michael; Zajc, William (May 2006). "The First Few Microseconds" (PDF). Scientific American.

- Singh, S. (2004). Big Bang: The origins of the universe. Fourth Estate. ISBN 0-00-716220-0.

- "Misconceptions about the Big Bang" (PDF). Scientific American. March 2005.

- Weinberg, S. (1993). *The First Three Minutes: A Modern View Of The Origin Of The Universe*. Basic Books. ISBN 0-465-02437-8.

1.15   External links

- big-bang model at *Encyclopædia Britannica*

- The Story of the Big Bang - STFC funded project explaining the history of the universe in easy-to-understand language

- Big Bang Cosmology WMAP

- The Big Bang - NASA Science

- Big bang model with animated graphics

- Cosmology at DMOZ
- Evidence for the Big Bang
- Universe
- A Cosmic History of the Universe

Chapter 2

Physical cosmology

This article is about the branch of astronomy. For other uses, see Cosmology.
"Cosmic Evolution" redirects here. For the book by Eric Chaisson, see Cosmic Evolution (book).

Physical cosmology is the study of the largest-scale structures and dynamics of the Universe and is concerned with fundamental questions about its origin, structure, evolution, and ultimate fate.[1] For most of human history, it was a branch of metaphysics and religion. Cosmology as a science originated with the Copernican principle, which implies that celestial bodies obey identical physical laws to those on Earth, and Newtonian mechanics, which first allowed us to understand those physical laws.

Physical cosmology, as it is now understood, began with the development in 1915 of Albert Einstein's general theory of relativity, followed by major observational discoveries in the 1920s: first, Edwin Hubble discovered that the universe contains a huge number of external galaxies beyond our own Milky Way; then, work by Vesto Slipher and others showed that the universe is expanding. These advances made it possible to speculate about the origin of the universe, and allowed the establishment of the Big Bang Theory, by Georges Lemaitre, as the leading cosmological model. A few researchers still advocate a handful of alternative cosmologies;[2] however, most cosmologists agree that the Big Bang theory explains the observations better.

Dramatic advances in observational cosmology since the 1990s, including the cosmic microwave background, distant supernovae and galaxy redshift surveys, have led to the development of a standard model of cosmology. This model requires the universe to contain large amounts of dark matter and dark energy whose nature is currently not well understood, but the model gives detailed predictions that are in excellent agreement with many diverse observations.[3]

Cosmology draws heavily on the work of many disparate areas of research in theoretical and applied physics. Areas relevant to cosmology include particle physics experiments and theory, theoretical and observational astrophysics, general relativity, quantum mechanics, and plasma physics.

2.1   Subject history

See also: Timeline of cosmology and List of cosmologists

Modern cosmology developed along tandem tracks of theory and observation. In 1916, Albert Einstein published his theory of general relativity, which provided a unified description of gravity as a geometric property of space and time.[4] At the time, Einstein believed in a static universe, but found that his original formulation of the theory did not permit it.[5] This is because masses distributed throughout the universe gravitationally attract, and move toward each other over time.[6] However, he realized that his equations permitted the introduction of a constant term which could counteract the attractive force of gravity on the cosmic scale. Einstein published his first paper on relativistic cosmology in 1917, in which he added this *cosmological constant* to his field equations in order to force them to model a static universe.[7] However, this so-called Einstein model is unstable to small perturbations—it will eventually start to expand or contract.[5] The Einstein model describes a static universe; space is finite and unbounded (analogous to the surface of a sphere, which has a finite area but no edges). It was later realized that Einstein's model was just one of a larger set of possibilities, all of which were consistent with general relativity and the cosmological principle. The cosmological solutions of general relativity were found by Alexander Friedmann in the early 1920s.[8] His equations describe the Friedmann–Lemaître–Robertson–Walker universe, which may expand or contract, and whose geometry may be open, flat, or closed.

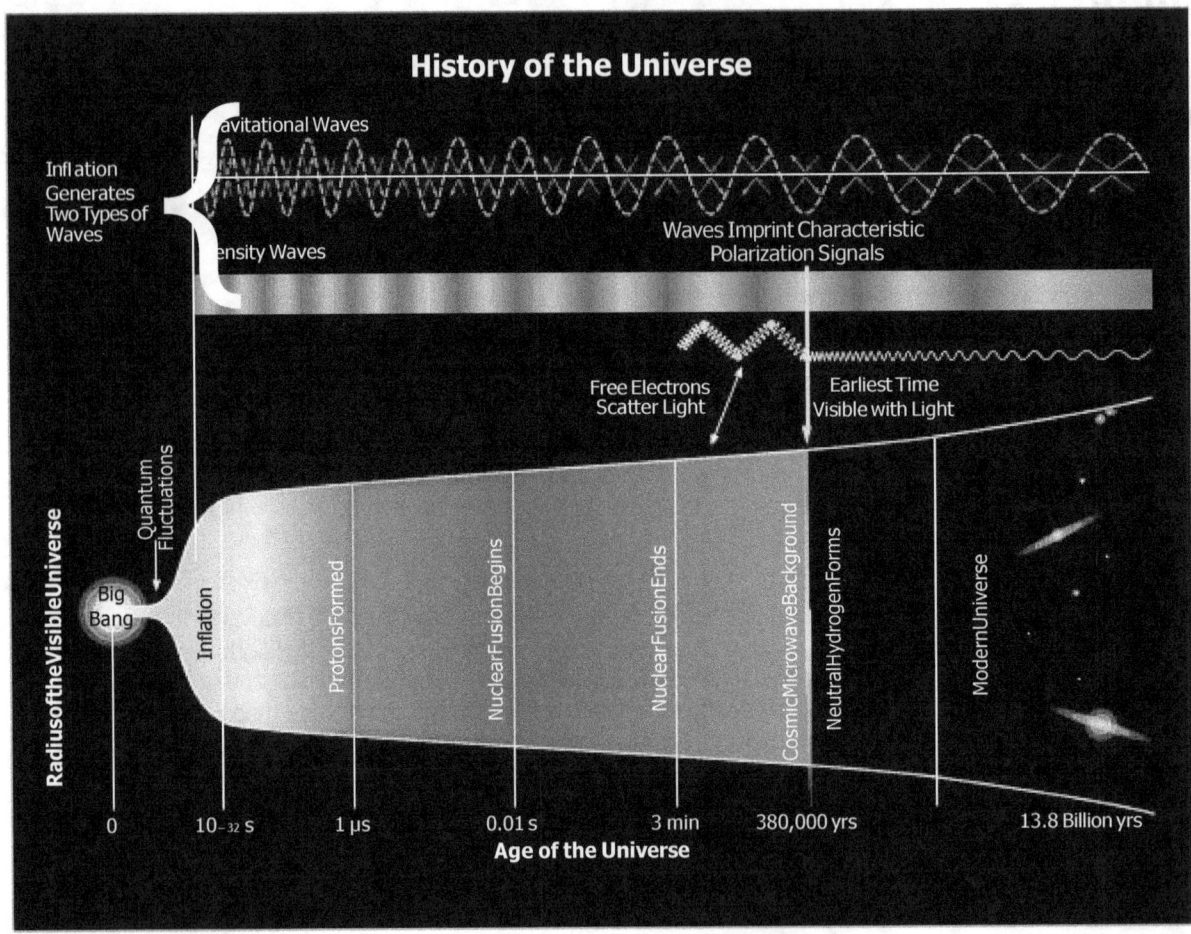

History of the Universe – gravitational waves are hypothesized to arise from cosmic inflation, a faster-than-light expansion just after the Big Bang (17 March 2014).[9][10][11]

In the 1910s, Vesto Slipher (and later Carl Wilhelm Wirtz) interpreted the red shift of spiral nebulae as a Doppler shift that indicated they were receding from Earth.[12][13] However, it is difficult to determine the distance to astronomical objects. One way is to compare the physical size of an object to its angular size, but a physical size must be assumed to do this. Another method is to measure the brightness of an object and assume an intrinsic luminosity, from which the distance may be determined using the inverse square law. Due to the difficulty of using these methods, they did not realize that the nebulae were actually galaxies outside our own Milky Way, nor did they speculate about the cosmological implications. In 1927, the Belgian Roman Catholic priest Georges Lemaître independently derived the Friedmann–Lemaître–Robertson–Walker equations and proposed, on the basis of the recession of spiral nebulae, that the universe began with the "explosion" of a "primeval atom"[14]—which was later called the Big Bang. In 1929, Edwin Hubble provided an observational basis for Lemaître's theory. Hubble showed that the spiral nebulae were galaxies by determining their distances using measurements of the brightness of Cepheid variable stars. He discovered a relationship between the redshift of a galaxy and its distance. He interpreted this as evidence that the galaxies are receding from Earth in every direction at speeds proportional to their distance.[15] This fact is now known as Hubble's law, though the numerical factor Hubble found relating recessional velocity and distance was off by a factor of ten, due to not knowing about the types of Cepheid variables.

Given the cosmological principle, Hubble's law suggested that the universe was expanding. Two primary explanations were proposed for the expansion. One was Lemaître's Big Bang theory, advocated and developed by George Gamow. The other explanation was Fred Hoyle's steady state model in which new matter is created as the galaxies move away from each other. In this model, the universe is roughly the same at any point in time.

For a number of years, support for these theories was evenly divided. However, the observational evidence began to support the idea that the universe evolved from a hot dense state. The discovery of the cosmic microwave background in 1965 lent strong support to the Big Bang model, and since the precise measurements of the cosmic microwave background by the Cosmic Background Explorer in the early 1990s, few cosmologists have seriously proposed other theories of the origin and evolution of the cosmos. One consequence of this is that in standard general relativity, the universe began with a singularity, as demonstrated by Roger Penrose and Stephen Hawking in the 1960s.

An alternative view to extend the Big Bang model, suggesting the universe had no beginning or singularity and the age of the universe is infinite, has been presented.[16][17][18]

2.2   Energy of the cosmos

Light chemical elements, primarily hydrogen and helium, were created in the Big Bang process *(see Nucleosynthesis)*. The small atomic nuclei combined into larger atomic nuclei to form heavier elements such as iron and nickel, which are more stable *(see Nuclear fusion)*. This caused a *later energy release*. Such reactions of nuclear particles inside stars continue to contribute to *sudden energy releases*, such as in nova stars. Gravitational collapse of matter into black holes is also thought to power the most energetic processes, generally seen at the centers of galaxies *(see Quasar and Active galaxy)*.

Cosmologists cannot explain all cosmic phenomena exactly, such as those related to the accelerating expansion of the universe, using conventional forms of energy. Instead, cosmologists propose a new form of energy called dark energy that permeates all space.[19] One hypothesis is that dark energy is the energy of virtual particles, which are believed to exist in a vacuum due to the uncertainty principle.

There is no clear way to define the total energy in the universe using the most widely accepted theory of gravity, general relativity. Therefore, it remains controversial whether the total energy is conserved in an expanding universe. For instance, each photon that travels through intergalactic space loses energy due to the redshift effect. This energy is not obviously transferred to any other system, so seems to be permanently lost. On the other hand, some cosmologists insist that energy is conserved in some sense; this follows the law of conservation of energy.[20]

Thermodynamics of the universe is a field of study that explores which form of energy dominates the cosmos – relativistic particles which are referred to as radiation, or non-relativistic particles referred to as matter. Relativistic particles are particles whose rest mass is zero or negligible compared to their kinetic energy, and so move at the speed of light or very close to it; non-relativistic particles have much higher rest mass than their energy and so move much slower than the speed of light.

As the universe expands, both matter and radiation in it become diluted. However, the energy densities of radiation and matter dilute at different rates. As a particular volume expands, mass energy density is changed only by the increase in volume, but the energy density of radiation is changed both by the increase in volume and by the increase in the wavelength of the photons that make it up. Thus the energy of radiation becomes a smaller part of the universe's total energy than that of matter as it expands. The very early universe is said to have been 'radiation dominated' and radiation controlled the deceleration of expansion. Later, as the average energy per photon becomes roughly 10 eV and lower, matter dictates the rate of deceleration and the universe is said to be 'matter dominated'. The intermediate case is not treated well analytically. As the expansion of the universe continues, matter dilutes even further and the cosmological constant becomes dominant, leading to an acceleration in the universe's expansion.

2.3   History of the universe

See also: Timeline of the Big Bang

The history of the universe is a central issue in cosmology. The history of the universe is divided into different periods called epochs, according to the dominant forces and processes in each period. The standard cosmological model is known as the Lambda-CDM model.

2.3.1 Equations of motion

Main article: Friedmann–Lemaître–Robertson–Walker metric

The equations of motion governing the universe as a whole are derived from general relativity with a small, positive cosmological constant.[21] The solution is an expanding universe; due to this expansion, the radiation and matter in the universe cool down and become diluted. At first, the expansion is slowed down by gravitation attracting the radiation and matter in the universe. However, as these become diluted, the cosmological constant becomes more dominant and the expansion of the universe starts to accelerate rather than decelerate. In our universe this happened billions of years ago.

2.3.2 Particle physics in cosmology

Main article: Particle physics in cosmology

Particle physics is important to the behavior of the early universe, because the early universe was so hot that the average energy density was very high. Because of this, scattering processes and decay of unstable particles are important in cosmology.

As a rule of thumb, a scattering or a decay process is cosmologically important in a certain cosmological epoch if the time scale describing that process is smaller than, or comparable to, the time scale of the expansion of the universe. The time scale that describes the expansion of the universe is $1/H$ with H being the Hubble constant, which itself actually varies with time. The expansion timescale $1/H$ is roughly equal to the age of the universe at that time.

2.3.3 Timeline of the Big Bang

Main article: Timeline of the Big Bang

Observations suggest that the universe began around 13.8 billion years ago.[22] Since then, the evolution of the universe has passed through three phases. The very early universe, which is still poorly understood, was the split second in which the universe was so hot that particles had energies higher than those currently accessible in particle accelerators on Earth. Therefore, while the basic features of this epoch have been worked out in the Big Bang theory, the details are largely based on educated guesses. Following this, in the early universe, the evolution of the universe proceeded according to known high energy physics. This is when the first protons, electrons and neutrons formed, then nuclei and finally atoms. With the formation of neutral hydrogen, the cosmic microwave background was emitted. Finally, the epoch of structure formation began, when matter started to aggregate into the first stars and quasars, and ultimately galaxies, clusters of galaxies and superclusters formed. The future of the universe is not yet firmly known, but according to the ΛCDM model it will continue expanding forever.

2.4 Areas of study

Below, some of the most active areas of inquiry in cosmology are described, in roughly chronological order. This does not include all of the Big Bang cosmology, which is presented in *Timeline of the Big Bang*.

2.4.1   Very early universe

The early, hot universe appears to be well explained by the Big Bang from roughly 10^{-33} seconds onwards. But there are several problems. One is that there is no compelling reason, using current particle physics, for the universe to be flat, homogeneous, and isotropic *(see the cosmological principle)*. Moreover, grand unified theories of particle physics suggest that there should be magnetic monopoles in the universe, which have not been found. These problems are resolved by a brief period of cosmic inflation, which drives the universe to flatness, smooths out anisotropies and inhomogeneities to the observed level, and exponentially dilutes the monopoles. The physical model behind cosmic inflation is extremely simple, but it has not yet been confirmed by particle physics, and there are difficult problems reconciling inflation and quantum field theory. Some cosmologists think that string theory and brane cosmology will provide an alternative to inflation.

Another major problem in cosmology is what caused the universe to contain far more matter than antimatter. Cosmologists can observationally deduce that the universe is not split into regions of matter and antimatter. If it were, there would be X-rays and gamma rays produced as a result of annihilation, but this is not observed. Therefore, some process in the early universe must have created a small excess of matter over antimatter, and this (currently not understood) process is called *baryogenesis*. Three required conditions for baryogenesis were derived by Andrei Sakharov in 1967, and requires a violation of the particle physics symmetry, called CP-symmetry, between matter and antimatter. However, particle accelerators measure too small a violation of CP-symmetry to account for the baryon asymmetry. Cosmologists and particle physicists look for additional violations of the CP-symmetry in the early universe that might account for the baryon asymmetry.

Both the problems of baryogenesis and cosmic inflation are very closely related to particle physics, and their resolution might come from high energy theory and experiment, rather than through observations of the universe.

2.4.2   Big bang theory

Main article: Big bang nucleosynthesis

Big Bang nucleosynthesis is the theory of the formation of the elements in the early universe. It finished when the universe was about three minutes old and its temperature dropped below that at which nuclear fusion could occur. Big Bang nucleosynthesis had a brief period during which it could operate, so only the very lightest elements were produced. Starting from hydrogen ions (protons), it principally produced deuterium, helium-4, and lithium. Other elements were produced in only trace abundances. The basic theory of nucleosynthesis was developed in 1948 by George Gamow, Ralph Asher Alpher, and Robert Herman. It was used for many years as a probe of physics at the time of the Big Bang, as the theory of Big Bang nucleosynthesis connects the abundances of primordial light elements with the features of the early universe. Specifically, it can be used to test the equivalence principle, to probe dark matter, and test neutrino physics. Some cosmologists have proposed that Big Bang nucleosynthesis suggests there is a fourth "sterile" species of neutrino.

Standard model of Big Bang cosmology

The ΛCDM (**Lambda cold dark matter**) or **Lambda-CDM** model is a parametrization of the Big Bang cosmological model in which the universe contains a cosmological constant, denoted by Lambda (Greek Λ), associated with dark energy, and cold dark matter (abbreviated **CDM**). It is frequently referred to as the **standard model** of Big Bang cosmology.

2.4.3   Cosmic microwave background

Main article: Cosmic microwave background
 The cosmic microwave background is radiation left over from decoupling after the epoch of recombination when neutral atoms first formed. At this point, radiation produced in the Big Bang stopped Thomson scattering from charged ions. The radiation, first observed in 1965 by Arno Penzias and Robert Woodrow Wilson, has a perfect thermal black-body spectrum. It has a temperature of 2.7 kelvins today and is isotropic to one part in 10^5. Cosmological perturbation theory, which describes the evolution of slight inhomogeneities in the early universe, has allowed cosmologists to precisely calculate the angular power spectrum of the radiation, and it has been measured by the recent satellite experiments (COBE

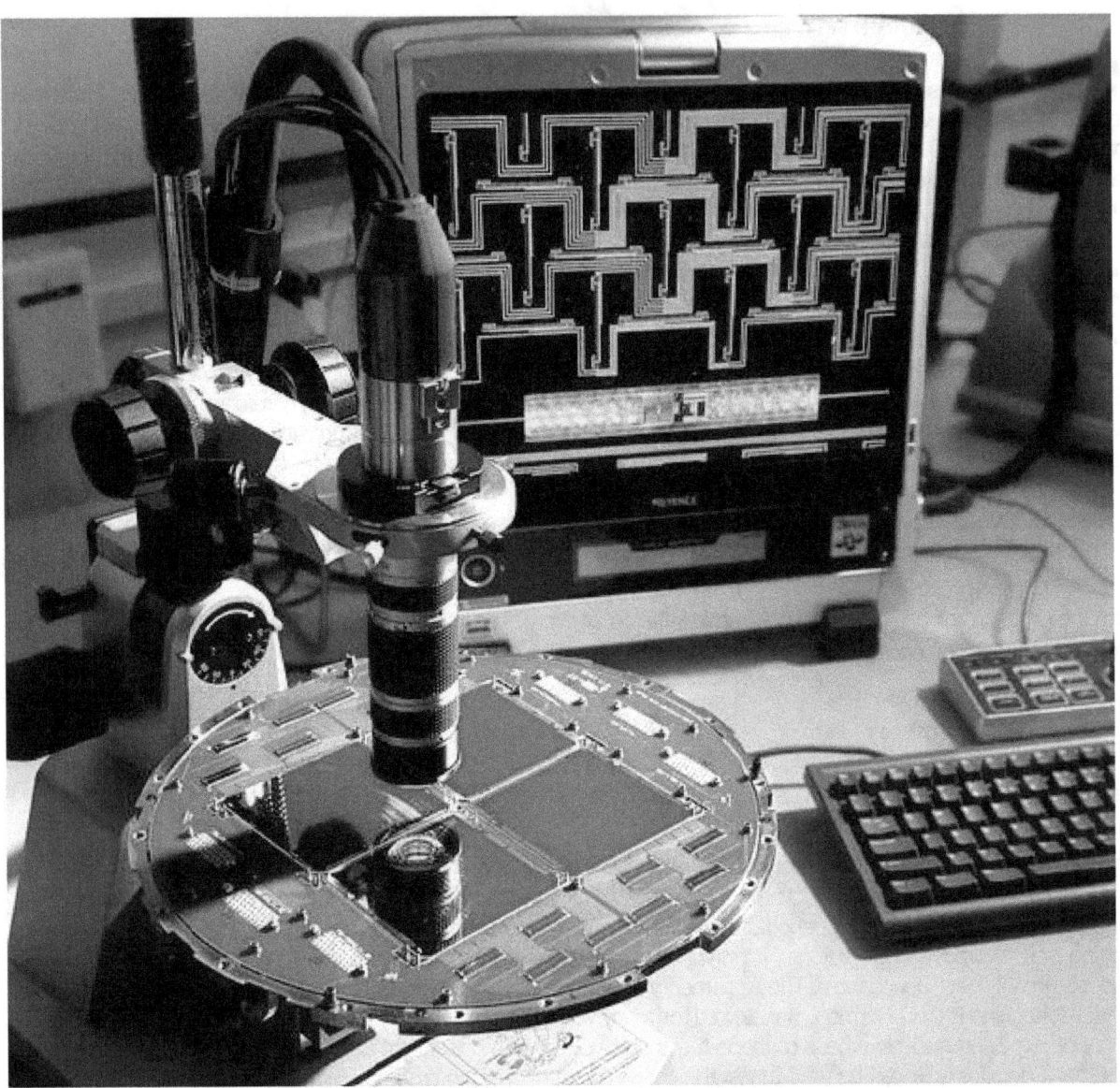

Evidence of gravitational waves in the infant universe may have been uncovered by the microscopic examination of the focal plane of the BICEP2 radio telescope.[9][10][11][23]

and WMAP) and many ground and balloon-based experiments (such as Degree Angular Scale Interferometer, Cosmic Background Imager, and Boomerang). One of the goals of these efforts is to measure the basic parameters of the Lambda-CDM model with increasing accuracy, as well as to test the predictions of the Big Bang model and look for new physics. The recent measurements made by WMAP, for example, have placed limits on the neutrino masses.

Newer experiments, such as QUIET and the Atacama Cosmology Telescope, are trying to measure the polarization of the cosmic microwave background. These measurements are expected to provide further confirmation of the theory as well as information about cosmic inflation, and the so-called secondary anisotropies, such as the Sunyaev-Zel'dovich effect and Sachs-Wolfe effect, which are caused by interaction between galaxies and clusters with the cosmic microwave background.

On 17 March 2014, astronomers at the Harvard–Smithsonian Center for Astrophysics announced the apparent detection of gravitational waves, which, if confirmed, may provide strong evidence for inflation and the Big Bang.[9][10][11][23] However, on 19 June 2014, lowered confidence in confirming the cosmic inflation findings was reported.[24][25][26]

2.4.4   Formation and evolution of large-scale structure

Main articles: Large-scale structure of the cosmos, Structure formation and Galaxy formation and evolution

Understanding the formation and evolution of the largest and earliest structures (i.e., quasars, galaxies, clusters and superclusters) is one of the largest efforts in cosmology. Cosmologists study a model of **hierarchical structure formation** in which structures form from the bottom up, with smaller objects forming first, while the largest objects, such as superclusters, are still assembling. One way to study structure in the universe is to survey the visible galaxies, in order to construct a three-dimensional picture of the galaxies in the universe and measure the matter power spectrum. This is the approach of the *Sloan Digital Sky Survey* and the 2dF Galaxy Redshift Survey.

Another tool for understanding structure formation is simulations, which cosmologists use to study the gravitational aggregation of matter in the universe, as it clusters into filaments, superclusters and voids. Most simulations contain only non-baryonic cold dark matter, which should suffice to understand the universe on the largest scales, as there is much more dark matter in the universe than visible, baryonic matter. More advanced simulations are starting to include baryons and study the formation of individual galaxies. Cosmologists study these simulations to see if they agree with the galaxy surveys, and to understand any discrepancy.

Other, complementary observations to measure the distribution of matter in the distant universe and to probe reionization include:

- The Lyman-alpha forest, which allows cosmologists to measure the distribution of neutral atomic hydrogen gas in the early universe, by measuring the absorption of light from distant quasars by the gas.

- The 21 centimeter absorption line of neutral atomic hydrogen also provides a sensitive test of cosmology

- Weak lensing, the distortion of a distant image by gravitational lensing due to dark matter.

These will help cosmologists settle the question of when and how structure formed in the universe.

2.4.5   Dark matter

Main article: Dark matter

Evidence from Big Bang nucleosynthesis, the cosmic microwave background and structure formation suggests that about 23% of the mass of the universe consists of non-baryonic dark matter, whereas only 4% consists of visible, baryonic matter. The gravitational effects of dark matter are well understood, as it behaves like a cold, non-radiative fluid that forms haloes around galaxies. Dark matter has never been detected in the laboratory, and the particle physics nature of dark matter remains completely unknown. Without observational constraints, there are a number of candidates, such as a stable supersymmetric particle, a weakly interacting massive particle, an axion, and a massive compact halo object. Alternatives to the dark matter hypothesis include a modification of gravity at small accelerations (MOND) or an effect from brane cosmology.

2.4.6   Dark energy

Main article: Dark energy

If the universe is flat, there must be an additional component making up 73% (in addition to the 23% dark matter and 4% baryons) of the energy density of the universe. This is called dark energy. In order not to interfere with Big Bang nucleosynthesis and the cosmic microwave background, it must not cluster in haloes like baryons and dark matter. There is strong observational evidence for dark energy, as the total energy density of the universe is known through constraints on the flatness of the universe, but the amount of clustering matter is tightly measured, and is much less than this. The case for dark energy was strengthened in 1999, when measurements demonstrated that the expansion of the universe has begun to gradually accelerate.

Apart from its density and its clustering properties, nothing is known about dark energy. *Quantum field theory* predicts a cosmological constant (CC) much like dark energy, but 120 orders of magnitude larger than that observed. Steven Weinberg and a number of string theorists *(see string landscape)* have invoked the 'weak anthropic principle': i.e. the reason that physicists observe a universe with such a small cosmological constant is that no physicists (or any life) could exist in a universe with a larger cosmological constant. Many cosmologists find this an unsatisfying explanation: perhaps because while the weak anthropic principle is self-evident (given that living observers exist, there must be at least one universe with a cosmological constant which allows for life to exist) it does not attempt to explain the context of that universe. For example, the weak anthropic principle alone does not distinguish between:

- Only one universe will ever exist and there is some underlying principle that constrains the CC to the value we observe.

- Only one universe will ever exist and although there is no underlying principle fixing the CC, we got lucky.

- Lots of universes exist (simultaneously or serially) with a range of CC values, and of course ours is one of the life-supporting ones.

Other possible explanations for dark energy include quintessence or a modification of gravity on the largest scales. The effect on cosmology of the dark energy that these models describe is given by the dark energy's equation of state, which varies depending upon the theory. The nature of dark energy is one of the most challenging problems in cosmology.

A better understanding of dark energy is likely to solve the problem of the ultimate fate of the universe. In the current cosmological epoch, the accelerated expansion due to dark energy is preventing structures larger than superclusters from forming. It is not known whether the acceleration will continue indefinitely, perhaps even increasing until a big rip, or whether it will eventually reverse.

2.4.7 Other areas of inquiry

Cosmologists also study:

- Whether primordial black holes were formed in our universe, and what happened to them.

- The GZK cutoff for high-energy cosmic rays, and whether it signals a failure of special relativity at high energies

- The equivalence principle, whether or not Einstein's general theory of relativity is the correct theory of gravitation, and if the fundamental laws of physics are the same everywhere in the universe.

- The increasing complexity of universal structures, an example being the progressively greater energy rate density. [27]

2.5 See also

- Hubble's law

- Illustris project

- List of cosmologists

- Photon

- Physical ontology

- String cosmology

- Universal Rotation Curve

2.6   References

[1] For an overview, see George FR Ellis (2006). "Issues in the Philosophy of Cosmology". In Jeremy Butterfield & John Earman. *Philosophy of Physics (Handbook of the Philosophy of Science) 3 volume set.* North Holland. pp. 1183*ff*. arXiv:astro-ph/0602280. ISBN 0-444-51560-7.

[2] An Open Letter to the Scientific Community as published in *New Scientist,* May 22, 2004

[3] Beringer, J.; et al. (Particle Data Group) (2012). "2013 Review of Particle Physics" (PDF). *Phys. Rev. D* **86**: 010001. Bibcode:2012PhRvD..86a0001B. doi:10.1103/PhysRevD.86.010001.

[4] "Nobel Prize Biography". *Nobel Prize Biography.* Nobel Prize. Retrieved 25 February 2011.

[5] Liddle, A. *An Introduction to Modern Cosmology.* Wiley. p. 51. ISBN 0-470-84835-9.

[6] Vilenkin, Alex (2007). *Many worlds in one : the search for other universes.* New York: Hill and Wang, A division of Farrar, Straus and Giroux. p. 19. ISBN 978-0-8090-6722-0.

[7] Jones, Mark; Lambourne, Robert (2004). *An introduction to galaxies and cosmology.* Milton Keynes Cambridge, UK New York: Open University Cambridge University Press. p. 228. ISBN 0-521-54623-0.

[8] Jones, Mark; Lambourne, Robert (2004). *An introduction to galaxies and cosmology.* Milton Keynes Cambridge, UK New York: Open University Cambridge University Press. p. 232. ISBN 0-521-54623-0.

[9] Staff (17 March 2014). "BICEP2 2014 Results Release". *National Science Foundation.* Retrieved 18 March 2014.

[10] Clavin, Whitney (17 March 2014). "NASA Technology Views Birth of the Universe". *NASA.* Retrieved 17 March 2014.

[11] Overbye, Dennis (17 March 2014). "Detection of Waves in Space Buttresses Landmark Theory of Big Bang". *New York Times.* Retrieved 17 March 2014.

[12] Slipher, V. M. (1922), Fox, Philip; Stebbins, Joel, eds., "Further Notes on Spectrographic Observations of Nebulae and Clusters", *Publications of the American Astronomical Society* **4**: 284–286, Bibcode:1922PAAS....4..284S

[13] Seitter, Waltraut C.; Duerbeck, Hilmar W. (1999), Egret, Daniel; Heck, Andre, eds., "Carl Wilhelm Wirtz – Pioneer in Cosmic Dimensions", *Harmonizing Cosmic Distance Scales in a Post-Hipparcos Era,* ASP Conference Series **167**: 237–242, Bibcode:1999ASPC..167..237S, ISBN 1-886733-88-0

[14] Lemaître, G. (1927), "Un Univers homogène de masse constante et de rayon croissant rendant compte de la vitesse radiale des nébuleuses extra-galactiques", *Annales de la Société Scientifique de Bruxelles* (in French) A**47**: 49–59, Bibcode:1927ASSB...47...49L

[15] Hubble, Edwin (March 1929), "A Relation between Distance and Radial Velocity among Extra-Galactic Nebulae", *Proceedings of the National Academy of Sciences of the United States of America* **15** (3): 168–173, Bibcode:1929PNAS...15..168H, doi:10.1073/pnas.15.3.168

[16] Ghose, Tia (26 February 2015). "Big Bang, Deflated? Universe May Have Had No Beginning". *Live Science.* Retrieved 28 February 2015.

[17] Ali, Ahmed Faraq (4 February 2015). "Cosmology from quantum potential". *Physics Letters B* **741**: 276–279. doi:10.1016/j.physletb.2014.12.057. Retrieved 28 February 2015.

[18] Das, Saurya; Bhaduri, Rajat K. (18 November 2014). "Dark matter and dark energy from Bose-Einstein condensate" (PDF). *arXiv.* Retrieved 28 February 2015.

[19] Science 20 June 2003:Vol. 300. no. 5627, pp. 1914 - 1918 Throwing Light on Dark Energy, Robert P. Kirshner. Retrieved December 2006

[20] e.g. Liddle, A. *An Introduction to Modern Cosmology.* Wiley. ISBN 0-470-84835-9. This argues cogently "Energy is always, always, always conserved."

[21] P. Ojeda; H. Rosu (June 2006). "Supersymmetry of FRW barotropic cosmologies". *Internat. J. Theoret. Phys.* (Springer) **45** (6): 1191–1196. arXiv:gr-qc/0510004. Bibcode:2006IJTP...45.1152R. doi:10.1007/s10773-006-9123-2.

[22] "Cosmic Detectives". The European Space Agency (ESA). 2013-04-02. Retrieved 2013-04-25.

[23] Overbye, Dennis (24 March 2014). "Ripples From the Big Bang". *New York Times*. Retrieved 24 March 2014.

[24] Overbye, Dennis (19 June 2014). "Astronomers Hedge on Big Bang Detection Claim". *New York Times*. Retrieved 20 June 2014.

[25] Amos, Jonathan (19 June 2014). "Cosmic inflation: Confidence lowered for Big Bang signal". *BBC News*. Retrieved 20 June 2014.

[26] Ade, P.A.R.; et al. (BICEP2 Collaboration) (19 June 2014). "Detection of B-Mode Polarization at Degree Angular Scales by BICEP2" (PDF). *Physical Review Letters* 112: 241101. arXiv:1403.3985. Bibcode:2014PhRvL.112x1101A. doi:10.1103/PhysRevLett.112.241101. PMID 24996078. Retrieved 20 June 2014.

[27] Chaisson, Eric (1987-01-01). "The life ERA: cosmic selection and conscious evolution". *Faculty Publications*.

2.7 Further reading

2.7.1 Popular

- Brian Greene (2005). *The Fabric of the Cosmos*. Penguin Books Ltd. ISBN 0-14-101111-4.

- Alan Guth (1997). *The Inflationary Universe: The Quest for a New Theory of Cosmic Origins*. Random House. ISBN 0-224-04448-6.

- Hawking, Stephen W. (1988). *A Brief History of Time: From the Big Bang to Black Holes*. Bantam Books, Inc. ISBN 0-553-38016-8.

- Hawking, Stephen W. (2001). *The Universe in a Nutshell*. Bantam Books, Inc. ISBN 0-553-80202-X.

- Ostriker, Jeremiah P.; Mitton, Simon (2013). *Heart of Darkness: Unraveling the mysteries of the invisible Universe*. Princeton, NJ: Princeton University Press. ISBN 978-0-691-13430-7.

- Simon Singh (2005). *Big Bang: The Origin of the Universe*. Fourth Estate. ISBN 0-00-716221-9.

- Steven Weinberg (1993) [First published 1978]. *The First Three Minutes*. Basic Books. ISBN 0-465-02437-8.

2.7.2 Textbooks

- Cheng, Ta-Pei (2005). *Relativity, Gravitation and Cosmology: a Basic Introduction*. Oxford and New York: Oxford University Press. ISBN 0-19-852957-0. Introductory cosmology and general relativity without the full tensor apparatus, deferred until the last part of the book.

- Dodelson, Scott (2003). *Modern Cosmology*. Academic Press. ISBN 0-12-219141-2. An introductory text, released slightly before the WMAP results.

- Grøn, Øyvind; Hervik, Sigbjørn (2007). *Einstein's General Theory of Relativity with Modern Applications in Cosmology*. New York: Springer. ISBN 978-0-387-69199-2.

- Harrison, Edward (2000). *Cosmology: the science of the universe*. Cambridge University Press. ISBN 0-521-66148-X. For undergraduates; mathematically gentle with a strong historical focus.

- Kutner, Marc (2003). *Astronomy: A Physical Perspective*. Cambridge University Press. ISBN 0-521-52927-1. An introductory astronomy text.

- Kolb, Edward; Michael Turner (1988). *The Early Universe*. Addison-Wesley. ISBN 0-201-11604-9. The classic reference for researchers.

- Liddle, Andrew (2003). *An Introduction to Modern Cosmology*. John Wiley. ISBN 0-470-84835-9. Cosmology without general relativity.

- Liddle, Andrew; David Lyth (2000). *Cosmological Inflation and Large-Scale Structure*. Cambridge. ISBN 0-521-57598-2. An introduction to cosmology with a thorough discussion of inflation.

- Mukhanov, Viatcheslav (2005). *Physical Foundations of Cosmology*. Cambridge University Press. ISBN 0-521-56398-4.

- Padmanabhan, T. (1993). *Structure formation in the universe*. Cambridge University Press. ISBN 0-521-42486-0. Discusses the formation of large-scale structures in detail.

- Peacock, John (1998). *Cosmological Physics*. Cambridge University Press. ISBN 0-521-42270-1. An introduction including more on general relativity and quantum field theory than most.

- Peebles, P. J. E. (1993). *Principles of Physical Cosmology*. Princeton University Press. ISBN 0-691-01933-9. Strong historical focus.

- Peebles, P. J. E. (1980). *The Large-Scale Structure of the Universe*. Princeton University Press. ISBN 0-691-08240-5. The classic work on large-scale structure and correlation functions.

- Rees, Martin (2002). *New Perspectives in Astrophysical Cosmology*. Cambridge University Press. ISBN 0-521-64544-1.

- Weinberg, Steven (1971). *Gravitation and Cosmology*. John Wiley. ISBN 0-471-92567-5. A standard reference for the mathematical formalism.

- Weinberg, Steven (2008). *Cosmology*. Oxford University Press. ISBN 0-19-852682-2.

- Benjamin Gal-Or, "Cosmology, Physics and Philosophy", Springer Verlag, 1981, 1983, 1987, ISBN 0-387-90581-2, ISBN 0-387-96526-2.

2.8 External links

2.8.1 From groups

- Cambridge Cosmology- from Cambridge University (public home page)

- Cosmology 101 - from the NASA WMAP group

- Center for Cosmological Physics. University of Chicago, Chicago.

- Origins, Nova Online - Provided by *PBS*.

2.8.2 From individuals

- Gale, George, "Cosmology: Methodological Debates in the 1930s and 1940s", *The Stanford Encyclopedia of Philosophy*, Edward N. Zalta (ed.)

- Madore, Barry F., "*Level 5 : A Knowledgebase for Extragalactic Astronomy and Cosmology*". Caltech and Carnegie. Pasadena, California, USA.

- Tyler, Pat, and Phil Newman "*Beyond Einstein*". Laboratory for High Energy Astrophysics (LHEA) NASA Goddard Space Flight Center.

- Wright, Ned. "*Cosmology tutorial and FAQ*". Division of Astronomy & Astrophysics, UCLA.

- George Musser (February 2004). "Four Keys to Cosmology". *Scientific American* (Scientific American). Retrieved 22 March 2015.

- Cliff Burgess; Fernando Quevedo (November 2007). "The Great Cosmic Roller-Coaster Ride". *Scientific American* (print). pp. 52–59. (subtitle) Could cosmic inflation be a sign that our universe is embedded in a far vaster realm?

Chapter 3

Observational cosmology

Observational cosmology is the study of the structure, the evolution and the origin of the universe through observation, using instruments such as telescopes and cosmic ray detectors.

3.1 Early observations

The science of physical cosmology as it is practiced today had its subject material defined in the years following the Shapley-Curtis debate when it was determined that the universe had a larger scale than the Milky Way galaxy. This was precipitated by observations that established the size and the dynamics of the cosmos that could be explained by Einstein's General Theory of Relativity. In its infancy, cosmology was a speculative science based on a very limited number of observations and characterized by a dispute between steady state theorists and promoters of Big Bang cosmology. It was not until the 1990s and beyond that the astronomical observations would be able to eliminate competing theories and drive the science to the "Golden Age of Cosmology" which was heralded by David Schramm at a National Academy of Sciences colloquium in 1992.[1]

3.1.1 Hubble's law and the cosmic distance ladder

Main articles: Hubble's law and cosmic distance ladder

Distance measurements in astronomy have historically been and continue to be confounded by considerable measurement uncertainty. In particular, while stellar parallax can be used to measure the distance to nearby stars, the observational limits imposed by the difficulty in measuring the minuscule parallaxes associated with objects beyond our galaxy meant that astronomers had to look for alternative ways to measure cosmic distances. To this end, a standard candle measurement for Cepheid variables was discovered by Henrietta Swan Leavitt in 1908 which would provide Edwin Hubble with the rung on the cosmic distance ladder he would need to determine the distance to spiral nebula. Hubble used the 100-inch Hooker Telescope at Mount Wilson Observatory to identify individual stars in those galaxies, and determine the distance to the galaxies by isolating individual Cepheids. This firmly established the spiral nebula as being objects well outside the Milky Way galaxy. Determining the distance to "island universes", as they were dubbed in the popular media, established the scale of the universe and settled the Shapley-Curtis debate once and for all.[2]

In 1927, by combining various measurements, including Hubble's distance measurements and Vesto Slipher's determinations of redshifts for these objects, Georges Lemaître was the first to estimate a constant of proportionality between galaxies' distances and what was termed their "recessional velocities", finding a value of about 600 km/s/Mpc.[3][4][5][6][7][8] He showed that this was theoretically expected in a universe model based on general relativity.[3] Two years later, Hubble showed that the relation between the distances and velocities was a positive correlation and had a slope of about 500 km/s/Mpc.[9] This correlation would come to be known as *Hubble's law* and would serve as the observational foundation for the expanding universe theories on which cosmology is still based. The publication of the observations by Slipher, Wirtz, Hubble and their colleagues and the acceptance by the theorists of their theoretical implications in light of Einstein's

39

Astronomer Edwin Hubble

General theory of relativity is considered the beginning of the modern science of cosmology.[10]

3.1.2   Nuclide abundances

Main articles: cosmochemistry and astrochemistry

Determination of the cosmic abundance of elements has a history dating back to early spectroscopic measurements of light from astronomical objects and the identification of emission and absorption lines which corresponded to particular electronic transitions in chemical elements identified on Earth. For example, the element Helium was first identified through its spectroscopic signature in the Sun before it was isolated as a gas on Earth.[11][12]

Computing relative abundances was achieved through corresponding spectroscopic observations to measurements of the elemental composition of meteorites.

3.1.3   Detection of the cosmic microwave background

Main article: Discovery of cosmic microwave background radiation
A cosmic microwave background was predicted in 1948 by George Gamow and Ralph Alpher, and by Alpher and Robert

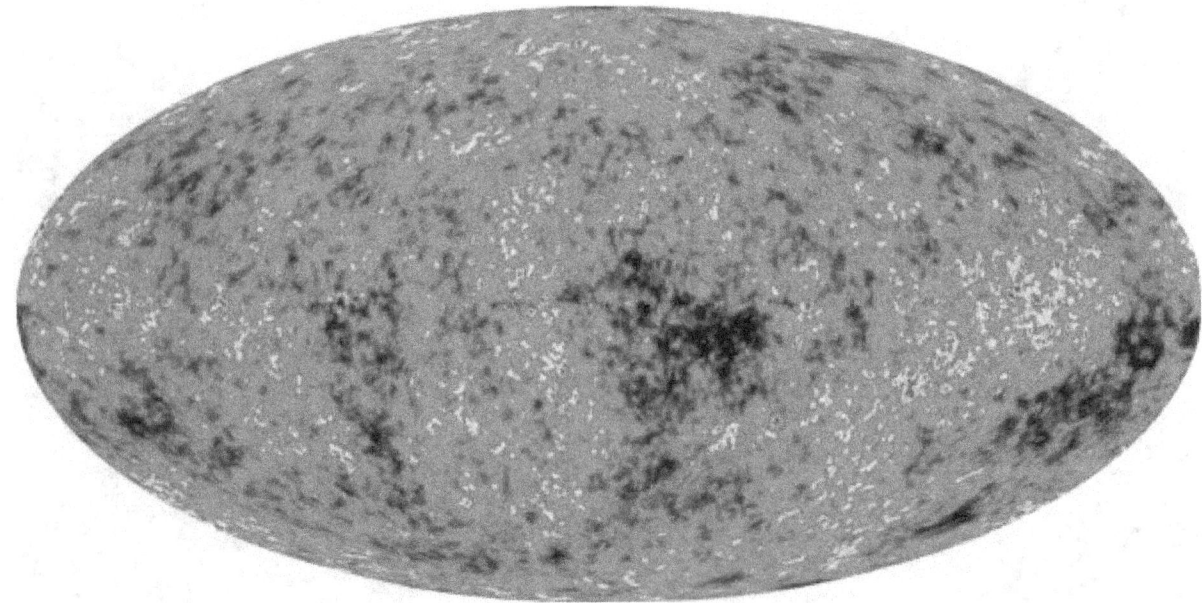

the CMB seen by WMAP

Herman as due to the hot big bang model. Moreover, Alpher and Herman were able to estimate the temperature,[13] but their results were not widely discussed in the community. Their prediction was rediscovered by Robert Dicke and Yakov Zel'dovich in the early 1960s with the first published recognition of the CMB radiation as a detectable phenomenon appeared in a brief paper by Soviet astrophysicists A. G. Doroshkevich and Igor Novikov, in the spring of 1964.[14] In 1964, David Todd Wilkinson and Peter Roll, Dicke's colleagues at Princeton University, began constructing a Dicke radiometer to measure the cosmic microwave background.[15] In 1965, Arno Penzias and Robert Woodrow Wilson at the Crawford Hill location of Bell Telephone Laboratories in nearby Holmdel Township, New Jersey had built a Dicke radiometer that they intended to use for radio astronomy and satellite communication experiments. Their instrument had an excess 3.5 K antenna temperature which they could not account for. After receiving a telephone call from Crawford Hill, Dicke famously quipped: "Boys, we've been scooped."[16] A meeting between the Princeton and Crawford Hill groups determined that the antenna temperature was indeed due to the microwave background. Penzias and Wilson received the 1978 Nobel Prize in Physics for their discovery.

3.2   Modern observations

Today, observational cosmology continues to test the predictions of theoretical cosmology and has led to the refinement of cosmological models. For example, the observational evidence for dark matter has heavily influenced theoretical modeling of structure and galaxy formation. When trying to calibrate the Hubble diagram with accurate supernova standard candles, observational evidence for dark energy was obtained in the late 1990s. These observations have been incorporated into a six-parameter framework known as the Lambda-CDM model which explains the evolution of the universe in terms of its constituent material. This model has subsequently been verified by detailed observations of the cosmic microwave background, especially through the WMAP experiment.

Included here are the modern observational efforts that have directly influenced cosmology.

3.2.1   Redshift surveys

Main article: Redshift survey

With the advent of automated telescopes and improvements in spectroscopes, a number of collaborations have been made to map the universe in redshift space. By combining redshift with angular position data, a redshift survey maps the 3D distribution of matter within a field of the sky. These observations are used to measure properties of the large-scale structure of the universe. The Great Wall, a vast supercluster of galaxies over 500 million light-years wide, provides a dramatic example of a large-scale structure that redshift surveys can detect.[17]

The first redshift survey was the CfA Redshift Survey, started in 1977 with the initial data collection completed in 1982.[18] More recently, the 2dF Galaxy Redshift Survey determined the large-scale structure of one section of the Universe, measuring z-values for over 220,000 galaxies; data collection was completed in 2002, and the final data set was released 30 June 2003.[19] (In addition to mapping large-scale patterns of galaxies, 2dF established an upper limit on neutrino mass.) Another notable investigation, the Sloan Digital Sky Survey (SDSS), is ongoing as of 2011 and aims to obtain measurements on around 100 million objects.[20] SDSS has recorded redshifts for galaxies as high as 0.4, and has been involved in the detection of quasars beyond $z = 6$. The DEEP2 Redshift Survey uses the Keck telescopes with the new "DEIMOS" spectrograph; a follow-up to the pilot program DEEP1, DEEP2 is designed to measure faint galaxies with redshifts 0.7 and above, and it is therefore planned to provide a complement to SDSS and 2dF.[21]

3.2.2   Cosmic microwave background experiments

Main article: Cosmic microwave background experiments

Subsequent to the discovery of the CMB, hundreds of cosmic microwave background experiments had been conducted to measure and characterize the signatures of the radiation. The most famous experiment is probably the NASA Cosmic Background Explorer (COBE) satellite that orbited in 1989–1996 and which detected and quantified the large-scale anisotropies at the limit of its detection capabilities. Inspired by the initial COBE results of an extremely isotropic and homogeneous background, a series of ground-based and balloon-based experiments quantified CMB anisotropies on smaller angular scales over the next decade. The primary goal of those experiments was to measure the angular scale of the first acoustic peak, for which COBE did not have sufficient resolution. The measurements were able to rule out cosmic strings as the leading theory of cosmic structure formation, and suggested cosmic inflation was the right theory. During the 1990s, the first peak was measured with increasing sensitivity and by 2000 the BOOMERanG experiment reported that the highest power fluctuations occur at scales of approximately one degree. Together with other cosmological data, these results implied that the geometry of the Universe is flat. A number of ground-based interferometers provided measurements of the fluctuations with higher accuracy over the next three years, including the Very Small Array, Degree Angular Scale Interferometer (DASI) and the Cosmic Background Imager (CBI). DASI made the first detection of the polarization of the CMB and the CBI provided the first E-mode spectrum with compelling evidence that it is out of phase with the T-mode spectrum.

In June 2001, NASA launched a second CMB space mission, WMAP, to make much more precise measurements of the large-scale anisotropies over the full sky. The first results from this mission, disclosed in 2003, were detailed measurements of the angular power spectrum to below degree scales, tightly constraining various cosmological parameters. The results are broadly consistent with those expected from cosmic inflation as well as various other competing theories, and are available in detail at NASA's data center for Cosmic Microwave Background (CMB) (see links below). Although WMAP provided very accurate measurements of the large angular-scale fluctuations in the CMB (structures about as large in the sky as the Moon), it did not have the angular resolution to measure the smaller scale fluctuations which had been observed using previous ground-based interferometers.

A third space mission, Planck, was launched in May 2009. Planck employs both HEMT radiometers and bolometer technology and measures the CMB anisotropies at a higher resolution than WMAP. Unlike the previous two space missions, Planck is a collaboration between NASA and the European Space Agency (ESA). Its detectors got a trial run at the Antarctic Viper telescope as ACBAR (Arcminute Cosmology Bolometer Array Receiver) experiment – which has produced the most precise measurements at small angular scales to date – and at the Archeops balloon telescope.

Additional ground-based instruments such as the South Pole Telescope in Antarctica and the proposed Clover Project, Atacama Cosmology Telescope and the QUIET telescope in Chile will provide additional data not available from satellite observations, possibly including the B-mode polarization.

3.2.3   Telescope observations

Radio

The brightest sources of low-frequency radio emission (10 MHz and 100 GHz) are radio galaxies which can be observed out to extremely high redshifts. These are subsets of the active galaxies that have extended features known as lobes and jets which extend away from the galactic nucleus distances on the order of megaparsecs. Because radio galaxies are so bright, astronomers have used them to probe extreme distances and early times in the evolution of the universe.

Infrared

Far infrared observations including submillimeter astronomy have revealed a number of sources at cosmological distances. With the exception of a few atmospheric windows, most of infrared light is blocked by the atmosphere, the observations generally take place from balloon or space-based instruments. Current observational experiments in the infrared include NICMOS, the Cosmic Origins Spectrograph, the Spitzer Space Telescope, the Keck Interferometer, the Stratospheric Observatory For Infrared Astronomy, and the Herschel Space Observatory. The next large space telescope planned by NASA, the James Webb Space Telescope will also explore in the infrared.

An additional infrared survey, the Two-Micron All Sky Survey, has also been very useful in revealing the distribution of galaxies, similar to other optical surveys described below.

Optical rays (visible to human eyes)

Optical light is still the primary means by which astronomy occurs, and in the context of cosmology, this means observing distant galaxies and galaxy clusters in order to learn about the large scale structure of the Universe as well as galaxy evolution. Redshift surveys have been a common means by which this has been accomplished with some of the most famous including the 2dF Galaxy Redshift Survey, the Sloan Digital Sky Survey, and the upcoming Large Synoptic Survey Telescope. These optical observations generally use either photometry or spectroscopy to measure the redshift of a galaxy and then, via Hubble's Law, determine its distance modulo redshift distortions due to peculiar velocities. Additionally, the position of the galaxies as seen on the sky in celestial coordinates can be used to gain information about the other two spatial dimensions.

Very deep observations (which is to say sensitive to dim sources) are also useful tools in cosmology. The Hubble Deep Field, Hubble Ultra Deep Field, Hubble Extreme Deep Field, and Hubble Deep Field South are all examples of this.

Ultraviolet

X-rays

See X-ray telescope.

Gamma-rays

3.2.4   Cosmic ray observations

3.3   Future observations

3.3.1   Cosmic neutrinos

Main article: Cosmic neutrino background

It is a prediction of the Big Bang model that the universe is filled with a neutrino background radiation, analogous to the cosmic microwave background radiation. The microwave background is a relic from when the universe was about 380,000 years old, but the neutrino background is a relic from when the universe was about two seconds old.

If this neutrino radiation could be observed, it would be a window into very early stages of the universe. Unfortunately, these neutrinos would now be very cold, and so they are effectively impossible to observe directly.

3.3.2   Gravitational waves

Main article: Cosmic gravitational wave background

3.4   See also

- Big bang
- Cosmic background radiation

3.5   References

[1] Arthur M. Sackler Colloquia of the National Academy of Sciences: Physical Cosmology; Irvine, California: March 27–28, 1992.

[2] "Island universe" is a reference to speculative ideas promoted by a variety of scholastic thinkers in the 18th and 19th centuries. The most famous early proponent of such ideas was philosopher Immanuel Kant who published a number of treatises on astronomy in addition to his more famous philosophical works. See Kant, I., 1755. *Allgemeine Naturgeschichte und Theorie des Himmels*, Part I, J.F. Peterson, Königsberg and Leipzig.

[3] Lemaître, G. (1927). "Un univers homogène de masse constante et de rayon croissant rendant compte de la vitesse radiale des nébuleuses extra-galactiques". *Annales de la Société Scientifique de Bruxelles A* 47: 49–56. Bibcode:1927ASSB...47...49L. Partially translated in Lemaître, G. (1931). "Expansion of the universe, A homogeneous universe of constant mass and increasing radius accounting for the radial velocity of extra-galactic nebulae". *Monthly Notices of the Royal Astronomical Society* 91: 483–490. Bibcode:1931MNRAS..91..483L. doi:10.1093/mnras/91.5.483.

[4] van den Bergh, S. (2011). "The Curious Case of Lemaitre's Equation No. 24". *Journal of the Royal Astronomical Society of Canada* 105 (4): 151. arXiv:1106.1195. Bibcode:2011JRASC.105..151V.

[5] Block, D. L. (2012). "Georges Lemaitre and Stiglers Law of Eponymy". In Holder, R. D.; Mitton, S. *Georges Lemaître: Life, Science and Legacy.* Astrophysics and Space Science Library **395**. pp. 89–96. arXiv:1106.3928. Bibcode:2012ASSL..395...89B. doi:10.1007/978-3-642-32254-9_8. ISBN 978-3-642-32253-2.

[6] Reich, E. S. (27 June 2011). "Edwin Hubble in translation trouble". *Nature News.* doi:10.1038/news.2011.385.

[7] Livio, M. (2011). "Lost in translation: Mystery of the missing text solved". *Nature* **479** (7372): 171. Bibcode:2011Natur.479..171L. doi:10.1038/479171a.

[8] Livio, M.; Riess, A. (2013). "Measuring the Hubble constant". *Physics Today* **66** (10): 41. Bibcode:2013PhT....66j..41L. doi:10.1063/PT.3.2148.

[9] Hubble, E. (1929). "A relation between distance and radial velocity among extra-galactic nebulae". *Proceedings of the National Academy of Sciences* **15** (3): 168–73. Bibcode:1929PNAS...15..168H. doi:10.1073/pnas.15.3.168. PMC 522427. PMID 16577160.

[10] This popular consideration is echoed in *Time Magazine's* listing for Edwin Hubble in their Time 100 list of most influential people of the 20th Century. Michael Lemonick recounts, "He discovered the cosmos, and in doing so founded the science of cosmology."

[11] *The Encyclopedia of the Chemical Elements,* page 256

[12] *Oxford English Dictionary* (1989), s.v. "helium". Retrieved December 16, 2006, from Oxford English Dictionary Online. Also, from quotation there: Thomson, W. (1872). *Rep. Brit. Assoc.* xcix: "Frankland and Lockyer find the yellow prominences to give a very decided bright line not far from D, but hitherto not identified with any terrestrial flame. It seems to indicate a new substance, which they propose to call Helium."

[13] G. Gamow, "The Origin of Elements and the Separation of Galaxies," *Physical Review* **74** (1948), 505. G. Gamow, "The evolution of the universe", *Nature* **162** (1948), 680. R. A. Alpher and R. Herman, "On the Relative Abundance of the Elements," *Physical Review* **74** (1948), 1577.

[14] A. A. Penzias. "The origin of elements." (PDF). *Nobel lecture.* Retrieved October 4, 2006.

[15] R. H. Dicke, "The measurement of thermal radiation at microwave frequencies", *Rev. Sci. Instrum.* **17**, 268 (1946). This basic design for a radiometer has been used in most subsequent cosmic microwave background experiments.

[16] A. A. Penzias and R. W. Wilson, "A Measurement of Excess Antenna Temperature at 4080 Mc/s," *Astrophysical Journal* **142** (1965), 419. R. H. Dicke, P. J. E. Peebles, P. G. Roll and D. T. Wilkinson, "Cosmic Black-Body Radiation," *Astrophysical Journal* **142** (1965), 414. The history is given in P. J. E. Peebles, *Principles of physical cosmology* (Princeton Univ. Pr., Princeton 1993).

[17] M. J. Geller & J. P. Huchra, *Science* **246**, 897 (1989). online

[18] See the official CfA website for more details.

[19] Shaun Cole; et al. (The 2dFGRS Collaboration) (2005). "The 2dF galaxy redshift survey: Power-spectrum analysis of the final dataset and cosmological implications". *Mon. Not. Roy. Astron. Soc.* **362**: 505–34. arXiv:astro-ph/0501174. Bibcode:2005MNRAS.362..505C. doi:10.1111/j.1365-2966.2005.09318.x. 2dF Galaxy Redshift Survey homepage

[20] SDSS Homepage

[21] Marc Davis; et al. (DEEP2 collaboration) (2002). "Science objectives and early results of the DEEP2 redshift survey". *Conference on Astronomical Telescopes and Instrumentation, Waikoloa, Hawaii, 22–28 August 2002.*

Chapter 4

Universe

For other uses, see Universe (disambiguation).

The **Universe** is all of time and space and its contents.[8][9][10][11] The Universe includes planets, stars, galaxies, the contents of intergalactic space, the smallest subatomic particles, and all matter and energy. The majority of matter and energy is most likely in the form of dark matter and dark energy.[12][13] The *observable universe* is about 28 billion parsecs (91 billion light-years) in diameter at the present time.[2] The size of the whole Universe is not known and may be infinite.[14] Observations and the development of physical theories have led to inferences about the composition and evolution of the Universe.

Throughout recorded history, cosmologies and cosmogonies, including scientific models, have been proposed to explain observations of the Universe. The earliest quantitative geocentric models were developed by ancient Greek philosophers and Indian philosophers.[15][16] Over the centuries, more precise astronomical observations led to Nicolaus Copernicus's heliocentric model of the Solar System and Johannes Kepler's improvement on that model with elliptical orbits, which was eventually explained by Isaac Newton's theory of gravity. Further observational improvements led to the realization that the Solar System is located in a galaxy composed of billions of stars, the Milky Way. It was subsequently discovered that our galaxy is just one of many. Observations of the distribution of these galaxies and their spectral lines have led to many of the theories of modern physical cosmology. The discovery in the early 20th century that galaxies are systematically redshifted suggested that the Universe is expanding, and the discovery of the cosmic microwave background radiation suggested that the Universe had a beginning.[17] Finally, observations in the late 1990s indicated the rate of the expansion of the Universe is increasing.[18]

The Big Bang theory is the prevailing cosmological model describing the development of the Universe. Space and time were created in the Big Bang, and these were imbued with a fixed amount of energy and matter; as space expands, the density of that matter and energy decreases. After the initial expansion, the Universe cooled sufficiently to allow the formation first of subatomic particles and later of simple atoms. Giant clouds of these primordial elements later coalesced through gravity to form stars. Assuming that the prevailing model is correct, the age of the Universe is measured to be 13.798 ± 0.037 billion years.[1][19]

There are many competing hypotheses about the ultimate fate of the Universe. Physicists and philosophers remain unsure about what, if anything, preceded the Big Bang. Many refuse to speculate, doubting that any information from any such prior state could ever be accessible. There are various multiverse hypotheses, in which some physicists have suggested that the Universe might be one among many universes that likewise exist.[20][21]

4.1   Definition

The Universe is customarily defined as everything that exists, everything that has existed, and everything that will exist.[22][23][24] According to our current understanding, the Universe consists of three constituents: spacetime, forms of energy, including electromagnetic radiation and matter; and the physical laws that relate them. The Universe also encompasses all of life, all of history, and some philosophers and scientists even suggest that it encompasses ideas such as mathematics.[25][26][27]

4.2 Etymology

The word *universe* derives from the Old French word *univers*, which in turn derives from the Latin word *universum*.[28] The Latin word was used by Cicero and later Latin authors in many of the same senses as the modern English word is used.[29] The Latin word derives from the poetic contraction *unvorsum* — first used by Lucretius in Book IV (line 262) of his *De rerum natura* (*On the Nature of Things*) — which connects *un*, *uni* (the combining form of *unus*, or "one") with *vorsum*, *versum* (a noun made from the perfect passive participle of *vertere*, meaning "something rotated, rolled, changed").[29]

An alternative interpretation of *unvorsum* is "everything rotated as one" or "everything rotated by one". In this sense, it may be considered a translation of an earlier Greek word for the *Universe*, περιφορά, (*periforá*, "circumambulation"), originally used to describe a course of a meal, the food being carried around the circle of dinner guests.[30] This Greek word refers to celestial spheres, an early Greek model of the Universe. Regarding Plato's Metaphor of the Sun, Aristotle suggests that the rotation of the sphere of fixed stars inspired by the prime mover,[31] motivates, in turn, terrestrial change via the Sun. Astronomical and physical measurements, such as the Foucault pendulum, demonstrate that the Earth rotates on its axis.

4.2.1 Synonyms

A term for the *Universe* in ancient Greece was τὸ πᾶν (*tò pán*, The All, Pan). Related terms were matter, (τὸ ὅλον, *tò hólon*, see also Hyle, lit. wood) and place (τὸ κενόν, *tò kenón*).[32][33] Other synonyms for the *Universe* among the ancient Greek philosophers included κόσμος (cosmos) and φύσις (meaning nature, Greek physis, from which we derive the word physics).[34] The same synonyms are found in Latin authors (*totum*, *mundus*, *natura*)[35] and survive in modern languages, e.g., the German words *Das All*, *Weltall*, and *Natur* for *Universe*. The same synonyms are found in English, such as everything (as in the theory of everything), the cosmos (as in cosmology), the world (as in the many-worlds interpretation), and nature (as in natural laws or natural philosophy).[36]

4.3 Chronology and the Big Bang

Main articles: Big Bang and Chronology of the Universe

The prevailing model for the evolution of the Universe is the Big Bang theory.[37][38] The Big Bang model states that the earliest state of the Universe was extremely hot and dense and that it subsequently expanded. The model is based on general relativity and on simplifying assumptions such as homogeneity and isotropy of space. A version of the model with a cosmological constant (Lambda) and cold dark matter, known as the Lambda-CDM model, is the the simplest model that provides a reasonably good account of various observations about the Universe. The Big Bang model accounts for observations such as the correlation of distance and redshift of galaxies, the ratio of the number of hydrogen to helium atoms, and the microwave radiation background.

The initial hot, dense state is called the Planck epoch, a brief period extending from time zero to one Planck time unit of approximately 10^{-43} seconds. During the Planck epoch, all types of matter and all types of energy were concentrated into a dense state, where gravitation is believed to have been as strong as the other fundamental forces, and all the forces may have been unified. Since the Planck epoch, the Universe has been expanding to its present form, possibly with a very brief period (less than 10^{-32} seconds) of cosmic inflation which caused the Universe to reach a much larger size almost instantaneously.[39] Several independent measurements support this theoretical expansion.

After the Planck epoch and inflation came the quark, hadron, and lepton epochs. Together, these epochs encompassd less than 10 seconds of time following the Big Bang. Other observations can be explained by combining the overall expansion of space with nuclear and atomic physics. As the Universe expands, the energy density of electromagnetic radiation decreases more quickly than does that of matter because the energy of a photon decreases with its wavelength. As the Universe expanded and cooled, elementary particles associated stably into ever larger combinations. Thus, in the early part of the matter-dominated era, stable protons and neutrons formed, which then associated into atomic nuclei. At this stage, matter in the Universe was mainly a hot, dense plasma of negatively charged electrons, neutral neutrinos and positive nuclei. This era, called the photon epoch, lasted about 380 thousand years. Nuclear reactions among nuclei led to the present abundances of lighter nuclei, particularly hydrogen, deuterium, and helium through a process known as Big Bang nucleosynthesis.

Eventually, at time known as recombination, electrons and nuclei formed stable atoms, which are transparent to most wavelengths of radiation. With photons decoupled from matter, the Universe entered the matter-dominated era. Light from this era could now travel freely, and it can still be seen in the Universe as the cosmic microwave background (CMB). After around 100 million years, the first stars formed; these were likely very massive, luminous, and responsible for the reionization of the Universe. Having no elements heavier than lithium, these stars also produced the first heavy elements through stellar nucleosynthesis.[40] The Universe also contains a mysterious energy called dark energy; the energy density of dark energy does not change over time. After about 9.8 billion years, the Universe had expanded sufficiently so that the density of matter was less than the density of dark energy, marking the beginning of the present dark-energy-dominated era.[41] In this era, the expansion of the Universe is accelerating due to dark energy.

4.4   Properties

Main articles: Observable universe, Age of the Universe and Metric expansion of space

The spacetime of the Universe is usually interpreted from a Euclidean perspective, with space as consisting of three dimensions, and time as consisting of one dimension, the "fourth dimension". By combining space and time into a single manifold called Minkowski space, physicists have simplified a large number of physical theories, as well as described in a more uniform way the workings of the Universe at both the supergalactic and subatomic levels.

The four dimensions of spacetime consist of events that are not absolutely defined spatially and temporally, but rather are known relative to the motion of an observer. Minkowski space first approximates the Universe without gravity; the pseudo-Riemannian manifolds of general relativity describe spacetime with matter and gravity. Some areas of theoretical physics, such as string theory, postulate the existence of additional dimensions.

Of the four fundamental interactions, gravitation is dominant at cosmological length scales; that is, the other three forces play a negligible role in determining structures at the level of galaxies and larger-scale structures. Gravity's effects are cumulative; by contrast, the effects of positive and negative charges tend to cancel one another, making electromagnetism relatively insignificant on cosmological length scales. The remaining two interactions, the weak and strong nuclear forces, decline very rapidly with distance; their effects are confined mainly to sub-atomic length scales.

The Universe appears to have much more matter than antimatter, an asymmetry possibly related to the observations of CP violation.[42] The Universe appears to have no net electric charge, and therefore gravity appears to be the dominant interaction on cosmological length scales. The Universe also appears to have neither net momentum nor angular momentum. The absence of net charge and momentum would follow from accepted physical laws (Gauss's law and the non-divergence of the stress-energy-momentum pseudotensor, respectively), if the Universe were finite.[43]

4.4.1   Shape

Main article: Shape of the Universe

The shape of the Universe is related to general relativity, which describes how spacetime is curved and bent by mass and energy. The curvature, topology or geometry of the Universe includes both local geometry in the observable universe and global geometry, which is possibly measurable. Investigations include, which 3-manifold corresponds to the spatial section in comoving coordinates of the four-dimensional spacetime of the Universe.{unclear} Cosmologists often work with a given space-like slice of spacetime called the comoving coordinates. In terms of observation, the section of spacetime

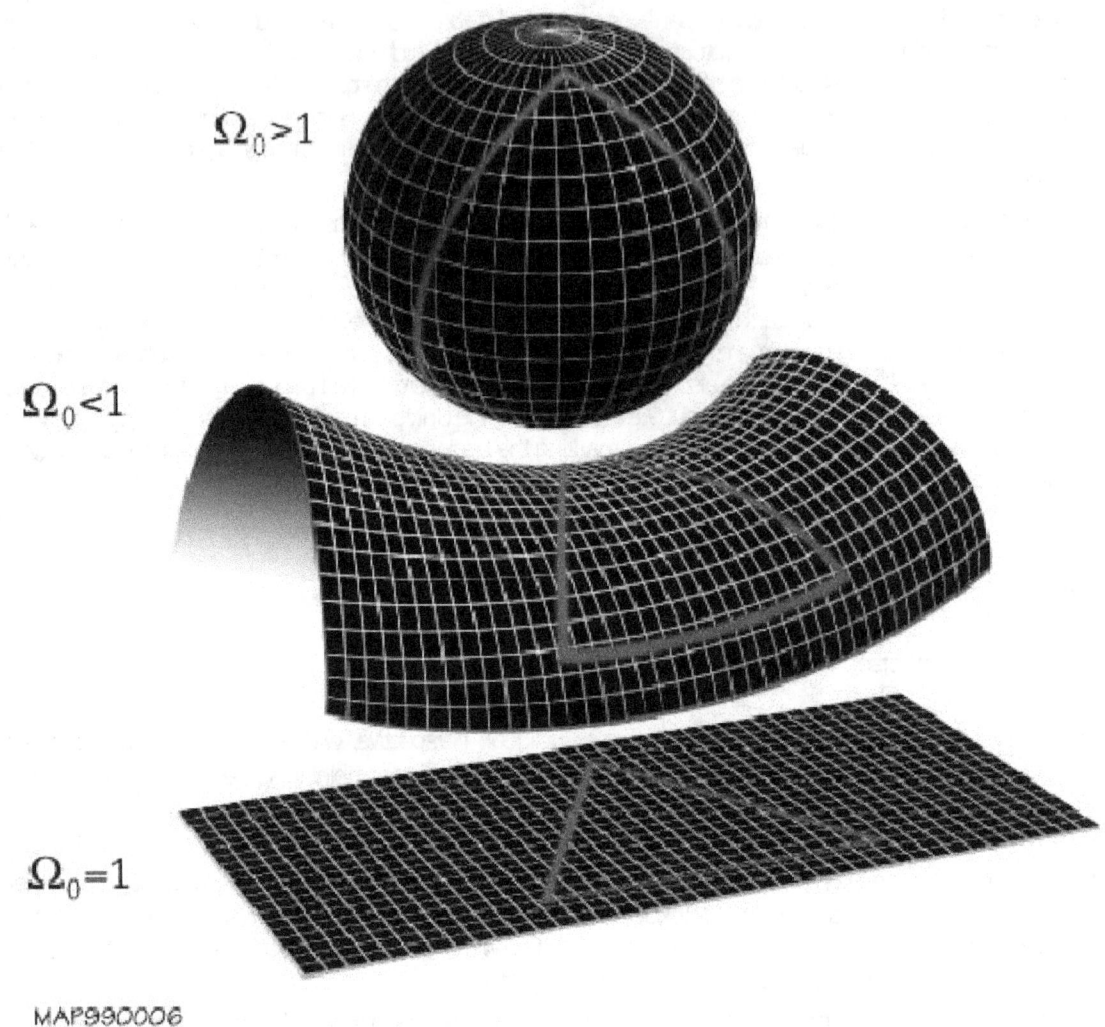

$\Omega_0 > 1$

$\Omega_0 < 1$

$\Omega_0 = 1$

MAP990006

The three possible options of the shape of the Universe.

that can be observed is the backward light cone, being the time it takes to reach a given observer within the cosmological horizon (particle horizon).

The particle horizon, the light horizon, or the cosmic light horizon, is the maximum distance from which particles can have traveled to the observer in the age of the Universe. An horizon represents the boundary between the observable and the unobservable region of the Universe.[44][45] The existence, properties, and significance of a cosmological horizon depend on the particular cosmological model.

Observational data suggest the cosmological topological of the Universe is infinite in extent with finite age, supported by the so-called Friedmann–Lemaître–Robertson–Walker (FLRW) models.[46][47][48][49] These FLRW models of space are consistent with the Wilkinson Microwave Anisotropy Probe (WMAP) and Planck maps of cosmic background radiation, thus supporting inflationary models and the standard model of cosmology, describing a flat, homogeneous universe dominated by dark matter and dark energy.[1][50]

4.4.2   Size and Regions

See also: Observable universe and Observational cosmology

According to a restrictive definition, the Universe is everything within our connected spacetime that could have a chance to interact with us and vice versa.[51] According to the general theory of relativity, some regions of space may never interact with ours even in the lifetime of the Universe due to the finite speed of light and the ongoing expansion of space. For example, radio messages sent from Earth may never reach some regions of space, even if the Universe were to exist forever: space may expand faster than light can traverse it.[52] Since we cannot observe space beyond the limitations of light, or any electromagnetic radiation, it is unknown whether the size of the Universe is finite or infinite.[14][53][54]

Distant regions of space are taken to exist and be part of reality as much as we are, yet we can never interact with them. The spatial region within which we can affect and be affected is the observable universe. The observable Universe depends on the location of the observer. By traveling, an observer can come into contact with a greater region of spacetime than an observer who remains still. Nevertheless, even the most rapid traveler will not be able to interact with all of space. Typically, the observable Universe is taken to mean the Universe observable from our vantage point in the Milky Way Galaxy.

The proper distance – the distance as would be measured at a specific time, including the present – between Earth and the edge of the observable universe is 46 billion light-years (14×10^9 pc), making the diameter of the observable universe about 91 billion light-years (28×10^9 pc). The distance the light from the edge of the observable universe has travelled is very close to the age of the Universe times the speed of light, 13.8 billion light-years (4.2×10^9 pc), but this does not represent the distance at any given time because the edge of the universe and the Earth have moved since further apart.[55] For comparison, the diameter of a typical galaxy is 30,000 light-years, and the typical distance between two neighboring galaxies is 3 million light-years.[56] As an example, the Milky Way Galaxy is roughly 100,000 light years in diameter,[57] and the nearest sister galaxy to the Milky Way, the Andromeda Galaxy, is located roughly 2.5 million light years away.[58]

4.4.3   Age and expansion

Main articles: Age of the universe and Metric expansion of space

Astronomers calculate the age of the Universe by assuming that the Lambda-CDM model accurately describes the evolution of the Universe from a very uniform, hot, dense primordial state to its present state and measuring the cosmological parameters which constitute the model. This model is well understood theoretically and supported by recent high-precision astronomical observations such as WMAP and Planck. Commonly, the set of observations fitted includes the cosmic microwave background anisotropy, the brightness/redshift relation for Type Ia supernovae, and large-scale galaxy clustering including the baryon acoustic oscillation feature. Other observations, such as the Hubble constant, the abundance of galaxy clusters, weak gravitational lensing and globular cluster ages, are generally consistent with these, providing a check of the model, but are less accurately measured at present. With the prior that the Lambda-CDM model is correct and the values of the parameters measured by various experiments including Planck, the best value of the age of the Universe as of 2015 is 13.798 ± 0.037 billion years.[1]

Over time, the Universe and its contents have evolved; for example, the relative population of quasars and galaxies has changed and space itself has expanded. This expansion accounts for how it is that scientists on Earth can observe the light from a galaxy 30 billion light years away, even if that light has traveled for only 13 billion years; the very space between them has expanded, and that is one of the tools used to calculate the age of the Universe. This expansion is consistent with the observation that the light from distant galaxies has been redshifted; the photons emitted have been stretched to longer wavelengths and lower frequency during their journey. The rate of this spatial expansion is accelerating, based on studies of Type Ia supernovae.

The more matter there is in the Universe, the stronger will be the gravitational pull among the matter. If the Universe were *too* dense then it would re-collapse into singularity. However, if the Universe contained too *little* matter then the expansion is accelerated greatly, thereby leaving no time for planets and planetary systems to form. After the Big Bang, the universe is continuously expanding. The rate of expansion is affected by the gravity among the matter present. Surprisingly, our universe has just the right mass density of about 5 protons per cubic meter which has allowed it to expand for last 13.8

billion years, giving time to form the universe as we see it today.[59]

There are dynamical forces acting on the particles in the Universe which affect the expansion rate. It was earlier expected that the Hubble Constant would be decreasing as time went on due to the influence of gravitational interactions in the Universe, and thus there is an additional observable quantity in the Universe called the deceleration parameter which cosmologists expected to be directly related to the matter density of the Universe. Surprisingly, the deceleration parameter was measured by two different groups to be less than zero (actually, consistent with −1) which implied that today Hubble's Constant is increasing as time goes on. Some cosmologists have whimsically called the effect associated with the "accelerating universe" the "cosmic jerk".[60] The 2011 Nobel Prize in Physics was given for the discovery of this phenomenon.[61]

4.4.4   Spacetime

Main articles: Spacetime and World line
See also: Lorentz transformation

Spacetimes are the arenas in which all physical events take place—an event is a point in spacetime specified by its time and place. For example, the motion of planets around the sun may be described in a particular type of spacetime, or the motion of light around a rotating star may be described in another type of spacetime. The basic elements of spacetime are events. In any given spacetime, an event is a unique position at a unique time. Because events are spacetime points, an example of an event in classical relativistic physics is (x, y, z, t), the location of an elementary (point-like) particle at a particular time. A spacetime itself can be viewed as the union of all events in the same way that a line is the union of all of its points, formally organized into a manifold, a space which can be described at small scales using coordinate systems.

The Universe appears to be a smooth spacetime continuum consisting of three spatial dimensions and one temporal (time) dimension. In spacetime, the displacement four-vector ΔR is given by the space displacement vector Δr and the time difference Δt between the events. The *spacetime interval*, also called *invariant interval*, is the interval between the two events, s^2

On the average, space is observed to be very nearly flat (close to zero curvature), meaning that Euclidean geometry is empirically true with high accuracy throughout most of the Universe.[62] Spacetime also appears to have a simply connected topology, in analogy with a sphere, at least on the length-scale of the observable Universe. However, present observations cannot exclude the possibilities that the Universe has more dimensions and that its spacetime may have a multiply connected global topology, in analogy with the cylindrical or toroidal topologies of two-dimensional spaces.[47][63]

4.5   Contents

See also: Galaxy formation and evolution, Galaxy cluster and Nebula

The Universe is composed of dark energy, dark matter, and ordinary matter. Ordinary baryonic matter, which includes atoms, stars, galaxies, and life, is only 4.9% of the contents. The present overall density of the this type of matter is very low, roughly 4.5×10^{-31} grams per cubic centimetre, corresponding to a density of the order of only one proton for every four cubic meters of volume.[4] The nature of both dark energy and dark matter is unknown. Dark matter, a mysterious form of matter that has not yet been identified, is 26.8% of the contents. Dark energy, which is the energy of empty space and that is causing the expansion of the Universe to accelerate, is the remaining 68.3% of the contents.[6][64][65]

Matter, dark matter, and dark energy are distributed homogeneously throughout the Universe over length scales longer than 300 million light-years or so.[66] However, on smaller length-scales, matter is observed to form clump hierarchically; many atoms are condensed into stars, most stars into galaxies, most galaxies into clusters, superclusters and, finally, the largest-scale structures such as the Sloan great wall. There are probably more than 100 billion (10^{11}) galaxies in the observable Universe.[67] Typical galaxies range from dwarfs with as few as ten million[68] (10^7) stars up to giants with one trillion[69] (10^{12}) stars. A 2010 study by astronomers estimated that the observable Universe contains 300 sextillion (3×10^{23}) stars.[70] Between the structures are voids, which are typically 10–150 Mpc (33 million–490 million ly) in

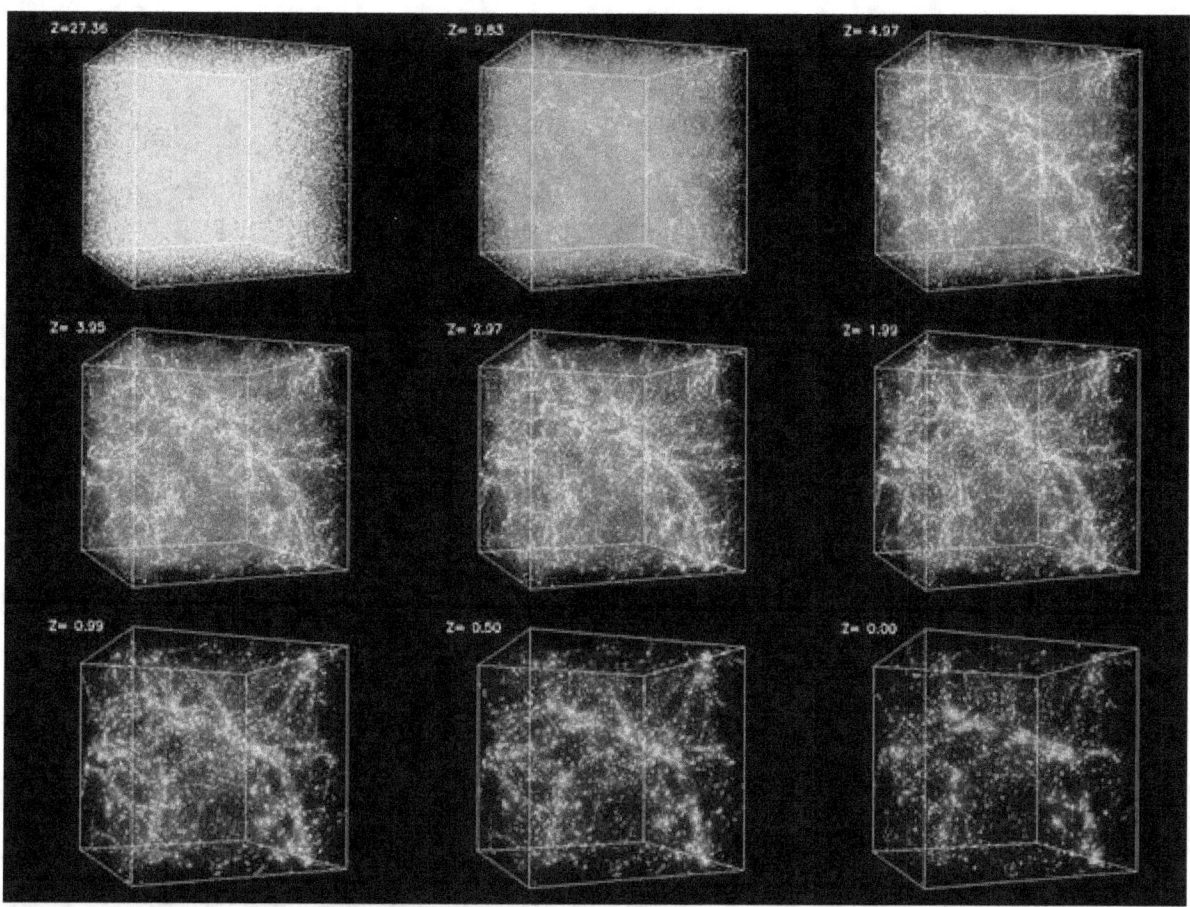

The formation of clusters and large-scale filaments in the Cold Dark Matter model with dark energy. The frames show the evolution of structures in a 43 million parsecs (or 140 million light years) box from redshift of 30 to the present epoch (upper left z=30 to lower right z=0).

diameter. The Milky Way is in the Local Group of galaxies, which in turn is in the Laniakea Supercluster.[71] This supercluster spans over 500 million light years, while the Local Group spans over 10 million light years.[72] In April 2015, astronomers announced the discovery of a supervoid, the largest known structure in the universe. This big hole is 1.8 billion ly (550 Mpc) across, characterized by its unusual emptiness.[73]

The observable matter of the Universe is also isotropic on large scales, meaning that no direction of observation appears to be different from any other.[74] The Universe is also bathed in a highly isotropic microwave radiation that corresponds to a thermal equilibrium blackbody spectrum of roughly 2.725 kelvin.[75] The hypothesis that the large-scale Universe is homogeneous and isotropic is known as the cosmological principle,[76] which is supported by astronomical observations.

4.5.1   Dark energy

Main article: Dark energy

An explanation for why the expansion of the Universe is accelerating remains elusive. It is often attributed to "dark energy", an unknown form of energy that is hypothesized to permeate space.[64] On a mass–energy equivalence basis, the density of dark energy (6.91×10^{-27} kg/m^3) is much less than the density of ordinary matter or dark matter within galaxies. However, in the present dark-energy era, it dominates the mass–energy of the universe because it is uniform across space.[77]

Two proposed forms for dark energy are the cosmological constant, a *constant* energy density filling space homogeneously,[78]

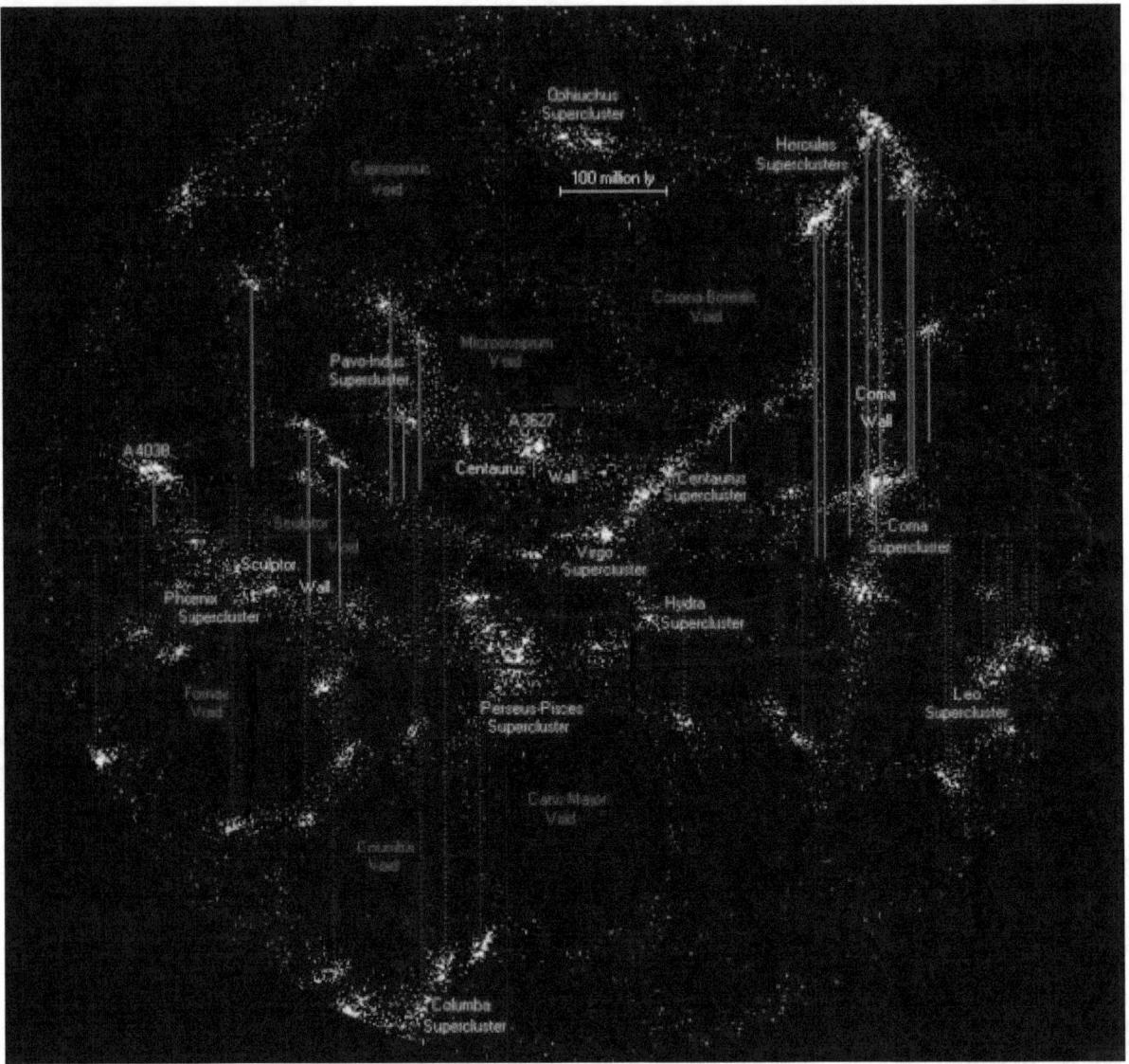

A map of the Superclusters and voids nearest to Earth

and scalar fields such as quintessence or moduli, *dynamic* quantities whose energy density can vary in time and space. Contributions from scalar fields that are constant in space are usually also included in the cosmological constant. The cosmological constant can be formulated to be equivalent to vacuum energy. Scalar fields having only a slight amont of spatial inhomogeneity can be difficult to distinguish from a cosmological constant.

4.5.2   Dark matter

Main article: Dark matter

Dark matter is a hypothetical kind of matter that cannot be seen with telescopes, but which accounts for most of the matter in the Universe. The existence and properties of dark matter are inferred from its gravitational effects on visible matter, radiation, and the large-scale structure of the Universe. Other than neutrinos, a form of hot dark matter, dark matter has not been detected directly, making it one of the greatest mysteries in modern astrophysics. Dark matter neither emits nor absorbs light or any other electromagnetic radiation at any significant level. Dark matter is estimated to constitute 26.8%

of the total mass–energy and 84.5% of the total matter in the Universe.[6][79]

4.5.3   Ordinary Matter

Main article: Matter

The remaining 4.9% of the mass–energy of the Universe is composed of matter made up of ordinary matter, that is atoms, ions, electrons and the objects they form. This matter includes stars, which produce nearly all of the light we see from galaxies, as well as interstellar gas in the interstellar and intergalactic media, planets, and all the objects from everyday life that we can bump into, touch or squeeze.[80]

Ordinary matter commonly exists in four states (or phases): solid, liquid and gas, and plasma. However, advances in experimental techniques have revealed other previously theoretical phases, such as Bose–Einstein condensates and fermionic condensates.

More fundamentally, matter is composed of two types of elementary particles: quarks and leptons.[81] For example, the neutron is formed of two down quarks and one up quark, while the proton is formed of two up quarks and one down quark. Atoms are made up of several neutrons and protons (two types of baryons), which have several electrons (a type of lepton) orbiting them. Because most of the mass of the atom is concentrated in the atomic nucleus, which is made up of baryons, astronomers often use the term *baryonic matter* to describe ordinary matter, although a small fraction of this baryonic matter is also composed of electrons. A focus on an elementary-particle view of matter also leads to new phases of matter, such as the quark–gluon plasma.[82]

The primordial protons and neutrons themselves were formed from the quark–gluon plasma during the Big Bang as it cooled below two trillion degrees. A few minutes afterwards, starting with only protons and neutrons, nuclei up to lithium and beryllium were formed, but the abundances of other elements dropped sharply with growing atomic mass. Some boron may have been formed at this time, but the process stopped before significant carbon could be formed. This process, known as Big Bang nucleosynthesis, essentially shut down after about 20 minutes, due to drops in temperature and density as the universe continued to expand. The subsequent nucleosynthesis of heavier elements required the extreme temperatures and pressures found within stars and supernovas, in the processes of stellar nucleosynthesis and supernova nucleosynthesis.

4.5.4   Particles

Main article: Particle physics

The ordinary matter in the Universe consists of particles described by particle physics. An elementary particle or fundamental particle is a particle whose substructure (domain of the bigger structure which shares the similar characteristics of the domain) is unknown, thus it is unknown whether it is composed of other particles.[83] Known elementary particles include the fundamental fermions (quarks, leptons, antiquarks, and antileptons), which generally are "matter particles" and "antimatter particles", as well as the fundamental bosons (gauge bosons and Higgs boson), which generally are "force particles" that mediate interactions among fermions.[83] A particle containing two or more elementary particles is a *composite particle*.

The Standard Model of particle physics suggests a universal set of physical laws and physical constants.[84] It concerns the electromagnetic, weak, and strong nuclear interactions, as well as classifying all the subatomic particles known. It was developed throughout the latter half of the 20th century as a collaborative effort of scientists around the world.[85] The current formulation was finalized in the mid-1970s upon experimental confirmation of the existence of quarks. Since then, discoveries of the top quark (1995), the tau neutrino (2000), and more recently the Higgs boson (2013), have given further credence to the Standard Model. Because of its success in explaining a wide variety of experimental results, the Standard Model is sometimes regarded as a "theory of almost everything".

Each particle generation is divided into two leptons and two quarks, both of which are fermions. Bosons are the other fundamental class of elementary particles. Additionally, there are hypothetical particles.

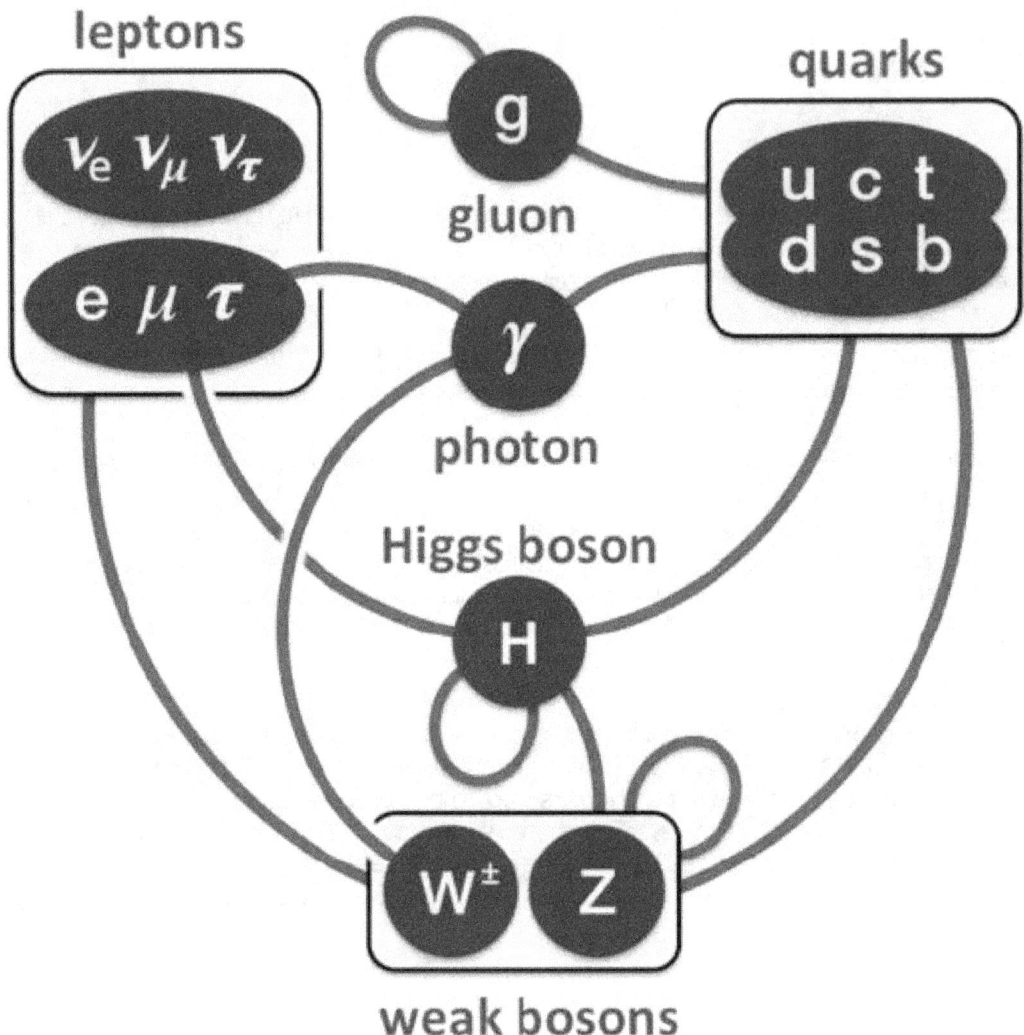

The tree-level interactions between elementary particles described in the Standard Model. Vertices (darkened circles) represent types of particles, and edges (blue arcs) connecting them represent interactions that can take place. The top row of vertices in the chart (leptons and quarks) are the matter particles; the second row of vertices (photon, W/Z, gluons) are the force mediating particles; and the bottom row is the Higgs boson. The elementary particles from which the Universe is constructed. Six leptons and six quarks comprise most of the matter; for example, the protons and neutrons of atomic nuclei are composed of quarks; the ubiquitous electron is a lepton. These particles interact via the gauge bosons shown in the middle row, each corresponding to a particular type of gauge symmetry. The Higgs boson is believed to confer mass on the particles with which it is connected. The graviton is not shown.

Hadrons

Main article: Hadron

A hadron is a composite particle made of quarks bound state by the strong force (in a similar way as molecules are held together by the electromagnetic force). From approximately 10^{-6} seconds after the Big Bang, when the temperature of the universe had fallen sufficiently to allow the quarks from the preceding quark epoch to bind together into hadrons, the mass of the universe was dominated by hadrons. This period is known as the hadron epoch. Initially the temperature was high enough to allow the formation of hadron/anti-hadron pairs, which kept matter and antimatter in thermal equilibrium.

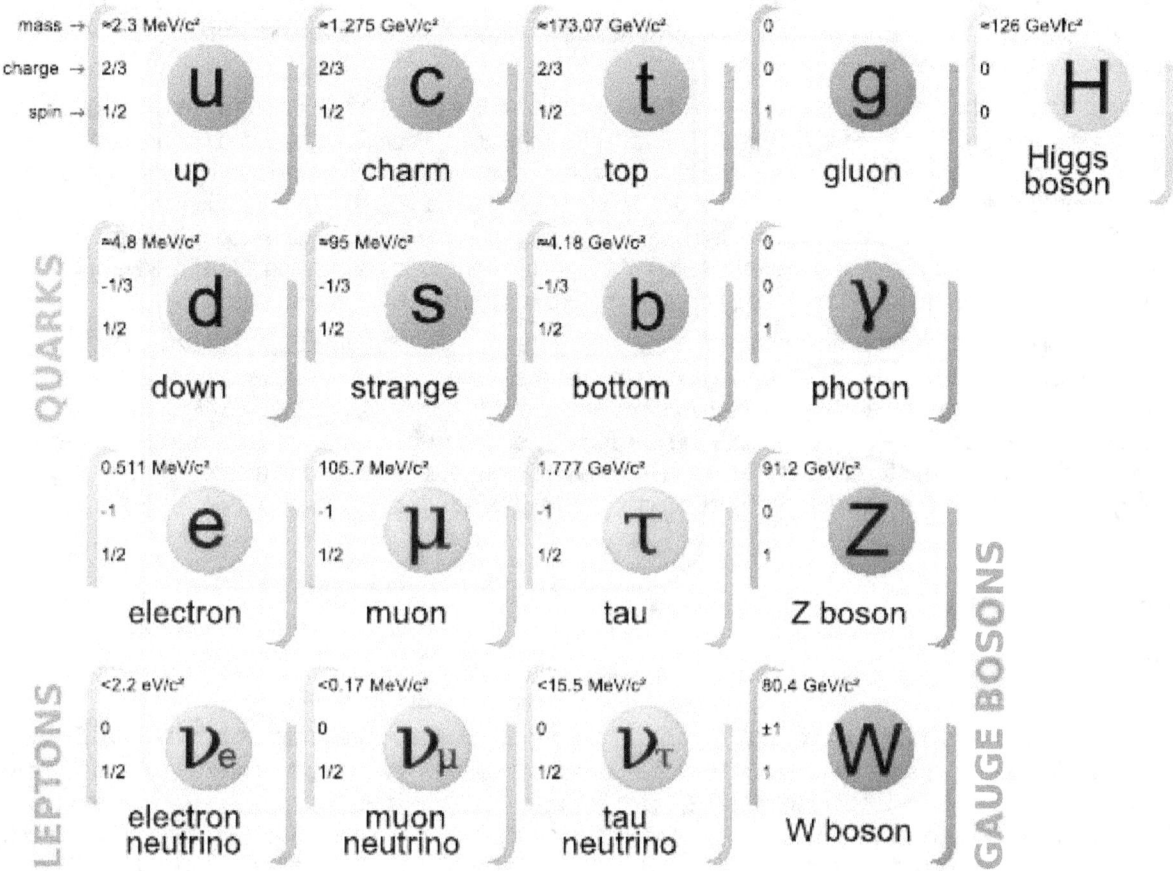

Six of the particles in the Standard Model are quarks (shown in purple). Each of the first three columns forms a generation of matter.

However, as the temperature of the Universe continued to fall, hadron/anti-hadron pairs were no longer produced. Most of the hadrons and anti-hadrons were then eliminated in annihilation reactions, leaving a small residual of hadrons. The elimination of anti-hadrons was completed by one second after the Big Bang, when the following lepton epoch began.

Hadrons are categorized into two families: baryons (such as protons and neutrons) made of three quarks and mesons (such as pions) made of one quark and one antiquark. Of the hadrons, protons are stable, and neutrons are bound within atomic nuclei are stable. Other hadrons are unstable under ordinary conditions; free neutrons decay with a half-life of about 611 seconds. Experimentally, hadron physics is studied by colliding protons or nuclei of heavy elements such as lead, and detecting the debris in the produced particle showers.

Leptons

Main article: Lepton

A lepton is an elementary, half-integer spin (spin $1/2$) particle that does not undergo strong interactions but is subject to the Pauli exclusion principle; no two leptons of the same species can be in exactly the same state at the same time.[86] Fermions differ from bosons, which obey Bose–Einstein statistics. The best known of all leptons is the electron, which governs nearly all of chemistry as it is found in atoms and is directly tied to all chemical properties. Two main classes of leptons exist: charged leptons (also known as the *electron-like* leptons), and neutral leptons (better known as neutrinos). The two leptons may be classified into one with electric charge –1 (electron-like) and one neutral (neutrino). Thus electrons are stable and the most common charged lepton in the Universe, whereas muons and taus can only be produced in high energy collisions (such as those involving cosmic rays and those carried out in particle accelerators).[87][88] Charged leptons can combine with other particles to form various composite particles such as atoms and positronium, while neutrinos rarely interact with anything, and are consequently rarely observed. Neutrinos of all generations stream throughout the Universe but rarely interact with normal matter.[89]

The lepton epoch was the period in the evolution of the early Universe in which the leptons dominated the mass of the Universe. It started roughly 1 second after the Big Bang, after the majority of hadrons and anti-hadrons annihilated each other at the end of the hadron epoch. During the lepton epoch the temperature of the Universe was still high enough to create lepton/anti-lepton pairs, so leptons and anti-leptons were in thermal equilibrium. Approximately 10 seconds after the Big Bang, the temperature of the Universe had fallen to the point where lepton/anti-lepton pairs were no longer created.[90] Most leptons and anti-leptons were then eliminated in annihilation reactions, leaving a small residue of leptons. The mass of the Universe was then dominated by photons as it entered the following photon epoch.

Leptons have various intrinsic properties, including electric charge, spin, and mass. Unlike quarks however, leptons are not subject to the strong interaction, but they are subject to the other three fundamental interactions: gravitation (best described at present by general relativity), electroweak (includes electromagnetism), and the weak interaction. During the quark epoch, the electroweak force split into the electromagnetic and weak forces. Interactions can be described by renormalized quantum field theory, and are mediated by gauge bosons that correspond to a particular type of gauge symmetry. A renormalized quantum field theory of general relativity has not yet been achieved.

Bosons

See also: Higgs boson

Bosons may be either elementary, like photons, or composite, like mesons. Bosons are characterized by Bose–Einstein statistics and all have integer spins. While most bosons are composite particles, and these are important in superfluidity and other applications of Bose–Einstein condensates, in the Standard Model there are five bosons which are elementary: photons, which carry the electromagnetic interaction; W and Z bosons, which carry the weak interaction; gluons, which carry the strong interaction, and the scalar boson Higgs Boson.[91] Additionally, the graviton is a hypothetical elementary particle not incorporated in the Standard Model. If it exists, a graviton must be a boson, and could conceivably be a gauge boson.

The Higgs boson or Higgs particle is an elementary particle in the Standard Model. Observations of the particle allows scientists to explore the Higgs field[92][93]—a fundamental field of crucial importance to particle physics theory, that unlike the more familiar electromagnetic field cannot be "turned off ", but instead takes a non-zero constant value almost everywhere.[93] The presence of this field, now believed to be confirmed, explains why some fundamental particles have mass even though the symmetries controlling their interactions should require them to be massless, and also answers several other long-standing puzzles in physics, such as the reason the weak force has a much shorter range than the electromagnetic force. The Higgs field can be detected through its excitations (i.e. Higgs particles), but these are extremely hard to produce and detect.

Photons

Main article: Photon epoch
See also: Photino

A photon is an elementary particle, the quantum of light and all other forms of electromagnetic radiation. It is the force carrier for the electromagnetic force, even when static via virtual photons. The effects of this force are easily observable at the microscopic and at the macroscopic level because the photon has zero rest mass; this allows long distance interactions. Like all elementary particles, photons are currently best explained by quantum mechanics and exhibit wave–particle duality, exhibiting properties of waves and of particles. For example, a single photon may be refracted by a lens or exhibit wave interference with itself but also act as a particle giving a definite result when its position is measured.

In the Standard Model, the photon is one of four gauge bosons in the electroweak interaction; the other three are denoted W^+, W^- and Z^0 and are responsible for the weak interaction. Unlike the photon, these gauge bosons have mass, owing to a mechanism that breaks their SU(2) gauge symmetry.

The photon epoch started after most leptons and anti-leptons were annihilated at the end of the lepton epoch, about 10 seconds after the Big Bang. Atomic nuclei were created in the process of nucleosynthesis which occurred during the first few minutes of the photon epoch. For the remainder of the photon epoch the universe contained a hot dense plasma of nuclei, electrons and photons. About 380,000 years after the Big Bang the temperature of the universe fell to the point where nuclei could combine with electrons to create neutral atoms. As a result, photons no longer interacted frequently with matter, the universe became transparent and the cosmic microwave background radiation was created and then structure formation took place.

In the Standard Model, photons and other elementary particles are described as a necessary consequence of physical laws having a certain symmetry at every point in spacetime. The intrinsic properties of particles, such as electric charge, mass and spin, are determined by the properties of this gauge symmetry.

4.6   Geometry

See also: Physics beyond the Standard Model and Elementary particle § Beyond the Standard Model

In modern physics, *space* and *time* are unified in a four-dimensional Minkowski continuum (a single interwoven continuum) called *Spacetime*, with the goal to create a phase space (dynamical system in which all possible states of a system are represented), whose metric treats the time dimension differently from the three spatial dimensions.

The geometry of 4-dimensional space is much more complex than that of 3-dimensional space, due to the extra degree of freedom (time, t). The set of points in Euclidean 4-space having the same distance R from a fixed point P_0 forms a hypersurface known as a 3-sphere (see also Hypersphere). The hypervolume of the enclosed space can be calculated with:

$$V = _{\frac{1}{2}} \pi^2 R^4$$

This is part of the Friedmann–Lemaître–Robertson–Walker metric in General relativity where R is substituted by function $R(t)$ with t meaning the cosmological age of the universe. Growing or shrinking R with time means expanding or collapsing universe, depending on the mass density inside.[94] Thus, after the singularity, space begun to expand, and models try to recapture past and future evolution of the Universe, and to understand the fundamental principles at work.

4.7   Theories of physics

See also: Supersymmetry

4.7.1   Special relativity

Main articles: Introduction to special relativity and Special relativity

Special relativity is the generally accepted physical theory regarding the relationship between space and time. It is based on two postulates: (1) that the laws of physics are invariant (i.e. identical) in all inertial systems (non-accelerating frames of reference); and (2) that the speed of light in a vacuum is the same for all observers, regardless of the motion of the light source. It was originally proposed in 1905 by Albert Einstein in the paper "On the Electrodynamics of Moving Bodies".[95] The inconsistency of Newtonian mechanics with Maxwell's equations of electromagnetism and the inability to discover Earth's motion through a luminiferous aether led to the development of special relativity, which corrects mechanics to handle situations involving motions nearing the speed of light. As of today, special relativity is the most accurate model of motion at any speed. Even so, Newtonian mechanics is still useful (due to its simplicity and high accuracy) as an approximation at small velocities relative to the speed of light.

Special relativity implies a wide range of consequences, which have been experimentally verified,[96] including length contraction, time dilation, relativistic mass, mass–energy equivalence, a universal speed limit, and relativity of simultaneity. It has replaced the conventional notion of an absolute universal time with the notion of a time that is dependent on reference frame and spatial position. Rather than an invariant time interval between two events, there is an invariant spacetime interval. Combined with other laws of physics, the two postulates of special relativity predict the equivalence of mass and energy, as expressed in the mass–energy equivalence formula $E = mc^2$, where c is the speed of light in vacuum.[97][98]

A defining feature of special relativity is the replacement of the Galilean transformations of Newtonian mechanics with the Lorentz transformations. Time and space cannot be defined separately from each other. Rather space and time are interwoven into a single continuum known as spacetime. Events that occur at the same time for one observer could occur at different times for another.

4.7.2 General relativity

Main articles: Introduction to general relativity, General relativity and Einstein's field equations

General relativity is the geometric theory of gravitation published by Albert Einstein in 1915[99] and the current description of gravitation in modern physics. It is the basis of current cosmological models of a consistently expanding universe. General relativity generalizes special relativity and Newton's law of universal gravitation, providing a unified description of gravity as a geometric property of space and time, or spacetime. In particular, the curvature of spacetime is directly related to the energy and momentum of whatever matter and radiation are present. The relation is specified by the Einstein field equations, a system of partial differential equations.

Some predictions of general relativity differ significantly from those of classical physics, especially concerning the passage of time, the geometry of space, the motion of bodies in free fall, and the propagation of light. Examples of such differences include gravitational time dilation, gravitational lensing, the gravitational redshift of light, and the gravitational time delay. The predictions of general relativity have been confirmed in all observations and experiments to date. Although general relativity is not the only relativistic theory of gravity, it is the simplest theory that is consistent with experiments and observations. However, unanswered questions remain, the most fundamental being how general relativity can be reconciled with the laws of quantum physics to produce a complete and self-consistent theory of quantum gravity.

Solving Einstein's field equations

See also: Big Bang and Ultimate fate of the Universe

In non-Cartesian (non-square) or curved coordinate systems, the Pythagorean theorem holds only on infinitesimal length scales and must be augmented with a more general metric tensor $g_{\mu\nu}$, which can vary from place to place and which describes the local geometry in the particular coordinate system. However, assuming the cosmological principle that the Universe is homogeneous and isotropic everywhere, every point in space is like every other point; hence, the metric tensor must be the same everywhere. That leads to a single form for the metric tensor, called the Friedmann–Lemaître–Robertson–Walker metric

$$ds^2 = -c^2 dt^2 + R(t)^2 \left(\frac{dr^2}{} + r^2 d\theta^2 + r^2 \sin^2 \theta d\phi^2 \right)$$

where (r, θ, φ) correspond to a spherical coordinate system. This metric has only two undetermined parameters: an overall length scale R that can vary with time, and a curvature index k that can be only 0, 1 or −1, corresponding to flat Euclidean geometry, or spaces of positive or negative curvature. In cosmology, solving for the history of the Universe is done by calculating R as a function of time, given k and the value of the cosmological constant Λ, which is a (small) parameter in

Einstein's field equations. The equation describing how R varies with time is known as the Friedmann equation, after its inventor, Alexander Friedmann.[100]

The solutions for $R(t)$ depend on k and Λ, but some qualitative features of such solutions are general. First and most importantly, the length scale R of the Universe can remain constant *only* if the Universe is perfectly isotropic with positive curvature ($k=1$) and has one precise value of density everywhere, as first noted by Albert Einstein. However, this equilibrium is unstable and because the Universe is known to be inhomogeneous on smaller scales, R must change, according to general relativity. When R changes, all the spatial distances in the Universe change in tandem; there is an overall expansion or contraction of space itself. This accounts for the observation that galaxies appear to be flying apart; the space between them is stretching. The stretching of space also accounts for the apparent paradox that two galaxies can be 40 billion light years apart, although they started from the same point 13.8 billion years ago[101] and never moved faster than the speed of light.

Second, all solutions suggest that there was a gravitational singularity in the past, when R goes to zero and matter and energy became infinitely dense. It may seem that this conclusion is uncertain because it is based on the questionable assumptions of perfect homogeneity and isotropy (the cosmological principle) and that only the gravitational interaction is significant. However, the Penrose–Hawking singularity theorems show that a singularity should exist for very general conditions. Hence, according to Einstein's field equations, R grew rapidly from an unimaginably hot, dense state that existed immediately following this singularity (when R had a small, finite value); this is the essence of the Big Bang model of the Universe. A common misconception is that the Big Bang model predicts that matter and energy exploded from a single point in space and time; that is false. Rather, space itself was created in the Big Bang and imbued with a fixed amount of energy and matter distributed uniformly throughout; as space expands (i.e., as $R(t)$ increases), the density of that matter and energy decreases.

Third, the curvature index k determines the sign of the mean spatial curvature of spacetime averaged over length scales greater than a billion light years. If $k=1$, the curvature is positive and the Universe has a finite volume. Such universes are often visualized as a three-dimensional sphere S^3 embedded in a four-dimensional space. Conversely, if k is zero or negative, the Universe *may* have infinite volume, depending on its overall topology. It may seem counter-intuitive that an infinite and yet infinitely dense Universe could be created in a single instant at the Big Bang when $R=0$, but exactly that is predicted mathematically when k does not equal 1. For comparison, an infinite plane has zero curvature but infinite area, whereas an infinite cylinder is finite in one direction and a torus is finite in both. A toroidal Universe could behave like a normal Universe with periodic boundary conditions, as seen in "wrap-around" video games such as *Asteroids*; a traveler crossing an outer "boundary" of space going *outwards* would reappear instantly at another point on the boundary moving *inwards*.

The ultimate fate of the Universe is still unknown, because it depends critically on the curvature index k and the cosmological constant Λ. If the Universe is sufficiently dense, k equals +1, meaning that its average curvature throughout is positive and the Universe will eventually recollapse in a Big Crunch, possibly starting a new Universe in a Big Bounce. Conversely, if the Universe is insufficiently dense, k equals 0 or −1 and the Universe will expand forever, cooling off and eventually becoming inhospitable for all life, as the stars die and all matter coalesces into black holes (the Big Freeze and the heat death of the Universe). As noted above, recent data suggests that the expansion speed of the Universe is not decreasing as originally expected, but increasing; if this continues indefinitely, the Universe will eventually rip itself to shreds (the Big Rip). Observationally, the Universe appears to have an overall density that is very close to the critical value between recollapse and eternal expansion; more careful astronomical observations are needed to resolve the question.

4.7.3 Multiverse hypothesis

Main articles: Multiverse, Many-worlds interpretation, Bubble universe theory and Parallel universe (fiction)
See also: Eternal inflation

Some speculative theories have proposed that our Universe is but one of a set of disconnected universes, collectively denoted as the multiverse, challenging or enhancing more limited definitions of the Universe.[20][102] Scientific multiverse models are distinct from concepts such as alternate planes of consciousness and simulated reality.

Max Tegmark developed a four-part classification scheme for the different types of multiverses that scientists have suggested in various problem domains. An example of such a model is the chaotic inflation model of the early universe.[103]

Another is the many-worlds interpretation of quantum mechanics. Parallel worlds are generated in a manner similar to quantum superposition and decoherence, with all states of the wave function being realized in separate worlds. Effectively, the multiverse evolves as a universal wavefunction. If the Big Bang that created our multiverse created an ensemble of multiverses, the wave function of the ensemble would be entangled in this sense.

The least controversial category of multiverse in Tegmark's scheme is Level I, which describes distant spacetime events "in our own universe", but suggests that statistical analysis exploiting the anthropic principle provides an opportunity to test multiverse theories in some cases. If space is infinite, or sufficiently large and uniform, identical instances of the history of Earth's entire Hubble volume occur every so often, simply by chance. Tegmark calculated our nearest so-called doppelgänger, is $10^{10^{115}}$ meters away from us (a double exponential function larger than a googolplex).[104][105] In principle, it would be impossible to scientifically verify an identical Hubble volume. However, it does follow as a fairly straightforward consequence from otherwise unrelated scientific observations and theories.

It is possible to conceive of disconnected spacetimes, each existing but unable to interact with one another.[104][106] An easily visualized metaphor is a group of separate soap bubbles, in which observers living on one soap bubble cannot interact with those on other soap bubbles, even in principle.[107] According to one common terminology, each "soap bubble" of spacetime is denoted as a *universe*, whereas our particular spacetime is denoted as *the Universe*,[20] just as we call our moon *the Moon*. The entire collection of these separate spacetimes is denoted as the multiverse.[20] With this terminology, different *Universes* are not causally connected to each other.[20] In principle, the other unconnected *Universes* may have different dimensionalities and topologies of spacetime, different forms of matter and energy, and different physical laws and physical constants, although such possibilities are purely speculative.[20] Others consider each of several bubbles created as part of chaotic inflation to be separate *Universes*, though in this model these universes all share a causal origin.[20]

4.7.4 Fine-tuned Universe

Main article: Fine-tuned Universe

The fine-tuned Universe is the proposition that the conditions that allow life in the Universe can only occur when certain universal fundamental physical constants lie within a very narrow range, so that if any of several fundamental constants were only slightly different, the Universe would be unlikely to be conducive to the establishment and development of matter, astronomical structures, elemental diversity, or life as it is understood.[108] The proposition is discussed among philosophers, scientists, theologians, and proponents and detractors of creationism.

4.8 Historical development

See also: Cosmology, Timeline of cosmology, Nicolaus Copernicus § Copernican system and Philosophiæ Naturalis Principia Mathematica § Beginnings of the Scientific Revolution

Historically, there have been many ideas of the cosmos (cosmologies) and its origin (cosmogonies). Theories of an impersonal Universe governed by physical laws were first proposed by the Greeks and Indians.[16] Ancient Chinese philosophy encompassed the notion of the Universe including both all of space and all of time.[109][110] Over the centuries, improvements in astronomical observations and theories of motion and gravitation led to ever more accurate descriptions of the Universe. The modern era of cosmology began with Albert Einstein's 1915 general theory of relativity, which made it possible to quantitatively predict the origin, evolution, and conclusion of the Universe as a whole. Most modern, accepted theories of cosmology are based on general relativity and, more specifically, the predicted Big Bang.[111]

4.8.1 Mythologies

Main articles: Creation myth, Creator deity and Religious cosmology

Many cultures have stories describing the origin of the world and universe. Cultures generally regard these stories as having some truth.[112] There are however many differing beliefs in how these stories apply amongst those believing in a supernatural origin, ranging from a god directly creating the Universe as it is now to a god just setting the "wheels in motion" (for example via mechanisms such as the big bang and evolution).

Creation stories may be roughly grouped into common types. In one type of story, the world is born from a world egg; such stories include the Finnish epic poem *Kalevala*, the Chinese story of Pangu or the Indian Brahmanda Purana. In related stories, the Universe is created by a single entity emanating or producing something by him- or herself, as in the Tibetan Buddhism concept of Adi-Buddha, the ancient Greek story of Gaia (Mother Earth), the Aztec goddess Coatlicue myth, the ancient Egyptian god Atum story, and the Judeo-Christian Genesis creation narrative in which the Abrahamic God created the Universe. In another type of story, the Universe is created from the union of male and female deities, as in the Maori story of Rangi and Papa. In other stories, the Universe is created by crafting it from pre-existing materials, such as the corpse of a dead god — as from Tiamat in the Babylonian epic *Enuma Elish* or from the giant Ymir in Norse mythology – or from chaotic materials, as in Izanagi and Izanami in Japanese mythology. In other stories, the Universe emanates from fundamental principles, such as Brahman and Prakrti, the creation myth of the Serers,[113] or the yin and yang of the Tao.

4.8.2   Philosophical models

Further information: Cosmology
See also: Pre-Socratic philosophy, Physics (Aristotle), Hindu cosmology, Islamic cosmology and Philosophy of space and time

The pre-Socratic Greek philosophers and Indian philosophers developed some of the earliest philosophical concepts of the Universe.[16][114] The earliest Greek philosophers noted that appearances can be deceiving, and sought to understand the underlying reality behind the appearances. In particular, they noted the ability of matter to change forms (e.g., ice to water to steam) and several philosophers proposed that all the physical materials in the world are different forms of a single primordial material, or *arche*. The first to do so was Thales, who proposed this material is water. Thales' student, Anaximander, proposed that everything came from the limitless *apeiron*. Anaximenes proposed air on account of its perceived attractive and repulsive qualities that cause the *arche* to condense or dissociate into different forms. Anaxagoras proposed the principle of *Nous* (Mind). Heraclitus proposed fire (and spoke of *logos*). Empedocles proposed the elements: earth, water, air and fire. His four-element model became very popular. Like Pythagoras, Plato believed that all things were composed of number, with Empedocles' elements taking the form of the Platonic solids. Democritus, and later philosophers—most notably Leucippus—proposed that the Universe is composed of indivisible atoms moving through void (vacuum). Aristotle did not believe that was feasible because air, like water, offers resistance to motion. Air will immediately rush in to fill a void, and moreover, without resistance, it would do so indefinitely fast.

Although Heraclitus argued for eternal change, his contemporary Parmenides made the radical suggestion that all change is an illusion, that the true underlying reality is eternally unchanging and of a single nature. Parmenides denoted this reality as τὸ ἐν (The One). Parmenides' idea seemed implausible to many Greeks, but his student Zeno of Elea challenged them with several famous paradoxes. Aristotle responded to these paradoxes by developing the notion of a potential countable infinity, as well as the infinitely divisible continuum. Unlike the eternal and unchanging cycles of time, he believed the world is bounded by the celestial spheres and that cumulative stellar magnitude is only finitely multiplicative.

The Indian philosopher Kanada, founder of the Vaisheshika school, developed a notion of atomism and proposed that light and heat were varieties of the same substance.[115] In the 5th century AD, the Buddhist atomist philosopher Dignāga proposed atoms to be point-sized, durationless, and made of energy. They denied the existence of substantial matter and proposed that movement consisted of momentary flashes of a stream of energy.[116]

The notion of temporal finitism was inspired by the doctrine of creation shared by the three Abrahamic religions: Judaism, Christianity and Islam. The Christian philosopher, John Philoponus, presented the philosophical arguments against the ancient Greek notion of an infinite past and future. Philoponus' arguments against an infinite past were used by the early Muslim philosopher, Al-Kindi (Alkindus); the Jewish philosopher, Saadia Gaon (Saadia ben Joseph); and the Muslim theologian, Al-Ghazali (Algazel).

4.8.3 Astronomical concepts

Main articles: History of astronomy and Timeline of astronomy
Astronomical models of the Universe were proposed soon after astronomy began with the Babylonian astronomers, who viewed the Universe as a flat disk floating in the ocean, and this forms the premise for early Greek maps like those of Anaximander and Hecataeus of Miletus.

Later Greek philosophers, observing the motions of the heavenly bodies, were concerned with developing models of the Universe based more profoundly on empirical evidence. The first coherent model was proposed by Eudoxus of Cnidos. According to Aristotle's physical interpretation of the model, celestial spheres eternally rotate with uniform motion around a stationary Earth. Normal matter is entirely contained within the terrestrial sphere.

De Mundo (composed before 250 BC or between 350 and 200 BC), stated, *Five elements, situated in spheres in five regions, the less being in each case surrounded by the greater — namely, earth surrounded by water, water by air, air by fire, and fire by ether — make up the whole Universe.*[117]

This model was also refined by Callippus and after concentric spheres were abandoned, it was brought into nearly perfect agreement with astronomical observations by Ptolemy. The success of such a model is largely due to the mathematical fact that any function (such as the position of a planet) can be decomposed into a set of circular functions (the Fourier modes). Other Greek scientists, such as the Pythagorean philosopher Philolaus, postulated (according to Stobaeus account) that at the center of the Universe was a "central fire" around which the Earth, Sun, Moon and Planets revolved in uniform circular motion.[118]

The Greek astronomer Aristarchus of Samos was the first known individual to propose a heliocentric model of the Universe. Though the original text has been lost, a reference in Archimedes' book *The Sand Reckoner* describes Aristarchus's heliocentric model. Archimedes wrote: (translated into English):

> "You, King Gelon, are aware the Universe is the name given by most astronomers to the sphere the center of which is the center of the Earth, while its radius is equal to the straight line between the center of the Sun and the center of the Earth. This is the common account as you have heard from astronomers. But Aristarchus has brought out a book consisting of certain hypotheses, wherein it appears, as a consequence of the assumptions made, that the Universe is many times greater than the Universe just mentioned. His hypotheses are that the fixed stars and the Sun remain unmoved, that the Earth revolves about the Sun on the circumference of a circle, the Sun lying in the middle of the orbit, and that the sphere of fixed stars, situated about the same center as the Sun, is so great that the circle in which he supposes the Earth to revolve bears such a proportion to the distance of the fixed stars as the center of the sphere bears to its surface"

Aristarchus thus believed the stars to be very far away, and saw this as the reason why stellar parallax had not been observed, that is, the stars had not been observed to move relative each other as the Earth moved around the Sun. The stars are in fact much farther away than the distance that was generally assumed in ancient times, which is why stellar parallax is only detectable with precision instruments. The geocentric model, consistent with planetary parallax, was assumed to be an explanation for the unobservability of the parallel phenomenon, stellar parallax. The rejection of the heliocentric view was apparently quite strong, as the following passage from Plutarch suggests (*On the Apparent Face in the Orb of the Moon*):

> "Cleanthes [a contemporary of Aristarchus and head of the Stoics] thought it was the duty of the Greeks to indict Aristarchus of Samos on the charge of impiety for putting in motion the Hearth of the Universe [i.e. the Earth], . . . supposing the heaven to remain at rest and the Earth to revolve in an oblique circle, while it rotates, at the same time, about its own axis"

The only other astronomer from antiquity known by name who supported Aristarchus's heliocentric model was Seleucus of Seleucia, a Hellenistic astronomer who lived a century after Aristarchus.[119][120][121] According to Plutarch, Seleucus was the first to prove the heliocentric system through reasoning, but it is not known what arguments he used. Seleucus' arguments for a heliocentric cosmology were probably related to the phenomenon of tides.[122] According to Strabo (1.1.9), Seleucus was the first to state that the tides are due to the attraction of the Moon, and that the height of the tides depends on the Moon's position relative to the Sun.[123] Alternatively, he may have proved heliocentricity by determining the constants of a geometric model for it, and by developing methods to compute planetary positions using this model, like what Nicolaus Copernicus later did in the 16th century.[124] During the Middle Ages, heliocentric models were also proposed by the Indian astronomer Aryabhata,[125] and by the Persian astronomers Albumasar[126] and Al-Sijzi.[127]

The Aristotelian model was accepted in the Western world for roughly two millennia, until Copernicus revived Aristarchus's perspective that the astronomical data could be explained more plausibly if the earth rotated on its axis and if the sun were placed at the center of the Universe.

As noted by Copernicus himself, the notion that the Earth rotates is very old, dating at least to Philolaus (c. 450 BC), Heraclides Ponticus (c. 350 BC) and Ecphantus the Pythagorean. Roughly a century before Copernicus, the Christian scholar Nicholas of Cusa also proposed that the Earth rotates on its axis in his book, *On Learned Ignorance* (1440).[128] Aryabhata (476–550 AD/CE), Brahmagupta (598–668), and Al-Sijzi,[129] also proposed that the Earth rotates on its axis. The first empirical evidence for the Earth's rotation on its axis, using the phenomenon of comets, was given by Tusi (1201–1274) and Ali Qushji (1403–1474).

This cosmology was accepted by Isaac Newton, Christiaan Huygens and later scientists.[130] Edmund Halley (1720)[131] and Jean-Philippe de Chéseaux (1744)[132] noted independently that the assumption of an infinite space filled uniformly with stars would lead to the prediction that the nighttime sky would be as bright as the Sun itself; this became known as Olbers' paradox in the 19th century.[133] Newton believed that an infinite space uniformly filled with matter would cause infinite forces and instabilities causing the matter to be crushed inwards under its own gravity.[130] This instability was clarified in 1902 by the Jeans instability criterion.[134] One solution to these paradoxes is the Charlier Universe, in which the matter is arranged hierarchically (systems of orbiting bodies that are themselves orbiting in a larger system, *ad infinitum*) in a fractal way such that the Universe has a negligibly small overall density; such a cosmological model had also been proposed earlier in 1761 by Johann Heinrich Lambert.[56][135] A significant astronomical advance of the 18th century was the realization by Thomas Wright, Immanuel Kant and others of nebulae.[131]

The modern era of physical cosmology began in 1917, when Albert Einstein first applied his general theory of relativity to model the structure and dynamics of the Universe.[136]

4.9 See also

- Cosmic Calendar (scaled down timeline)
- Cosmic latte
- Dyson's eternal intelligence
- Esoteric cosmology
- False vacuum
- Fine-tuned Universe
- Hindu cosmology
- Illustris project
- Jain cosmology
- Kardashev scale
- Nucleocosmochronology
- Non-standard cosmology
- Rare Earth hypothesis
- Vacuum genesis
- World view
- Zero-energy Universe

4.10 References

[1] Planck collaboration (2014). "Planck 2013 results. XVI. Cosmological parameters". *Astronomy & Astrophysics.* arXiv:1303.5076. Bibcode:2014A&A...571A..16P. doi:10.1051/0004-6361/201321591.

[2] Itzhak Bars; John Terning (2009). *Extra Dimensions in Space and Time.* Springer. pp. 27ff. ISBN 978-0-387-77637-8. Retrieved 2011-05-01.

[3] Paul Davies (2006). *The Goldilocks Enigma.* First Mariner Books. p. 43ff. ISBN 978-0-618-59226-5. Retrieved 2013-07-01.

[4] NASA/WMAP Science Team (24 January 2014). "Universe 101: What is the Universe Made Of?". NASA. Retrieved 2015-02-17.

[5] Fixsen, D. J. (2009). "The Temperature of the Cosmic Microwave Background". *The Astrophysical Journal* **707** (2): 916–920. arXiv:0911.1955. Bibcode:2009ApJ...707..916F. doi:10.1088/0004-637X/707/2/916.

[6] Sean Carroll, Ph.D., Cal Tech, 2007, The Teaching Company, *Dark Matter, Dark Energy: The Dark Side of the Universe,* Guidebook Part 1 pages 1 and 3, Accessed Oct. 7, 2013, "...only 5% of the Universe is made of ordinary matter, with 25 percent being some kind of unseen dark matter and a full 70% being a smoothly distributed dark energy..."

[7] NASA/WMAP Science Team (24 January 2014). "Universe 101: Will the Universe expand forever?". NASA. Retrieved 16 April 2015.

[8] *Universe. Webster's New World College Dictionary, Wiley Publishing, Inc.* 2010.

[9] "Universe". *Dictionary.com.* Retrieved 2012-09-21.

[10] "Universe". *Merriam-Webster Dictionary.* Retrieved 2012-09-21.

[11] Zeilik, Michael; Gregory, Stephen A. (1998). *Introductory Astronomy & Astrophysics* (4th ed.). Saunders College Publishing. ISBN 0030062284. The totality of all space and time; all that is, has been, and will be.

[12] *The American Heritage Dictionary of the English Language* (4th ed.). Houghton Mifflin Harcourt Publishing Company. 2010.

[13] *Cambridge Advanced Learner's Dictionary.*

[14] Brian Greene (2011). *The Hidden Reality.* Alfred A. Knopf.

[15] Dold-Samplonius, Yvonne (2002). *From China to Paris: 2000 Years Transmission of Mathematical Ideas.* Franz Steiner Verlag.

[16] Thomas F. Glick; Steven Livesey; Faith Wallis. *Medieval Science Technology and Medicine: An Encyclopedia.* Routledge.

[17] Hawking, Stephen (1988). *A Brief History of Time.* Bantam Books. p. 125. ISBN 0-553-05340-X.

[18] "The Nobel Prize in Physics 2011". Retrieved 16 April 2015.

[19] "Planck reveals an almost perfect universe". *Planck.* ESA. 2013-03-21. Retrieved 2013-03-21.

[20] Ellis, George F.R.; U. Kirchner; W.R. Stoeger (2004). "Multiverses and physical cosmology". *Monthly Notices of the Royal Astronomical Society* **347** (3): 921–936. arXiv:astro-ph/0305292. Bibcode:2004MNRAS.347..921E. doi:10.1111/j.1365-2966.2004.07261.x.

[21] Palmer, Jason. (2011-08-03) BBC News – 'Multiverse' theory suggested by microwave background. Retrieved 2011-11-28.

[22] Paul Copan; William Lane Craig (2004). *Creation Out of Nothing: A Biblical, Philosophical, and Scientific Exploration.* Baker Academic. p. 220. ISBN 9780801027338.

[23] Alexander Bolonkin (November 2011). *Universe, Human Immortality and Future Human Evaluation.* Elsevier. pp. 3–. ISBN 978-0-12-415801-6.

[24] Duco A. Schreuder (3 December 2014). *Vision and Visual Perception.* Archway Publishing. pp. 135–. ISBN 978-1-4808-1294-9.

[25] Tegmark, Max. "The Mathematical Universe". *Foundations of Physics* **38** (2): 101–150. arXiv:0704.0646. Bibcode:2008FoPh...38..101T. doi:10.1007/s10701-007-9186-9. a short version of which is available at *Shut up and calculate.* (in reference to David Mermin's famous quote "shut up and calculate"

[26] Jim Holt (2012). *Why Does the World Exist?*. Liveright Publishing. p. 308.

[27] Timothy Ferris (1997). *The Whole Shebang: A State-of-the-Universe(s) Report.* Simon & Schuster. p. 400.

[28] *The Compact Edition of the Oxford English Dictionary*, volume II, Oxford: Oxford University Press, 1971, p. 3518.

[29] Lewis, C. T. and Short, S (1879) *A Latin Dictionary*, Oxford University Press, ISBN 0-19-864201-6, pp. 1933, 1977–1978.

[30] Liddell & Scott (1968, p. 1392)

[31] Mary Louise Gill; Pierre Pellegrin (9 February 2009). *A Companion to Ancient Philosophy*. John Wiley & Sons. pp. 369–. ISBN 978-1-4051-7825-9.

[32] Liddell & Scott (1968, pp. 1345–1346)

[33] Yonge, Charles Duke (1870). *An English-Greek lexicon*. New York: American Book Company. p. 567.

[34] Liddell & Scott (1968, pp. 985, 1964)

[35] Lewis, C. T.; Short, S (1879). *A Latin Dictionary*. Oxford University Press. pp. 1881–1882, 1175, 1189–1190. ISBN 0-19-864201-6.

[36] *The Compact Edition of the Oxford English Dictionary* II. Oxford: Oxford University Press. 1971. pp. 909, 569, 3821–3822, 1900. ISBN 978-0198611172.

[37] Joseph Silk (2009). *Horizons of Cosmology*. Templeton Pressr. p. 208.

[38] Simon Singh (2005). *Big Bang: The Origin of the Universe*. Harper Perennial. p. 560.

[39] C. Sivaram (1986). "Evolution of the Universe through the Planck epoch". *Astrophysics & Space Science* 125: 189. Bibcode:1986Ap&SS.125..189S. doi:10.1007/BF00643984.

[40] Richard B. Larson and Volker Bromm (March 2002). "The First Stars in the Universe". *Scientific American*.

[41] Ryden, Barbara, "Introduction to Cosmology", 2006, eqn. 6.33

[42] "Antimatter". Particle Physics and Astronomy Research Council. October 28, 2003. Retrieved 2006-08-10.

[43] Landau and Lifshitz, p. 361.

[44] Edward Robert Harrison (2000). *Cosmology: the science of the universe*. Cambridge University Press. pp. 447–. ISBN 978-0-521-66148-5. Retrieved 1 May 2011.

[45] Andrew R. Liddle; David Hilary Lyth (13 April 2000). *Cosmological inflation and large-scale structure*. Cambridge University Press. pp. 24–. ISBN 978-0-521-57598-0. Retrieved 1 May 2011.

[46] Will the Universe expand forever?, WMAP website at NASA.

[47] Luminet, Jean-Pierre; Weeks, Jeffrey R.; Riazuelo, Alain; Lehoucq, Roland; Uzan, Jean-Philippe (2003-10-09). "Dodecahedral space topology as an explanation for weak wide-angle temperature correlations in the cosmic microwave background". *Nature* 425 (6958): 593–5. arXiv:astro-ph/0310253. Bibcode:2003Natur.425..593L. doi:10.1038/nature01944. PMID 14534579.

[48] Roukema, Boudewijn; Zbigniew Buliński; Agnieszka Szaniewska; Nicolas E. Gaudin (2008). "A test of the Poincare dodecahedral space topology hypothesis with the WMAP CMB data". *Astronomy and Astrophysics* 482 (3): 747. arXiv:0801.0006. Bibcode:2008A&A...482..747L. doi:10.1051/0004-6361:20078777.

[49] Aurich, Ralf; Lustig, S.; Steiner, F.; Then, H. (2004). "Hyperbolic Universes with a Horned Topology and the CMB Anisotropy". *Classical and Quantum Gravity* 21 (21): 4901–4926. arXiv:astro-ph/0403597. Bibcode:2004CQGra..21.4901A. doi:10.1088/0264-9381/21/21/010.

[50] "Planck reveals 'almost perfect' universe". *Michael Banks*. Physics World. 2013-03-21. Retrieved 2013-03-21.

[51] McCall, Storrs. *A Model of the Universe: Space-time, Probability, and Decision*. Oxford University. p. 23.

[52] Michio Kaku (11 March 2008). *Physics of the Impossible: A Scientific Exploration into the World of Phasers, Force Fields, Teleportation, and Time Travel*. Knopf Doubleday Publishing Group. pp. 202–. ISBN 978-0-385-52544-2.

[53] "How can space travel faster than the speed of light?". *Vannesa Janek.* Universe Today. 20 February 2015. Retrieved 6 June 2015.

[54] "Is faster-than-light travel or communication possible? Section: Expansion of the Universe". *Philip Gibbs.* 1997. Retrieved 6 June 2015.

[55] Christopher Crockett (February 20, 2013). "What is a light-year?". *EarthSky.*

[56] Rindler, p. 196.

[57] Christian, Eric; Samar, Safi-Harb. "How large is the Milky Way?". Retrieved 2007-11-28.

[58] I. Ribas; C. Jordi; F. Vilardell; E.L. Fitzpatrick; R.W. Hilditch; F. Edward Guinan (2005). "First Determination of the Distance and Fundamental Properties of an Eclipsing Binary in the Andromeda Galaxy". *Astrophysical Journal* 635 (1): L37–L40. arXiv:astro-ph/0511045. Bibcode:2005ApJ...635L..37R. doi:10.1086/499161.
McConnachie, A. W.; Irwin, M. J.; Ferguson, A. M. N.; Ibata, R. A.; Lewis, G. F.; Tanvir, N. (2005). "Distances and metallicities for 17 Local Group galaxies". *Monthly Notices of the Royal Astronomical Society* 356 (4): 979–997. arXiv:astro-ph/0410489. Bibcode:2005MNRAS.356..979M. doi:10.1111/j.1365-2966.2004.08514.x.

[59] Sean Carroll and Michio Kaku (2014). *How the Universe Works 3.* End of the Universe. Discovery Channel.

[60] Overbye, Dennis (October 11, 2003). "A 'Cosmic Jerk' That Reversed the Universe". *New York Times.*

[61] The Nobel Prize in Physics 2011

[62] WMAP Mission: Results – Age of the Universe. Map.gsfc.nasa.gov. Retrieved 2011-11-28.

[63] Luminet, Jean-Pierre; Boudewijn F. Roukema (1999). "Topology of the Universe: Theory and Observations". *Proceedings of Cosmology School held at Cargese, Corsica, August 1998.* arXiv:astro-ph/9901364.

[64] Peebles, P. J. E. & Ratra, Bharat (2003). "The cosmological constant and dark energy". *Reviews of Modern Physics* 75 (2): 559–606. arXiv:astro-ph/0207347. Bibcode:2003RvMP...75..559P. doi:10.1103/RevModPhys.75.559.

[65] "First Planck results: the Universe is still weird and interesting".

[66] Mandolesi, N.; Calzolari, P.; Cortiglioni, S.; Delpino, F.; Sironi, G.; Inzani, P.; Deamici, G.; Solheim, J. -E.; Berger, L.; Partridge, R. B.; Martenis, P. L.; Sangree, C. H.; Harvey, R. C. (1986). "Large-scale homogeneity of the Universe measured by the microwave background". *Nature* 319 (6056): 751. doi:10.1038/319751a0.

[67] Mackie, Glen (February 1, 2002). "To see the Universe in a Grain of Taranaki Sand". Swinburne University. Retrieved 2006-12-20.

[68] "Unveiling the Secret of a Virgo Dwarf Galaxy". ESO. 2000-05-03. Retrieved 2007-01-03.

[69] "Hubble's Largest Galaxy Portrait Offers a New High-Definition View". NASA. 2006-02-28. Retrieved 2007-01-03.

[70] Vergano, Dan (1 December 2010). "Universe holds billions more stars than previously thought". *USA Today.* Retrieved 2010-12-14.

[71] Earth's new address: 'Solar System, Milky Way, Laniakea' / Nature

[72] http://www.universetoday.com/30286/local-group/

[73] "Astronomers discover largest known structure in the universe is ...   a big hole". The Guardian. 20 April 2015.

[74] Hinshaw, Gary (November 29, 2006). "New Three Year Results on the Oldest Light in the Universe". NASA WMAP. Retrieved 2006-08-10.

[75] Hinshaw, Gary (December 15, 2005). "Tests of the Big Bang: The CMB". NASA WMAP. Retrieved 2007-01-09.

[76] Rindler, p. 202.

[77] "Dark Energy". *Hyperphysics.* Retrieved January 4, 2014.

[78] Carroll, Sean (2001). "The cosmological constant". *Living Reviews in Relativity* 4. Retrieved 2006-09-28.

[79] "Planck captures portrait of the young Universe, revealing earliest light". University of Cambridge. 21 March 2013. Retrieved 21 March 2013.

[80] P. Davies (1992). *The New Physics: A Synthesis.* Cambridge University Press. p. 1. ISBN 0-521-43831-4.

[81] G. 't Hooft (1997). *In search of the ultimate building blocks.* Cambridge University Press. p. 6. ISBN 0-521-57883-3.

[82] "RHIC Scientists Serve Up "Perfect" Liquid" (Press release). Brookhaven National Laboratory. 18 April 2005. Retrieved 2009-09-15.

[83] Sylvie Braibant; Giorgio Giacomelli; Maurizio Spurio (2012). *Particles and Fundamental Interactions: An Introduction to Particle Physics* (2nd ed.). Springer. pp. 1–3. ISBN 978-94-007-2463-1.

[84] Strobel, Nick (May 23, 2001). "The Composition of Stars". Astronomy Notes. Retrieved 2007-01-04. "Have physical constants changed with time?". Astrophysics (Astronomy Frequently Asked Questions). Retrieved 2007-01-04.

[85] R. Oerter (2006). *The Theory of Almost Everything: The Standard Model, the Unsung Triumph of Modern Physics* (Kindle ed.). Penguin Group. p. 2. ISBN 0-13-236678-9.

[86] "Lepton (physics)". *Encyclopædia Britannica*. Retrieved 2010-09-29.

[87] Harari, H. (1977). "Beyond charm". In Balian, R.; Llewellyn-Smith, C.H. *Weak and Electromagnetic Interactions at High Energy, Les Houches, France, Jul 5- Aug 14, 1976*. Les Houches Summer School Proceedings **29**. North-Holland. p. 613.

[88] Harari H. (1977). "Three generations of quarks and leptons" (PDF). In E. van Goeler; Weinstein R. (eds.). *Proceedings of the XII Rencontre de Moriond*. p. 170. SLAC-PUB-1974.

[89] "Experiment confirms famous physics model" (Press release). MIT News Office. 18 April 2007.

[90] The Timescale of Creation

[91] Veltman, Martinus (2003). *Facts and Mysteries in Elementary Particle Physics.* World Scientific. ISBN 981-238-149-X.

[92] Onyisi, P. (23 October 2012). "Higgs boson FAQ". University of Texas ATLAS group. Retrieved 2013-01-08.

[93] Strassler, M. (12 October 2012). "The Higgs FAQ 2.0". *ProfMattStrassler.com*. Retrieved 2013-01-08. [Q] Why do particle physicists care so much about the Higgs particle?
[A] Well, actually, they don't. What they really care about is the Higgs *field*, because it is *so* important. [emphasis in original]

[94] Ray d'Inverno (1992), *Introducing Einstein's Relativity*, Clarendon Press, chp. 22.8 *Geometry of 3-spaces of constant curvature*, p.319ff, ISBN 0-19-859653-7

[95] Albert Einstein (1905) "*Zur Elektrodynamik bewegter Körper*", Annalen der Physik 17: 891; English translation On the Electrodynamics of Moving Bodies by George Barker Jeffery and Wilfrid Perrett (1923); Another English translation On the Electrodynamics of Moving Bodies by Megh Nad Saha (1920).

[96] Tom Roberts and Siegmar Schleif (October 2007). "What is the experimental basis of Special Relativity?". *Usenet Physics FAQ*. Retrieved 2008-09-17.

[97] Albert Einstein (2001). *Relativity: The Special and the General Theory* (Reprint of 1920 translation by Robert W. Lawson ed.). Routledge. p. 48. ISBN 0-415-25384-5.

[98] Richard Phillips Feynman (1998). *Six Not-so-easy Pieces: Einstein's relativity, symmetry, and space-time* (Reprint of 1995 ed.). Basic Books. p. 68. ISBN 0-201-32842-9.

[99] O'Connor, J.J. and E.F. Robertson (1996), "General relativity". *Mathematical Physics index*, School of Mathematics and Statistics, University of St. Andrews, Scotland, May, 1996. Retrieved 2015-02-04.

[100] Friedmann A. (1922). "Über die Krümmung des Raumes". *Zeitschrift für Physik* **10** (1): 377–386. Bibcode:1922ZPhy...10..377F. doi:10.1007/BF01332580.

[101] "Cosmic Detectives". The European Space Agency (ESA). 2013-04-02. Retrieved 2013-04-15.

[102] Munitz MK (1959). "One Universe or Many?". *Journal of the History of Ideas* **12** (2): 231–255. doi:10.2307/2707516. JSTOR 2707516.

[103] Linde A. (1986). "Eternal chaotic inflation". *Mod. Phys. Lett.* **A1** (2): 81–85. Bibcode:1986MPLA....1...81L. doi:10.1142/S0217732386000129. Linde A. (1986). "Eternally existing self-reproducing chaotic inflationary Universe" (PDF). *Phys. Lett.* **B175** (4): 395–400. Bibcode:1986PhLB..175..395L. doi:10.1016/0370-2693(86)90611-8. Retrieved 2011-03-17.

[104] Tegmark M. (2003). "Parallel universes. Not just a staple of science fiction, other universes are a direct implication of cosmological observations". *Scientific American* **288** (5): 40–51. doi:10.1038/scientificamerican0503-40. PMID 12701329.

[105] Tegmark, Max (2003). J. D. Barrow; P.C.W. Davies; C.L. Harper, eds. "Parallel Universes". *Scientific American: "Science and Ultimate Reality: from Quantum to Cosmos", honoring John Wheeler's 90th birthday* (Cambridge University Press): 2131. arXiv:astro-ph/0302131. Bibcode:2003astro.ph..2131T. doi:10.1038/scientificamerican0503-40.

[106] Ellis G. F (2011). "Does the Multiverse Really Exist?". *Scientific American* **305** (2): 38–43. doi:10.1038/scientificamerican0811-38.

[107] Clara Moskowitz (August 12, 2011). "Weird! Our Universe May Be a 'Multiverse,' Scientists Say". *livescience*.

[108] Mark Isaak (ed.) (2005). "CI301: The Anthropic Principle". *Index to Creationist Claims*. TalkOrigins Archive. Retrieved 2007-10-31.

[109] Gernet, J. (1993–1994). "Space and time: Science and religion in the encounter between China and Europe". *Chinese Science* **11**. pp. 93–102.

[110] Ng, Tai (2007). "III.3". *Chinese Culture, Western Culture: Why Must We Learn from Each Other?*. iUniverse, Inc.

[111] Blandford R. D. "A century of general relativity: Astrophysics and cosmology". *Science* **347** (6226): 103–108. Bibcode:2015Sci...347.1103B. doi:10.1126/science.aaa4033.

[112] Leeming, David A. (2010). *Creation Myths of the World*. ABC-CLIO. p. xvii. ISBN 978-1-59884-174-9. In common usage the word 'myth' refers to narratives or beliefs that are untrue or merely fanciful; the stories that make up national or ethnic mythologies describe characters and events that common sense and experience tell us are impossible. Nevertheless, all cultures celebrate such myths and attribute to them various degrees of literal or symbolic *truth*.

[113] (Henry Gravrand, "La civilisation Sereer -Pangool") [in] Universität Frankfurt am Main, Frobenius-Institut, Deutsche Gesellschaft für Kulturmorphologie, Frobenius Gesellschaft, "Paideuma: Mitteilungen zur Kulturkunde, Volumes 43–44", F. Steiner (1997), pp. 144–5, ISBN 3515028420

[114] B. Young, Louise. *The Unfinished Universe*. Oxford University Press. p. 21.

[115] Will Durant, *Our Oriental Heritage*.

> "Two systems of Hindu thought propound physical theories suggestively similar to those of Greece. Kanada, founder of the Vaisheshika philosophy, held that the world is composed of atoms as many in kind as the various elements. The Jains more nearly approximated to Democritus by teaching that all atoms were of the same kind, producing different effects by diverse modes of combinations. Kanada believed light and heat to be varieties of the same substance; Udayana taught that all heat comes from the Sun; and Vachaspati, like Newton, interpreted light as composed of minute particles emitted by substances and striking the eye."

[116] Stcherbatsky, F. Th. (1930, 1962), *Buddhist Logic*, Volume 1, p. 19, Dover, New York:

> "The Buddhists denied the existence of substantial matter altogether. Movement consists for them of moments, it is a staccato movement, momentary flashes of a stream of energy... "Everything is evanescent",... says the Buddhist, because there is no stuff... Both systems [Sānkhya, and later Indian Buddhism] share in common a tendency to push the analysis of existence up to its minutest, last elements which are imagined as absolute qualities, or things possessing only one unique quality. They are called "qualities" (*guna-dharma*) in both systems in the sense of absolute qualities, a kind of atomic, or intra-atomic, energies of which the empirical things are composed. Both systems, therefore, agree in denying the objective reality of the categories of Substance and Quality,... and of the relation of Inference uniting them. There is in Sānkhya philosophy no separate existence of qualities. What we call quality is but a particular manifestation of a subtle entity. To every new unit of quality corresponds a subtle quantum of matter which is called *guna*, "quality", but represents a subtle substantive entity. The same applies to early Buddhism where all qualities are substantive... or, more precisely, dynamic entities, although they are also called *dharmas* ('qualities')."

[117] Aristotle; Forster, E. S. (Edward Seymour), 1879-1950; Dobson, J. F. (John Frederic), 1875-1947 (1914). *De Mundo*. p. 2.

[118] Boyer, C. (1968) *A History of Mathematics*. Wiley, p. 54.

[119] Neugebauer, Otto E. (1945). "The History of Ancient Astronomy Problems and Methods". *Journal of Near Eastern Studies* 4
       (1): 1–38. doi:10.1086/370729. JSTOR 595168. the Chaldaean Seleucus from Seleucia

[120] Sarton, George (1955). "Chaldaean Astronomy of the Last Three Centuries B. C". *Journal of the American Oriental Society*
       75 (3): 166–173 (169). doi:10.2307/595168. JSTOR 595168. the heliocentrical astronomy invented by Aristarchos of Samos
       and still defended a century later by Seleucos the Babylonian

[121] William P. D. Wightman (1951, 1953), *The Growth of Scientific Ideas*, Yale University Press p. 38, where Wightman calls him
       Seleukos the Chaldean.

[122] Lucio Russo, *Flussi e riflussi*, Feltrinelli, Milano, 2003, ISBN 88-07-10349-4.

[123] Bartel (1987, p. 527)

[124] Bartel (1987, pp. 527–9)

[125] Bartel (1987, pp. 529–34)

[126] Bartel (1987, pp. 534–7)

[127] Nasr, Seyyed H. (1993) [1964]. *An Introduction to Islamic Cosmological Doctrines* (2nd ed.). 1st edition by Harvard University
       Press, 2nd edition by State University of New York Press. pp. 135–6. ISBN 0-7914-1515-5.

[128] Misner, Thorne and Wheeler, p. 754.

[129] Ālī, Ema Ākabara. *Science in the Quran* 1. Malik Library. p. 218.

[130] Misner, Thorne and Wheeler, p. 755–756.

[131] Misner, Thorne and Wheeler, p. 756.

[132] de Cheseaux JPL (1744). *Traité de la Comète*. Lausanne. pp. 223ff.. Reprinted as Appendix II in Dickson FP (1969). *The
       Bowl of Night: The Physical Universe and Scientific Thought*. Cambridge, MA: M.I.T. Press. ISBN 978-0-262-54003-2.

[133] Olbers HWM (1826). "Unknown title". *Bode's Jahrbuch* 111.. Reprinted as Appendix I in Dickson FP (1969). *The Bowl of
       Night: The Physical Universe and Scientific Thought*. Cambridge, MA: M.I.T. Press. ISBN 978-0-262-54003-2.

[134] Jeans, J. H. (1902). "The Stability of a Spherical Nebula" (PDF). *Philosophical Transactions of the Royal Society A* 199 (312–
       320): 1–53. Bibcode:1902RSPTA.199....1J. doi:10.1098/rsta.1902.0012. JSTOR 90845. Retrieved 2011-03-17.

[135] Misner, Thorne and Wheeler, p. 757.

[136] Einstein, A (1917). "Kosmologische Betrachtungen zur allgemeinen Relativitätstheorie". *Preussische Akademie der Wis-
       senschaften, Sitzungsberichte*. 1917. (part 1): 142–152.

4.11   Bibliography

- Bartel, Leendert van der Waerden (1987). "The Heliocentric System in Greek, Persian and Hindu Astronomy". *Annals of the New York Academy of Sciences* 500 (1): 525–545. Bibcode:1987NYASA.500..525V. doi:10.1111/j.1749-
  6632.1987.tb37224.x.

- Landau, Lev, Lifshitz, E.M. (1975). *The Classical Theory of Fields (Course of Theoretical Physics)* 2 (revised 4th
  English ed.). New York: Pergamon Press. pp. 358–397. ISBN 978-0-08-018176-9.

- Liddell, H. G. & Scott, R. (1968). *A Greek-English Lexicon*. Oxford University Press. ISBN 0-19-864214-8.

- Misner, C.W., Thorne, Kip, Wheeler, J.A. (1973). *Gravitation*. San Francisco: W. H. Freeman. pp. 703–816.
  ISBN 978-0-7167-0344-0.

- Rindler, W. (1977). *Essential Relativity: Special, General, and Cosmological*. New York: Springer Verlag. pp.
  193–244. ISBN 0-387-10090-3.

4.12   Further reading

- Weinberg, S. (1993). *The First Three Minutes: A Modern View of the Origin of the Universe* (2nd updated ed.). New York: Basic Books. ISBN 978-0-465-02437-7. OCLC 28746057. For lay readers.

- Nussbaumer, Harry; Bieri, Lydia; Sandage, Allan (2009). *Discovering the Expanding Universe*. Cambridge University Press. ISBN 978-0-521-51484-2.

4.13   External links

- Is there a hole in the Universe? at HowStuffWorks

- Cosmology FAQ

- The Dark Side and the Bright Side of the Universe Princeton University, Shirley Ho

- Multiple Big Bangs

4.13.1   Videos

- Discovery Channel, Stephen Hawking's Universe

- Cosmography of the Local Universe at irfu.cea.fr (17:35) (arXiv)

- The Known Universe by the American Museum of Natural History

- Understand The Size Of The Universe – by Powers of Ten

- A Flight Through the Universe – BerkeleyLab

- 3D Universe Hayden Planetarium

- Browser visualization of nearby stars (Chrome browser)

High-precision test of general relativity by the Cassini space probe (artist's impression): radio signals sent between the Earth and the probe (green wave) are delayed by the warping of space and time (blue lines) due to the Sun's mass.

Depiction of a multiverse of seven "bubble" universes, which are separate spacetime continua, each having different physical laws, physical constants, and perhaps even different numbers of dimensions or topologies.

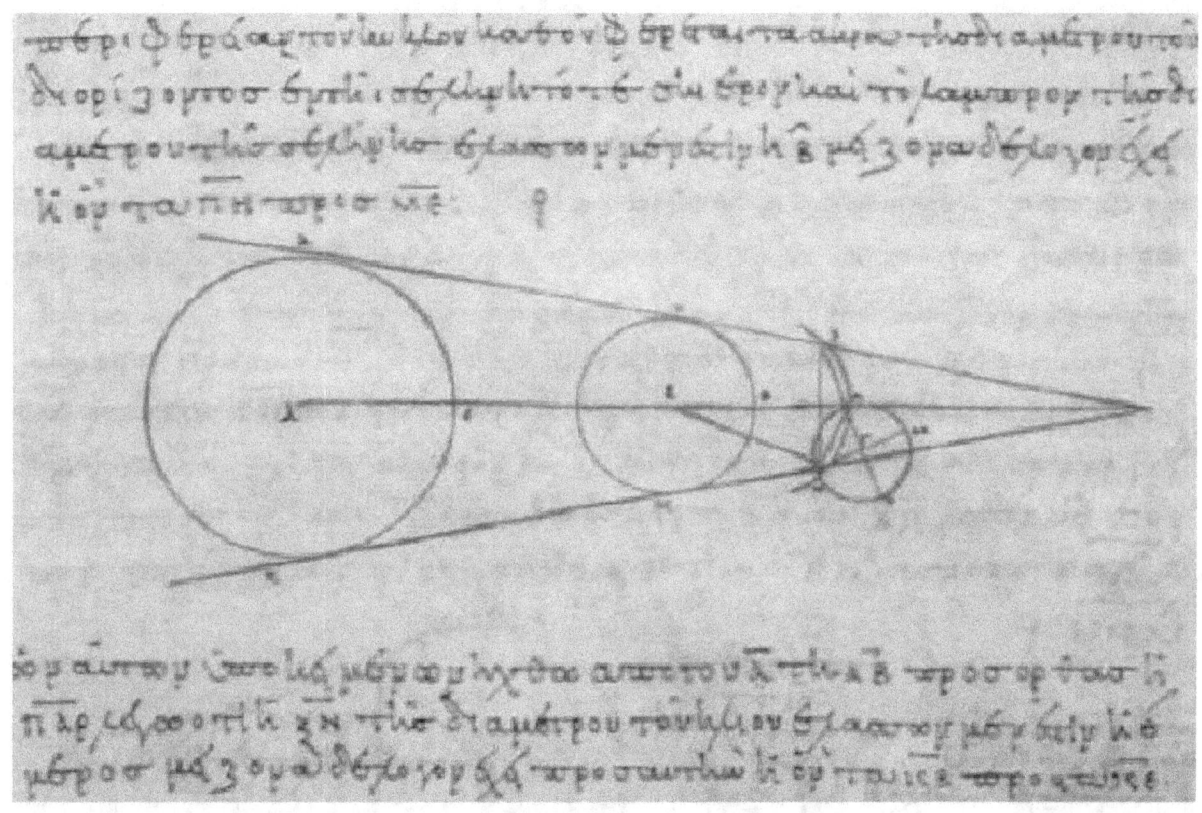

Aristarchus's 3rd century BCE calculations on the relative sizes of from left the Sun, Earth and Moon, from a 10th-century AD Greek copy

Flammarion engraving, Paris 1888

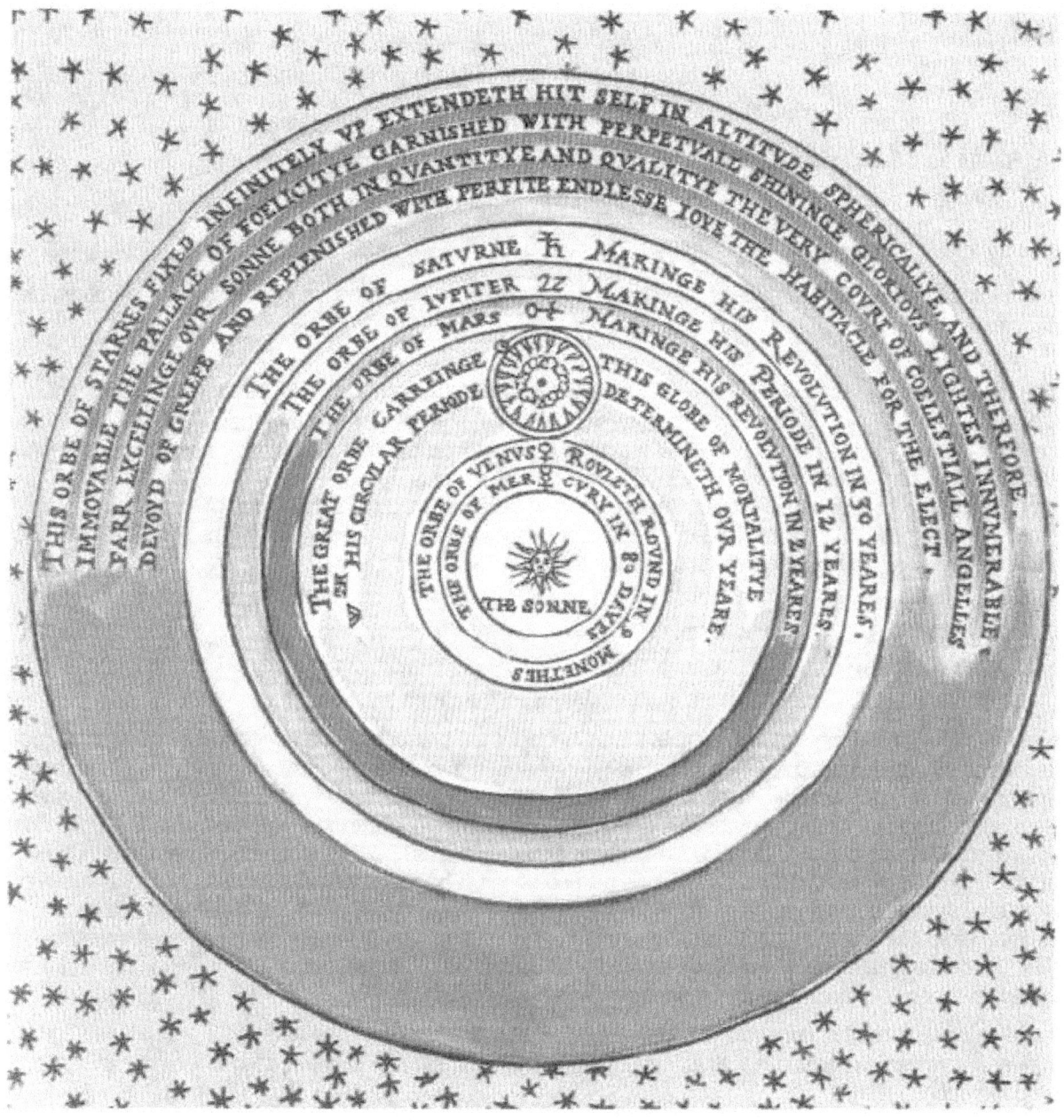

Model of the Copernican Universe by Thomas Digges in 1576, with the amendment that the stars are no longer confined to a sphere, but spread uniformly throughout the space surrounding the planets.

Chapter 5

Age of the universe

This article is about scientific estimates of the age of the universe. For religious and other non-scientific estimates, see Dating creation.

In physical cosmology, the **age of the universe** is the time elapsed since the Big Bang. The current measurement of the age of the universe is 13.798±0.037 billion years ((13.798±0.037)×10^9 years) within the Lambda-CDM concordance model.[1] The uncertainty of 37 million years has been obtained by the agreement of a number of scientific research projects, such as microwave background radiation measurements by the Planck satellite, the Wilkinson Microwave Anisotropy Probe and other probes. Measurements of the cosmic background radiation give the cooling time of the universe since the Big Bang,[2] and measurements of the expansion rate of the universe can be used to calculate its approximate age by extrapolating backwards in time.

5.1   Explanation

The Lambda-CDM concordance model describes the evolution of the universe from a very uniform, hot, dense primordial state to its present state over a span of about 13.8 billion years[3] of cosmological time. This model is well understood theoretically and strongly supported by recent high-precision astronomical observations such as WMAP. In contrast, theories of the origin of the primordial state remain very speculative. If one extrapolates the Lambda-CDM model backward from the earliest well-understood state, it quickly (within a small fraction of a second) reaches a singularity called the "Big Bang singularity". This singularity is not understood as having a physical significance in the usual sense, but it is convenient to quote times measured "since the Big Bang" even though they do not correspond to a physically measurable time. For example, "10^{-6} seconds after the Big Bang" is a well-defined era in the universe's evolution. If one referred to the same era as "13.8 billion years minus 10^{-6} seconds ago", the precision of the meaning would be lost because the minuscule latter time interval is swamped by uncertainty in the former.

Though the universe might in theory have a longer history, the International Astronomical Union[4] presently use "age of the universe" to mean the duration of the Lambda-CDM expansion, or equivalently the elapsed time since the Big Bang in the current observable universe.

5.2   Observational limits

Since the universe must be at least as old as the oldest thing in it, there are a number of observations which put a lower limit on the age of the universe; these include the temperature of the coolest white dwarfs, which gradually cool as they age, and the dimmest turnoff point of main sequence stars in clusters (lower-mass stars spend a greater amount of time on the main sequence, so the lowest-mass stars that have evolved off of the main sequence set a minimum age).

5.3 Cosmological parameters

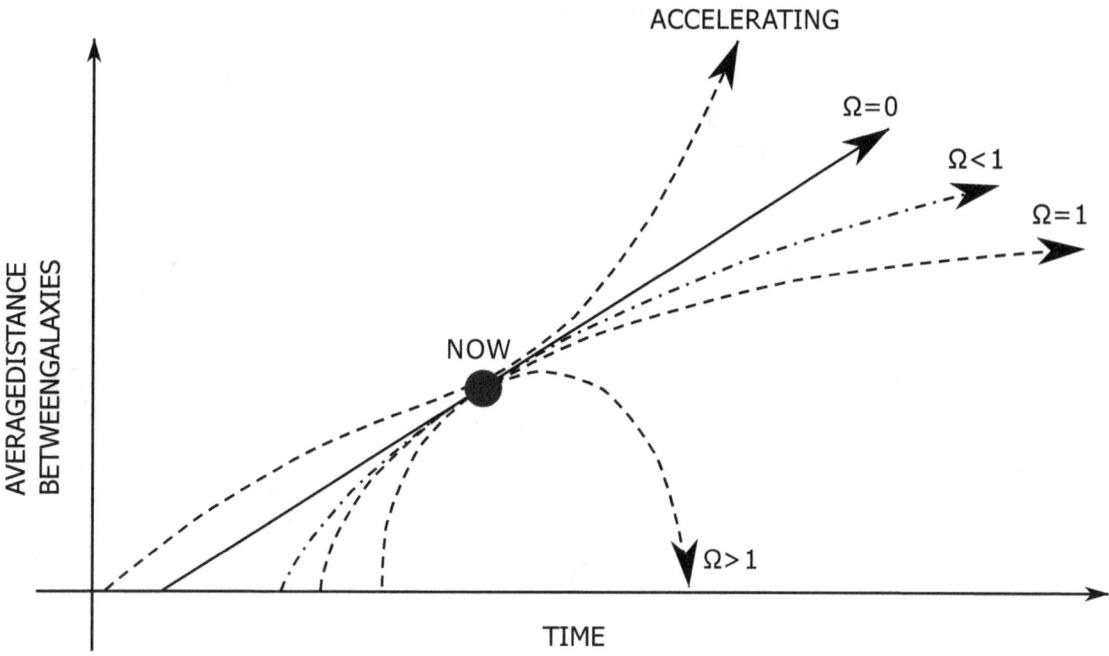

The age of the universe can be determined by measuring the Hubble constant today and extrapolating back in time with the observed value of density parameters (Ω). Before the discovery of dark energy, it was believed that the universe was matter-dominated, and so Ω on this graph corresponds to Ω . Note that the accelerating universe has the greatest age, while the Big Crunch universe has the least age.

The problem of determining the age of the universe is closely tied to the problem of determining the values of the cosmological parameters. Today this is largely carried out in the context of the ΛCDM model, where the universe is assumed to contain normal (baryonic) matter, cold dark matter, radiation (including both photons and neutrinos), and a cosmological constant. The fractional contribution of each to the current energy density of the universe is given by the density parameters Ωm, Ωr, and $\Omega \Lambda$. The full ΛCDM model is described by a number of other parameters, but for the purpose of computing its age these three, along with the Hubble parameter H_0 , are the most important.

If one has accurate measurements of these parameters, then the age of the universe can be determined by using the Friedmann equation. This equation relates the rate of change in the scale factor $a(t)$ to the matter content of the universe. Turning this relation around, we can calculate the change in time per change in scale factor and thus calculate the total age of the universe by integrating this formula. The age t_0 is then given by an expression of the form

$$t_0 = \frac{1}{H_0}\, F\,(\Omega_r, \Omega_m, \Omega_\Lambda \dots)$$

where H_0 is the Hubble parameter and the function F depends only on the fractional contribution to the universe's energy content that comes from various components. The first observation that one can make from this formula is that it is the Hubble parameter that controls that age of the universe, with a correction arising from the matter and energy content. So a rough estimate of the age of the universe comes from the Hubble time, the inverse of the Hubble parameter. With a value for H_0 around 68 km/s/Mpc, the Hubble time evaluates to $1/H_0$ = 14.4 billion years.[5]

To get a more accurate number, the correction factor F must be computed. In general this must be done numerically, and the results for a range of cosmological parameter values are shown in the figure. For the Planck values $(\Omega m, \Omega \Lambda)$ = (0.3086, 0.6914), shown by the box in the upper left corner of the figure, this correction factor is about F = 0.956. For a flat universe without any cosmological constant, shown by the star in the lower right corner, F = $^2/_3$ is much smaller and thus the universe is younger for a fixed value of the Hubble parameter. To make this figure, Ωr is held constant (roughly

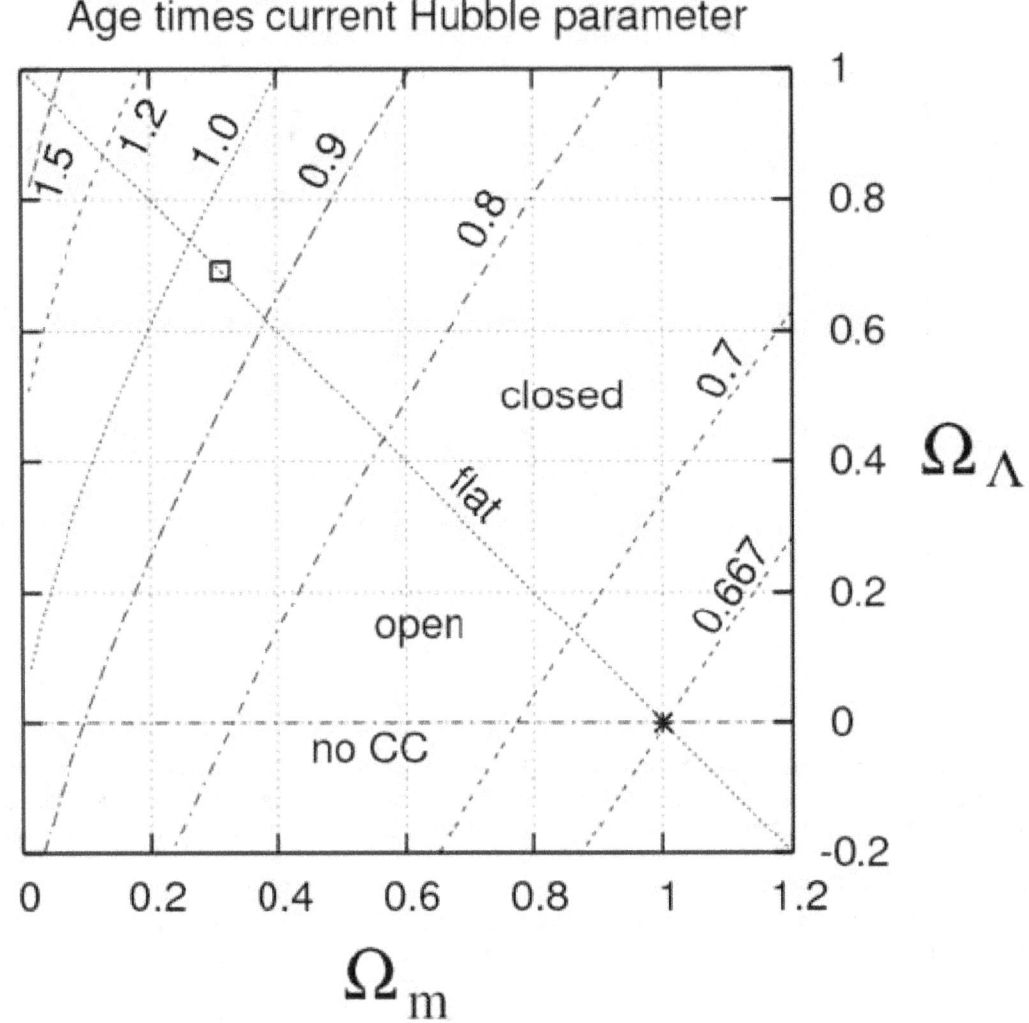

The value of the age correction factor, F, is shown as a function of two cosmological parameters: the current fractional matter density Ω and cosmological constant density $\Omega\Lambda$. The best-fit values of these parameters are shown by the box in the upper left; the matter-dominated universe is shown by the star in the lower right.

equivalent to holding the CMB temperature constant) and the curvature density parameter is fixed by the value of the other three.

Apart from the Planck satellite, the Wilkinson Microwave Anisotropy Probe (WMAP) was instrumental in establishing an accurate age of the universe, though other measurements must be folded in to gain an accurate number. CMB measurements are very good at constraining the matter content Ωm[6] and curvature parameter Ωk.[7] It is not as sensitive to $\Omega\Lambda$ directly,[7] partly because the cosmological constant becomes important only at low redshift. The most accurate determinations of the Hubble parameter H_0 come from Type Ia supernovae. Combining these measurements leads to the generally accepted value for the age of the universe quoted above.

The cosmological constant makes the universe "older" for fixed values of the other parameters. This is significant, since before the cosmological constant became generally accepted, the Big Bang model had difficulty explaining why globular clusters in the Milky Way appeared to be far older than the age of the universe as calculated from the Hubble parameter and a matter-only universe.[8][9] Introducing the cosmological constant allows the universe to be older than these clusters, as well as explaining other features that the matter-only cosmological model could not.[10]

5.4   WMAP

NASA's Wilkinson Microwave Anisotropy Probe (WMAP) project's nine-year data release in 2012 estimated the age of the universe to be $(13.772\pm0.059)\times10^9$ years (13.772 billion years, with an uncertainty of plus or minus 59 million years).[2]

However, this age is based on the assumption that the project's underlying model is correct; other methods of estimating the age of the universe could give different ages. Assuming an extra background of relativistic particles, for example, can enlarge the error bars of the WMAP constraint by one order of magnitude.[11]

This measurement is made by using the location of the first acoustic peak in the microwave background power spectrum to determine the size of the decoupling surface (size of the universe at the time of recombination). The light travel time to this surface (depending on the geometry used) yields a reliable age for the universe. Assuming the validity of the models used to determine this age, the residual accuracy yields a margin of error near one percent.[12]

5.5   Planck

In 2013, the European Space Agency's Planck spacecraft team estimated the age of the universe to be 13.82 billion years,[13][14][15][16] slightly higher but within the uncertainties of the earlier number derived from the WMAP data. By combining the Planck data with previous missions, the best combined estimate of the age of the universe is $(13.798\pm0.037)\times10^9$ years old.[1]

5.6   Assumption of strong priors

Calculating the age of the universe is accurate only if the assumptions built into the models being used to estimate it are also accurate. This is referred to as strong priors and essentially involves stripping the potential errors in other parts of the model to render the accuracy of actual observational data directly into the concluded result. Although this is not a valid procedure in all contexts (as noted in the accompanying caveat: "based on the fact we have assumed the underlying model we used is correct"), the age given is thus accurate to the specified error (since this error represents the error in the instrument used to gather the raw data input into the model).

The age of the universe based on the best fit to Planck 2013 data alone is 13.813±0.058 billion years (the other estimate of 13.798±0.037 billion years uses Gaussian priors based on earlier estimates from other studies to determine the combined uncertainty). This number represents the first accurate "direct" measurement of the age of the universe (other methods typically involve Hubble's law and the age of the oldest stars in globular clusters, etc.). It is possible to use different methods for determining the same parameter (in this case – the age of the universe) and arrive at different answers with no overlap in the "errors". To best avoid the problem, it is common to show two sets of uncertainties; one related to the actual measurement and the other related to the systematic errors of the model being used.

An important component to the analysis of data used to determine the age of the universe (e.g. from Planck) therefore is to use a Bayesian statistical analysis, which normalizes the results based upon the priors (i.e. the model).[12] This quantifies any uncertainty in the accuracy of a measurement due to a particular model used.[18][19]

5.7   History

Main article: Cosmic age problem

In the 18th century, the concept that the age of the Earth was millions, if not billions, of years began to appear. However, most scientists throughout the 19th century and into the first decades of the 20th century presumed that the universe itself was Steady State and eternal, with maybe stars coming and going but no changes occurring at the largest scale known at the time.

The first scientific theories indicating that the age of the universe might be finite were the studies of thermodynamics, formalized in the mid-19th century. The concept of entropy dictates that if the universe (or any other closed system) were infinitely old, then everything inside would be at the same temperature, and thus there would be no stars and no life. No scientific explanation for this contradiction was put forth at the time. In 1915 Albert Einstein published the theory of general relativity[20] and in 1917 constructed the first cosmological model based on his theory. In order to remain consistent with a steady state universe, Einstein added what was later called a cosmological constant to his equations. However, already in 1922, also using Einstein's theory, Alexander Friedman, and independently five years later Georges Lemaître, showed that the universe cannot be static and must be either expanding or contracting. Einstein's model of a static universe was in addition proved unstable by Arthur Eddington.

The first direct observational hint that the universe has a finite age came from the observations of 'recession velocities', mostly by Vesto Slipher, combined with distances to the 'nebulae' (galaxies) by Edwin Hubble in a work published in 1929.[21] Earlier in the 20th century, Hubble and others resolved individual stars within certain nebulae, thus determining that they were galaxies, similar to, but external to, our Milky Way Galaxy. In addition, these galaxies were very large and very far away. Spectra taken of these distant galaxies showed a red shift in their spectral lines presumably caused by the Doppler effect, thus indicating that these galaxies were moving away from the Earth. In addition, the farther away these galaxies seemed to be (the dimmer they appeared to us) the greater was their redshift, and thus the faster they seemed to be moving away. This was the first direct evidence that the universe is not static but expanding. The first estimate of the age of the universe came from the calculation of when all of the objects must have started speeding out from the same point. Hubble's initial value for the universe's age was very low, as the galaxies were assumed to be much closer than later observations found them to be.

The first reasonably accurate measurement of the rate of expansion of the universe, a numerical value now known as the Hubble constant, was made in 1958 by astronomer Allan Sandage.[22] His measured value for the Hubble constant came very close to the value range generally accepted today.

However Sandage, like Einstein, did not believe his own results at the time of discovery. His value for the age of the universe was too short to reconcile with the 25-billion-year age estimated at that time for the oldest known stars. Sandage and other astronomers repeated these measurements numerous times, attempting to reduce the Hubble constant and thus increase the resulting age for the universe. Sandage even proposed new theories of cosmogony to explain this discrepancy. This issue was finally resolved by improvements in the theoretical models used for estimating the ages of stars. As of 2013, using the latest models for stellar evolution, the estimated age of the oldest known star is 14.46±0.8 billion years.[23]

The discovery of microwave cosmic background radiation announced in 1965[24] finally brought an effective end to the remaining scientific uncertainty over the expanding universe. The recently launched space probes WMAP, launched in 2001, and Planck, launched in 2009, produced data that determines the Hubble constant and the age of the universe independent of galaxy distances, removing the largest source of error.[12]

More recently, in February 2015, an alternative view to extend the Big Bang model was presented that suggests the Universe had no beginning or singularity and that the age of the Universe may be infinite.[25][26][27]

5.8 See also

- Age of the Earth

- Anthropic principle

- Cosmic age problem

- Cosmic Calendar (age of universe scaled to a single year)

- Cosmology

- Dark Ages Radio Explorer (DARE)

- Hubble Deep Field

- Illustris project

- Metric expansion of space

- Multiverse

- Observable universe

- Red shift observations in astronomy

- Static universe

- *The First Three Minutes* (1977 book by Steven Weinberg).

5.9   References

[1] Planck Collaboration (2014). "Planck 2013 results. I. Overview of products and scientific results". *Astronomy & Astrophysics* **571**: 1. arXiv:1303.5062. Bibcode:2014A&A...571A...1P. doi:10.1051/0004-6361/201321529.

[2] Bennett, C.L. et al. (2013). "Nine-Year Wilkinson Microwave Anisotropy Probe (WMAP) Observations: Final Maps and Results". arXiv:1212.5225 [astro-ph.CO].

[3] "Cosmic Detectives". European Space Agency. 2 April 2013. Retrieved 2013-04-15.

[4] Chang, K. (9 March 2008). "Gauging Age of Universe Becomes More Precise". *The New York Times.*

[5] Liddle, A. R. (2003). *An Introduction to Modern Cosmology* (2nd ed.). Wiley. p. 57. ISBN 0-470-84835-9.

[6] Hu, W. "Animation: Matter Content Sensitivity. The matter-radiation ratio is raised while keeping all other parameters fixed.". University of Chicago. Archived from the original on 23 February 2008. Retrieved 2008-02-23.

[7] Hu, W. "Animation: Angular diameter distance scaling with curvature and lambda". University of Chicago. Archived from the original on 23 February 2008. Retrieved 2008-02-23.

[8] "Globular Star Clusters". SEDS. 1 July 2011. Archived from the original on 24 February 2008. Retrieved 2013-07-19.

[9] Iskander, E. (11 January 2006). "Independent age estimates". University of British Columbia. Archived from the original on 6 March 2008. Retrieved 2008-02-23.

[10] Ostriker, J. P.; Steinhardt, P. J. (1995). "Cosmic Concordance". arXiv:astro-ph/9505066.

[11] de Bernardis, F.; Melchiorri, A.; Verde, L.; Jimenez, R. (2008). "The Cosmic Neutrino Background and the Age of the Universe". *Journal of Cosmology and Astroparticle Physics* **2008** (3): 20. arXiv:0707.4170. Bibcode:2008JCAP...03..020D. doi:10.1088/1475-7516/2008/03/020.

[12] Spergel, D. N. et al. (2003). "First-Year Wilkinson Microwave Anisotropy Probe (WMAP) Observations: Determination of Cosmological Parameters". *The Astrophysical Journal Supplement Series* **148** (1): 175–194. arXiv:astro-ph/0302209. Bibcode:2003ApJS..148..175S. doi:10.1086/377226.

[13] Staff (21 March 2013). "Planck Reveals An Almost Perfect Universe". European Space Agency. Retrieved 2013-03-21.

[14] Clavin, W.; Harrington, J. D. (21 March 2013). "Planck Mission Brings Universe Into Sharp Focus". NASA. Retrieved 2013-03-21.

[15] Overbye, D. (21 March 2013). "An Infant Universe, Born Before We Knew". *New York Times.* Retrieved 2013-03-21.

[16] Boyle, A. (21 March 2013). "Planck probe's cosmic 'baby picture' revises universe's vital statistics". *NBC News.* Retrieved 2013-03-21.

[17] Planck collaboration (2014). "Planck 2013 results. XVI. Cosmological parameters". *Astronomy & Astrophysics* **571**: 16. arXiv:1303.5076. Bibcode:2014A&A...571A..16P. doi:10.1051/0004-6361/201321591.

[18] Loredo, T. J. (1992). "The Promise of Bayesian Inference for Astrophysics" (PDF). In Feigelson, E. D.; Babu, G. J. *Statistical Challenges in Modern Astronomy*. Springer-Verlag. pp. 275–297. Bibcode:1992scma.conf..275L. doi:10.1007/978-1-4613-9290-3_31. ISBN 978-1-4613-9292-7.

[19] Colistete, R.; Fabris, J. C.; Concalves, S. V. B. (2005). "Bayesian Statistics and Parameter Constraints on the Generalized Chaplygin Gas Model Using SNe ia Data". *International Journal of Modern Physics D* 14 (5): 775–796. arXiv:astro-ph/0409245. Bibcode:2005IJMPD..14..775C. doi:10.1142/S0218271805006729.

[20] Einstein, A. (1915). "Zur allgemeinen Relativitätstheorie". *Sitzungsberichte der Königlich Preußischen Akademie der Wissenschaften* (in German): 778–786. Bibcode:1915SPAW...... 778E.

[21] Hubble, E. (1929). "A relation between distance and radial velocity among extra-galactic nebulae". *Proceedings of the National Academy of Sciences* 15 (3): 168–173. Bibcode:1929PNAS...15..168H. doi:10.1073/pnas.15.3.168. PMC 522427. PMID 16577160.

[22] Sandage, A. R. (1958). "Current Problems in the Extragalactic Distance Scale". *The Astrophysical Journal* 127 (3): 513–526. Bibcode:1958ApJ...127..513S. doi:10.1086/146483.

[23] Bond, H. E.; Nelan, E. P.; Vandenberg, D. A.; Schaefer, G. H.; Harmer, D. (2013). "HD 140283: A Star in the Solar Neighborhood that Formed Shortly After the Big Bang". *The Astrophysical Journal* 765 (12): L12. arXiv:1302.3180. Bibcode:2013ApJ...765L..12B. doi:10.1088/2041-8205/765/1/L12.

[24] Penzias, A. A.; Wilson, R .W. (1965). "A Measurement of Excess Antenna Temperature at 4080 Mc/s". *The Astrophysical Journal* 142: 419–421. Bibcode:1965ApJ...142..419P. doi:10.1086/148307.

[25] Ghose, Tia (26 February 2015). "Big Bang, Deflated? Universe May Have Had No Beginning". *Live Science*. Retrieved 28 February 2015.

[26] Das, Saurya; Bhaduri, Rajat K. (18 November 2014). "Dark matter and dark energy from Bose-Einstein condensate" (PDF). *arXiv*. Retrieved 28 February 2015.

[27] Ali, Ahmed Faraq (4 February 2015). "Cosmology from quantum potential". *Physics Letters B* 741: 276–279. Retrieved 28 February 2015.

5.10 External links

- Ned Wright's Cosmology Tutorial

- Wright, Edward L. (2 July 2005). "Age of the Universe".

- Wayne Hu's cosmological parameter animations

- Ostriker; Steinhardt (1995). "Cosmic Concordance". arXiv:astro-ph/9505066.

- SEDS page on "Globular Star Clusters"

- Douglas Scott "Independent Age Estimates"

- KryssTal "The Scale of the Universe" Space and Time scaled for the beginner.

- iCosmos: Cosmology Calculator (With Graph Generation)

- The Expanding Universe (American Institute of Physics)

Chapter 6

Chronology of the universe

See also: Timeline of the formation of the Universe

The **chronology of the universe** describes the history and future of the universe according to Big Bang cosmology, the

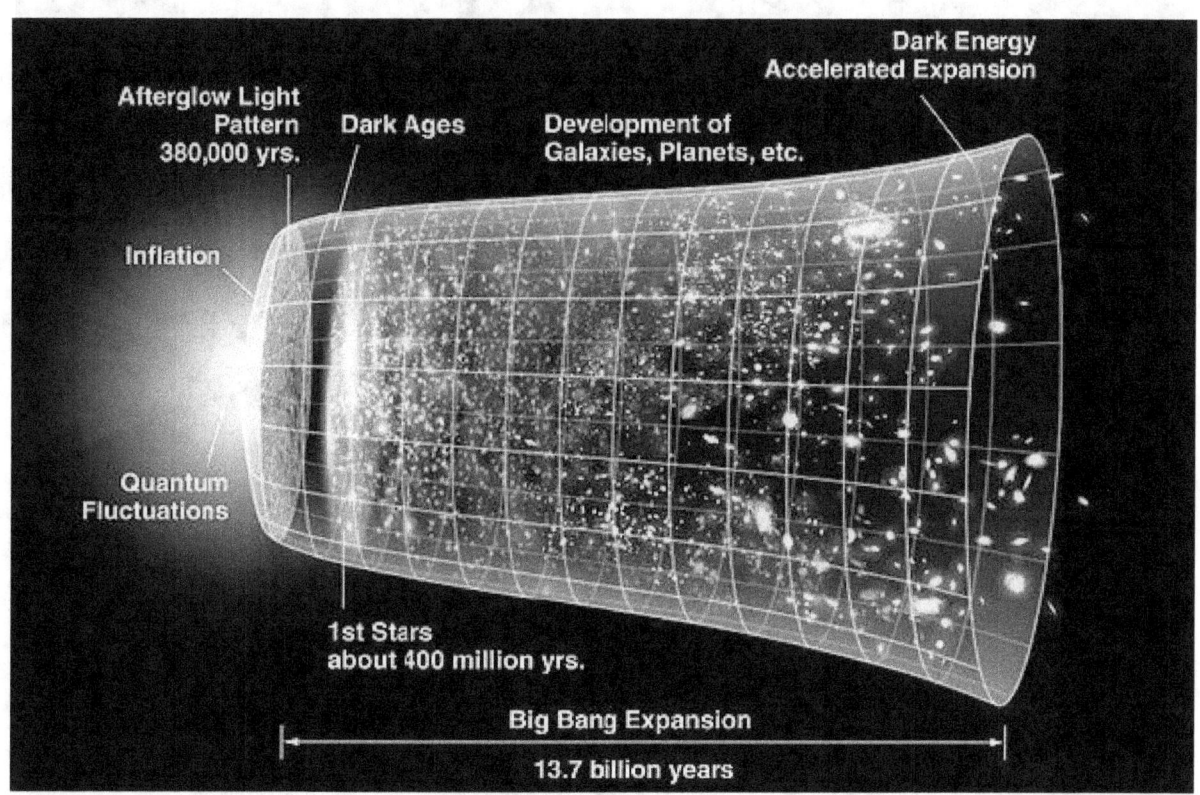

Diagram of Evolution of the universe from the Big Bang (left) - to the present.

prevailing scientific model of how the universe developed over time from the Planck epoch, using the cosmological time parameter of comoving coordinates. The model of the universe's expansion is known as the Big Bang. As of 2013, this expansion is estimated to have begun 13.798 ± 0.037 billion years ago.[1] It is convenient to divide the evolution of the universe so far into three phases.

6.1 Summary

In the first phase, the very earliest universe was so hot, or energetic, that initially no matter particles existed or could exist perhaps only fleetingly. According to prevailing scientific theories it was at this time that the forces we see around us today merged into one unified force. Space-time itself expanded during an inflationary epoch due to the immensity of the energies involved. Gradually the immense energies cooled – still to a temperature inconceivably hot compared to any we see around us now, but sufficiently to allow forces to gradually undergo symmetry breaking, a kind of repeated condensation from one status quo to another, leading finally to the separation of the strong force from the electroweak force and the first particles.

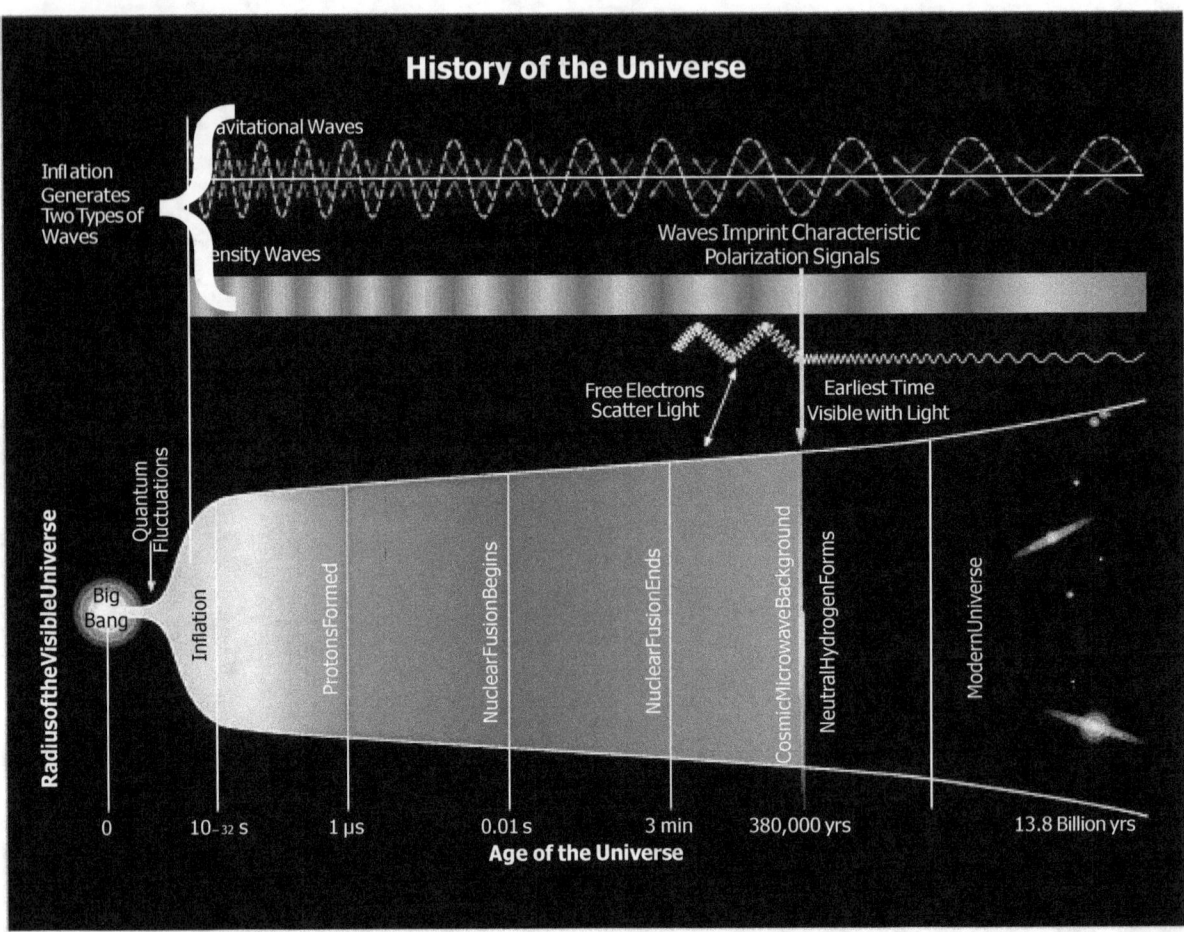

History of the Universe - gravitational waves are hypothesized to arise from cosmic inflation, a faster-than-light expansion just after the Big Bang (17 March 2014).[2][3][4]

In the second phase, this quark–gluon plasma universe then cooled further, the current fundamental forces we know take their present forms through further symmetry breaking – notably the breaking of electroweak symmetry – and the full range of complex and composite particles we see around us today became possible, leading to a gravitationally dominated universe, the first neutral atoms (~ 80% hydrogen), and the cosmic microwave background radiation we can detect today. Modern high energy particle physics theories are satisfactory at these energy levels, and so physicists believe they have a good understanding of this and subsequent development of the fundamental universe around us. Because of these changes, space had also become largely transparent to light and other electromagnetic energy, rather than "foggy", by the end of this phase.

The third phase started after a short dark age with a universe whose fundamental particles and forces were as we know them, and witnessed the emergence of large scale stable structures, such as the earliest stars, quasars, galaxies, clusters of galaxies and superclusters, and the development of these to create the kind of universe we see today. Some researchers call the development of all this physical structure over billions of years "cosmic evolution". Other, more interdisciplinary, researchers refer to "cosmic evolution" as the entire scenario of growing complexity from big bang to humankind, thereby incorporating biology and culture into a grand unified view of all complex systems in the universe to date.[5]

Beyond the present day, scientists anticipate that the Earth will cease to be able to support life in about a billion years, and will be drawn into the Sun in about 5 billion years. On a far longer timescale, the Stelliferous Era will end as stars eventually die and fewer are born to replace them, leading to a darkening universe. Various theories suggest a number of subsequent possibilities. If particles such as protons are unstable then eventually matter may evaporate into low level energy in a kind of entropy related heat death. Alternatively the universe may collapse in a big crunch, although current data shows the rate of expansion is still increasing. If this is correct then it may end in a "big freeze" as matter and energy become very thinly spread and cool down. Alternative suggestions include a false vacuum catastrophe or a Big Rip as possible ends to the universe.

6.2   Very early universe

All ideas concerning the very early universe (cosmogony) are speculative. No accelerator experiments have yet probed energies of sufficient magnitude to provide any experimental insight into the behavior of matter at the energy levels that prevailed during this period. Proposed scenarios differ radically. Some examples are the Hartle–Hawking initial state, string landscape, brane inflation, string gas cosmology, and the ekpyrotic universe. Some of these are mutually compatible, while others are not.

6.2.1   Planck epoch

0 to 10^{-43} second after the Big Bang

Main article: Planck epoch

The Planck epoch is an era in traditional (non-inflationary) big bang cosmology wherein the temperature was so high that the four fundamental forces—electromagnetism, gravitation, weak nuclear interaction, and strong nuclear interaction—were one fundamental force. Little is understood about physics at this temperature; different hypotheses propose different scenarios. Traditional big bang cosmology predicts a gravitational singularity before this time, but this theory relies on general relativity and is expected to break down due to quantum effects.

In inflationary cosmology, times before the end of inflation (roughly 10^{-32} second after the Big Bang) do not follow the traditional big bang timeline.

6.2.2   Grand unification epoch

Between 10^{-43} second and 10^{-36} second after the Big Bang[6]

Main article: Grand unification epoch

As the universe expanded and cooled, it crossed transition temperatures at which forces separate from each other. These are phase transitions much like condensation and freezing. The grand unification epoch began when gravitation separated from the other forces of nature, which are collectively known as gauge forces. The non-gravitational physics in this epoch would be described by a so-called grand unified theory (GUT). The grand unification epoch ended when the GUT forces further separate into the strong and electroweak forces.

6.2.3   Electroweak epoch

Between 10^{-36} second (or the end of inflation) and 10^{-32} second after the Big Bang[6]

Main article: Electroweak epoch

According to traditional big bang cosmology, the Electroweak epoch began 10^{-36} second after the Big Bang, when the temperature of the universe was low enough (10^{28} K) to separate the strong force from the electroweak force (the name for the unified forces of electromagnetism and the weak interaction). In inflationary cosmology, the electroweak epoch ends when the inflationary epoch begins, at roughly 10^{-32} second.

Inflationary epoch

Unknown duration, ending 10^{-32}(?) second after the Big Bang

Main article: Inflationary epoch

Cosmic inflation was an era of accelerating expansion produced by a hypothesized field called the inflaton, which would have properties similar to the Higgs field and dark energy. While decelerating expansion would magnify deviations from homogeneity, making the universe more chaotic, accelerating expansion would make the universe more homogeneous. A sufficiently long period of inflationary expansion in the past could explain the high degree of homogeneity that is observed in the universe today at large scales, even if the state of the universe before inflation was highly disordered.

Inflation ended when the inflaton field decayed into ordinary particles in a process called "reheating", at which point ordinary Big Bang expansion began. The time of reheating is usually quoted as a time "after the Big Bang". This refers to the time that would have passed in traditional (non-inflationary) cosmology between the Big Bang singularity and the universe dropping to the same temperature that was produced by reheating, even though, in inflationary cosmology, the traditional Big Bang did not occur.

According to the simplest inflationary models, inflation ended at a temperature corresponding to roughly 10^{-32} second after the Big Bang. As explained above, this does not imply that the inflationary era lasted less than 10^{-32} second. In fact, in order to explain the observed homogeneity of the universe, the duration must be longer than 10^{-32} second. In inflationary cosmology, the earliest meaningful time "after the Big Bang" is the time of the end of inflation.

On March 17, 2014, astrophysicists of the BICEP2 collaboration announced the detection of inflationary gravitational waves in the B-mode power spectrum which was interpreted as clear experimental evidence for the theory of inflation.[2][3][4][7][8][9] However, on June 19, 2014, lowered confidence in confirming the cosmic inflation findings was reported [8][10][11] and finally, on February 2, 2015, a joint analysis of data from BICEP2/Keck and Planck satellite concluded that the statistical "significance [of the data] is too low to be interpreted as a detection of primordial B-modes" and can be attributed mainly to polarized dust in the Milky Way.[12][13][14][15]

Baryogenesis

Main article: Baryogenesis

There is currently insufficient observational evidence to explain why the universe contains far more baryons than antibaryons. A candidate explanation for this phenomenon must allow the Sakharov conditions to be satisfied at some time after the end of cosmological inflation. While particle physics suggests asymmetries under which these conditions are met, these asymmetries are too small empirically to account for the observed baryon-antibaryon asymmetry of the universe.

6.3 Early universe

After cosmic inflation ends, the universe is filled with a quark–gluon plasma. From this point onwards the physics of the early universe is better understood, and less speculative.

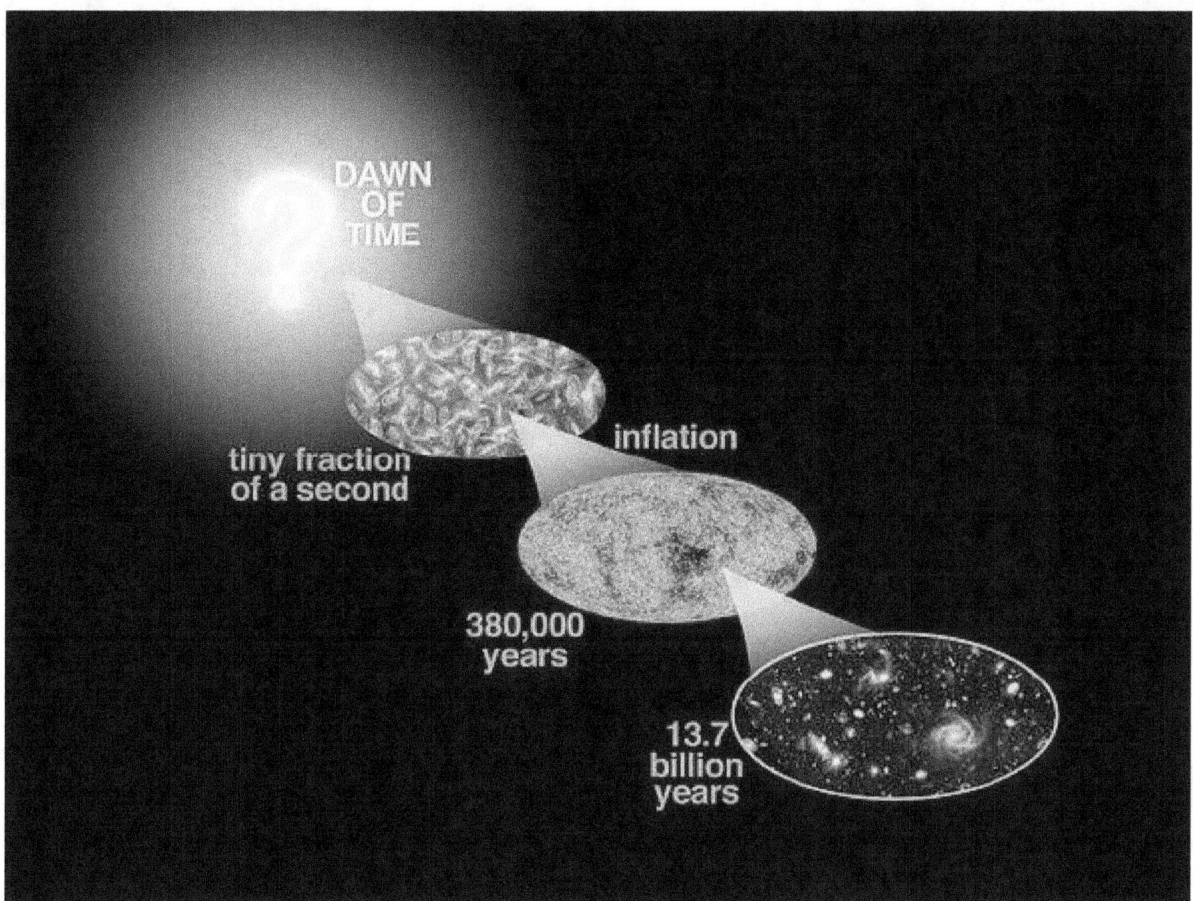

Cosmic History

6.3.1   Supersymmetry breaking (speculative)

Main article: Supersymmetry breaking

If supersymmetry is a property of our universe, then it must be broken at an energy that is no lower than 1 TeV, the electroweak symmetry scale. The masses of particles and their superpartners would then no longer be equal, which could explain why no superpartners of known particles have ever been observed.

6.3.2   Electroweak symmetry breaking and the quark epoch

Between 10^{-12} second and 10^{-6} second after the Big Bang

Main articles: Electroweak symmetry breaking and Quark epoch

As the universe's temperature falls below a certain very high energy level, it is believed that the Higgs field spontaneously acquires a vacuum expectation value, which breaks electroweak gauge symmetry. This has two related effects:

1. The weak force and electromagnetic force, and their respective bosons (the W and Z bosons and photon) manifest differently in the present universe, with different ranges;

2. Via the Higgs mechanism, all elementary particles interacting with the Higgs field become massive, having been massless at higher energy levels.

At the end of this epoch, the fundamental interactions of gravitation, electromagnetism, the strong interaction and the weak interaction have now taken their present forms, and fundamental particles have mass, but the temperature of the universe is still too high to allow quarks to bind together to form hadrons.

6.3.3 Hadron epoch

Between 10^{-6} second and 1 second after the Big Bang

Main article: Hadron epoch

The quark–gluon plasma that composes the universe cools until hadrons, including baryons such as protons and neutrons, can form. At approximately 1 second after the Big Bang neutrinos decouple and begin traveling freely through space. This cosmic neutrino background, while unlikely to ever be observed in detail since the neutrino energies are very low, is analogous to the cosmic microwave background that was emitted much later. (See above regarding the quark–gluon plasma, under the String Theory epoch.) However, there is strong indirect evidence that the cosmic neutrino background exists, both from Big Bang nucleosynthesis predictions of the helium abundance, and from anisotropies in the cosmic microwave background

6.3.4 Lepton epoch

Between 1 second and 10 seconds after the Big Bang

Main article: Lepton epoch

The majority of hadrons and anti-hadrons annihilate each other at the end of the hadron epoch, leaving leptons and anti-leptons dominating the mass of the universe. Approximately 10 seconds after the Big Bang the temperature of the universe falls to the point at which new lepton/anti-lepton pairs are no longer created and most leptons and anti-leptons are eliminated in annihilation reactions, leaving a small residue of leptons.[16]

6.3.5 Photon epoch

Between 10 seconds and 380,000 years after the Big Bang

Main article: Photon epoch

After most leptons and anti-leptons are annihilated at the end of the lepton epoch the energy of the universe is dominated by photons. These photons are still interacting frequently with charged protons, electrons and (eventually) nuclei, and continue to do so for the next 380,000 years.

Nucleosynthesis

Between 3 minutes and 20 minutes after the Big Bang[17]

Main article: Big Bang nucleosynthesis

During the photon epoch the temperature of the universe falls to the point where atomic nuclei can begin to form. Protons (hydrogen ions) and neutrons begin to combine into atomic nuclei in the process of nuclear fusion. Free neutrons combine with protons to form deuterium. Deuterium rapidly fuses into helium-4. Nucleosynthesis only lasts for about seventeen minutes, since the temperature and density of the universe has fallen to the point where nuclear fusion cannot continue. By this time, all neutrons have been incorporated into helium nuclei. This leaves about three times more hydrogen than helium-4 (by mass) and only trace quantities of other light nuclei.

Matter domination

70,000 years after the Big Bang

At this time, the densities of non-relativistic matter (atomic nuclei) and relativistic radiation (photons) are equal. The Jeans length, which determines the smallest structures that can form (due to competition between gravitational attraction and pressure effects), begins to fall and perturbations, instead of being wiped out by free-streaming radiation, can begin to grow in amplitude.

According to ΛCDM, at this stage, cold dark matter dominates, paving the way for gravitational collapse to amplify the tiny inhomogeneities left by cosmic inflation, making dense regions denser and rarefied regions more rarefied. However, because present theories as to the nature of dark matter are inconclusive, there is as yet no consensus as to its origin at earlier times, as currently exist for baryonic matter.

Recombination

ca. 377,000 years after the Big Bang

Main article: Recombination (cosmology)
Hydrogen and helium *atoms* begin to form as the density of the universe falls. This is thought to have occurred about

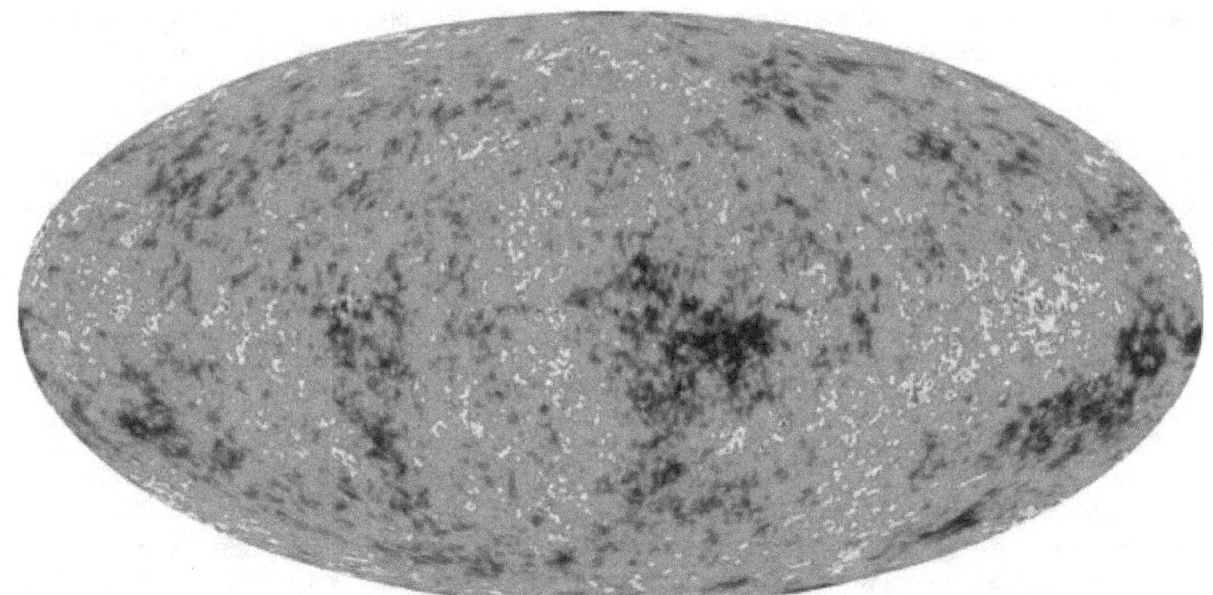

9 year WMAP data (2012) shows the cosmic microwave background radiation variations throughout the universe from our perspective, though the actual variations are much smoother than the diagram suggests.[18][19]

377,000 years after the Big Bang.[20] Hydrogen and helium are at the beginning ionized, i.e., no electrons are bound to the nuclei, which (containing positively charged protons) are therefore electrically charged (+1 and +2 respectively). As the universe cools down, the electrons get captured by the ions, forming electrically neutral atoms. This process is relatively fast (and faster for the helium than for the hydrogen), and is known as recombination.[21] At the end of recombination, most of the protons in the universe are bound up in neutral atoms. Therefore, the photons' mean free path becomes effectively infinite and the photons can now travel freely (see Thomson scattering): the universe has become transparent. This cosmic event is usually referred to as *decoupling*.

The photons present at the time of decoupling are the same photons that we see in the cosmic microwave background (CMB) radiation, after being greatly cooled by the expansion of the universe. Around the same time, existing pressure waves within the electron-baryon plasma — known as baryon acoustic oscillations — became embedded in the distribution of matter as it condensed, giving rise to a very slight preference in distribution of large scale objects. Therefore the cosmic microwave background is a picture of the universe at the end of this epoch including the tiny fluctuations generated during inflation (see diagram), and the spread of objects such as galaxies in the universe is an indication of the scale and size of the universe as it developed over time.[22]

Habitable epoch

See also: Abiogenesis

The chemistry of life may have begun shortly after the Big Bang, 13.8 billion years ago, during a habitable epoch when the Universe was only 10-17 million years old.[23][24][25]

Dark Ages

See also: Hydrogen line

Before decoupling occurred, most of the photons in the universe were interacting with electrons and protons in the photon–baryon fluid. The universe was opaque or "foggy" as a result. There was light but not light we can now observe through telescopes. The baryonic matter in the universe consisted of ionized plasma, and it only became neutral when it gained free electrons during "recombination", thereby releasing the photons creating the CMB. When the photons were released (or decoupled) the universe became transparent. At this point the only radiation emitted was the 21 cm spin line of neutral hydrogen. There is currently an observational effort underway to detect this faint radiation, as it is in principle an even more powerful tool than the cosmic microwave background for studying the early universe. The Dark Ages are currently thought to have lasted between 150 million to 800 million years after the Big Bang. The October 2010 discovery of UDFy-38135539, the first observed galaxy to have existed during the following reionization epoch, gives us a window into these times. The galaxy earliest in this period observed and thus also the most distant galaxy ever observed is currently on the record of Leiden University's Richard J. Bouwens and Garth D. Illingsworth from UC Observatories/Lick Observatory. They found the galaxy UDFj-39546284 to be at a time some 480 million years after the Big Bang or about halfway through the Cosmic Dark Ages at a distance of about 13.2 billion light-years. More recently, the UDFj-39546284 galaxy was found to be around "380 million years" after the Big Bang and at a distance of 13.37 billion light-years.[26]

6.4 Structure formation

See also: Large-scale structure of the cosmos and Structure formation

Structure formation in the big bang model proceeds hierarchically, with smaller structures forming before larger ones. The first structures to form are quasars, which are thought to be bright, early active galaxies, and population III stars. Before this epoch, the evolution of the universe could be understood through linear cosmological perturbation theory: that is, all structures could be understood as small deviations from a perfect homogeneous universe. This is computationally relatively easy to study. At this point non-linear structures begin to form, and the computational problem becomes much more difficult, involving, for example, N-body simulations with billions of particles.

6.4.1 Reionization

150 million to 1 billion years after the Big Bang

See also: Reionization and 21 centimeter radiation

The first stars and quasars form from gravitational collapse. The intense radiation they emit reionizes the surrounding universe. From this point on, most of the universe is composed of plasma.

The Hubble Ultra Deep Fields often showcase galaxies from an ancient era that tell us what the early Stelliferous Age was like.

6.4.2   Formation of stars

See also: Star formation

The first stars, most likely Population III stars, form and start the process of turning the light elements that were formed in the Big Bang (hydrogen, helium and lithium) into heavier elements. However, as yet there have been no observed Population III stars, and understanding of them is currently based on computational models of their formation and evolution. Fortunately observations of the Cosmic Microwave Background radiation can be used to date when star formation began in earnest. Analysis of such observations made by the European Space Agency's Planck telescope, as reported by BBC News in early February, 2015, concludes that the first generation of stars lit up 560 million years after the Big Bang. [27] [28]

Another Hubble image shows an infant galaxy forming nearby, which means this happened very recently on the cosmological timescale. This shows that new galaxy formation in the universe is still occurring.

6.4.3 Formation of galaxies

See also: Galaxy formation and evolution

Large volumes of matter collapse to form a galaxy. Population II stars are formed early on in this process, with Population I stars formed later.

Johannes Schedler's project has identified a quasar CFHQS 1641+3755 at 12.7 billion light-years away,[29] when the universe was just 7% of its present age.

On July 11, 2007, using the 10-metre Keck II telescope on Mauna Kea, Richard Ellis of the California Institute of Technology at Pasadena and his team found six star forming galaxies about 13.2 billion light years away and therefore created when the universe was only 500 million years old.[30] Only about 10 of these extremely early objects are currently known.[31] More recent observations have shown these ages to be shorter than previously indicated. The most distant galaxy observed as of October 2013 has been reported to be 13.1 billion light years away.[32]

The Hubble Ultra Deep Field shows a number of small galaxies merging to form larger ones, at 13 billion light years, when the universe was only 5% its current age.[33] This age estimate is now believed to be slightly shorter.[32]

Based upon the emerging science of nucleocosmochronology, the Galactic thin disk of the Milky Way is estimated to have been formed 8.8 ± 1.7 billion years ago.[34]

6.4.4   Formation of groups, clusters and superclusters

See also: Large-scale structure of the cosmos

Gravitational attraction pulls galaxies towards each other to form groups, clusters and superclusters.

6.4.5   Formation of the Solar System

9 billion years after the Big Bang

Main article: Formation and evolution of the Solar System

The Solar System began forming about 4.6 billion years ago, or about 9 billion years after the Big Bang. A fragment of a molecular cloud made mostly of hydrogen and traces of other elements began to collapse, forming a large sphere in the center which would become the Sun, as well as a surrounding disk. The surrounding accretion disk would coalesce into a multitude of smaller objects that would become planets, asteroids, and comets. The Sun is a late-generation star, and the Solar System incorporates matter created by previous generations of stars.

6.4.6   Today

13.8 billion years after the Big Bang

The Big Bang is estimated to have occurred about 13.8 billion years ago.[35] Since the expansion of the universe appears to be accelerating, its large-scale structure is likely to be the largest structure that will ever form in the universe. The present accelerated expansion prevents any more inflationary structures entering the horizon and prevents new gravitationally bound structures from forming.

6.5   Ultimate fate of the universe

Main article: Ultimate fate of the universe

As with interpretations of what happened in the very early universe, advances in fundamental physics are required before it will be possible to know the ultimate fate of the universe with any certainty. Below are some of the main possibilities.

6.5.1   Fate of the Solar System: 1 to 5 billion years

Main articles: Formation and evolution of the Solar System § Future, Stability of the Solar System, Future of the Earth § Solar evolution and Red giant § The Sun as a red giant

Over a timescale of a billion years or more, the Earth and Solar System are unstable. Earth's existing biosphere is expected to vanish in about a billion years, as the Sun's heat production gradually increases to the point that liquid water and life are unlikely;[36] the Earth's magnetic fields, axial tilt and atmosphere are subject to long term change; and the Solar System

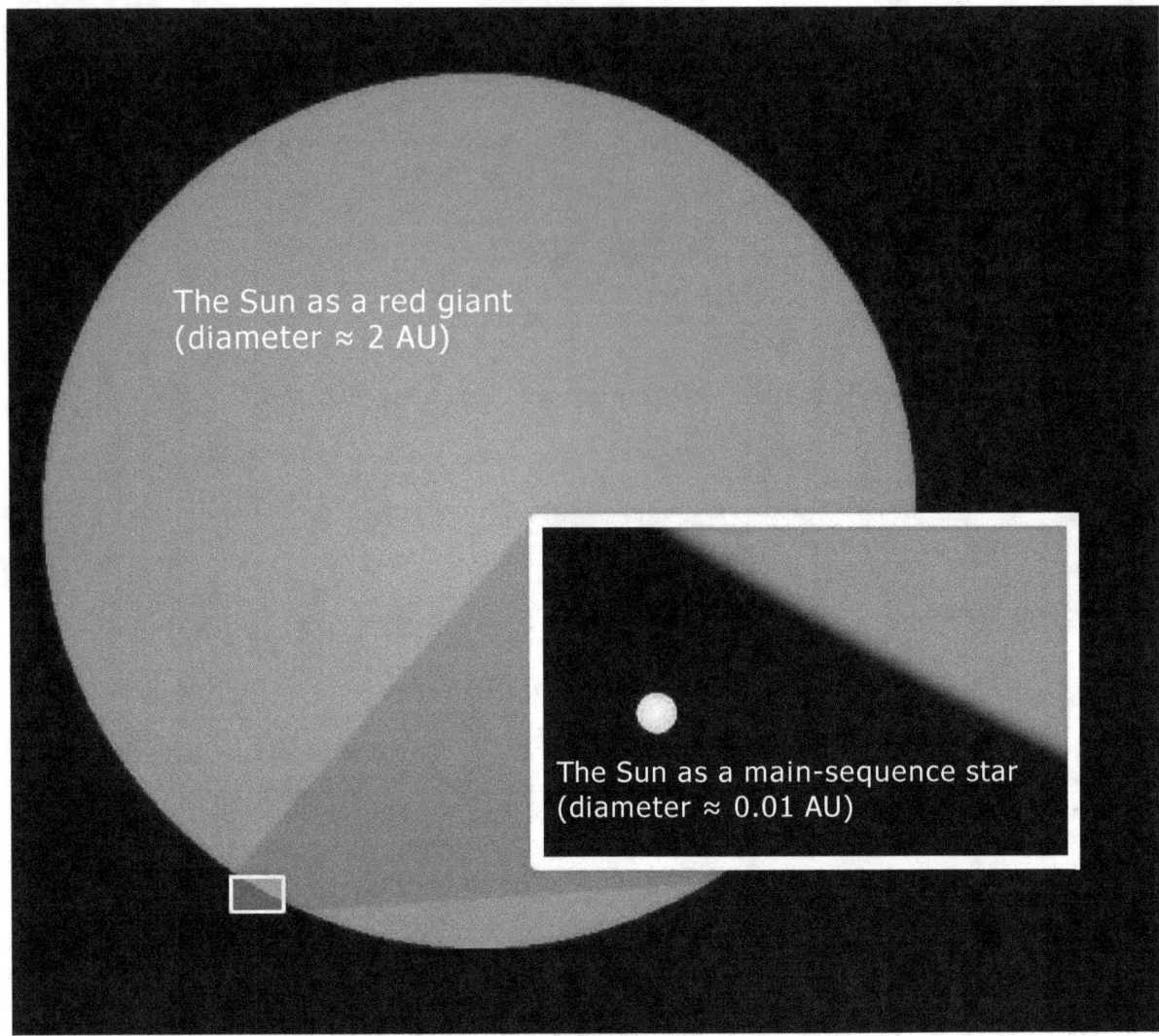

Relative size of our Sun as it is now (inset) compared to its estimated future size as a red giant

itself is chaotic over million- and billion-year timescales;[37] Eventually in around 5.4 billion years from now, the core of the Sun will become hot enough to trigger hydrogen fusion in its surrounding shell.[36] This will cause the outer layers of the star to expand greatly, and the star will enter a phase of its life in which it is called a red giant.[38][39] Within 7.5 billion years, the Sun will have expanded to a radius of 1.2 AU—256 times its current size, and studies announced in 2008 show that due to tidal interaction between Sun and Earth, Earth would actually fall back into a lower orbit, and get engulfed and incorporated inside the Sun before the Sun reaches its largest size, despite the Sun losing about 38% of its mass.[40] The Sun itself will continue to exist for many billions of years, passing through a number of phases, and eventually ending up as a long-lived white dwarf. Eventually, after billions more years, the Sun will finally cease to shine altogether, becoming a black dwarf.[41]

6.5.2   Big Rip: ≥20 billion years from now

See also: Big Rip

This scenario is possible only if the energy density of dark energy actually increases without limit over time. Such dark energy is called phantom energy and is unlike any known kind of energy. In this case, the expansion rate of the universe will increase without limit. Gravitationally bound systems, such as clusters of galaxies, galaxies, and ultimately the Solar System will be torn apart. Eventually the expansion will be so rapid as to overcome the electromagnetic forces holding molecules and atoms together. Finally even atomic nuclei will be torn apart and the universe as we know it will end in an unusual kind of gravitational singularity. At the time of this singularity, the expansion rate of the universe will reach infinity, so that any and all forces (no matter how strong) that hold composite objects together (no matter how closely) will be overcome by this expansion, literally tearing everything apart.

6.5.3   Big Crunch: $\geq 10^2$ billion years from now

See also: Big Crunch

If the energy density of dark energy were negative or the universe were closed, then it would be possible that the expansion of the universe would reverse and the universe would contract towards a hot, dense state. This is a required element of oscillatory universe scenarios, such as the cyclic model, although a Big Crunch does not necessarily imply an oscillatory universe. Current observations suggest that this model of the universe is unlikely to be correct, and the expansion will continue or even accelerate.

6.5.4   Big Freeze: $\geq 10^5$ billion years from now

Main articles: Future of an expanding universe and Heat death of the universe

This scenario is generally considered to be the most likely, as it occurs if the universe continues expanding as it has been. Over a time scale on the order of 10^{14} years or less, existing stars burn out, stars cease to be created, and the universe goes dark.[42], §IID. Over a much longer time scale in the eras following this, the galaxy evaporates as the stellar remnants comprising it escape into space, and black holes evaporate via Hawking radiation.[42], §III, §IVG. In some grand unified theories, proton decay after at least 10^{34} years will convert the remaining interstellar gas and stellar remnants into leptons (such as positrons and electrons) and photons. Some positrons and electrons will then recombine into photons.[42], §IV, §VF. In this case, the universe has reached a high-entropy state consisting of a bath of particles and low-energy radiation. It is not known however whether it eventually achieves thermodynamic equilibrium.[42], §VIB, VID.

6.5.5   Heat Death: 10^{1000} years from now

See also: Heat death of the universe

The heat death is a possible final state of the universe, estimated at after 10^{1000} years, in which it has "run down" to a state of no thermodynamic free energy to sustain motion or life. In physical terms, it has reached maximum entropy (because of this, the term "entropy" has often been confused with Heat Death, to the point of entropy being labelled as the "force killing the universe"). The hypothesis of a universal heat death stems from the 1850s ideas of William Thomson (Lord Kelvin)[43] who extrapolated the theory of heat views of mechanical energy loss in nature, as embodied in the first two laws of thermodynamics, to universal operation.

6.5.6   Vacuum metastability event

See also: False vacuum

If our universe is in a very long-lived false vacuum, it is possible that a small region of the universe will tunnel into a lower energy state (see Bubble nucleation). If this happens, all structures within will be destroyed instantaneously and the region will expand at near light speed, bringing destruction without any forewarning.

6.6   See also

- Cosmic Calendar (age of universe scaled to a single year)
- Cyclic model
- Dark-energy-dominated era
- Dyson's eternal intelligence
- Entropy (arrow of time)
- Graphical timeline from Big Bang to Heat Death
- Graphical timeline of the Big Bang
- Graphical timeline of the Stelliferous Era
- Illustris project
- Matter-dominated era
- Radiation-dominated era
- Timeline of the far future
- Ultimate fate of the universe

6.7   References

[1] Planck collaboration (2013). "Planck 2013 results. XVI. Cosmological parameters". *Submitted to Astronomy & Astrophysics* **571**: A16. arXiv:1303.5076. Bibcode:2014A&A...571A..16P. doi:10.1051/0004-6361/201321591.

[2] Staff (17 March 2014). "BICEP2 2014 Results Release". *National Science Foundation*. Retrieved 18 March 2014.

[3] Clavin, Whitney (17 March 2014). "NASA Technology Views Birth of the Universe". *NASA*. Retrieved 17 March 2014.

[4] Overbye, Dennis (17 March 2014). "Detection of Waves in Space Buttresses Landmark Theory of Big Bang". *The New York Times*. Retrieved 17 March 2014.

[5] Chaisson, E., (2001). *Cosmic Evolution: The Rise of Complexity in Nature*, Harvard University Press, ISBN 0-674-00987-8; see also Cosmic Evolution

[6] Ryden B: "Introduction to Cosmology", pg. 196 Addison-Wesley 2003

[7] Overbye, Dennis (March 24, 2014). "Ripples From the Big Bang". *New York Times*. Retrieved March 24, 2014.

[8] Ade, P.A.R. (BICEP2 Collaboration) et al. (June 19, 2014). "Detection of B-Mode Polarization at Degree Angular Scales by BI-CEP2" (PDF). *Physical Review Letters* **112**: 241101. arXiv:1403.3985. Bibcode:2014PhRvL.112x1101A. doi:10.1103/PhysRevLett.112.241101. PMID 24996078. Retrieved June 20, 2014.

[9] http://www.math.columbia.edu/~{}woit/wordpress/?p=6865

[10] Overbye, Dennis (June 19, 2014). "Astronomers Hedge on Big Bang Detection Claim". *New York Times*. Retrieved June 20, 2014.

[11] Amos, Jonathan (June 19, 2014). "Cosmic inflation: Confidence lowered for Big Bang signal". *BBC News*. Retrieved June 20, 2014.

[12] BICEP2/Keck, Planck Collaborations (2015). "A Joint Analysis of BICEP2/Keck Array and Planck Data (Provisionally accepted by PRL)". *arXiv*. arXiv:1502.00612v1. Retrieved 13 February 2015.

[13] Clavin, Whitney (30 January 2015). "Gravitational Waves from Early Universe Remain Elusive". *NASA*. Retrieved 30 January 2015.

[14] Overbye, Dennis (30 January 2015). "Speck of Interstellar Dust Obscures Glimpse of Big Bang". *New York Times*. Retrieved 31 January 2015.

[15] "Gravitational waves from early universe remain elusive". *Science Daily*. 31 January 2015. Retrieved 3 February 2015.

[16] The Timescale of Creation

[17] Detailed timeline of Big Bang nucleosynthesis processes

[18] Gannon, Megan (December 21, 2012). "New 'Baby Picture' of Universe Unveiled". Space.com. Retrieved December 21, 2012.

[19] Bennett, C.L.; Larson, L.; Weiland, J.L.; Jarosk, N.; Hinshaw, N.; Odegard, N.; Smith, K.M.; Hill, R.S.; Gold, B.; Halpern, M.; Komatsu, E.; Nolta, M.R.; Page, L.; Spergel, D.N.; Wollack, E.; Dunkley, J.; Kogut, A.; Limon, M.; Meyer, S.S.; Tucker, G.S.; Wright, E.L. (December 20, 2012). "Nine-Year Wilkinson Microwave Anisotropy Probe (WMAP) Observations: Final Maps and Results". *The Astrophysical Journal Supplement Series* 208: 20. arXiv:1212.5225. Bibcode:2013ApJS..208...20B. doi:10.1088/0067-0049/208/2/20. Retrieved December 22, 2012.

[20] Hinshaw, G. et al. (2009). "Five-Year Wilkinson Microwave Anisotropy Probe (WMAP) Observations: Data Processing, Sky Maps, and Basic Results" (PDF). *Astrophysical Journal Supplement* 180 (2): 225–245. arXiv:0803.0732. Bibcode:2009ApJS..180..225H. doi:10.1088/0067-0049/180/2/225.

[21] Mukhanov, V: "Physical foundations of Cosmology", pg. 120, Cambridge 2005

[22] Amos, Jonathan (2012-11-13). "Quasars illustrate dark energy's roller coaster ride". *BBC News*. Retrieved 13 November 2012.

[23] Loeb, Abraham (October 2014). "The Habitable Epoch of the Early Universe". *International Journal of Astrobiology* 13 (04): 337–339. arXiv:1312.0613. Bibcode:2014IJAsB..13..337L. doi:10.1017/S1473550414000196. Retrieved 15 December 2014.

[24] Loeb, Abraham (2 December 2013). "The Habitable Epoch of the Early Universe" (PDF). *Arxiv*. arXiv:1312.0613v3. Retrieved 15 December 2014.

[25] Dreifus, Claudia (2 December 2014). "Much-Discussed Views That Go Way Back - Avi Loeb Ponders the Early Universe, Nature and Life". *New York Times*. Retrieved 3 December 2014.

[26] Wall, Mike (December 12, 2012). "Ancient Galaxy May Be Most Distant Ever Seen". Space.com. Retrieved December 12, 2012.

[27] *Ferreting Out The First Stars*; physorg.com

[28]

[29] APOD: 2007 September 6 - Time Tunnel

[30] "New Scientist" 14 July 2007

[31] HET Helps Astronomers Learn Secrets of One of Universe's Most Distant Objects

[32] Scientists confirm most distant galaxy ever

[33] APOD: 2004 March 9 – The Hubble Ultra Deep Field

[34] Eduardo F. del Peloso a1a, Licio da Silva a1, Gustavo F. Porto de Mello and Lilia I. Arany-Prado (2005), "The age of the Galactic thin disk from Th/Eu nucleocosmochronology: extended sample" (Proceedings of the International Astronomical Union (2005), 1: 485–486 Cambridge University Press)

[35] "Cosmic Detectives". The European Space Agency (ESA). 2013-04-02. Retrieved 2013-04-15.

[36] K. P. Schroder, Robert Connon Smith (2008). "Distant future of the Sun and Earth revisited". *Monthly Notices of the Royal Astronomical Society* 386 (1): 155–163. arXiv:0801.4031. Bibcode:2008MNRAS.386..155S. doi:10.1111/j.1365-2966.2008.13022.x.

[37] J. Laskar (1994). "Large-scale chaos in the solar system". *Astronomy and Astrophysics* 287: L9–L12. Bibcode:1994A&A...287L...9L.

[38] Zeilik & Gregory 1998, p. 320–321.

[39] "Introduction to Cataclysmic Variables (CVs)". *NASA Goddard Space Center*. 2006. Retrieved 2006-12-29.

[40] Palmer, Jason (22 February 2008). "Hope dims that Earth will survive Sun's death". *New Scientist*.

[41] G. Fontaine, P. Brassard, P. Bergeron (2001). "The Potential of White Dwarf Cosmochronology". *Publications of the Astronomical Society of the Pacific* 113 (782): 409–435. Bibcode:2001PASP..113..409F. doi:10.1086/319535. Retrieved 2008-05-11.

[42] A dying universe: the long-term fate and evolution of astrophysical objects, Fred C. Adams and Gregory Laughlin, *Reviews of Modern Physics* 69, #2 (April 1997), pp. 337–372. Bibcode: 1997RvMP...69..337A. doi:10.1103/RevModPhys.69.337.

[43] Thomson, William. (1851). "On the Dynamical Theory of Heat, with numerical results deduced from Mr Joule's equivalent of a Thermal Unit, and M. Regnault's Observations on Steam." Excerpts. [§§1-14 & §§99-100], *Transactions of the Royal Society of Edinburgh*, March, 1851; and *Philosophical Magazine* IV. 1852, [from *Mathematical and Physical Papers*, vol. i, art. XLVIII, pp. 174]

6.8 External links

- PBS Online (2000). From the Big Bang to the End of the Universe – The Mysteries of Deep Space Timeline. Retrieved March 24, 2005.

- Schulman, Eric (1997). The History of the Universe in 200 Words or Less. Retrieved March 24, 2005.

- Space Telescope Science Institute Office of Public Outreach (2005). Home of the Hubble Space Telescope. Retrieved March 24, 2005.

- Fermilab graphics (see "Energy time line from the Big Bang to the present" and "History of the Universe Poster")

- Exploring Time from Planck time to the lifespan of the Universe

- Cosmic Evolution is a multi-media web site that explores the cosmic-evolutionary scenario from big bang to humankind.

- Astronomers' first detailed hint of what was going on less than a trillionth of a second after time began

- The Universe Adventure

- Cosmology FAQ, Professor Edward L. Wright, UCLA

- Sean Carroll on the arrow of time (Part 1), *The origin of the universe and the arrow of time*, Sean Carroll, video, CHAST 2009, Templeton, Faculty of science, University of Sydney, November 2009, TED.com

- A Universe From Nothing, video, Lawrence Krauss, AAI 2009, YouTube.com

- Once Upon A Universe - Story of the Universe told in 13 chapters. Science communication site supported by STFC.

- Cosmic Evolution through Time - an interactive timeline explains the main events in the history of our Universe

Chapter 7

Planck epoch

In physical cosmology, the **Planck epoch** (or **Planck era**) is the earliest period of time in the history of the universe, from zero to approximately 10^{-43} seconds (Planck time). It is believed that, due to the extraordinarily small scale of the universe at the time, quantum effects of gravity dominated physical interactions. During this period, approximately 13.79 billion years ago, gravitation is believed to have been as strong as the other fundamental forces, and all the forces may have been unified. Inconceivably hot and dense, the state of the universe during the Planck epoch was unstable. As it expanded and cooled, the familiar manifestations of the fundamental forces arose through a process known as symmetry breaking.

Modern cosmology now suggests that the Planck epoch may have inaugurated a period of unification, known as the grand unification epoch, and that symmetry breaking then quickly led to the era of cosmic inflation, the Inflationary epoch, during which the universe greatly expanded in scale over a very short period of time.[1]

7.1   Theoretical ideas

As there presently exists no widely accepted framework for how to combine quantum mechanics with relativistic gravity, science is not currently able to make predictions about events occurring over intervals shorter than the Planck time or distances shorter than one Planck length, the distance light travels in one Planck time—about 1.616×10^{-35} meters. Without an understanding of quantum gravity, a theory unifying quantum mechanics and relativistic gravity, the physics of the Planck epoch are unclear, and the exact manner in which the fundamental forces were unified, and how they came to be separate entities, is still poorly understood. Three of the four forces have been successfully integrated in a common framework, but gravity remains problematic. If quantum effects are ignored, the universe starts from a singularity with an infinite density. This conclusion could change when quantum gravity is taken into account. String theory and loop quantum gravity are leading candidates for a theory of unification, which have yielded meaningful insights already, but work in noncommutative geometry and other fields also holds promise for our understanding of the very beginning.

7.2   Experiments exploring this time

Experimental data casting light on this cosmological epoch has been scant or non-existent until now, but recent results from the WMAP probe have allowed scientists to test hypotheses about the universe's first trillionth of a second (although the cosmic microwave background radiation observed by WMAP originated when the universe was already several hundred thousand years old). Although this interval is still orders of magnitude longer than the Planck time, other experiments currently coming online including the Planck Surveyor probe, promise to push back our 'cosmic clock' further to reveal quite a bit more about the very first moments of our universe's history, hopefully giving us some insight into the Planck epoch itself. Data from particle accelerators provides meaningful insight into the early universe as well. Experiments with the Relativistic Heavy Ion Collider have allowed physicists to determine that the quark–gluon plasma (an early phase of matter) behaved more like a liquid than a gas, and the Large Hadron Collider at CERN will probe still earlier phases of matter, but no accelerator (current or planned) will be capable of probing the Planck scale directly.

7.3 See also

- Big Bang
- Planck particle
- Quantum gravity
- Chronology of the universe
- Unified field theory
- Planck scale

7.4 Footnotes

[1] Edward W. Kolb; Michael S. Turner (1994). *The Early Universe*. Basic Books. p. 447. ISBN 978-0-201-62674-2. Retrieved 10 April 2010.

7.5 References

- *A Brief History of the Universe*.
- *The Planck Epoch*.
- *Genesis I: The Planck Epoch*.
- *Evolution of the Universe through the Planck Epoch*.

7.6 External links

- The Planck Era from U of Tennessee Astrophysics pages
- The Planck Era from U of Oregon Cosmology pages
- The Planck Era by Sten Odenwald from Astronomy Cafe
- The Plank Epoch by professor James Schombert 390
- The Planck Era - definition from U of Ottawa's Astronomy Knowledge Base

Chapter 8

Cosmic microwave background

"CMB" redirects here. For other uses, see CMB (disambiguation).

The **cosmic microwave background (CMB)** is the thermal radiation left over from the time of recombination in Big Bang cosmology. In older literature, the CMB is also variously known as cosmic microwave background radiation (CMBR) or "relic radiation." The CMB is a cosmic background radiation that is fundamental to observational cosmology because it is the oldest light in the universe, dating to the epoch of recombination. With a traditional optical telescope, the space between stars and galaxies (the *background*) is completely dark. However, a sufficiently sensitive radio telescope shows a faint background glow, almost exactly the same in all directions, that is not associated with any star, galaxy, or other object. This glow is strongest in the microwave region of the radio spectrum. The accidental discovery of CMB in 1964 by American radio astronomers Arno Penzias and Robert Wilson[1][2] was the culmination of work initiated in the 1940s, and earned the discoverers the 1978 Nobel Prize.

> *The CMB is a snapshot of the oldest light in our Universe, imprinted on the sky when the Universe was just 380,000 years old. It shows tiny temperature fluctuations that correspond to regions of slightly different densities, representing the seeds of all future structure: the stars and galaxies of today.*[3]

The CMB is well explained as radiation left over from an early stage in the development of the universe, and its discovery is considered a landmark test of the Big Bang model of the universe. When the universe was young, before the formation of stars and planets, it was denser, much hotter, and filled with a uniform glow from a white-hot fog of hydrogen plasma. As the universe expanded, both the plasma and the radiation filling it grew cooler. When the universe cooled enough, protons and electrons combined to form neutral atoms. These atoms could no longer absorb the thermal radiation, and so the universe became transparent instead of being an opaque fog.[4] Cosmologists refer to the time period when neutral atoms first formed as the *recombination epoch*, and the event shortly afterwards when photons started to travel freely through space rather than constantly being scattered by electrons and protons in plasma is referred to as photon decoupling. The photons that existed at the time of photon decoupling have been propagating ever since, though growing fainter and less energetic, since the expansion of space causes their wavelength to increase over time (and wavelength is inversely proportional to energy according to Planck's relation). This is the source of the alternative term *relic radiation*. The *surface of last scattering* refers to the set of points in space at the right distance from us so that we are now receiving photons originally emitted from those points at the time of photon decoupling.

Precise measurements of the CMB are critical to cosmology, since any proposed model of the universe must explain this radiation. The CMB has a thermal black body spectrum at a temperature of 2.72548±0.00057 K.[5] The spectral radiance dEv/dv peaks at 160.2 GHz, in the microwave range of frequencies. (Alternatively if spectral radiance is defined as dEλ/dλ then the peak wavelength is 1.063 mm.) The glow is very nearly uniform in all directions, but the tiny residual variations show a very specific pattern, the same as that expected of a fairly uniformly distributed hot gas that has expanded to the current size of the universe. In particular, the spectral radiance at different angles of observation in the sky contains small anisotropies, or irregularities, which vary with the size of the region examined. They have been measured in detail, and match what would be expected if small thermal variations, generated by quantum fluctuations of matter in a very tiny space, had expanded to the size of the observable universe we see today. This is a very active field of study, with scientists seeking both better data (for example, the Planck spacecraft) and better interpretations of the initial conditions of expansion. Although many different processes might produce the general form of a black body spectrum, no model other than the Big Bang has yet explained the fluctuations. As a result, most cosmologists consider the Big Bang model of the universe to be the best explanation for the CMB.

The high degree of uniformity throughout the observable universe and its faint but measured anisotropy lend strong support for the Big Bang model in general and the ΛCDM ("Lambda Cold Dark Matter") model in particular. Moreover, the fluctuations are coherent on angular scales that are larger than the apparent cosmological horizon at recombination. Either such coherence is acausally fine-tuned, or cosmic inflation occurred.[6][7]

8.1 Features

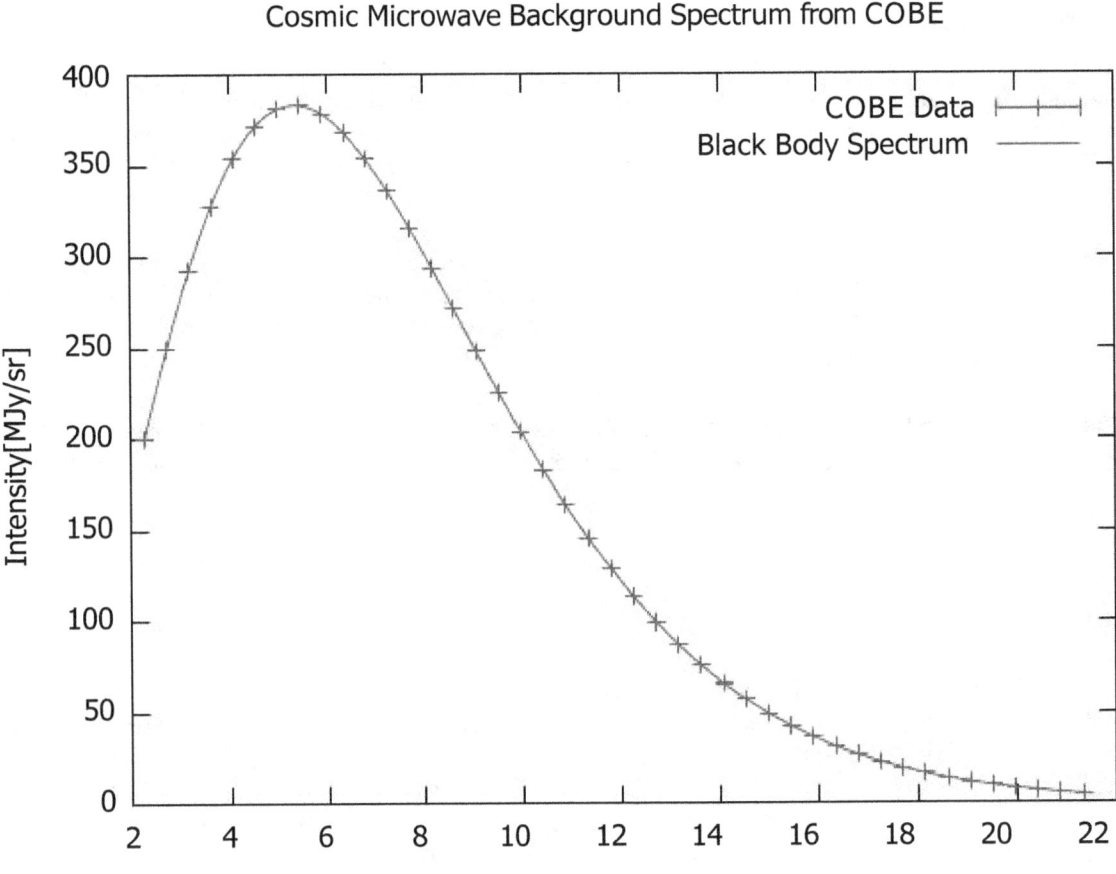

Graph of cosmic microwave background spectrum measured by the FIRAS instrument on the COBE, the most precisely measured black body spectrum in nature.[8] The error bars are too small to be seen even in an enlarged image, and it is impossible to distinguish the observed data from the theoretical curve.

The cosmic microwave background radiation is an emission of uniform, black body thermal energy coming from all parts of the sky. The radiation is isotropic to roughly one part in 100,000: the root mean square variations are only 18 µK,[9] after subtracting out a dipole anisotropy from the Doppler shift of the background radiation. The latter is caused by the peculiar velocity of the Earth relative to the comoving cosmic rest frame as the planet moves at some 371 km/s towards the constellation Leo. The CMB dipole as well as aberration at higher multipoles have been measured, consistent with galactic motion.[10]

In the Big Bang model for the formation of the universe, Inflationary Cosmology predicts that after about 10^{-37} seconds[11] the nascent universe underwent exponential growth that smoothed out nearly all inhomogeneities. The remaining inhomogeneities were caused by quantum fluctuations in the inflaton field that caused the inflation event.[12] After 10^{-6} seconds, the early universe was made up of a hot, interacting plasma of photons, electrons, and baryons. As the universe expanded, adiabatic cooling caused the energy density of the plasma to decrease until it became favorable for electrons to combine with protons, forming hydrogen atoms. This recombination event happened when the temperature was around 3000 K or when the universe was approximately 379,000 years old.[13] At this point, the photons no longer interacted with the now electrically neutral atoms and began to travel freely through space, resulting in the decoupling of matter and radiation.[14]

The color temperature of the ensemble of decoupled photons has continued to diminish ever since; now down to 2.7260±0.0013 K,[5] it will continue to drop as the universe expands. The intensity of the radiation also corresponds to black-body radiation at 2.726 K because red-shifted black-body radiation is just like black-body radiation at a lower temperature. According to the Big Bang model, the radiation from the sky we measure today comes from a spherical surface called *the surface of last scattering*. This represents the set of locations in space at which the decoupling event is estimated to have occurred[15] and at a point in time such that the photons from that distance have just reached observers. Most of the radiation energy in the universe is in the cosmic microwave background,[16] making up a fraction of roughly 6×10^{-5} of the total density of the universe.[17]

Two of the greatest successes of the Big Bang theory are its prediction of the almost perfect black body spectrum and its detailed prediction of the anisotropies in the cosmic microwave background. The CMB spectrum has become the most precisely measured black body spectrum in nature.[8]

Density of energy for CMB is 0.25 eV/cm^{3}[18] (4.005×10^{-14} J/m^3) or (400–500 photons/cm^{3}[19]).

8.2   History

See also: Discovery of cosmic microwave background radiation

The cosmic microwave background was first predicted in 1948 by Ralph Alpher, and Robert Herman.[20][21][22] Alpher and Herman were able to estimate the temperature of the cosmic microwave background to be 5 K, though two years later they re-estimated it at 28 K. This high estimate was due to a mis-estimate of the Hubble constant by Alfred Behr, which could not be replicated and was later abandoned for the earlier estimate. Although there were several previous estimates of the temperature of space, these suffered from two flaws. First, they were measurements of the *effective* temperature of space and did not suggest that space was filled with a thermal Planck spectrum. Next, they depend on our being at a special spot at the edge of the Milky Way galaxy and they did not suggest the radiation is isotropic. The estimates would yield very different predictions if Earth happened to be located elsewhere in the universe.[23]

The 1948 results of Alpher and Herman were discussed in many physics settings through about 1955, when both left the Applied Physics Laboratory at Johns Hopkins University. The mainstream astronomical community, however, was not intrigued at the time by cosmology. Alpher and Herman's prediction was rediscovered by Yakov Zel'dovich in the early 1960s, and independently predicted by Robert Dicke at the same time. The first published recognition of the CMB radiation as a detectable phenomenon appeared in a brief paper by Soviet astrophysicists A. G. Doroshkevich and Igor Novikov, in the spring of 1964.[24] In 1964, David Todd Wilkinson and Peter Roll, Dicke's colleagues at Princeton University, began constructing a Dicke radiometer to measure the cosmic microwave background.[25] In 1964, Arno Penzias and Robert Woodrow Wilson at the Crawford Hill location of Bell Telephone Laboratories in nearby Holmdel Township, New Jersey had built a Dicke radiometer that they intended to use for radio astronomy and satellite communication experiments. On 20 May 1964 they made their first measurement clearly showing the presence of the microwave background,[26] with their instrument having an excess 4.2K antenna temperature which they could not account for. After receiving a telephone call from Crawford Hill, Dicke famously quipped: "Boys, we've been scooped."[1][27][28] A meeting between the Princeton and Crawford Hill groups determined that the antenna temperature was indeed due to the microwave background. Penzias and Wilson received the 1978 Nobel Prize in Physics for their discovery.[29]

The interpretation of the cosmic microwave background was a controversial issue in the 1960s with some proponents of the steady state theory arguing that the microwave background was the result of scattered starlight from distant galaxies.[30]

Using this model, and based on the study of narrow absorption line features in the spectra of stars, the astronomer Andrew McKellar wrote in 1941: "It can be calculated that the 'rotational temperature' of interstellar space is 2 K."[31] However, during the 1970s the consensus was established that the cosmic microwave background is a remnant of the big bang. This was largely because new measurements at a range of frequencies showed that the spectrum was a thermal, black body spectrum, a result that the steady state model was unable to reproduce.[32]

The Holmdel Horn Antenna on which Penzias and Wilson discovered the cosmic microwave background.

Harrison, Peebles, Yu and Zel'dovich realized that the early universe would have to have inhomogeneities at the level of 10^{-4} or 10^{-5}.[33][34][35] Rashid Sunyaev later calculated the observable imprint that these inhomogeneities would have on the cosmic microwave background.[36] Increasingly stringent limits on the anisotropy of the cosmic microwave background were set by ground based experiments during the 1980s. RELIKT-1, a Soviet cosmic microwave background anisotropy experiment on board the Prognoz 9 satellite (launched 1 July 1983) gave upper limits on the large-scale anisotropy. The NASA COBE mission clearly confirmed the primary anisotropy with the Differential Microwave Radiometer instrument, publishing their findings in 1992.[37][38] The team received the Nobel Prize in physics for 2006 for this discovery.

Inspired by the COBE results, a series of ground and balloon-based experiments measured cosmic microwave background anisotropies on smaller angular scales over the next decade. The primary goal of these experiments was to measure the scale of the first acoustic peak, which COBE did not have sufficient resolution to resolve. This peak corresponds to large scale density variations in the early universe that are created by gravitational instabilities, resulting in acoustical oscillations in the plasma.[39] The first peak in the anisotropy was tentatively detected by the Toco experiment and the result was confirmed by the BOOMERanG and MAXIMA experiments.[40][41][42] These measurements demonstrated that the geometry of the universe is approximately flat, rather than curved.[43] They ruled out cosmic strings as a major component of cosmic structure formation and suggested cosmic inflation was the right theory of structure formation.[44]

The second peak was tentatively detected by several experiments before being definitively detected by WMAP, which has also tentatively detected the third peak.[45] As of 2010, several experiments to improve measurements of the polarization and the microwave background on small angular scales are ongoing. These include DASI, WMAP, BOOMERanG, QUaD, Planck spacecraft, Atacama Cosmology Telescope, South Pole Telescope and the QUIET telescope.

8.2.1 Timeline

Thermal (non-microwave background) temperature predictions

- 1896 – Charles Édouard Guillaume estimates the "radiation of the stars" to be 5.6K.[46]

- 1926 – Sir Arthur Eddington estimates the non-thermal radiation of starlight in the galaxy "... by the formula $E = \sigma T^4$ the effective temperature corresponding to this density is 3.18° absolute ...   black body".[47]

- 1930s – Cosmologist Erich Regener calculates that the non-thermal spectrum of cosmic rays in the galaxy has an effective temperature of 2.8 K

- 1931 – Term *microwave* first used in print: "When trials with wavelengths as low as 18 cm. were made known, there was undisguised surprise+that the problem of the micro-wave had been solved so soon." Telegraph & Telephone Journal XVII. 179/1

- 1934 – Richard Tolman shows that black-body radiation in an expanding universe cools but remains thermal

- 1938 – Nobel Prize winner (1920) Walther Nernst reestimates the cosmic ray temperature as 0.75K

- 1941 – Andrew McKellar was attempting to measure the average temperature of the interstellar medium, and used the excitation of CN doublet lines to measure that the "eff ective temperature of space" (the average bolometric temperature) is about 2.3 K [31][48]

- 1946 – Robert Dicke predicts "... radiation from cosmic matter" at <20 K, but did not refer to background radiation [49]

- 1946 – George Gamow calculates a temperature of 50 K (assuming a 3-billion year old universe),[50] commenting it "... is in reasonable agreement with the actual temperature of interstellar space", but does not mention background radiation.[51]

- 1953 – Erwin Finlay-Freundlich in support of his tired light theory, derives a blackbody temperature for inter-galactic space of 2.3K [52] with comment from Max Born suggesting radio astronomy as the arbitrator between expanding and infinite cosmologies.

Microwave background radiation predictions

- 1946 – George Gamow calculates a temperature of 50 K (assuming a 3-billion year old universe),[50] commenting it "... is in reasonable agreement with the actual temperature of interstellar space", but does not mention background radiation.

- 1948 – Ralph Alpher and Robert Herman estimate "the temperature in the universe" at 5 K. Although they do not specifically mention microwave background radiation, it may be inferred.[53]

- 1949 – Ralph Alpher and Robert Herman re-re-estimate the temperature at 28 K.

- 1953 – George Gamow estimates 7 K.[49]

- 1956 – George Gamow estimates 6 K.[49]

- 1955 – Émile Le Roux of the Nançay Radio Observatory, in a sky survey at λ = 33 cm, reported a near-isotropic background radiation of 3 kelvins, plus or minus 2.[49]

- 1957 – Tigran Shmaonov reports that "the absolute effective temperature of the radioemission background ... is 4±3 K".[54] It is noted that the "measurements showed that radiation intensity was independent of either time or direction of observation ... it is now clear that Shmaonov did observe the cosmic microwave background at a wavelength of 3.2 cm".[55][56]

- 1960s – Robert Dicke re-estimates a microwave background radiation temperature of 40 K [49][57]

- 1964 – A. G. Doroshkevich and Igor Dmitrievich Novikov publish a brief paper suggesting microwave searches for the black-body radiation predicted by Gamow, Alpher, and Herman, where they name the CMB radiation phenomenon as detectable.[58]

- 1964–65 – Arno Penzias and Robert Woodrow Wilson measure the temperature to be approximately 3 K. Robert Dicke, James Peebles, P. G. Roll, and D. T. Wilkinson interpret this radiation as a signature of the big bang.

- 1966 – Rainer K. Sachs and Arthur M. Wolfe theoretically predict microwave background fluctuation amplitudes created by gravitational potential variations between observers and the last scattering surface (see Sachs-Wolfe effect)

- 1968 – Martin Rees and Dennis Sciama theoretically predict microwave background fluctuation amplitudes created by photons traversing time-dependent potential wells

- 1969 – R. A. Sunyaev and Yakov Zel'dovich study the inverse Compton scattering of microwave background photons by hot electrons (see Sunyaev-Zel'dovich effect)

- 1983 – Researchers from the Cambridge Radio Astronomy Group and the Owens Valley Radio Observatory first detect the Sunyaev-Zel'dovich effect from clusters of galaxies

- 1983 – RELIKT-1 Soviet CMB anisotropy experiment was launched.

- 1990 – FIRAS on the Cosmic Background Explorer (COBE) satellite measures the black body form of the CMB spectrum with exquisite precision, and shows that the microwave background has a nearly perfect black-body spectrum and thereby strongly constrains the density of the intergalactic medium.

- January 1992 – Scientists that analysed data from the RELIKT-1 report the discovery of anisotropy in the cosmic microwave background at the Moscow astrophysical seminar.[59]

- 1992 – Scientists that analysed data from COBE DMR report the discovery of anisotropy in the cosmic microwave background.[60]

- 1995 – The Cosmic Anisotropy Telescope performs the first high resolution observations of the cosmic microwave background.

- 1999 – First measurements of acoustic oscillations in the CMB anisotropy angular power spectrum from the TOCO, BOOMERANG, and Maxima Experiments. The BOOMERanG experiment makes higher quality maps at intermediate resolution, and confirms that the universe is "flat".

- 2002 – Polarization discovered by DASI.[61]

- 2003 – E-mode polarization spectrum obtained by the CBI.[62] The CBI and the Very Small Array produces yet higher quality maps at high resolution (covering small areas of the sky).

- 2003 – The WMAP spacecraft produces an even higher quality map at low and intermediate resolution of the whole sky (WMAP provides *no* high-resolution data, but improves on the intermediate resolution maps from BOOMERanG).

- 2004 – E-mode polarization spectrum obtained by the CBI.[63]

- 2004 – The Arcminute Cosmology Bolometer Array Receiver produces a higher quality map of the high resolution structure not mapped by WMAP.

- 2005 – The Arcminute Microkelvin Imager and the Sunyaev-Zel'dovich Array begin the first surveys for very high redshift clusters of galaxies using the Sunyaev-Zel'dovich effect.

- 2005 – Ralph A. Alpher is awarded the National Medal of Science for his groundbreaking work in nucleosynthesis and prediction that the universe expansion leaves behind background radiation, thus providing a model for the Big Bang theory.

- 2006 – The long-awaited three-year WMAP results are released, confirming previous analysis, correcting several points, and including polarization data.

- 2006 – Two of COBE's principal investigators, George Smoot and John Mather, received the Nobel Prize in Physics in 2006 for their work on precision measurement of the CMBR.

- 2006-2011 – Improved measurements from WMAP, new supernova surveys ESSENCE and SNLS, and baryon acoustic oscillations from SDSS and WiggleZ, continue to be consistent with the standard Lambda-CDM model.

- 2014 – On March 17, 2014, astrophysicists of the BICEP2 collaboration announced the detection of inflationary gravitational waves in the B-mode power spectrum, which if confirmed, would provide clear experimental evidence for the theory of inflation.[64][65][66][67][68][69] However, on 19 June 2014, lowered confidence in confirming the cosmic inflation findings was reported.[68][70][71]

8.3 Relationship to the Big Bang

The cosmic microwave background radiation and the cosmological redshift-distance relation are together regarded as the best available evidence for the Big Bang theory. Measurements of the CMB have made the inflationary Big Bang theory the Standard Model of Cosmology.[72] The discovery of the CMB in the mid-1960s curtailed interest in alternatives such as the steady state theory.[73]

The CMB essentially confirms the Big Bang theory. In the late 1940s Alpher and Herman reasoned that if there was a big bang, the expansion of the universe would have stretched and cooled the high-energy radiation of the very early universe into the microwave region and down to a temperature of about 5 K. They were slightly off with their estimate, but they had exactly the right idea. They predicted the CMB. It took another 15 years for Penzias and Wilson to stumble into discovering that the microwave background was actually there.[74]

The CMB gives a snapshot of the universe when, according to standard cosmology, the temperature dropped enough to allow electrons and protons to form hydrogen atoms, thus making the universe transparent to radiation. When it originated some 380,000 years after the Big Bang—this time is generally known as the "time of last scattering" or the period of recombination or decoupling—the temperature of the universe was about 3000 K. This corresponds to an energy of about 0.25 eV, which is much less than the 13.6 eV ionization energy of hydrogen.[75]

Since decoupling, the temperature of the background radiation has dropped by a factor of roughly 1,100[76] due to the expansion of the universe. As the universe expands, the CMB photons are redshifted, making the radiation's temperature inversely proportional to a parameter called the universe's scale length. The temperature T_r of the CMB as a function of redshift, z, can be shown to be proportional to the temperature of the CMB as observed in the present day (2.725 K or 0.235 meV):[77]

$$T_r = 2.725(1 + z)$$

For details about the reasoning that the radiation is evidence for the Big Bang, see Cosmic background radiation of the Big Bang.

8.3.1 Primary anisotropy

The anisotropy of the cosmic microwave background is divided into two types: primary anisotropy, due to effects which occur at the last scattering surface and before; and secondary anisotropy, due to effects such as interactions of the background radiation with hot gas or gravitational potentials, which occur between the last scattering surface and the observer.

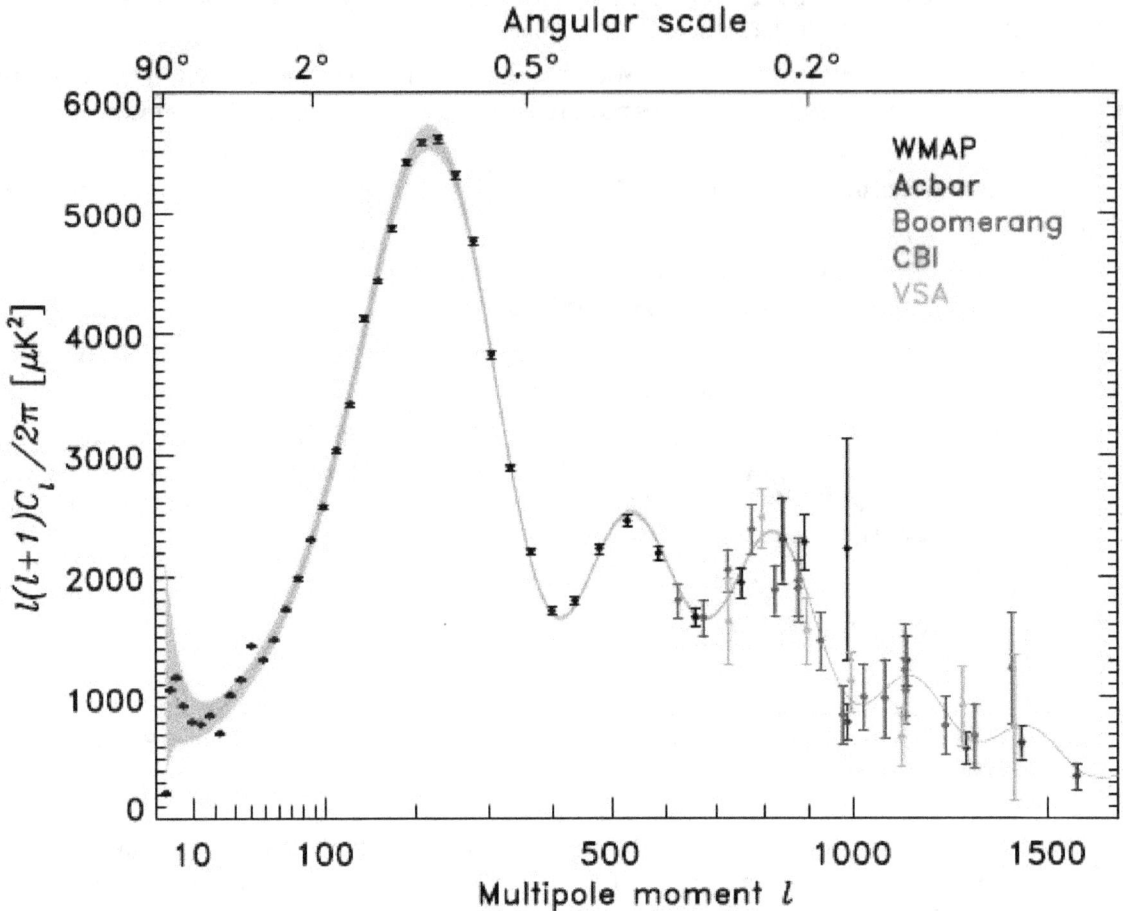

The power spectrum of the cosmic microwave background radiation temperature anisotropy in terms of the angular scale (or multipole moment). The data shown comes from the WMAP (2006), Acbar (2004) Boomerang (2005), CBI (2004), and VSA (2004) instruments. Also shown is a theoretical model (solid line).

The structure of the cosmic microwave background anisotropies is principally determined by two effects: acoustic oscillations and diffusion damping (also called collisionless damping or Silk damping). The acoustic oscillations arise because of a conflict in the photon–baryon plasma in the early universe. The pressure of the photons tends to erase anisotropies, whereas the gravitational attraction of the baryons—moving at speeds much slower than light—makes them tend to collapse to form dense haloes. These two effects compete to create acoustic oscillations which give the microwave background its characteristic peak structure. The peaks correspond, roughly, to resonances in which the photons decouple when a particular mode is at its peak amplitude.

The peaks contain interesting physical signatures. The angular scale of the first peak determines the curvature of the universe (but not the topology of the universe). The next peak—ratio of the odd peaks to the even peaks—determines the reduced baryon density.[78] The third peak can be used to get information about the dark matter density.[79]

The locations of the peaks also give important information about the nature of the primordial density perturbations. There are two fundamental types of density perturbations—called *adiabatic* and *isocurvature*. A general density perturbation is a mixture of both, and different theories that purport to explain the primordial density perturbation spectrum predict different mixtures.

- Adiabatic density perturbations

   the fractional additional density of each type of particle (baryons, photons ...) is the same. That is, if at one place there is 1% more energy in baryons than average, then at that place there is also 1% more energy in photons (and 1% more energy in neutrinos) than average. Cosmic inflation predicts that the primordial perturbations are adiabatic.

- Isocurvature density perturbations

  in each place the sum (over different types of particle) of the fractional additional densities is zero. That is, a perturbation where at some spot there is 1% more energy in baryons than average, 1% more energy in photons than average, and 2% *less* energy in neutrinos than average, would be a pure isocurvature perturbation. Cosmic strings would produce mostly isocurvature primordial perturbations.

The CMB spectrum can distinguish between these two because these two types of perturbations produce different peak locations. Isocurvature density perturbations produce a series of peaks whose angular scales (l-values of the peaks) are roughly in the ratio 1:3:5:..., while adiabatic density perturbations produce peaks whose locations are in the ratio 1:2:3:...[80] Observations are consistent with the primordial density perturbations being entirely adiabatic, providing key support for inflation, and ruling out many models of structure formation involving, for example, cosmic strings.

Collisionless damping is caused by two effects, when the treatment of the primordial plasma as fluid begins to break down:

- the increasing mean free path of the photons as the primordial plasma becomes increasingly rarefied in an expanding universe
- the finite depth of the last scattering surface (LSS), which causes the mean free path to increase rapidly during decoupling, even while some Compton scattering is still occurring.

These effects contribute about equally to the suppression of anisotropies at small scales, and give rise to the characteristic exponential damping tail seen in the very small angular scale anisotropies.

The depth of the LSS refers to the fact that the decoupling of the photons and baryons does not happen instantaneously, but instead requires an appreciable fraction of the age of the universe up to that era. One method of quantifying how long this process took uses the *photon visibility function* (PVF). This function is defined so that, denoting the PVF by P(t), the probability that a CMB photon last scattered between time t and t+dt is given by P(t)dt.

The maximum of the PVF (the time when it is most likely that a given CMB photon last scattered) is known quite precisely. The first-year WMAP results put the time at which P(t) is maximum as 372,000 years.[81] This is often taken as the "time" at which the CMB formed. However, to figure out how *long* it took the photons and baryons to decouple, we need a measure of the width of the PVF. The WMAP team finds that the PVF is greater than half of its maximum value (the "full width at half maximum", or FWHM) over an interval of 115,000 years. By this measure, decoupling took place over roughly 115,000 years, and when it was complete, the universe was roughly 487,000 years old.

8.3.2   Late time anisotropy

Since the CMB came into existence, it has apparently been modified by several subsequent physical processes, which are collectively referred to as late-time anisotropy, or secondary anisotropy. When the CMB photons became free to travel unimpeded, ordinary matter in the universe was mostly in the form of neutral hydrogen and helium atoms. However, observations of galaxies today seem to indicate that most of the volume of the intergalactic medium (IGM) consists of ionized material (since there are few absorption lines due to hydrogen atoms). This implies a period of reionization during which some of the material of the universe was broken into hydrogen ions.

The CMB photons are scattered by free charges such as electrons that are not bound in atoms. In an ionized universe, such charged particles have been liberated from neutral atoms by ionizing (ultraviolet) radiation. Today these free charges are at sufficiently low density in most of the volume of the universe that they do not measurably affect the CMB. However, if the IGM was ionized at very early times when the universe was still denser, then there are two main effects on the CMB:

1. Small scale anisotropies are erased. (Just as when looking at an object through fog, details of the object appear fuzzy.)

2. The physics of how photons are scattered by free electrons (Thomson scattering) induces polarization anisotropies on large angular scales. This broad angle polarization is correlated with the broad angle temperature perturbation.

Both of these effects have been observed by the WMAP spacecraft, providing evidence that the universe was ionized at very early times, at a redshift more than 17. The detailed provenance of this early ionizing radiation is still a matter of scientific debate. It may have included starlight from the very first population of stars (population III stars), supernovae when these first stars reached the end of their lives, or the ionizing radiation produced by the accretion disks of massive black holes.

The time following the emission of the cosmic microwave background—and before the observation of the first stars—is semi-humorously referred to by cosmologists as the dark age, and is a period which is under intense study by astronomers (See 21 centimeter radiation).

Two other effects which occurred between reionization and our observations of the cosmic microwave background, and which appear to cause anisotropies, are the Sunyaev–Zel'dovich effect, where a cloud of high-energy electrons scatters the radiation, transferring some of its energy to the CMB photons, and the Sachs–Wolfe effect, which causes photons from the Cosmic Microwave Background to be gravitationally redshifted or blueshifted due to changing gravitational fields.

8.4 Polarization

The cosmic microwave background is polarized at the level of a few microkelvin. There are two types of polarization, called E-modes and B-modes. This is in analogy to electrostatics, in which the electric field (*E*-field) has a vanishing curl and the magnetic field (*B*-field) has a vanishing divergence. The E-modes arise naturally from Thomson scattering in a heterogeneous plasma. The B-modes are not sourced by standard scalar type perturbations. Instead they can be created by two mechanisms: the first one is by gravitational lensing of E-modes, which has been measured by the South Pole Telescope in 2013;[82] the second one is from gravitational waves arising from cosmic inflation. Detecting the B-modes is extremely difficult, particularly as the degree of foreground contamination is unknown, and the weak gravitational lensing signal mixes the relatively strong E-mode signal with the B-mode signal.[83]

8.4.1 E-modes

E-modes were first seen in 2002 by the Degree Angular Scale Interferometer (DASI).

8.4.2 B-modes

Cosmologists predict two types of B-modes, the first generated during cosmic inflation shortly after the big bang,[84][85][86] and the second generated by gravitational lensing at later times.[87]

Primordial gravitational waves

Primordial gravitational waves are gravitational waves that could be observed in the polarisation of the cosmic microwave background and having their origin in the early universe. Models of cosmic inflation predict that such gravitational waves should appear; thus, their detection supports the theory of inflation, and their strength can confirm and exclude different models of inflation. It is the result of three things: inflationary expansion of space itself, reheating after inflation, and turbulent fluid mixing of matter and radiation. [88]

On 17 March 2014 it was announced that the BICEP2 instrument had detected the first type of B-modes, consistent with inflation and gravitational waves in the early universe at the level of $r = 0.20+0.07$
-0.05, which is the amount of power present in gravitational waves compared to the amount of power present in other scalar density perturbations in the very early universe. Had this been confirmed it would have provided strong evidence of cosmic inflation and the Big Bang,[64][65][66][67][89][90][91] but on 19 June 2014, considerably lowered confidence in

confirming the findings was reported[68][68][70][70][71][71] and on 19 September 2014 new results of the Planck experiment reported that the results of BICEP2 can be fully attributed to cosmic dust.[92][93]

Paul Steinhardt is skeptical, suggesting that light scattering from cosmic dust and synchrotron radiation from electrons, both in the Milky Way Galaxy, could have caused the readings.[94]

Gravitational lensing

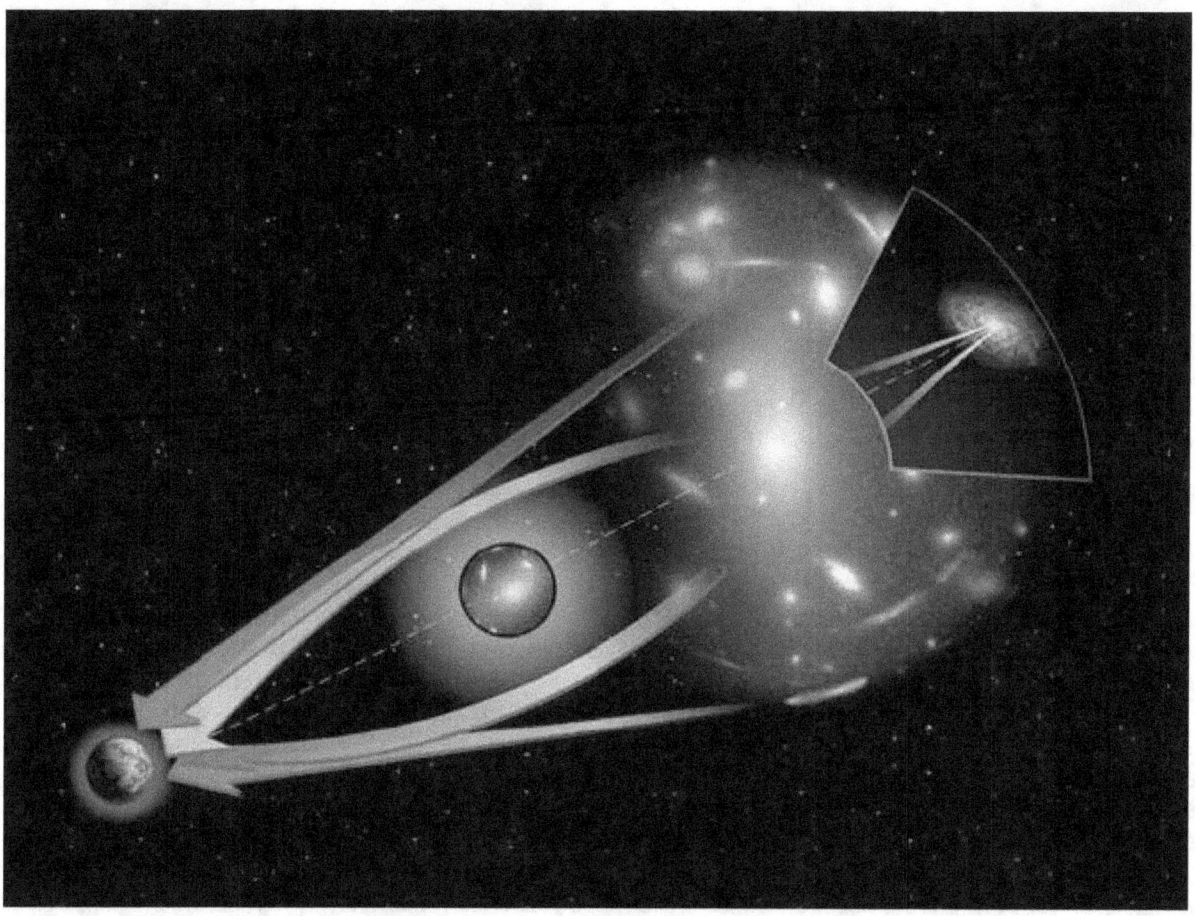

This artist's impression shows how light from the early universe is deflected by the gravitational lensing effect of massive cosmic structures forming B-modes as it travels across the universe. (Credit: ESA)

The second type of B-modes was discovered in 2013 using the South Pole Telescope with help from the Herschel Space Observatory.[95] This discovery may help test theories on the origin of the universe. Scientists are using data from the Planck mission by the European Space Agency, to gain a better understanding of these waves.[96][97][98]

In October 2014, a measurement of the B-mode polarization at 150 GHz was published by the POLARBEAR experiment.[99] Compared to BICEP2, POLARBEAR focuses on a smaller patch of the sky and is less susceptible to dust effects. The team reported that POLARBEAR's measured B-mode polarization was of cosmological origin (and not just due to dust) at a 97.2% confidence level.[100]

8.5   Microwave background observations

Main article: List of cosmic microwave background experiments

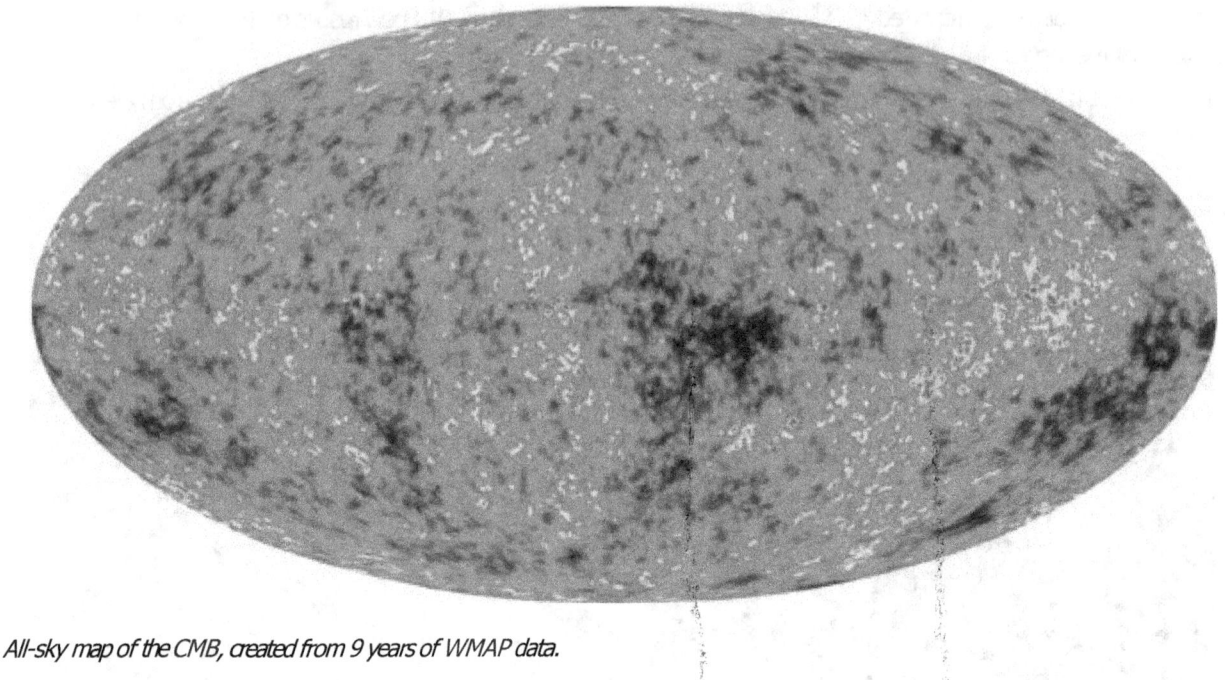

All-sky map of the CMB, created from 9 years of WMAP data.

Subsequent to the discovery of the CMB, hundreds of cosmic microwave background experiments have been conducted to measure and characterize the signatures of the radiation. The most famous experiment is probably the NASA Cosmic Background Explorer (COBE) satellite that orbited in 1989–1996 and which detected and quantified the large scale anisotropies at the limit of its detection capabilities. Inspired by the initial COBE results of an extremely isotropic and homogeneous background, a series of ground- and balloon-based experiments quantified CMB anisotropies on smaller angular scales over the next decade. The primary goal of these experiments was to measure the angular scale of the first acoustic peak, for which COBE did not have sufficient resolution. These measurements were able to rule out cosmic strings as the leading theory of cosmic structure formation, and suggested cosmic inflation was the right theory. During the 1990s, the first peak was measured with increasing sensitivity and by 2000 the BOOMERanG experiment reported that the highest power fluctuations occur at scales of approximately one degree. Together with other cosmological data, these results implied that the geometry of the universe is flat. A number of ground-based interferometers provided measurements of the fluctuations with higher accuracy over the next three years, including the Very Small Array, Degree Angular Scale Interferometer (DASI), and the Cosmic Background Imager (CBI). DASI made the first detection of the polarization of the CMB and the CBI provided the first E-mode polarization spectrum with compelling evidence that it is out of phase with the T-mode spectrum.

In June 2001, NASA launched a second CMB space mission, WMAP, to make much more precise measurements of the large scale anisotropies over the full sky. WMAP used symmetric, rapid-multi-modulated scanning, rapid switching radiometers to minimize non-sky signal noise.[76] The first results from this mission, disclosed in 2003, were detailed measurements of the angular power spectrum at a scale of less than one degree, tightly constraining various cosmological parameters. The results are broadly consistent with those expected from cosmic inflation as well as various other competing theories, and are available in detail at NASA's data bank for Cosmic Microwave Background (CMB) (see links below). Although WMAP provided very accurate measurements of the large scale angular fluctuations in the CMB (structures about as broad in the sky as the moon), it did not have the angular resolution to measure the smaller scale fluctuations which had been observed by former ground-based interferometers.

A third space mission, the ESA (European Space Agency) Planck Surveyor, was launched in May 2009 and is currently performing an even more detailed investigation. Planck employs both HEMT radiometers and bolometer technology and will measure the CMB at a smaller scale than WMAP. Its detectors were trialled in the Antarctic Viper telescope as ACBAR (Arcminute Cosmology Bolometer Array Receiver) experiment—which has produced the most precise measurements at small angular scales to date—and in the Archeops balloon telescope.

On 21 March 2013, the European-led research team behind the Planck cosmology probe released the mission's all-sky

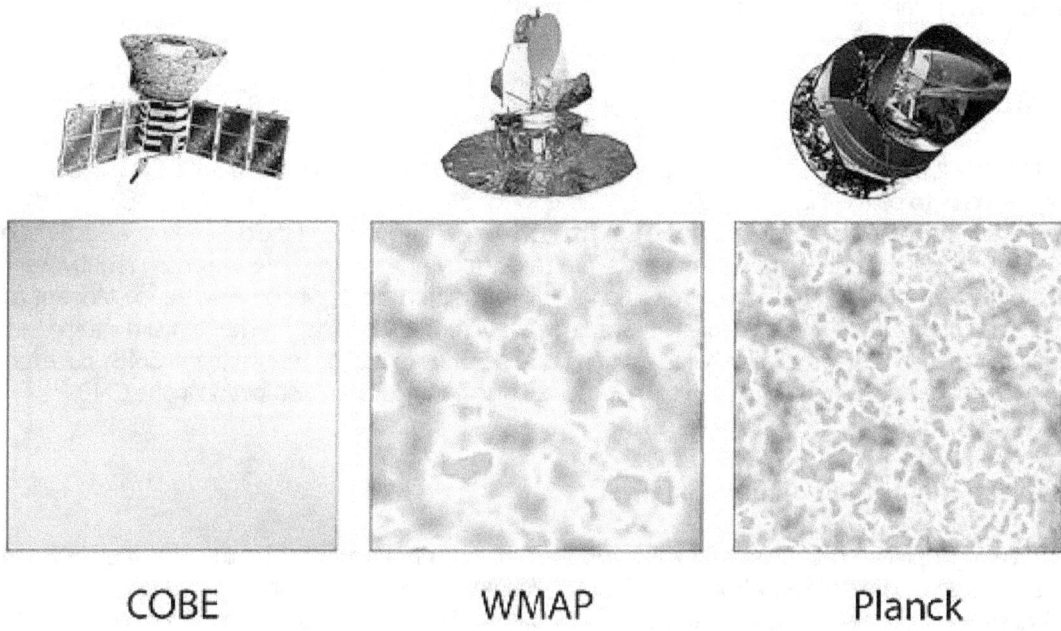

COBE   WMAP   Planck

Comparison of CMB results from COBE, WMAP and Planck – March 21, 2013.

map (565x318 jpeg, 3600x1800 jpeg) of the cosmic microwave background.[101][102] The map suggests the universe is slightly older than researchers thought. According to the map, subtle fluctuations in temperature were imprinted on the deep sky when the cosmos was about 370,000 years old. The imprint reflects ripples that arose as early, in the existence of the universe, as the first nonillionth of a second. Apparently, these ripples gave rise to the present vast cosmic web of galaxy clusters and dark matter. According to the team, the universe is 13.798 ± 0.037 billion years old,[103] and contains 4.9% ordinary matter, 26.8% dark matter and 68.3% dark energy. Also, the Hubble constant was measured to be 67.80 ± 0.77 (km/s)/Mpc.[101][104][105][106]

Additional ground-based instruments such as the South Pole Telescope in Antarctica and the proposed Clover Project, Atacama Cosmology Telescope and the QUIET telescope in Chile will provide additional data not available from satellite observations, possibly including the B-mode polarization.

8.6   Data reduction and analysis

Raw CMBR data from the space vehicle (i.e. WMAP) contain foreground effects that completely obscure the fine-scale structure of the cosmic microwave background. The fine-scale structure is superimposed on the raw CMBR data but is too small to be seen at the scale of the raw data. The most prominent of the foreground effects is the dipole anisotropy caused by the Sun's motion relative to the CMBR background. The dipole anisotropy and others due to Earth's annual motion relative to the Sun and numerous microwave sources in the galactic plane and elsewhere must be subtracted out to reveal the extremely tiny variations characterizing the fine-scale structure of the CMBR background.

The detailed analysis of CMBR data to produce maps, an angular power spectrum, and ultimately cosmological parameters is a complicated, computationally difficult problem. Although computing a power spectrum from a map is in principle a simple Fourier transform, decomposing the map of the sky into spherical harmonics, in practice it is hard to take the effects of noise and foreground sources into account. In particular, these foregrounds are dominated by galactic emissions such as Bremsstrahlung, synchrotron, and dust that emit in the microwave band; in practice, the galaxy has to be removed, resulting in a CMB map that is not a full-sky map. In addition, point sources like galaxies and clusters represent another source of foreground which must be removed so as not to distort the short scale structure of the CMB power spectrum.

Constraints on many cosmological parameters can be obtained from their effects on the power spectrum, and results are often calculated using Markov Chain Monte Carlo sampling techniques.

8.6.1   CMBR dipole anisotropy

From the CMB data it is seen that our local group of galaxies (the galactic cluster that includes the Solar System's Milky Way Galaxy) appears to be moving at 627±22 km/s relative to the reference frame of the CMB (also called the CMB rest frame, or the frame of reference in which there is no motion through the CMB) in the direction of galactic longitude l = 276°±3° , b = 30°±3° .[107][108] This motion results in an anisotropy of the data (CMB appearing slightly warmer in the direction of movement than in the opposite direction).[109] From a theoretical point of view, the existence of a CMB rest frame breaks Lorentz invariance even in empty space far away from any galaxy.[110] The standard interpretation of this temperature variation is a simple velocity red shift and blue shift due to motion relative to the CMB, but alternative cosmological models can explain some fraction of the observed dipole temperature distribution in the CMB. [111]

8.6.2   Low multipoles and other anomalies

With the increasingly precise data provided by WMAP, there have been a number of claims that the CMB exhibits anomalies, such as very large scale anisotropies, anomalous alignments, and non-Gaussian distributions.[112][113][114][115] The most longstanding of these is the low-l multipole controversy. Even in the COBE map, it was observed that the quadrupole (l = 2, spherical harmonic) has a low amplitude compared to the predictions of the Big Bang. In particular, the quadrupole and octupole (l = 3) modes appear to have an unexplained alignment with each other and with both the ecliptic plane and equinoxes,[116][117][118] an alignment sometimes referred to as the *axis of evil*.[113] A number of groups have suggested that this could be the signature of new physics at the greatest observable scales; other groups suspect systematic errors in the data.[119][120][121] Ultimately, due to the foregrounds and the cosmic variance problem, the greatest modes will never be as well measured as the small angular scale modes. The analyses were performed on two maps that have had the foregrounds removed as far as possible: the "internal linear combination" map of the WMAP collaboration and a similar map prepared by Max Tegmark and others.[45][76][122] Later analyses have pointed out that these are the modes most susceptible to foreground contamination from synchrotron, dust, and Bremsstrahlung emission, and from experimental uncertainty in the monopole and dipole. A full Bayesian analysis of the WMAP power spectrum demonstrates that the quadrupole prediction of Lambda-CDM cosmology is consistent with the data at the 10% level and that the observed octupole is not remarkable.[123] Carefully accounting for the procedure used to remove the foregrounds from the full sky map further reduces the significance of the alignment by ~5%.[124][125][126][127]

Recent observations with the Planck telescope, which is very much more sensitive than WMAP and has a larger angular resolution, confirm the observation of the axis of evil. Since two different instruments recorded the same anomaly, instrumental error (but not foreground contamination) appears to be ruled out.[128] Coincidence is a possible explanation, chief scientist from WMAP, Charles L. Bennett suggested coincidence and human psychology were involved, *"I do think there is a bit of a psychological effect; people want to find unusual things."*[129]

8.7   Future evolution

Assuming the universe keeps expanding and it does not suffer a Big Crunch, a Big Rip, or another similar fate, the cosmic microwave background will continue redshifting until it will no longer be detectable,[130] and will be overtaken first by the one produced by starlight, and later by the background radiation fields of processes that are assumed will take place in the far future of the universe.[131], §VD.

8.8   In popular culture

- In the *Stargate Universe* TV series, an Ancient spaceship, *Destiny*, was built to study patterns in the CMBR which indicate that the universe as we know it might have been created by some form of sentient intelligence.[132]

- In *Wheelers*, a novel by Ian Stewart & Jack Cohen, CMBR is explained as the encrypted transmissions of an ancient civilization. This allows the Jovian "blimps" to have a society older than the currently-observed age of the universe.

- In *The Three-Body Problem*, a novel by Liu Cixin, CMBR becomes observable to the naked eye due to interference from an alien civilization.

8.9   See also

- Physical cosmology

- Observational cosmology

- Gravitational wave background

- Cosmic gravitational wave background

- Observation history of galaxies

- Lambda-CDM model

- Heat death of the universe

- Computational packages for Cosmologists

8.10   References

[1] Penzias, A. A.; Wilson, R. W. (1965). "A Measurement of Excess Antenna Temperature at 4080 Mc/s". *The Astrophysical Journal* **142** (1): 419–421. Bibcode:1965ApJ...142..419P. doi:10.1086/148307.

[2] Smoot Group (28 March 1996). "The Cosmic Microwave Background Radiation". Lawrence Berkeley Lab. Retrieved 2008-12-11.

[3] "Planck reveals an almost perfect Universe". Max Planck Gesellschaft. 21 March 2013. Retrieved 2013-06-03.

[4] Kaku, M. (2014). "First Second of the Big Bang". *How the Universe Works*. Discovery Science.

[5] Fixsen, D. J. (2009). "The Temperature of the Cosmic Microwave Background". *The Astrophysical Journal* **707** (2): 916–920. arXiv:0911.1955. Bibcode:2009ApJ...707..916F. doi:10.1088/0004-637X/707/2/916.

[6] Dodelson, S. (2003). "Coherent Phase Argument for Inflation". *AIP Conference Proceedings* **689**: 184–196. arXiv:hep-ph/0309057. Bibcode:2003AIPC..689..184D. doi:10.1063/1.1627736.

[7] Baumann, D. (2011). "The Physics of Inflation" (PDF). University of Cambridge. Retrieved 2015-05-09.

[8] White, M. (1999). "Anisotropies in the CMB". *Proceedings of the Los Angeles Meeting, DPF 99*. UCLA. arXiv:astro-ph/9903232. Bibcode:1999dpf..conf.... W.

[9] Wright, E.L. (2004). "Theoretical Overview of Cosmic Microwave Background Anisotropy". In W. L. Freedman. *Measuring and Modeling the Universe*. Carnegie Observatories Astrophysics Series. Cambridge University Press. p. 291. arXiv:astro-ph/0305591. ISBN 0-521-75576-X.

[10] The Planck Collaboration, *Planck 2013 results. XXVII. Doppler boosting of the CMB: Eppur si muove*, arXiv:1303.5087, Bibcode:2014A&A...571A..27P, doi:10.1051/0004-6361/201321556

[11] Guth, A. H. (1998). *The Inflationary Universe: The Quest for a New Theory of Cosmic Origins*. Basic Books. p. 186. ISBN 978-0201328400. OCLC 35701222.

[12] Cirigliano, D.; de Vega, H.J.; Sanchez, N. G. (2005). "Clarifying inflation models: The precise inflationary potential from effective field theory and the WMAP data". *Physical Review D* **71** (10): 77–115. arXiv:astro-ph/0412634. Bibcode:2005PhRvD..71j3518C. doi:10.1103/PhysRevD.71.103518.

[13] Abbott, B. (2007). "Microwave (WMAP) All-Sky Survey". Hayden Planetarium. Retrieved 2008-01-13.

[14] Gawiser, E.; Silk, J. (2000). "The cosmic microwave background radiation". *Physics Reports*. 333–334: 245–267. arXiv:astro-ph/0002044. Bibcode:2000PhR...333..245G. doi:10.1016/S0370-1573(00)00025-9.

[15] Smoot, G. F. (2006). "Cosmic Microwave Background Radiation Anisotropies: Their Discovery and Utilization". *Nobel Lecture*. Nobel Foundation. Retrieved 2008-12-22.

[16] Hobson, M.P.; Efstathiou, G.; Lasenby, A.N. (2006). *General Relativity: An Introduction for Physicists*. Cambridge University Press. p. 388. ISBN 0-521-82951-8.

[17] Unsöld, A.; Bodo, B. (2002). *The New Cosmos, An Introduction to Astronomy and Astrophysics* (5th ed.). Springer–Verlag. p. 485. ISBN 3-540-67877-8.

[18] Confrontation of Cosmological Theories with Observational Data, M. S. Longair, page 144

[19] Cosmology II: The thermal history of the Universe, Ruth Durrer

[20] Gamow, G. (1948). "The Origin of Elements and the Separation of Galaxies". *Physical Review* 74 (4): 505–506. Bibcode:1948PhRv...74..505G. doi:10.1103/PhysRev.74.505.2.

[21] Gamow, G. (1948). "The evolution of the universe". *Nature* 162 (4122): 680–682. Bibcode:1948Natur.162..680G. doi:10.1038/162680a0. PMID 18893719.

[22] Alpher, R. A.; Herman, R. C. (1948). "On the Relative Abundance of the Elements". *Physical Review* 74 (12): 1737–1742. Bibcode:1948PhRv...74.1737A. doi:10.1103/PhysRev.74.1737.

[23] Assis, A. K. T.; Neves, M. C. D. (1995). "History of the 2.7 K Temperature Prior to Penzias and Wilson" (PDF) (3). pp. 79–87. but see also Wright, E. L. (2006). "Eddington's Temperature of Space". UCLA. Retrieved 2008-12-11.

[24] Penzias, A. A. (2006). "The origin of elements" (PDF). *Nobel lecture*. Nobel Foundation. Retrieved 2006-10-04.

[25] Dicke, R. H. (1946). "The Measurement of Thermal Radiation at Microwave Frequencies". *Review of Scientific Instruments* 17 (7): 268–275. Bibcode:1946RScI...17..268D. doi:10.1063/1.1770483. PMID 20991753. This basic design for a radiometer has been used in most subsequent cosmic microwave background experiments.

[26] The Cosmic Microwave Background Radiation (Nobel Lecture) by Robert Wilson 8 Dec 1978, p. 474

[27] Dicke, R. H. et al. (1965). "Cosmic Black-Body Radiation". *Astrophysical Journal* 142: 414–419. Bibcode:1965ApJ...142..414D. doi:10.1086/148306.

[28] The history is given in Peebles, P. J. E (1993). *Principles of Physical Cosmology*. Princeton University Press. pp. 139–148. ISBN 0-691-01933-9.

[29] "The Nobel Prize in Physics 1978". Nobel Foundation. 1978. Retrieved 2009-01-08.

[30] Narlikar, J. V.; Wickramasinghe, N. C. (1967). "Microwave Background in a Steady State Universe". *Nature* 216 (5110): 43–44. Bibcode:1967Natur.216...43N. doi:10.1038/216043a0.

[31] McKellar, A.; Kan-Mitchell, June; Conti, Peter S. (1941). "Molecular Lines from the Lowest States of Diatomic Molecules Composed of Atoms Probably Present in Interstellar Space". *Publications of the Dominion Astrophysical Observatory (Victoria, BC)* 7 (6): 251–272.

[32] Peebles, P. J. E. et al. (1991). "The case for the relativistic hot big bang cosmology". *Nature* 352 (6338): 769–776. Bibcode:1991Natur.352..769P. doi:10.1038/352769a0.

[33] Harrison, E. R. (1970). "Fluctuations at the threshold of classical cosmology". *Physical Review D* 1 (10): 2726–2730. Bibcode:1970PhRvD...1.2726H. doi:10.1103/PhysRevD.1.2726.

[34] Peebles, P. J. E.; Yu, J. T. (1970). "Primeval Adiabatic Perturbation in an Expanding Universe". *Astrophysical Journal* 162: 815–836. Bibcode:1970ApJ...162..815P. doi:10.1086/150713.

[35] Zeldovich, Y. B. (1972). "A hypothesis, unifying the structure and the entropy of the Universe". *Monthly Notices of the Royal Astronomical Society* 160 (7–8): 1P–4P. doi:10.1016/S0026-0576(07)80178-4.

[36] Doroshkevich, A. G.; Zel'Dovich, Y. B.; Syunyaev, R. A. (1978) [12–16 September 1977]. "Fluctuations of the microwave background radiation in the adiabatic and entropic theories of galaxy formation". In Longair, M. S.; Einasto, J. *The large scale structure of the universe; Proceedings of the Symposium.* Tallinn, Estonian SSR: Dordrecht, D. Reidel Publishing Co. pp. 393–404. Bibcode:1978IAUS...79..393S. While this is the first paper to discuss the detailed observational imprint of density inhomogeneities as anisotropies in the cosmic microwave background, some of the groundwork was laid in Peebles and Yu, above.

[37] Smooth, G. F. et al. (1992). "Structure in the COBE differential microwave radiometer first-year maps". *Astrophysical Journal Letters* 396 (1): L1–L5. Bibcode:1992ApJ...396L...1S. doi:10.1086/186504.

[38] Bennett, C.L. et al. (1996). "Four-Year COBE DMR Cosmic Microwave Background Observations: Maps and Basic Results". *Astrophysical Journal Letters* 464: L1–L4. arXiv:astro-ph/9601067. Bibcode:1996ApJ...464L...1B. doi:10.1086/310075.

[39] Grupen, C. et al. (2005). *Astroparticle Physics.* Springer. pp. 240–241. ISBN 3-540-25312-2.

[40] Miller, A. D. et al. (1999). "A Measurement of the Angular Power Spectrum of the Microwave Background Made from the High Chilean Andes". *Astrophysical Journal* 521 (2): L79–L82. arXiv:astro-ph/9905100. Bibcode:1999ApJ...521L..79T. doi:10.1086/312197.

[41] Melchiorri, A. et al. (2000). "A Measurement of Ω from the North American Test Flight of Boomerang". *Astrophysical Journal* 536 (2): L63–L66. arXiv:astro-ph/9911445. Bibcode:2000ApJ...536L..63M. doi:10.1086/312744.

[42] Hanany, S. et al. (2000). "MAXIMA-1: A Measurement of the Cosmic Microwave Background Anisotropy on Angular Scales of 10′–5°". *Astrophysical Journal* 545 (1): L5–L9. arXiv:astro-ph/0005123. Bibcode:2000ApJ...545L...5H. doi:10.1086/317322.

[43] de Bernardis, P. et al. (2000). "A flat Universe from high-resolution maps of the cosmic microwave background radiation". *Nature* 404 (6781): 955–959. arXiv:astro-ph/0004404. Bibcode:2000Natur.404..955D. doi:10.1038/35010035. PMID 10801117.

[44] Pogosian, L. et al. (2003). "Observational constraints on cosmic string production during brane inflation". *Physical Review D* 68 (2): 023506. arXiv:hep-th/0304188. Bibcode:2003PhRvD..68b3506P. doi:10.1103/PhysRevD.68.023506.

[45] Hinshaw, G.; (WMAP collaboration); Bennett, C. L.; Bean, R.; Doré, O.; Greason, M. R.; Halpern, M.; Hill, R. S.; Jarosik, N.; Kogut, A.; Komatsu, E.; Limon, M.; Odegard, N.; Meyer, S. S.; Page, L.; Peiris, H. V.; Spergel, D. N.; Tucker, G. S.; Verde, L.; Weiland, J. L.; Wollack, E.; Wright, E. L. et al. (2007). "Three-year Wilkinson Microwave Anisotropy Probe (WMAP) observations: temperature analysis". *Astrophysical Journal (Supplement Series)* 170 (2): 288–334. arXiv:astro-ph/0603451. Bibcode:2007ApJS..170..288H. doi:10.1086/513698.

[46] Guillaume, C.-É., 1896, *La Nature* 24, series 2, p. 234, cited in "History of the 2.7 K Temperature Prior to Penzias and Wilson" (PDF)

[47] Eddington, A., The Internal Constitution of the Stars, cited in "History of the 2.7 K Temperature Prior to Penzias and Wilson" (PDF)

[48] Weinberg, S. (1972). *Oxford Astronomy Encyclopedia.* John Wiley & Sons. p. 514. ISBN 0-471-92567-5.

[49] Kragh, H. (1999). *Cosmology and Controversy: The Historical Development of Two Theories of the Universe.* ISBN 0-691-00546-X. "In 1946, Robert Dicke and coworkers at MIT tested equipment that could test a cosmic microwave background of intensity corresponding to about 20K in the microwave region. However, they did not refer to such a background, but only to 'radiation from cosmic matter'. Also, this work was unrelated to cosmology and is only mentioned because it suggests that by 1950, detection of the background radiation might have been technically possible, and also because of Dicke's later role in the discovery". See also Dicke, R. H. et al. (1946). "Atmospheric Absorption Measurements with a Microwave Radiometer". *Physical Review* 70 (5–6): 340–348. Bibcode:1946PhRv...70..340D. doi:10.1103/PhysRev.70.340.

[50] George Gamow, *The Creation Of The Universe* p.50 (Dover reprint of revised 1961 edition) ISBN 0-486-43868-6

[51] Gamow, G. (2004) [1961]. *Cosmology and Controversy: The Historical Development of Two Theories of the Universe.* Courier Dover Publications. p. 40. ISBN 978-0-486-43868-9.

[52] Erwin Finlay-Freundlich, "Ueber die Rotverschiebung der Spektrallinien" (1953) *Contributions from the Observatory, University of St. Andrews;* no. 4, p. 96–102. Finlay-Freundlich also gave two extreme values of 1.9K and 6.0K in Finlay-Freundlich, E.: 1954, "Red shifts in the spectra of celestial bodies", Phil. Mag., Vol. 45, pp. 303–319.

[53] Helge Kragh, Cosmology and Controversy: The Historical Development of Two Theories of the Universe (1999) ISBN 0-691-00546-X. "Alpher and Herman first calculated the present temperature of the decoupled primordial radiation in 1948, when they reported a value of 5 K. Although it was not mentioned either then or in later publications that the radiation is in the microwave region, this follows immediately from the temperature ...   Alpher and Herman made it clear that what they had called "the temperature in the univerese" the previous year referred to a blackbody distributed background radiation quite different from sunliight".

[54] Shmaonov, T. A. (1957). "Commentary". *Pribory i Tekhnika Eksperimenta* (in Russian) **1**: 83. doi:10.1016/S0890-5096(06)60772-3.

[55] It is noted that the "measurements showed that radiation intensity was independent of either time or direction of observation ... it is now clear that Shmaonov did observe the cosmic microwave background at a wavelength of 3.2cm"

[56] Naselsky, P. D.; Novikov, D.I.; Novikov, I. D. (2006). *The Physics of the Cosmic Microwave Background.* ISBN 0-521-85550-0.

[57] Helge Kragh, Cosmology and Controversy: The Historical Development of Two Theories of the Universe

[58] Doroshkevich, A. G.; Novikov, I.D. (1964). "Mean Density of Radiation in the Metagalaxy and Certain Problems in Relativistic Cosmology". *Soviet Physics Doklady* **9** (23): 4292–4298. Bibcode:1999EnST...33.4292W. doi:10.1021/es990537g.

[59] *Nobel Prize In Physics: Russia's Missed Opportunities,* RIA Novosti, Nov 21, 2006

[60] Sanders, R.; Kahn, J. (13 October 2006). "UC Berkeley, LBNL cosmologist George F. Smoot awarded 2006 Nobel Prize in Physics". UC Berkeley News. Retrieved 2008-12-11.

[61] Kovac, J.M. et al. (2002). "Detection of polarization in the cosmic microwave background using DASI". *Nature* **420** (6917): 772–787. arXiv:astro-ph/0209478. Bibcode:2002Natur.420..772K. doi:10.1038/nature01269. PMID 12490941.

[62] Readhead, A. C. S. et al. (2004). "Polarization Observations with the Cosmic Background Imager". *Science* **306** (5697): 836–844. arXiv:astro-ph/0409569. Bibcode:2004Sci...306..836R. doi:10.1126/science.1105598. PMID 15472038.

[63] A. Readhead et al., "Polarization observations with the Cosmic Background Imager", Science 306, 836-844 (2004).

[64] Staff (March 17, 2014). "BICEP2 2014 Results Release". *National Science Foundation.* Retrieved March 18, 2014.

[65] Clavin, Whitney (March 17, 2014). "NASA Technology Views Birth of the Universe". *NASA.* Retrieved March 17, 2014.

[66] Overbye, Dennis (March 17, 2014). "Space Ripples Reveal Big Bang's Smoking Gun". *The New York Times.* Retrieved March 17, 2014.

[67] Overbye, Dennis (March 24, 2014). "Ripples From the Big Bang". *New York Times.* Retrieved March 24, 2014.

[68] Ade, P.A.R. (BICEP2 Collaboration) et al. (June 19, 2014). "Detection of B-Mode Polarization at Degree Angular Scales by BICEP2" (PDF). *Physical Review Letters* **112**: 241101. arXiv:1403.3985. Bibcode:2014PhRvL.112x1101A. doi:10.1103/PhysRevLett.112.241101. PMID 24996078. Retrieved June 20, 2014.

[69] http://www.math.columbia.edu/~{}woit/wordpress/?p=6865

[70] Overbye, Dennis (June 19, 2014). "Astronomers Hedge on Big Bang Detection Claim". *New York Times.* Retrieved June 20, 2014.

[71] Amos, Jonathan (June 19, 2014). "Cosmic inflation: Confidence lowered for Big Bang signal". *BBC News.* Retrieved June 20, 2014.

[72] Scott, D. (2005). "The Standard Cosmological Model". arXiv:astro-ph/0510731 [astro-ph].

[73] Durham, Frank; Purrington, Robert D. (1983). *Frame of the universe: a history of physical cosmology.* Columbia University Press. pp. 193–209. ISBN 0-231-05393-2.

[74] Assis, A. K. T.; Paulo, São; Neves, M. C. D. (July 1995). "History of the 2.7 K Temperature Prior to Penzias and Wilson" (PDF). *Apeiron* **2** (3): 79–87.

[75] Brandenberger, Robert H. (1995). "Formation of Structure in the Universe". p. 8159. arXiv:astro-ph/9508159. Bibcode:1995astro.ph..8159B.

[76] Bennett, C. L.; (WMAP collaboration); Hinshaw, G.; Jarosik, N.; Kogut, A.; Limon, M.; Meyer, S. S.; Page, L.; Spergel, D. N.; Tucker, G. S.; Wollack, E.; Wright, E. L.; Barnes, C.; Greason, M. R.; Hill, R. S.; Komatsu, E.; Nolta, M. R.; Odegard, N.; Peiris, H. V.; Verde, L.; Weiland, J. L. et al. (2003). "First-year Wilkinson Microwave Anisotropy Probe (WMAP) observations: preliminary maps and basic results". *Astrophysical Journal (Supplement Series)* 148: 1–27. arXiv:astro-ph/0302207. Bibcode:2003ApJS..148....1B. doi:10.1086/377253. This paper warns, "the statistics of this internal linear combination map are complex and inappropriate for most CMB analyses."

[77] Noterdaeme, P.; Petitjean, P.; Srianand, R.; Ledoux, C.; López, S. (February 2011). "The evolution of the cosmic microwave background temperature. Measurements of TCMB at high redshift from carbon monoxide excitation". *Astronomy and Astrophysics* 526: L7. arXiv:1012.3164. Bibcode:2011A&A...526L...7N. doi:10.1051/0004-6361/201016140.

[78] Wayne Hu. "Baryons and Inertia".

[79] Wayne Hu. "Radiation Driving Force".

[80] Hu, W.; White, M. (1996). "Acoustic Signatures in the Cosmic Microwave Background". *Astrophysical Journal* 471: 30–51. arXiv:astro-ph/9602019. Bibcode:1996ApJ...471...30H. doi:10.1086/177951.

[81] WMAP Collaboration; Verde, L.; Peiris, H. V.; Komatsu, E.; Nolta, M. R.; Bennett, C. L.; Halpern, M.; Hinshaw, G. et al. (2003). "First-Year Wilkinson Microwave Anisotropy Probe (WMAP) Observations: Determination of Cosmological Parameters". *Astrophysical Journal Supplement Series* 148 (1): 175–194. arXiv:astro-ph/0302209. Bibcode:2003ApJS..148..175S. doi:10.1086/377226.

[82] Hanson, D. et al. (2013). "Detection of B-mode polarization in the Cosmic Microwave Background with data from the South Pole Telescope". *Physical Review Letters* 111 (14). arXiv:1307.5830. Bibcode:2013PhRvL.111n1301H. doi:10.1103/PhysRevLett.111.141301.

[83] Lewis, A.; Challinor, A. (2006). "Weak gravitational lensing of the CMB". *Physics Reports* 429: 1–65. arXiv:astro-ph/0601594. Bibcode:2006PhR...429... 1L. doi:10.1016/j.physrep.2006.03.002.

[84] Seljak, U. (June 1997). "Measuring Polarization in the Cosmic Microwave Background". *Astrophysical Journal* 482: 6–16. arXiv:astro-ph/9608131. Bibcode:1997ApJ...482... 6S. doi:10.1086/304123.

[85] Seljak, U.; Zaldarriaga M. (March 17, 1997). "Signature of Gravity Waves in the Polarization of the Microwave Background". *Phys. Rev.Lett.* 78 (11): 2054–2057. arXiv:astro-ph/9609169. Bibcode:1997PhRvL..78.2054S. doi:10.1103/PhysRevLett.78.2054.

[86] Kamionkowski, M.; Kosowsky A. & Stebbins A. (March 17, 1997). "A Probe of Primordial Gravity Waves and Vorticity". *Phys. Rev.Lett.* 78 (11): 2058–2061. arXiv:astro-ph/9609132. Bibcode:1997PhRvL..78.2058K. doi:10.1103/PhysRevLett.78.2058.

[87] Zaldarriaga, M.; Seljak U. (July 15, 1998). "Gravitational lensing effect on cosmic microwave background polarization". *Physical Review D.* 2 58. arXiv:astro-ph/9803150. Bibcode:1998PhRvD..58b3003Z. doi:10.1103/PhysRevD.58.023003.

[88] "Scientists Report Evidence for Gravitational Waves in Early Universe". Retrieved 2007-06-20.

[89] "Gravitational waves: have US scientists heard echoes of the big bang?". The Guardian. 2014-03-14. Retrieved 2014-03-14.

[90] 'BICEP2 I: Detection Of B-mode Polarization at Degree Angular Scales' on arXiv

[91] "Space Ripples Reveal Big Bang's Smoking Gun". March 17, 2014.

[92] Planck Collaboration Team (19 September 2014). "Planck intermediate results. XXX. The angular power spectrum of polarized dust emission at intermediate and high Galactic latitudes". arXiv:1409.5738.

[93] Overbye, Dennis (22 September 2014). "Study Confirms Criticism of Big Bang Finding". *New York Times.* Retrieved 22 September 2014.

[94] "Big Bang research blunder leaves multiverse theory in ruins, theoretical physicist claims". *The Independent.*

[95] "Polarization detected in Big Bang's echo". *Nature News & Comment.*

[96] ESA Planck (Oct 22, 2013). "Planck Space Mission". Retrieved Oct 23, 2013.

[97] NASA/Jet Propulsion Laboratory (October 22, 2013). "Long-sought pattern of ancient light detected". *ScienceDaily.* Retrieved October 23, 2013.

[98] Hanson, D. et al. (Sep 30, 2013). "Detection of B-Mode Polarization in the Cosmic Microwave Background with Data from the South Pole Telescope". *Physical Review Letters.* 14 111. arXiv:1307.5830. Bibcode:2013PhRvL.111n1301H. doi:10.1103/PhysRevLett.111.141301.

[99] The Polarbear Collaboration (October 2014). "A Measurement of the Cosmic Microwave Background B-Mode Polarization Power Spectrum at Sub-Degree Scales with POLARBEAR" (PDF). *The Astrophysical Journal* 794: 171. arXiv:1403.2369. Bibcode:2014ApJ...794..171T. doi:10.1088/0004-637X/794/2/171. Retrieved November 16, 2014.

[100] "POLARBEAR project offers clues about origin of universe's cosmic growth spurt". *Christian Science Monitor.* October 21, 2014.

[101] Clavin, Whitney; Harrington, J.D. (21 March 2013). "Planck Mission Brings Universe Into Sharp Focus". *NASA.* Retrieved 21 March 2013.

[102] Staff (21 March 2013). "Mapping the Early Universe". *New York Times.* Retrieved 23 March 2013.

[103] Planck collaboration; Ade, P. A. R.; Aghanim, N.; Armitage-Caplan, C.; Arnaud, M.; Ashdown, M.; Atrio-Barandela, F.; Aumont, J.; Baccigalupi, C.; Banday, A. J.; Barreiro, R. B.; Bartelmann, M.; Bartlett, J. G.; Battaner, E.; Benabed, K.; Benoît, A.; Benoit-Lévy, A.; Bernard, J. -P.; Bersanelli, M.; Bielewicz, P.; Bobin, J.; Bock, J. J.; Bonaldi, A.; Bond, J. R.; Borrill, J.; Bouchet, F. R.; Boulanger, F.; Bowyer, J. W.; Bridges, M.; Bucher, M. (2013). "Planck 2013 results. I. Overview of products and scientific results". arXiv:1303.5062 [astro-ph.CO].

[104] Staff (21 March 2013). "Planck Reveals An Almost Perfect Universe". *ESA.* Retrieved 21 March 2013.

[105] Overbye, Dennis (21 March 2013). "An Infant Universe, Born Before We Knew". *New York Times.* Retrieved 21 March 2013.

[106] Boyle, Alan (21 March 2013). "Planck probe's cosmic 'baby picture' revises universe's vital statistics". *NBC News.* Retrieved 21 March 2013.

[107] Kogut, A.; Lineweaver, C.; Smoot, G. F.; Bennett, C. L.; Banday, A.; Boggess, N. W.; Cheng, E. S.; De Amici, G.; Fixsen, D. J.; Hinshaw, G.; Jackson, P. D.; Janssen, M.; Keegstra, P.; Loewenstein, K.; Lubin, P.; Mather, J. C.; Tenorio, L.; Weiss, R.; Wilkinson, D. T.; Wright, E. L. (1993). "Dipole Anisotropy in the COBE Differential Microwave Radiometers First-Year Sky Maps". *Astrophysical Journal* 419: 1–6. arXiv:astro-ph/9312056. Bibcode:1993ApJ...419... 1K. doi:10.1086/173453.

[108] Aghanim, N.; Armitage-Caplan, C. et al. (2013). "Planck 2013 results. XXVII. Doppler boosting of the CMB: Eppur si muove". *Astronomy & Astrophysics* 571 (27): A27. arXiv:1303.5087. Bibcode:2014A&A...571A..27P. doi:10.1051/0004-6361/201321556.

[109] http://antwrp.gsfc.nasa.gov/apod/ap090906.html

[110] http://iopscience.iop.org/1126-6708/2005/07/029/

[111] Inoue, K. T.; Silk, J. (2007). "Local Voids as the Origin of Large-Angle Cosmic Microwave Background Anomalies: The Effect of a Cosmological Constant". *Astrophysical Journal* 664 (2): 650–659. arXiv:astro-ph/0612347. Bibcode:2007ApJ...664..650I. doi:10.1086/517603.

[112] Rossmanith, G.; Räth, C.; Banday, A. J.; Morfill, G. (2009). "Non-Gaussian Signatures in the five-year WMAP data as identified with isotropic scaling indices". *Monthly Notices of the Royal Astronomical Society* 399 (4): 1921–1933. arXiv:0905.2854. Bibcode:2009MNRAS.399.1921R. doi:10.1111/j.1365-2966.2009.15421.x.

[113] Schild, R. E.; Gibson, C. H. (2008). "Goodness in the Axis of Evil". arXiv:0802.3229 [astro-ph].

[114] Bernui, A.; Mota, B.; Rebouças, M. J.; Tavakol, R. (2005). "Mapping the large-scale anisotropy in the WMAP data". *Astronomy and Astrophysics* 464 (2): 479–485. arXiv:astro-ph/0511666. Bibcode:2007A&A...464..479B. doi:10.1051/0004-6361:20065585.

[115] Jaffe, T.R.; Banday, A. J.; Eriksen, H. K.; Górski, K. M.; Hansen, F. K. (2005). "Evidence of vorticity and shear at large angular scales in the WMAP data: a violation of cosmological isotropy?". *The Astrophysical Journal* 629: L1–L4. arXiv:astro-ph/0503213. Bibcode:2005ApJ...629L...1J. doi:10.1086/444454.

[116] de Oliveira-Costa, A.; Tegmark, Max; Zaldarriaga, Matias; Hamilton, Andrew (2004). "The significance of the largest scale CMB fluctuations in WMAP". *Physical Review D* 69 (6): 063516. arXiv:astro-ph/0307282. Bibcode:2004PhRvD..69f3516D. doi:10.1103/PhysRevD.69.063516.

[117] Schwarz, D. J.; Starkman, Glenn D. et al. (2004). "Is the low-/ microwave background cosmic?". *Physical Review Letters* **93** (22): 221301. arXiv:astro-ph/0403353. Bibcode:2004PhRvL..93v1301S. doi:10.1103/PhysRevLett.93.221301.

[118] Bielewicz, P.; Gorski, K. M.; Banday, A. J. (2004). "Low-order multipole maps of CMB anisotropy derived from WMAP". *Monthly Notices of the Royal Astronomical Society* **355** (4): 1283–1302. arXiv:astro-ph/0405007. Bibcode:2004MNRAS.355.1283B. doi:10.1111/j.1365-2966.2004.08405.x.

[119] Liu, Hao; Li, Ti-Pei (2009). "Improved CMB Map from WMAP Data". arXiv:0907.2731v3 [astro-ph].

[120] Sawangwit, Utane; Shanks, Tom (2010). "Lambda-CDM and the WMAP Power Spectrum Beam Profile Sensitivity". arXiv:1006.1270v1 [astro-ph].

[121] Liu, Hao et al. (2010). "Diagnosing Timing Error in WMAP Data". arXiv:1009.2701v1 [astro-ph].

[122] Tegmark, M.; de Oliveira-Costa, A.; Hamilton, A. (2003). "A high resolution foreground cleaned CMB map from WMAP". *Physical Review D* **68** (12): 123523. arXiv:astro-ph/0302496. Bibcode:2003PhRvD..68l3523T. doi:10.1103/PhysRevD.68.123523. This paper states, "Not surprisingly, the two most contaminated multipoles are [the quadrupole and octupole], which most closely trace the galactic plane morphology."

[123] O'Dwyer, I.; Eriksen, H. K.; Wandelt, B. D.; Jewell, J. B.; Larson, D. L.; Górski, K. M.; Banday, A. J.; Levin, S.; Lilje, P. B. (2004). "Bayesian Power Spectrum Analysis of the First-Year Wilkinson Microwave Anisotropy Probe Data". *Astrophysical Journal Letters* **617** (2): L99–L102. arXiv:astro-ph/0407027. Bibcode:2004ApJ...617L..99O. doi:10.1086/427386.

[124] Slosar, A.; Seljak, U. (2004). "Assessing the effects of foregrounds and sky removal in WMAP". *Physical Review D* **70** (8): 083002. arXiv:astro-ph/0404567. Bibcode:2004PhRvD..70h3002S. doi:10.1103/PhysRevD.70.083002.

[125] Bielewicz, P.; Eriksen, H. K.; Banday, A. J.; Górski, K. M.; Lilje, P. B. (2005). "Multipole vector anomalies in the first-year WMAP data: a cut-sky analysis". *Astrophysical Journal* **635** (2): 750–60. arXiv:astro-ph/0507186. Bibcode:2005ApJ...635..750B. doi:10.1086/497263.

[126] Copi, C.J.; Huterer, Dragan; Schwarz, D. J.; Starkman, G. D. (2006). "On the large-angle anomalies of the microwave sky". *Monthly Notices of the Royal Astronomical Society* **367**: 79–102. arXiv:astro-ph/0508047. Bibcode:2006MNRAS.367...79C. doi:10.1111/j.1365-2966.2005.09980.x.

[127] de Oliveira-Costa, A.; Tegmark, M. (2006). "CMB multipole measurements in the presence of foregrounds". *Physical Review D* **74** (2): 023005. arXiv:astro-ph/0603369. Bibcode:2006PhRvD..74b3005D. doi:10.1103/PhysRevD.74.023005.

[128] Planck shows almost perfect cosmos – plus axis of evil

[129] Found: Hawking's initials written into the universe

[130] Krauss, Lawrence M.; Scherrer, Robert J. (2007). "The return of a static universe and the end of cosmology". *General Relativity and Gravitation* **39** (10): 1545–1550. arXiv:0704.0221. Bibcode:2007GReGr..39.1545K. doi:10.1007/s10714-007-0472-9.

[131] Adams, Fred C.; Laughlin, Gregory (1997). "A dying universe: The long-term fate and evolution of astrophysical objects". *Reviews of Modern Physics* **69** (2): 337–372. arXiv:astro-ph/9701131. Bibcode:1997RvMP...69..337A. doi:10.1103/RevModPhys.69.337.

[132] Cosmic Rebirth Encoded in Background Radiation?

8.11 External links

- CMBR Theme on arxiv.org

- Audio: Fraser Cain and Dr. Pamela Gay – Astronomy Cast. The Big Bang and Cosmic Microwave Background – October 2006

- Visualization of the CMB data from the Planck mission

- Copeland, Ed. "CMBR: Cosmic Microwave Background Radiation". *Sixty Symbols*. Brady Haran for the University of Nottingham.

Chapter 9

Observable universe

The **observable universe** consists of the galaxies and other matter that can, in principle, be observed from Earth at the present time because light and other signals from these objects has had time to reach the Earth since the beginning of the cosmological expansion. Assuming the universe is isotropic, the distance to the edge of the observable universe is roughly the same in every direction. That is, the observable universe is a spherical volume (a ball) centered on the observer. Every location in the Universe has its own observable universe, which may or may not overlap with the one centered on Earth.

The word *observable* used in this sense does not depend on whether modern technology actually permits detection of radiation from an object in this region (or indeed on whether there is any radiation to detect). It simply indicates that it is possible *in principle* for light or other signals from the object to reach an observer on Earth. In practice, we can see light only from as far back as the time of photon decoupling in the recombination epoch. That is when particles were first able to emit photons that were not quickly re-absorbed by other particles. Before then, the Universe was filled with a plasma that was opaque to photons.

The surface of last scattering is the collection of points in space at the exact distance that photons from the time of photon decoupling just reach us today. These are the photons we detect today as cosmic microwave background radiation (CMBR). However, with future technology, it may be possible to observe the still older relic neutrino background, or even more distant events via gravitational waves (which also should move at the speed of light). Sometimes astrophysicists distinguish between the *visible* universe, which includes only signals emitted since recombination—and the *observable* universe, which includes signals since the beginning of the cosmological expansion (the Big Bang in traditional cosmology, the end of the inflationary epoch in modern cosmology). According to calculations, the *comoving distance* (current proper distance) to particles from the CMBR, which represent the radius of the visible universe, is about 14.0 billion parsecs (about 45.7 billion light years), while the comoving distance to the edge of the observable universe is about 14.3 billion parsecs (about 46.6 billion light years),[7] about 2% larger.

The best estimate of the age of the universe as of 2013 is 13.798±0.037 billion years[5] but due to the expansion of space humans are observing objects that were originally much closer but are now considerably farther away (as defined in terms of cosmological proper distance, which is equal to the comoving distance at the present time) than a static 13.8 billion light-years distance.[8] It is estimated that the diameter of the observable universe is about 28 gigaparsecs (93 billion light-years, 8.8×10^{26} metres or 5.5×10^{23} miles),[9] putting the edge of the observable universe at about 46–47 billion light-years away.[10][11]

9.1   The Universe versus the observable universe

Some parts of the Universe are too far away for the light emitted since the Big Bang to have had enough time to reach Earth, so these portions of the Universe lie outside the observable universe. In the future, light from distant galaxies will have had more time to travel, so additional regions will become observable. However, due to Hubble's law regions sufficiently distant from us are expanding away from us faster than the speed of light (special relativity prevents nearby objects in the same local region from moving faster than the speed of light with respect to each other, but there is no such constraint for distant objects when the space between them is expanding; see uses of the proper distance for a discussion) and furthermore the expansion rate appears to be accelerating due to dark energy. Assuming dark energy remains constant (an unchanging cosmological constant), so that the expansion rate of the Universe continues to accelerate, there is a "future visibility limit" beyond which objects will *never* enter our observable universe at any time in the infinite future, because light emitted by objects outside that limit would never reach us. (A subtlety is that, because the Hubble parameter is decreasing with time, there can be cases where a galaxy that is receding from us just a bit faster than light does emit a signal that reaches us eventually[11][12]). This future visibility limit is calculated at a comoving distance of 19 billion parsecs (62 billion light years) assuming the Universe will keep expanding forever, which implies the number of galaxies that we can ever theoretically observe in the infinite future (leaving aside the issue that some may be impossible to observe in practice due to redshift, as discussed in the following paragraph) is only larger than the number currently observable by a factor of 2.36.[7]

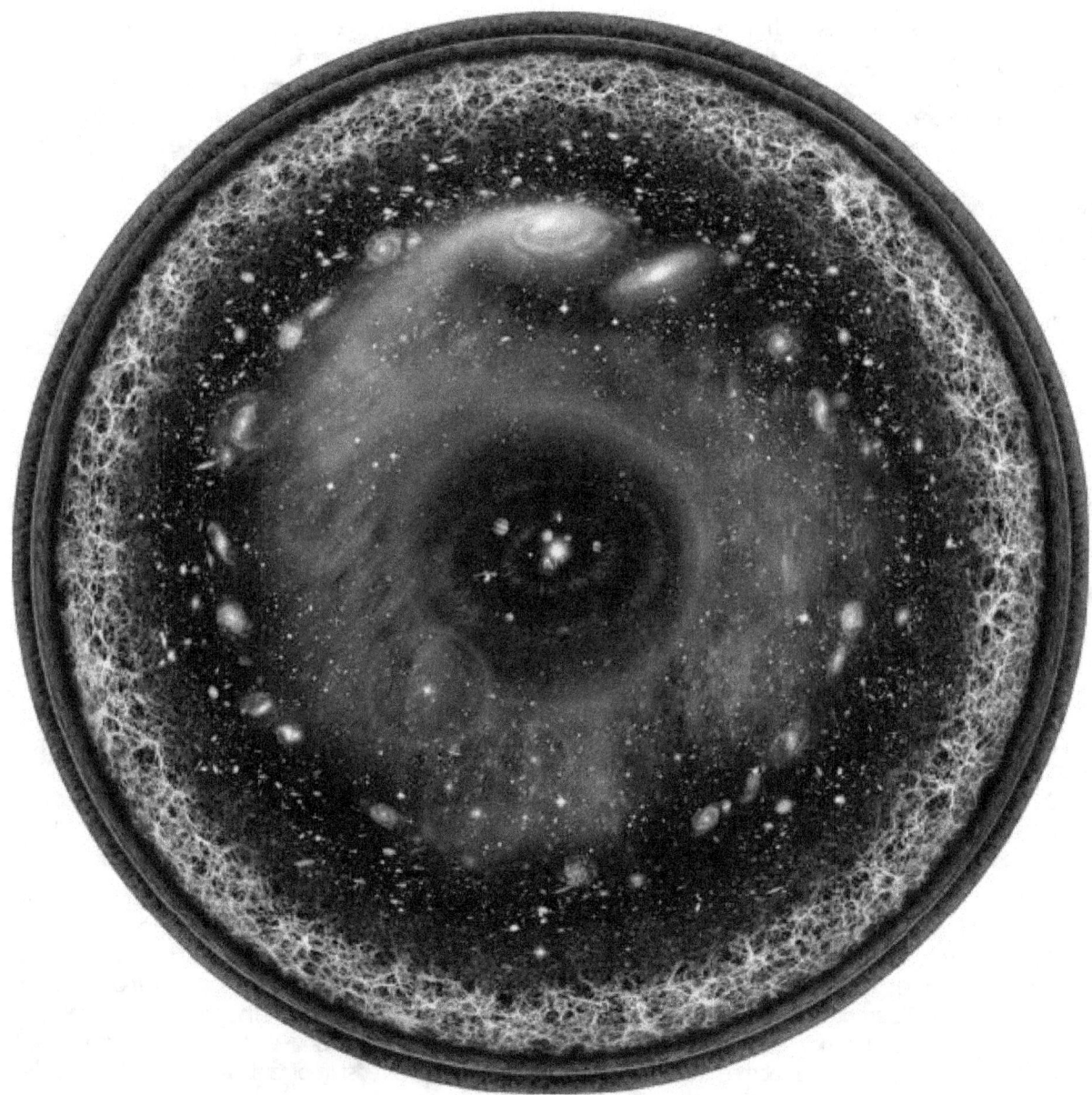

Artist's logarithmic scale conception of the observable universe with the Solar System at the center, inner and outer planets, Kuiper belt, Oort cloud, Alpha Centauri, Perseus Arm, Milky Way galaxy, Andromeda galaxy, nearby galaxies, Cosmic Web, Cosmic microwave radiation and the Big Bang's invisible plasma on the edge.

Though in principle more galaxies will become observable in the future, in practice an increasing number of galaxies will become extremely redshifted due to ongoing expansion, so much so that they will seem to disappear from view and become invisible.[13][14][15] An additional subtlety is that a galaxy at a given comoving distance is defined to lie within the "observable universe" if we can receive signals emitted by the galaxy at any age in its past history (say, a signal sent from the galaxy only 500 million years after the Big Bang), but because of the Universe's expansion, there may be some later age at which a signal sent from the same galaxy can *never* reach us at any point in the infinite future (so for example we might never see what the galaxy looked like 10 billion years after the Big Bang),[16] even though it remains at the same comoving distance (comoving distance is defined to be constant with time—unlike proper distance, which is used to define recession velocity due to the expansion of space), which is less than the comoving radius of the observable universe. This fact can be used to define a type of cosmic event horizon whose distance from us changes over time. For example, the current distance to this horizon is about 16 billion light years, meaning that a signal from an event happening *at present* can eventually reach us in the future if the event is less than 16 billion light years away, but the signal will never reach us if the event is more than 16 billion light years away.[11]

Both popular and professional research articles in cosmology often use the term "universe" to mean "observable universe". This can be justified on the grounds that we can never know anything by direct experimentation about any part of the Universe that is causally disconnected from us, although many credible theories require a total universe much larger than the observable universe. No evidence exists to suggest that the boundary of the observable universe constitutes a boundary on the Universe as a whole, nor do any of the mainstream cosmological models propose that the Universe has any physical boundary in the first place, though some models propose it could be finite but unbounded, like a higher-dimensional analogue of the 2D surface of a sphere that is finite in area but has no edge. It is plausible that the galaxies within our observable universe represent only a minuscule fraction of the galaxies in the Universe. According to the theory of cosmic inflation and its founder, Alan Guth, if it is assumed that inflation began about 10^{-37} seconds after the Big Bang, then with the plausible assumption that the size of the Universe at this time was approximately equal to the speed of light times its age, that would suggest that at present the entire universe's size is at least 3×10^{23} times larger than the size of the observable universe.[17] There are also lower estimates claiming that the entire universe is in excess of 250 times larger than the observable universe.[18]

If the Universe is finite but unbounded, it is also possible that the Universe is *smaller* than the observable universe. In this case, what we take to be very distant galaxies may actually be duplicate images of nearby galaxies, formed by light that has circumnavigated the Universe. It is difficult to test this hypothesis experimentally because different images of a galaxy would show different eras in its history, and consequently might appear quite different. Bielewicz et al.[19] claims to establish a lower bound of 27.9 gigaparsecs (91 billion light-years) on the diameter of the last scattering surface (since this is only a lower bound, the paper leaves open the possibility that the whole universe is much larger, even infinite). This value is based on matching-circle analysis of the WMAP 7 year data. This approach has been disputed.[20]

9.2 Size

The comoving distance from Earth to the edge of the observable universe is about 14 gigaparsecs (46 billion light years or 4.3×10^{26} meters) in any direction. The observable universe is thus a sphere with a diameter of about 29 gigaparsecs[21] (93 Gly or 8.8×10^{26} m).[22] Assuming that space is roughly flat, this size corresponds to a comoving volume of about 1.3×10^4 Gpc^3 (4.1×10^5 Gly^3 or 3.5×10^{80} m^3).

The figures quoted above are distances *now* (in cosmological time), not distances *at the time the light was emitted*. For example, the cosmic microwave background radiation that we see right now was emitted at the time of photon decoupling, estimated to have occurred about 380000 years after the Big Bang,[23][24] which occurred around 13.8 billion years ago. This radiation was emitted by matter that has, in the intervening time, mostly condensed into galaxies, and those galaxies are now calculated to be about 46 billion light-years from us.[7][11] To estimate the distance to that matter at the time the light was emitted, we may first note that according to the Friedmann–Lemaître–Robertson–Walker metric, which is used to model the expanding universe, if at the present time we receive light with a redshift of z, then the scale factor at the time the light was originally emitted is given by[25][26]

$$a(t) = \frac{1}{}$$

WMAP nine-year results combined with other measurements give the redshift of photon decoupling as $z = 1091.64 \pm 0.47$,[27]

Hubble Ultra-Deep Field image of a region of the observable universe (equivalent sky area size shown in bottom left corner), near the constellation Fornax. Each spot is a galaxy, consisting of billions of stars. The light from the smallest, most red-shifted galaxies originated nearly 14 billion years ago.

which implies that the scale factor at the time of photon decoupling would be $^{1}/_{1092.64}$. So if the matter that originally emitted the oldest CMBR photons has a *present* distance of 46 billion light years, then at the time of decoupling when the photons were originally emitted, the distance would have been only about 42 *million* light-years.

9.2.1   Misconceptions on its size

Many secondary sources have reported a wide variety of incorrect figures for the size of the visible universe. Some of these figures are listed below, with brief descriptions of possible reasons for misconceptions about them.

13.8 billion light-years The age of the universe is estimated to be 13.8 billion years. While it is commonly understood that nothing can accelerate to velocities equal to or greater than that of light, it is a common misconception that

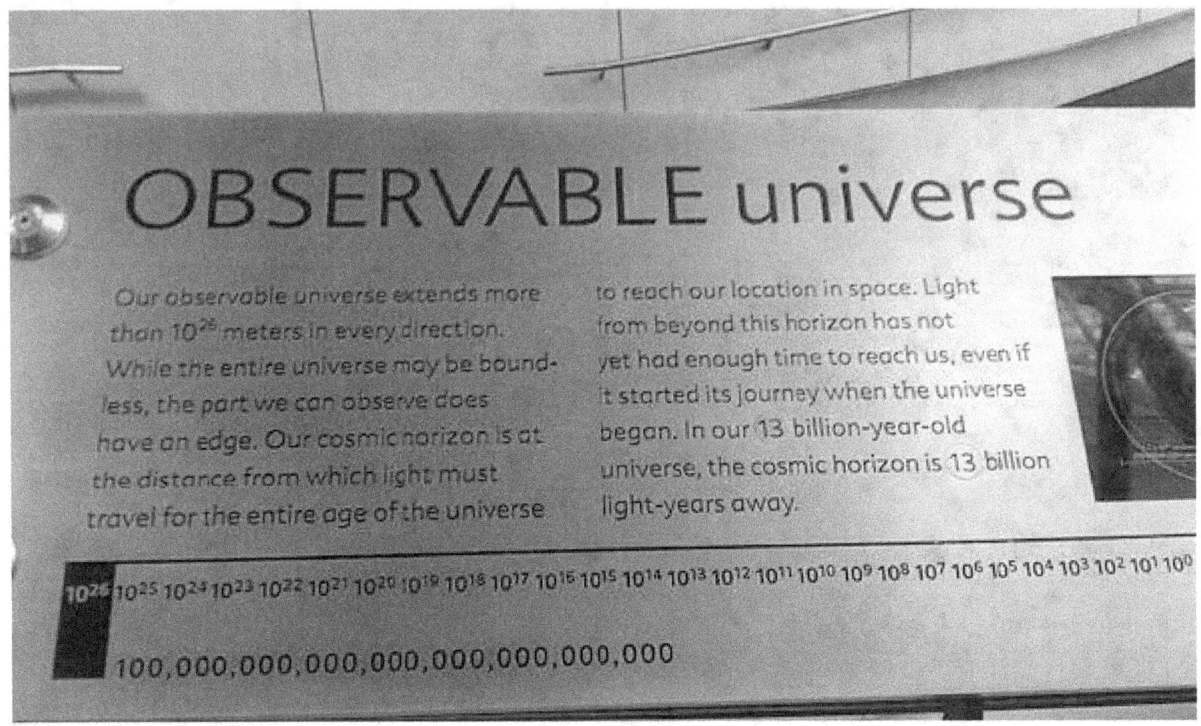

An example of one of the most common misconceptions about the size of the observable universe. Despite the fact that the universe is 13.8 billion years old, the distance to the edge of the observable universe is not 13.8 billion light-years, because the universe is expanding. This plaque appears at the Rose Center for Earth and Space in New York City.

the radius of the observable universe must therefore amount to only 13.8 billion light-years. This reasoning would only make sense if the flat, static Minkowski spacetime conception under special relativity were correct. In the real universe, spacetime is curved in a way that corresponds to the expansion of space, as evidenced by Hubble's law. Distances obtained as the speed of light multiplied by a cosmological time interval have no direct physical significance.[28]

15.8 billion light-years This is obtained in the same way as the 13.8 billion light year figure, but starting from an incorrect age of the universe that the popular press reported in mid-2006.[29][30] For an analysis of this claim and the paper that prompted it, see the following reference at the end of this article.[31]

27.6 billion light-years This is a diameter obtained from the (incorrect) radius of 13.8 billion light-years.

78 billion light-years In 2003, Cornish et al.[32] found this lower bound for the diameter of the *whole* universe (not just the observable part), if we postulate that the universe is finite in size due to its having a nontrivial topology,[33][34] with this lower bound based on the estimated current distance between points that we can see on opposite sides of the cosmic microwave background radiation (CMBR). If the whole universe is smaller than this sphere, then light has had time to circumnavigate it since the big bang, producing multiple images of distant points in the CMBR, which would show up as patterns of repeating circles.[35] Cornish et al. looked for such an effect at scales of up to 24 gigaparsecs (78 Gly or 7.4×10^{26} m) and failed to find it, and suggested that if they could extend their search to all possible orientations, they would then "be able to exclude the possibility that we live in a universe smaller than 24 Gpc in diameter". The authors also estimated that with "lower noise and higher resolution CMB maps (from WMAP's extended mission and from Planck), we will be able to search for smaller circles and extend the limit to ~28 Gpc."[32] This estimate of the maximum lower bound that can be established by future observations corresponds to a radius of 14 gigaparsecs, or around 46 billion light years, about the same as the figure for the radius of the visible universe (whose radius is defined by the CMBR sphere) given in the opening section. A 2012 preprint by most of the same authors as the Cornish et al. paper has extended the current lower bound to a diameter of 98.5% the diameter of the CMBR sphere, or about 26 Gpc.[36]

156 billion light-years This figure was obtained by doubling 78 billion light-years on the assumption that it is a radius.[37]

Since 78 billion light-years is already a diameter (the original paper by Cornish et al. says, "By extending the search to all possible orientations, we will be able to exclude the possibility that we live in a universe smaller than 24 Gpc in diameter," and 24 Gpc is 78 billion light years),[32] the doubled figure is incorrect. This figure was very widely reported.[37][38][39] A press release from Montana State University – Bozeman, where Cornish works as an astrophysicist, noted the error when discussing a story that had appeared in *Discover* magazine, saying "*Discover* mistakenly reported that the universe was 156 billion light-years wide, thinking that 78 billion was the radius of the universe instead of its diameter."[40]

180 billion light-years This estimate combines the erroneous 156-billion-light-year figure with evidence that the M33 Galaxy is actually fifteen percent farther away than previous estimates and that, therefore, the Hubble constant is fifteen percent smaller.[41] The 180-billion figure is obtained by adding 15% to 156 billion light years.

9.3 Large-scale structure

Sky surveys and mappings of the various wavelength bands of electromagnetic radiation (in particular 21-cm emission) have yielded much information on the content and character of the universe's structure. The organization of structure appears to follow as a hierarchical model with organization up to the scale of superclusters and filaments. Larger than this, there seems to be no continued structure, a phenomenon that has been referred to as the *End of Greatness*.[42]

9.3.1 Walls, filaments, nodes, and voids

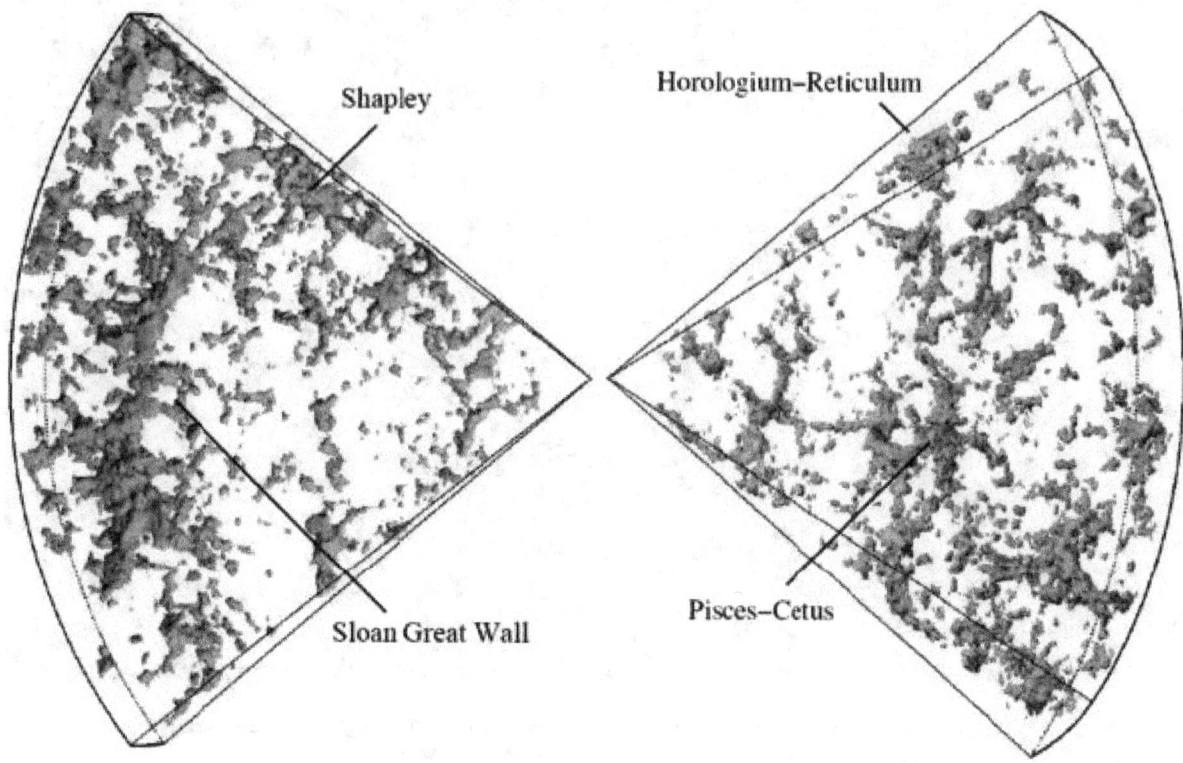

DTFE reconstruction of the inner parts of the 2dF Galaxy Redshift Survey

The organization of structure arguably begins at the stellar level, though most cosmologists rarely address astrophysics on that scale. Stars are organized into galaxies, which in turn form galaxy groups, galaxy clusters, superclusters, sheets, walls and filaments, which are separated by immense voids, creating a vast foam-like structure sometimes called the "cosmic web". Prior to 1989, it was commonly assumed that virialized galaxy clusters were the largest structures in existence, and that they were distributed more or less uniformly throughout the Universe in every direction. However, since the early 1980s, more and more structures have been discovered. In 1983, Adrian Webster identified the Webster LQG, a large quasar group consisting of 5 quasars. The discovery was the first identification of a large-scale structure, and has expanded the information about the known grouping of matter in the Universe. In 1987, Robert Brent Tully identified the Pisces–Cetus Supercluster Complex, the galaxy filament in which the Milky Way resides. It is about 1 billion light years across. That same year, an unusually large region with no galaxies has been discovered, the Giant Void, which measures 1.3 billion light years across. Based on redshift survey data, in 1989 Margaret Geller and John Huchra discovered the "Great Wall",[43] a sheet of galaxies more than 500 million light-years long and 200 million wide, but only 15 million light-years thick. The existence of this structure escaped notice for so long because it requires locating the position of galaxies in three dimensions, which involves combining location information about the galaxies with distance information from redshifts. Two years later, astronomers Roger G. Clowes and Luis E. Campusano discovered the Clowes–Campusano LQG, a large quasar group measuring two billion light years at its widest point, and was the largest known structure in the Universe at the time of its announcement. In April 2003, another large-scale structure was discovered, the Sloan Great Wall. In August 2007, a possible supervoid was detected in the constellation Eridanus.[44] It coincides with the 'CMB cold spot', a cold region in the microwave sky that is highly improbable under the currently favored cosmological model. This supervoid could cause the cold spot, but to do so it would have to be improbably big, possibly a billion light-years across, almost as big as the Giant Void mentioned above.

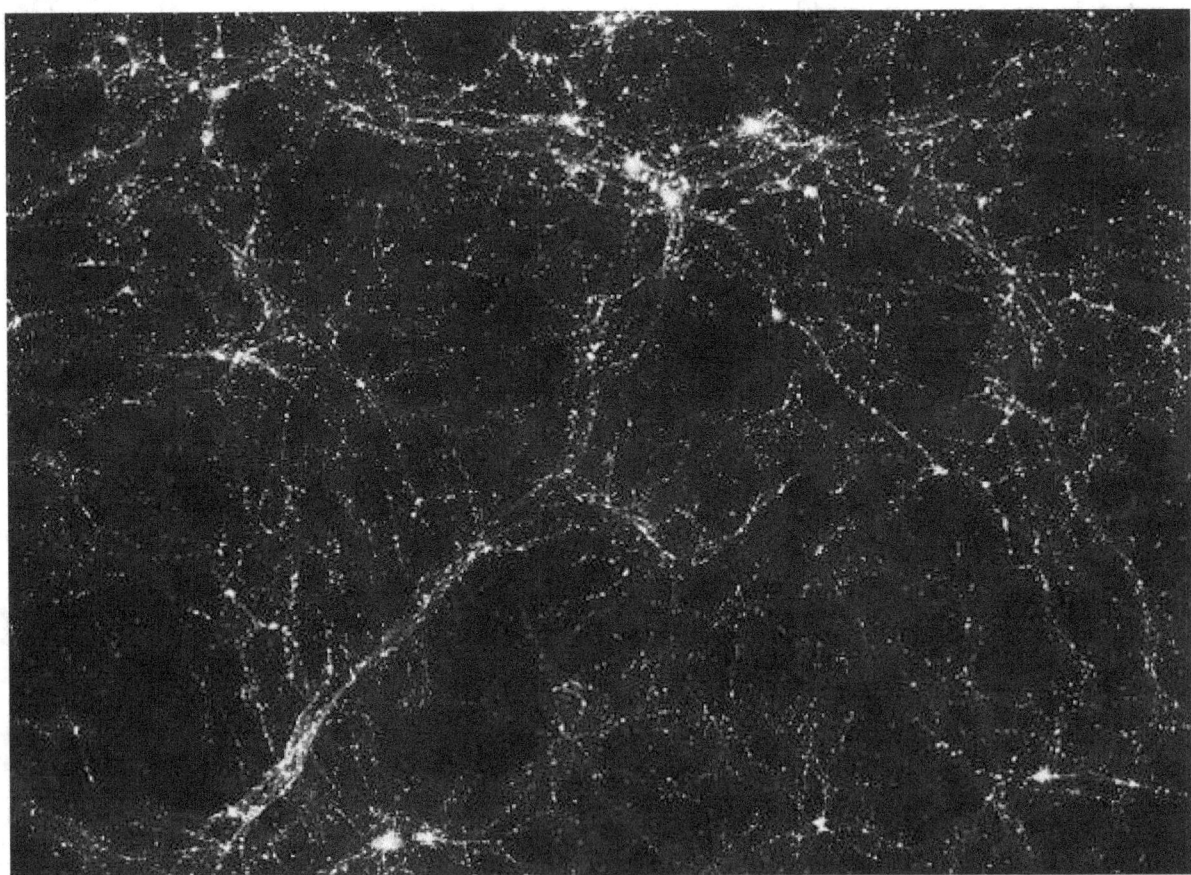

Image (computer simulated) of an area of space more than 50 million light years across, presenting a possible large-scale distribution of light sources in the universe - precise relative contributions of galaxies and quasars are unclear.

Another large-scale structure is the Newfound Blob, a collection of galaxies and enormous gas bubbles that measures about 200 million light years across.

In recent studies the Universe appears as a collection of giant bubble-like voids separated by sheets and filaments of galaxies, with the superclusters appearing as occasional relatively dense nodes. This network is clearly visible in the

2dF Galaxy Redshift Survey. In the figure, a three-dimensional reconstruction of the inner parts of the survey is shown, revealing an impressive view of the cosmic structures in the nearby universe. Several superclusters stand out, such as the Sloan Great Wall.

In 2011, a large quasar group was discovered, U1.11, measuring about 2.5 billion light years across. On January 11, 2013, another large quasar group, the Huge-LQG, was discovered, which was measured to be four billion light-years across, the largest known structure in the Universe that time.[45] In November 2013 astronomers discovered the Hercules–Corona Borealis Great Wall,[46][47] an even bigger structure twice as large as the former. It was defined by mapping of gamma-ray bursts.[46][48]

9.3.2   End of Greatness

The *End of Greatness* is an observational scale discovered at roughly 100 Mpc (roughly 300 million lightyears) where the lumpiness seen in the large-scale structure of the universe is homogenized and isotropized in accordance with the Cosmological Principle.[42] At this scale, no pseudo-random fractalness is apparent.[49] The superclusters and filaments seen in smaller surveys are randomized to the extent that the smooth distribution of the Universe is visually apparent. It was not until the redshift surveys of the 1990s were completed that this scale could accurately be observed.[42]

9.3.3   Observations

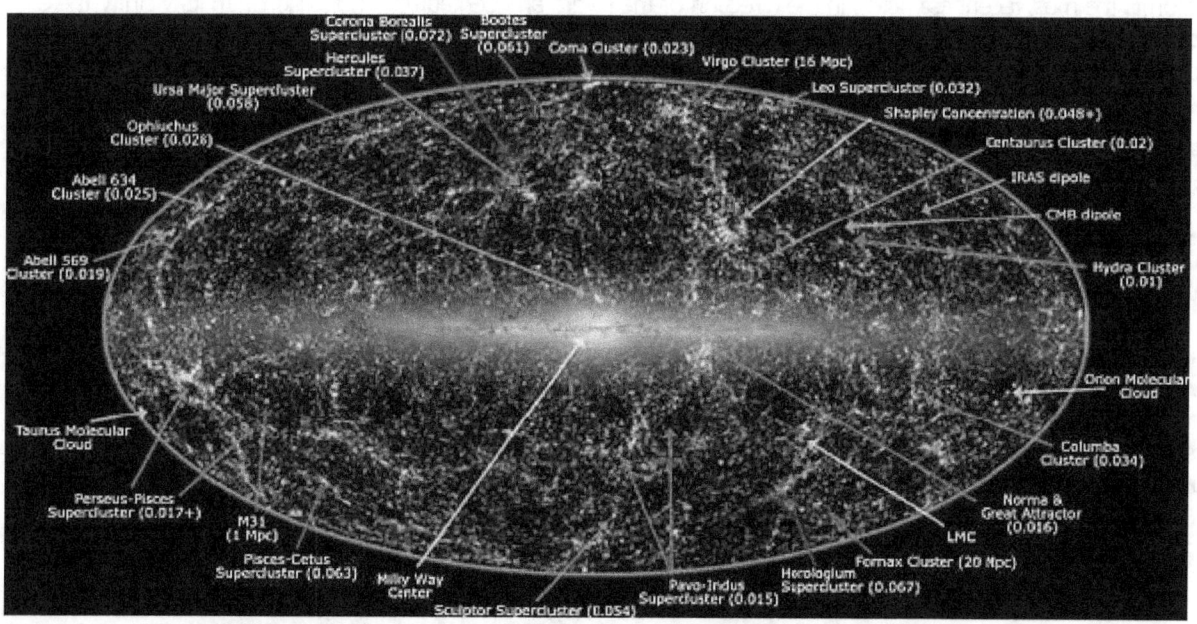

"Panoramic view of the entire near-infrared sky reveals the distribution of galaxies beyond the Milky Way. The image is derived from the 2MASS Extended Source Catalog (XSC)—more than 1.5 million galaxies, and the Point Source Catalog (PSC)–nearly 0.5 billion Milky Way stars. The galaxies are color-coded by 'redshift' obtained from the UGC, CfA, Tully NBGC, LCRS, 2dF, 6dFGS, and SDSS surveys (and from various observations compiled by the NASA Extragalactic Database), or photo-metrically deduced from the K band (2.2 um). Blue are the nearest sources (z < 0.01); green are at moderate distances (0.01 < z < 0.04) and red are the most distant sources that 2MASS resolves (0.04 < z < 0.1). The map is projected with an equal area Aitoff in the Galactic system (Milky Way at center)."[50]

Another indicator of large-scale structure is the 'Lyman-alpha forest'. This is a collection of absorption lines that appear in the spectra of light from quasars, which are interpreted as indicating the existence of huge thin sheets of intergalactic (mostly hydrogen) gas. These sheets appear to be associated with the formation of new galaxies.

Caution is required in describing structures on a cosmic scale because things are often different from how they appear. Gravitational lensing (bending of light by gravitation) can make an image appear to originate in a different direction from its real source. This is caused when foreground objects (such as galaxies) curve surrounding spacetime (as predicted by general relativity), and deflect passing light rays. Rather usefully, strong gravitational lensing can sometimes magnify distant galaxies, making them easier to detect. Weak lensing (gravitational shear) by the intervening universe in general also subtly changes the observed large-scale structure. As of 2004, measurements of this subtle shear showed considerable promise as a test of cosmological models.

The large-scale structure of the Universe also looks different if one only uses redshift to measure distances to galaxies. For example, galaxies behind a galaxy cluster are attracted to it, and so fall towards it, and so are slightly blueshifted (compared to how they would be if there were no cluster) On the near side, things are slightly redshifted. Thus, the environment of the cluster looks a bit squashed if using redshifts to measure distance. An opposite effect works on the galaxies already within a cluster: the galaxies have some random motion around the cluster center, and when these random motions are converted to redshifts, the cluster appears elongated. This creates a "*finger of God*"—the illusion of a long chain of galaxies pointed at the Earth.

9.3.4   Cosmography of our cosmic neighborhood

At the centre of the Hydra-Centaurus Supercluster, a gravitational anomaly called the Great Attractor affects the motion of galaxies over a region hundreds of millions of light-years across. These galaxies are all redshifted, in accordance with Hubble's law. This indicates that they are receding from us and from each other, but the variations in their redshift are sufficient to reveal the existence of a concentration of mass equivalent to tens of thousands of galaxies.

The Great Attractor, discovered in 1986, lies at a distance of between 150 million and 250 million light-years (250 million is the most recent estimate), in the direction of the Hydra and Centaurus constellations. In its vicinity there is a preponderance of large old galaxies, many of which are colliding with their neighbours, or radiating large amounts of radio waves.

In 1987 astronomer R. Brent Tully of the University of Hawaii's Institute of Astronomy identified what he called the Pisces-Cetus Supercluster Complex, a structure one billion light years long and 150 million light years across in which, he claimed, the Local Supercluster was embedded.[51][52]

9.4   Mass of ordinary matter

The mass of the known Universe is often quoted as 10^{50} tonnes or 10^{53} kg.[3] In this context, mass refers to ordinary matter and includes the interstellar medium (ISM) and the intergalactic medium (IGM). However, it excludes dark matter and dark energy. Three calculations substantiate this quoted value for the mass of ordinary matter in the Universe: Estimates based on critical density, extrapolations from number of stars, and estimates based on steady-state. The calculations obviously assume a finite universe.

9.4.1   Estimates based on critical density

Critical Density is the energy density where the expansion of the Universe is poised between continued expansion and collapse.[53] Observations of the cosmic microwave background from the Wilkinson Microwave Anisotropy Probe suggest that the spatial curvature of the Universe is very close to zero, which in current cosmological models implies that the value of the density parameter must be very close to a certain critical density value. At this condition, the calculation for ρ_c critical density, is:[54]

$$\rho_c = \frac{3H_0^2}{8\pi G}$$

where G is the gravitational constant. From The European Space Agency's Planck Telescope results: H_0, is 67.15 kilometers per second per mega parsec. This gives a critical density of 0.85×10^{-26} kg/m^3 (commonly quoted as about 5 hydrogen atoms per cubic meter). This density includes four significant types of energy/mass: ordinary matter (4.8%), neutrinos (0.1%), cold dark matter (26.8%), and dark energy (68.3%).[5] Note that although neutrinos are defined as particles like electrons, they are listed separately because they are difficult to detect and so different from ordinary matter. Thus, the density of ordinary matter is 4.8% of the total critical density calculated or 4.08×10^{-28} kg/m^3. To convert this density to mass we must multiply by volume, a value based on the radius of the "observable universe". Since the

Universe has been expanding for 13.8 billion years, the comoving distance (radius) is now about 46.6 billion light years. Thus, volume ($^4/_3\pi r^3$) equals 3.58×10^{80} m^3 and mass of ordinary matter equals density (4.08×10^{-28} kg/m^3) times volume (3.58×10^{80} m^3) or 1.46×10^{53} kg.

9.4.2 Extrapolation from number of stars

There is no way to know exactly the number of stars, but from current literature, the range of 10^{22} to 10^{24} is normally quoted.[55][56][57][58] One way to substantiate this range is to estimate the number of galaxies and multiply by the number of stars in an average galaxy. The 2004 Hubble Ultra-Deep Field image contains an estimated 10000 galaxies.[59] The patch of sky in this area, is 3.4 arc minutes on each side. For a relative comparison, it would require over 50 of these images to cover the full moon. If this area is typical for the entire sky, there are over 100 billion galaxies in the Universe.[60] More recently, in 2012, Hubble scientists produced the Hubble Extreme Deep Field image which showed slightly more galaxies for a comparable area.[61] However, in order to compute the number of stars based on these images, we would need additional assumptions: the percent of both large and dwarf galaxies; and, their average number of stars. Thus, a reasonable option is to assume 100 billion average galaxies and 100 billion stars per average galaxy. This results in 10^{22} stars. Next, we need average star mass which can be calculated from the distribution of stars in the Milky Way. Within the Milky Way, if a large number of stars are counted by spectral class, 73% are class M stars which contain only 30% of the Sun's mass. Considering mass and number of stars in each spectral class, the average star is 51.5% of the Sun's mass.[62] The Sun's mass is 2×10^{30} kg. so a reasonable number for the mass of an average star in the Universe is 10^{30} kg. Thus, the mass of all stars equals the number of stars (10^{22}) times an average mass of star (10^{30} kg) or 10^{52} kg. The next calculation adjusts for Interstellar Medium (ISM) and Intergalactic Medium (IGM). ISM is material between stars: gas (mostly hydrogen) and dust. IGM is material between galaxies, mostly hydrogen. Ordinary matter (protons, neutrons and electrons) exists in ISM and IGM as well as in stars. In the reference, "The Cosmic Energy Inventory", the percentage of each part is defined: stars = 5.9%, Interstellar Medium (ISM) = 1.7%, and Intergalactic Medium (IGM) = 92.4%.[63] Thus, to extrapolate the mass of the Universe from the star mass, divide the 10^{52} kg mass calculated for stars by 5.9%. The result is 1.7×10^{53} kg for all the ordinary matter.

9.4.3 Estimates based on steady-state universe

Sir Fred Hoyle calculated the mass of an observable steady-state universe using the formula:[64]

$$\frac{4}{H}$$

which can also be stated as [65]

$$\frac{c^3}{2GH}$$

Here H = Hubble constant, ρ = Hoyle's value for the density, G = gravitational constant, and c = speed of light.

This calculation yields approximately 0.92×10^{53} kg; however, this represents all energy/matter and is based on the Hubble volume (the volume of a sphere with radius equal to the Hubble length of about 13.8 billion light years). The critical density calculation above was based on the comoving distance radius of 46.6 billion light years. Thus, the Hoyle equation mass/energy result must be adjusted for increased volume. The comoving distance radius gives a volume about 39 times greater (46.7 cubed divided by 13.8 cubed). However, as volume increases, ordinary matter and dark matter would not increase; only dark energy increases with volume. Thus, assuming ordinary matter, neutrinos, and dark matter are 31.7% of the total mass/energy, and dark energy is 68.3%, the amount of total mass/energy for the steady-state calculation would be: mass of ordinary matter and dark matter (31.7% times 0.92×10^{53} kg) plus the mass of dark energy ((68.3% times 0.92×10^{53} kg) times increased volume (39)). This equals: 2.48×10^{54} kg. As noted above for the Critical Density method, ordinary matter is 4.8% of all energy/matter. If the Hoyle result is multiplied by this percent, the result for ordinary matter is 1.20×10^{53} kg.

9.4.4   Comparison of results

In summary, the three independent calculations produced reasonably close results: 1.46×10^{53}, 1.7×10^{53}, and 1.20×10^{53} kg. The average is 1.45×10^{53} kg.

The key assumptions using the Extrapolation from Star Mass method were the number of stars (10^{22}) and the percentage of ordinary matter in stars (5.9%). The key assumptions using the Critical Density method were the comoving distance radius of the Universe (46.6 billion light years) and the percentage of ordinary matter in all matter (4.8%). The key assumptions using the Hoyle steady-state method were the comoving distance radius and the percentage of dark energy in all mass (68.3%). Both the Critical Density and the Hoyle steady-state equations also used the Hubble constant (67.15 (km/s)/Mpc).

9.5   Matter content — number of atoms

Main article: cosmic abundance of elements

Assuming the mass of ordinary matter is about 1.45×10^{53} kg (reference previous section) and assuming all atoms are hydrogen atoms (which in reality make up about 74% of all atoms in our galaxy by mass, see Abundance of the chemical elements), calculating the estimated total number of atoms in the Universe is straightforward. Divide the mass of ordinary matter by the mass of a hydrogen atom (1.45×10^{53} kg divided by 1.67×10^{-27} kg). The result is approximately 10^{80} hydrogen atoms. The chemistry of life may have begun shortly after the Big Bang, 13.8 billion years ago, during a habitable epoch when the Universe was only 10–17 million years old.[66][67][68] According to the panspermia hypothesis, microscopic life—distributed by meteoroids, asteroids and other small Solar System bodies—may exist throughout the Universe.[69] Though life is confirmed only on the Earth, many think that extraterrestrial life is not only plausible, but probable or inevitable.[70][71]

9.6   Most distant objects

The most distant astronomical object yet announced as of January 2011 is a galaxy candidate classified UDFj-39546284. In 2009, a gamma ray burst, GRB 090423, was found to have a redshift of 8.2, which indicates that the collapsing star that caused it exploded when the Universe was only 630 million years old.[72] The burst happened approximately 13 billion years ago,[73] so a distance of about 13 billion light years was widely quoted in the media (or sometimes a more precise figure of 13.035 billion light years),[72] though this would be the "light travel distance" (*see* Distance measures (cosmology)) rather than the "proper distance" used in both Hubble's law and in defining the size of the observable universe (cosmologist Ned Wright argues against the common use of light travel distance in astronomical press releases on this page, and at the bottom of the page offers online calculators that can be used to calculate the current proper distance to a distant object in a flat universe based on either the redshift z or the light travel time). The proper distance for a redshift of 8.2 would be about 9.2 Gpc,[74] or about 30 billion light years. Another record-holder for most distant object is a galaxy observed through and located beyond Abell 2218, also with a light travel distance of approximately 13 billion light years from Earth, with observations from the Hubble telescope indicating a redshift between 6.6 and 7.1, and observations from Keck telescopes indicating a redshift towards the upper end of this range, around 7.[75] The galaxy's light now observable on Earth would have begun to emanate from its source about 750 million years after the Big Bang.[76]

9.7   Horizons

Main article: cosmological horizon

The limit of observability in our universe is set by a set of cosmological horizons which limit, based on various physical constraints, the extent to which we can obtain information about various events in the Universe. The most famous horizon is the particle horizon which sets a limit on the precise distance that can be seen due to the finite age of the Universe. Additional horizons are associated with the possible future extent of observations (larger than the particle horizon owing to the expansion of space), an "optical horizon" at the surface of last scattering, and associated horizons with the surface of last scattering for neutrinos and gravitational waves.

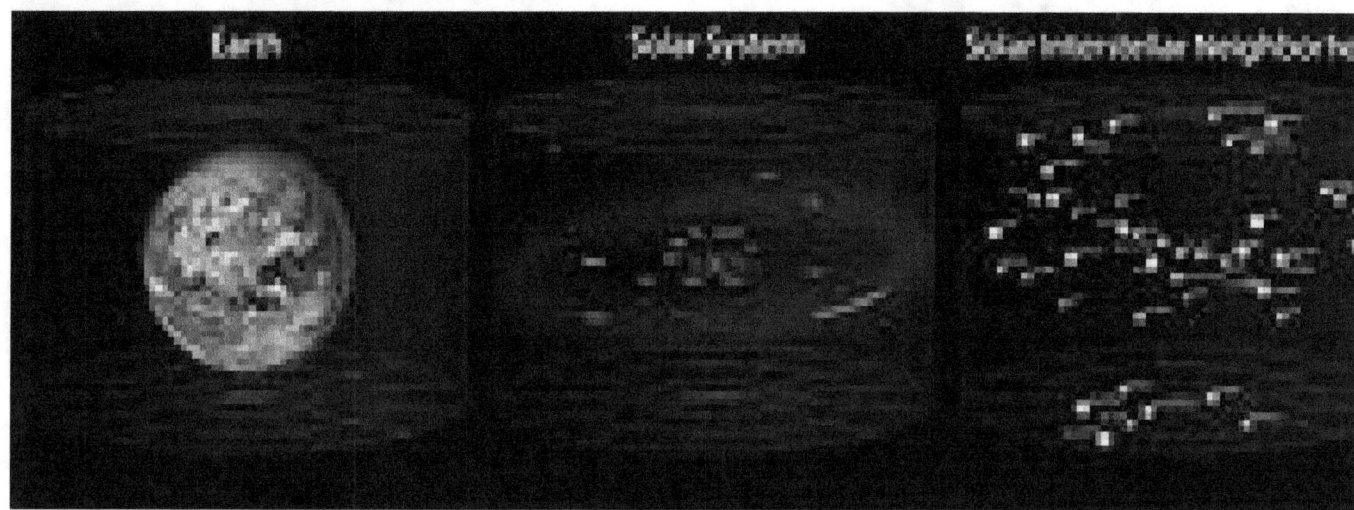

A diagram of our location in the observable universe. (*Click here for an alternate image.*)

9.8   See also

- Big Bang
- Bolshoi Cosmological Simulation
- Causality (physics)
- Chronology of the universe
- Dark flow
- Event horizon of the universe
- Hubble volume
- Illustris project
- Multiverse
- Orders of magnitude (length)
- Timeline of the Big Bang

9.9   References

[1] Itzhak Bars; John Terning (November 2009). *Extra Dimensions in Space and Time*. Springer. pp. 27–. ISBN 978-0-387-77637-8. Retrieved 2011-05-01.

[2] http://www.wolframalpha.com/input/?i=volume+universe

[3] Paul Davies (2006). *The Goldilocks Enigma*. First Mariner Books. p. 43–. ISBN 978-0-618-59226-5. Retrieved 1 July 2013.

[4] http://map.gsfc.nasa.gov/universe/uni_matter.html January 13, 2015

[5] Planck collaboration (2013). "Planck 2013 results. XVI. Cosmological parameters". arXiv:1303.5076 [astro-ph.CO].

[6] Fixsen, D. J. (December 2009). "The Temperature of the Cosmic Microwave Background". *The Astrophysical Journal* **707** (2): 916–920. arXiv:0911.1955. Bibcode:2009ApJ...707..916F. doi:10.1088/0004-637X/707/2/916.

[7] Gott III, J. Richard; Mario Jurić; David Schlegel; Fiona Hoyle et al. (2005). "A Map of the Universe" (PDF). *The Astrophysics Journal* **624** (2): 463. arXiv:astro-ph/0310571. Bibcode:2005ApJ...624..463G. doi:10.1086/428890.

[8] Davis, Tamara M.; Charles H. Lineweaver (2004). "Expanding Confusion: common misconceptions of cosmological horizons and the superluminal expansion of the universe". *Publications of the Astronomical Society of Australia* **21** (1): 97. arXiv:astro-ph/0310808. Bibcode:2004PASA...21...97D. doi:10.1071/AS03040.

[9] Itzhak Bars; John Terning (November 2009). *Extra Dimensions in Space and Time.* Springer. pp. 27–. ISBN 978-0-387-77637-8. Retrieved 1 May 2011.

[10] Frequently Asked Questions in Cosmology. Astro.ucla.edu. Retrieved on 2011-05-01.

[11] Lineweaver, Charles; Tamara M. Davis (2005). "Misconceptions about the Big Bang". Scientific American.

[12] Is the universe expanding faster than the speed of light? (see the last two paragraphs)

[13] Krauss, Lawrence M.; Robert J. Scherrer (2007). "The Return of a Static Universe and the End of Cosmology". *General Relativity and Gravitation* **39** (10): 1545–1550. arXiv:0704.0221. Bibcode:2007GReGr..39.1545K. doi:10.1007/s10714-007-0472-9.

[14] Using Tiny Particles To Answer Giant Questions. Science Friday, 3 Apr 2009. According to the transcript, Brian Greene makes the comment "And actually, in the far future, everything we now see, except for our local galaxy and a region of galaxies will have disappeared. The entire universe will disappear before our very eyes, and it's one of my arguments for actually funding cosmology. We've got to do it while we have a chance."

[15] See also Faster than light#Universal expansion and Future of an expanding universe#Galaxies outside the Local Supercluster are no longer detectable.

[16] Loeb, Abraham (2002). "The Long-Term Future of Extragalactic Astronomy". *Physical Review D* **65** (4). arXiv:astro-ph/0107568. Bibcode:2002PhRvD..65d7301L. doi:10.1103/PhysRevD.65.047301.

[17] Alan H. Guth (17 March 1998). *The inflationary universe: the quest for a new theory of cosmic origins.* Basic Books. pp. 186–. ISBN 978-0-201-32840-0. Retrieved 1 May 2011.

[18] Universe Could be 250 Times Bigger Than What is Observable - by Vanessa D'Amico on February 8, 2011 http://www.universetoday.com/83167/universe-could-be-250-times-bigger-than-what-is-observable/

[19] Bielewicz, P.; Banday, A. J.; Gorski, K. M. (2013). "Constraints on the Topology of the Universe". arXiv:1303.4004 [astro-ph.CO].

[20] Mota; Reboucas; Tavakol (2010). "Observable circles-in-the-sky in flat universes". arXiv:1007.3466 [astro-ph.CO].

[21] "WolframAlpha". Retrieved 29 November 2011.

[22] "WolframAlpha". Retrieved 29 November 2011.

[23] "Seven-Year Wilson Microwave Anisotropy Probe (WMAP) Observations: Sky Maps, Systematic Errors, and Basic Results" (PDF). nasa.gov. Retrieved 2010-12-02. (see p. 39 for a table of best estimates for various cosmological parameters)

[24] Abbott, Brian (May 30, 2007). "Microwave (WMAP) All-Sky Survey". Hayden Planetarium. Retrieved 2008-01-13.

[25] Paul Davies (28 August 1992). *The new physics.* Cambridge University Press. pp. 187–. ISBN 978-0-521-43831-5. Retrieved 1 May 2011.

[26] V. F. Mukhanov (2005). *Physical foundations of cosmology.* Cambridge University Press. pp. 58–. ISBN 978-0-521-56398-7. Retrieved 1 May 2011.

[27] Bennett, C. L.; Larson, D.; Weiland, J. L.; Jarosik, N. et al. (1 October 2013). "Nine-year Wilkinson Microwave Anisotropy Probe (WMAP) Observations: Final Maps and Results". *The Astrophysical Journal Supplement Series* **208** (2): 20. arXiv:1212.5225. Bibcode:2013ApJS..208...20B. doi:10.1088/0067-0049/208/2/20.

[28] Ned Wright, "Why the Light Travel Time Distance should not be used in Press Releases".

[29] Universe Might be Bigger and Older than Expected. Space.com (2006-08-07). Retrieved on 2011-05-01.

[30] Big bang pushed back two billion years – space – 04 August 2006 – New Scientist. Space.newscientist.com. Retrieved on 2011-05-01.

[31] Edward L. Wright, "An Older but Larger Universe?"

[32] Cornish; Spergel; Starkman; Eiichiro Komatsu (May 2004) [October 2003 (arXiv)]. "Constraining the Topology of the Universe". *Phys. Rev. Lett.* **92** (20). arXiv:astro-ph/0310233. Bibcode:2004PhRvL..92t1302C. doi:10.1103/PhysRevLett.92.201302. 201302.

[33] Levin, Janna. "In space, do all roads lead to home?". plus.maths.org. Retrieved 2012-08-15.

[34] http://cosmos.phy.tufts.edu/~{}zirbel/ast21/sciam/IsSpaceFinite.pdf

[35] Bob Gardner's "Topology, Cosmology and Shape of Space" Talk, Section 7. Etsu.edu. Retrieved on 2011-05-01.

[36] Vaudrevange; Starkmanl; Cornish; Spergel. "Constraints on the Topology of the Universe: Extension to General Geometries". arXiv:1206.2939. Bibcode:2012PhRvD..86h3526V. doi:10.1103/PhysRevD.86.083526.

[37] SPACE.com – Universe Measured: We're 156 Billion Light-years Wide!

[38] Roy, Robert. (2004-05-24) New study super-sizes the universe – Technology & science – Space – Space.com – msnbc.com. MSNBC. Retrieved on 2011-05-01.

[39] "Astronomers size up the Universe". *BBC News.* 2004-05-28. Retrieved 2010-05-20.

[40] "MSU researcher recognized for discoveries about universe". 2004-12-21. Retrieved 2011-02-08.

[41] Space.com – Universe Might be Bigger and Older than Expected

[42] Robert P Kirshner (2002). *The Extravagant Universe: Exploding Stars, Dark Energy and the Accelerating Cosmos.* Princeton University Press. p. 71. ISBN 0-691-05862-8.

[43] M. J. Geller; J. P. Huchra (1989). "Mapping the universe.". *Science* **246** (4932): 897–903. Bibcode:1989Sci...246..897G. doi:10.1126/science.246.4932.897. PMID 17812575.

[44] Biggest void in space is 1 billion light years across – space – 24 August 2007 – New Scientist. Space.newscientist.com. Retrieved on 2011-05-01.

[45] Wall, Mike (2013-01-11). "Largest structure in universe discovered". Fox News.

[46] Horváth, I; Hakkila, Jon; Bagoly, Z. (2014). "Possible structure in the GRB sky distribution at redshift two". arXiv:1401.0533. Bibcode:2014A&A...561L..12H. doi:10.1051/0004-6361/201323020.

[47] Horvath, I.; Hakkila, J.; Bagoly, Z. (2013). "The largest structure of the Universe, defined by Gamma-Ray Bursts". arXiv:1311.1104 [astro-ph.CO].

[48] Klotz, Irene (2013-11-19). "Universe's Largest Structure is a Cosmic Conundrum". *Discovery.*

[49] LiveScience.com, "The Universe Isn't a Fractal, Study Finds", Natalie Wolchover,22 August 2012

[50] 1Jarrett, T. H. (2004). "Large Scale Structure in the Local Universe: The 2MASS Galaxy Catalog". *Publications of the Astronomical Society of Australia* **21** (4): 396. arXiv:astro-ph/0405069. Bibcode:2004PASA...21..396J. doi:10.1071/AS04050.

[51] Massive Clusters of Galaxies Defy Concepts of the Universe N.Y. Times Tue. November 10, 1987:

[52] Map of the Pisces-Cetus Supercluster Complex:

[53] Michio Kaku (2005). *Parallel Worlds.* Anchor Books. p. 385. ISBN 978-1-4000-3372-0. Retrieved 1 July 2013.

[54] Bernard F. Schutz (2003). *Gravity from the ground up*. Cambridge University Press. pp. 361–. ISBN 978-0-521-45506-0. Retrieved 1 May 2011.

[55] "Astronomers count the stars". BBC News. July 22, 2003. Retrieved 2006-07-18.

[56] "trillions-of-earths-could-be-orbiting-300-sextillion-stars"

[57] van Dokkum, Pieter G.; Charlie Conroy (2010). "A substantial population of low-mass stars in luminous elliptical galaxies". *Nature* **468** (7326): 940–942. arXiv:1009.5992. Bibcode:2010Natur.468..940V. doi:10.1038/nature09578. PMID 21124316.

[58] "How many stars?"

[59] [url= http://hubblesite.org/newscenter/archive/releases/2004/28/text/]| NASA, Hubble News Release STSci - 2004-7

[60] James R Johnson. *Comprehending the Cosmos, a Macro View of the Universe*. p. 36. ISBN 978-1-477-64969-5. Retrieved 1 July 2013.

[61] "Hubble Goes to the eXtreme to Assemble Farthest Ever View of the Universe" (Press release). 25 September 2012. Retrieved 1 July 2013.

[62] James R Johnson. *Comprehending the Cosmos, a Macro View of the Universe*. p. 34. ISBN 978-1-477-64969-5. Retrieved 1 July 2013.

[63] Fukugita, Masataka; Peebles, P. J. E. (2004). "The Cosmic Energy Inventory". *Astrophysical Journal* **616** (2): 643–668. arXiv:astro-ph/0406095. Bibcode:2004ApJ...616..643F. doi:10.1086/425155.

[64] Helge Kragh (1999-02-22). "Chapter 5". *Cosmology and Controversy: The Historical Development of Two Theories of the Universe*. Princeton University Press. p. 212. ISBN 0-691-00546-X.

[65] Valev, Dimitar (2010). "Estimation of the total mass and energy of the universe". arXiv:1004.1035 [physics.gen-ph].

[66] Loeb, Abraham (October 2014). "The Habitable Epoch of the Early Universe". *International Journal of Astrobiology* **13** (04): 337–339. arXiv:1312.0613. Bibcode:2014IJAsB..13..337L. doi:10.1017/S1473550414000196. Retrieved 15 December 2014.

[67] Loeb, Abraham (2 December 2013). "The Habitable Epoch of the Early Universe" (PDF). *Arxiv*. arXiv:1312.0613v3. Retrieved 15 December 2014.

[68] Dreifus, Claudia (2 December 2014). "Much-Discussed Views That Go Way Back - Avi Loeb Ponders the Early Universe, Nature and Life". *New York Times*. Retrieved 3 December 2014.

[69] Rampelotto, P.H. (2010). "Panspermia: A Promising Field Of Research" (PDF). *Astrobiology Science Conference*. Retrieved 3 December 2014.

[70] Race, Margaret S.; Randolph, Richard O. (2002). "The need for operating guidelines and a decision making framework applicable to the discovery of non-intelligent extraterrestrial life". *Advances in Space Research* **30** (6): 1583–1591. Bibcode:2002AdSpR..30.1583R. doi:10.1016/S0273-1177(02)00478-7. ISSN 0273-1177. There is growing scientific confidence that the discovery of extraterrestrial life in some form is nearly inevitable

[71] Cantor, Matt (15 February 2009). "Alien Life 'Inevitable': Astronomer". *newser*. Archived from the original on 3 May 2013. Retrieved 3 May 2013. Scientists now believe there could be as many habitable planets in the cosmos as there are stars, and that makes life's existence elsewhere "inevitable" over billions of years, says one.

[72] New Gamma-Ray Burst Smashes Cosmic Distance Record – NASA Science. Science.nasa.gov. Retrieved on 2011-05-01.

[73] More Observations of GRB 090423, the Most Distant Known Object in the Universe. Universetoday.com (2009-10-28). Retrieved on 2011-05-01.

[74] Meszaros, Attila et al. (2009). "Impact on cosmology of the celestial anisotropy of the short gamma-ray bursts". *Baltic Astronomy* **18**: 293–296. arXiv:1005.1558. Bibcode:2009BaltA..18..293M.

[75] Hubble and Keck team up to find farthest known galaxy in the Universe|Press Releases|ESA/Hubble. Spacetelescope.org (2004-02-15). Retrieved on 2011-05-01.

[76] MSNBC: "Galaxy ranks as most distant object in cosmos"

9.10   Further reading

- Vicent J. Martínez; Jean-Luc Starck; Enn Saar; David L. Donoho et al. (2005). "Morphology Of The Galaxy Distribution From Wavelet Denoising". *The Astrophysical Journal* 634 (2): 744–755. arXiv:astro-ph/0508326. Bibcode:2005ApJ...634..744M. doi:10.1086/497125.

- Mureika, J. R. & Dyer, C. C. (2004). "Review: Multifractal Analysis of Packed Swiss Cheese Cosmologies". *General Relativity and Gravitation* 36 (1): 151–184. arXiv:gr-qc/0505083. Bibcode:2004GReGr..36..151M. doi:10.1023/B:GERG.0000006699.45969.49.

- Gott, III, J. R. et al. (May 2005). "A Map of the Universe". *The Astrophysical Journal* 624 (2): 463–484. arXiv:astro-ph/0310571. Bibcode:2005ApJ...624..463G. doi:10.1086/428890.

- F. Sylos Labini; M. Montuori & L. Pietronero (1998). "Scale-invariance of galaxy clustering". *Physics Reports* 293 (1): 61–226. arXiv:astro-ph/9711073. Bibcode:1998PhR...293...61S. doi:10.1016/S0370-1573(97)00044-6.

9.11   External links

- Calculating the total mass of ordinary matter in the universe, what you always wanted to know

- "Millennium Simulation" of structure forming Max Planck Institute of Astrophysics, Garching, Germany

- Visualisations of large-scale structure: animated spins of groups, clusters, filaments and voids, identified in SDSS data by MSPM (Sydney Institute for Astronomy)

- The Sloan Great Wall: Largest Known Structure? on APOD

- Cosmology FAQ

- Forming Galaxies Captured In The Young Universe By Hubble, VLT & Spitzer

- NASA featured Images and Galleries

- Star Survey reaches 70 sextillion

- Animation of the cosmic light horizon

- Inflation and the Cosmic Microwave Background by Charles Lineweaver

- Logarithmic Maps of the Universe

- List of publications of the 2dF Galaxy Redshift Survey

- List of publications of the 6dF Galaxy Redshift and peculiar velocity survey

- The Universe Within 14 Billion Light Years—NASA Atlas of the Universe (note—this map only gives a rough cosmographical estimate of the expected distribution of superclusters within the observable universe; very little actual mapping has been done beyond a distance of one billion light years):

- Video: "The Known Universe", from the American Museum of Natural History

- NASA/IPAC Extragalactic Database

- Cosmography of the Local Universe at irfu.cea.fr (17:35) (arXiv)

- What is the size of the universe? — Astronoo

Chapter 10

Hubble's law

Hubble's law is the name for the observation in physical cosmology that:

1. Objects observed in deep space (extragalactic space, ~10 megaparsecs or more) are found to have a Doppler shift interpretable as relative velocity away from the Earth;

2. This Doppler-shift-measured velocity, of various galaxies receding from the Earth, is approximately proportional to their distance from the Earth for galaxies up to a few hundred megaparsecs away.[1][2]

Hubble's law is considered the first observational basis for the expansion of the universe and today serves as one of the pieces of evidence most often cited in support of the Big Bang model.[3] The motion of astronomical objects due solely to this expansion is known as the **Hubble flow.**[4]

Although widely attributed to Edwin Hubble, the law was first derived from the general relativity equations by Georges Lemaître in a 1927 article where he proposed the expansion of the universe and suggested an estimated value of the rate of expansion, now called the **Hubble constant.**[5][6][7][8][9][10] Two years later Edwin Hubble confirmed the existence of that law and determined a more accurate value for the constant that now bears his name.[11] Hubble inferred the recession velocity of the objects from their redshifts, many of which were earlier measured and related to velocity by Vesto Slipher in 1917.[12]

The law is often expressed by the equation $v = H_0 D$, with H_0 the constant of proportionality (Hubble constant) between the "proper distance" D to a galaxy (which can change over time, unlike the comoving distance) and its velocity v (i.e. the derivative of proper distance with respect to cosmological time coordinate; see *Uses of the proper distance* for some discussion of the subtleties of this definition of 'velocity'). The SI unit of H_0 is s^{-1} but it is most frequently quoted in (km/s)/Mpc, thus giving the speed in km/s of a galaxy 1 megaparsec (3.09×10^{19} km) away. The reciprocal of H_0 is the Hubble time.

10.1   Observed values

10.2   Discovery

A decade before Hubble made his observations, a number of physicists and mathematicians had established a consistent theory of the relationship between space and time by using Einstein's field equations of general relativity. Applying the most general principles to the nature of the universe yielded a dynamic solution that conflicted with the then-prevailing notion of a static universe.

10.2.1   FLRW equations

In 1922, Alexander Friedmann derived his Friedmann equations from Einstein's field equations, showing that the Universe might expand at a rate calculable by the equations.[27] The parameter used by Friedmann is known today as the scale factor which can be considered as a scale invariant form of the proportionality constant of Hubble's law. Georges Lemaître independently found a similar solution in 1927. The Friedmann equations are derived by inserting the metric for a homogeneous and isotropic universe into Einstein's field equations for a fluid with a given density and pressure. This idea of an expanding spacetime would eventually lead to the Big Bang and Steady State theories of cosmology.

10.2.2   Lemaitre's Equation

In 1927, two years before Hubble published his own article, the Belgian priest and astronomer Georges Lemaître was the first to publish research deriving what is now known as Hubble's Law. Unfortunately, for reasons unknown, "all discussions of radial velocities and distances (and the very first empirical determination of "H") were omitted".[28] It is speculated that these omissions were deliberate. According to the Canadian astronomer Sidney van den Bergh, "The 1927 discovery of the expansion of the Universe by Lemaitre was published in French in a low-impact journal. In the 1931 high-impact English translation of this article a critical equation was changed by omitting reference to what is now known as the Hubble constant. That the section of the text of this paper dealing with the expansion of the Universe was also deleted from that English translation suggests a deliberate omission by the unknown translator."[29]

10.2.3   Shape of the universe

Before the advent of modern cosmology, there was considerable talk about the size and shape of the universe. In 1920, the famous Shapley-Curtis debate took place between Harlow Shapley and Heber D. Curtis over this issue. Shapley argued for a small universe the size of the Milky Way galaxy and Curtis argued that the Universe was much larger. The issue was resolved in the coming decade with Hubble's improved observations.

10.2.4   Cepheid variable stars outside of the Milky Way

Edwin Hubble did most of his professional astronomical observing work at Mount Wilson Observatory, the world's most powerful telescope at the time. His observations of Cepheid variable stars in spiral nebulae enabled him to calculate the distances to these objects. Surprisingly, these objects were discovered to be at distances which placed them well outside the Milky Way. They continued to be called "nebulae" and it was only gradually that the term "galaxies" took over.

10.2.5   Combining redshifts with distance measurements

The parameters that appear in Hubble's law: velocities and distances, are not directly measured. In reality we determine, say, a supernova brightness, which provides information about its distance, and the redshift $z = \Delta\lambda/\lambda$ of its spectrum of radiation. Hubble correlated brightness and parameter z.

Combining his measurements of galaxy distances with Vesto Slipher and Milton Humason's measurements of the redshifts associated with the galaxies, Hubble discovered a rough proportionality between redshift of an object and its distance. Though there was considerable scatter (now known to be caused by peculiar velocities – the 'Hubble flow' is used to refer to the region of space far enough out that the recession velocity is larger than local peculiar velocities), Hubble was able to plot a trend line from the 46 galaxies he studied and obtain a value for the Hubble constant of 500 km/s/Mpc (much higher than the currently accepted value due to errors in his distance calibrations). (See cosmic distance ladder for details.)

At the time of discovery and development of Hubble's law, it was acceptable to explain redshift phenomenon as a Doppler shift in the context of special relativity, and use the Doppler formula to associate redshift z with velocity. Today, the velocity-distance relationship of Hubble's law is viewed as a theoretical result with velocity to be connected with observed redshift not by the Doppler effect, but by a cosmological model relating recessional velocity to the expansion of the

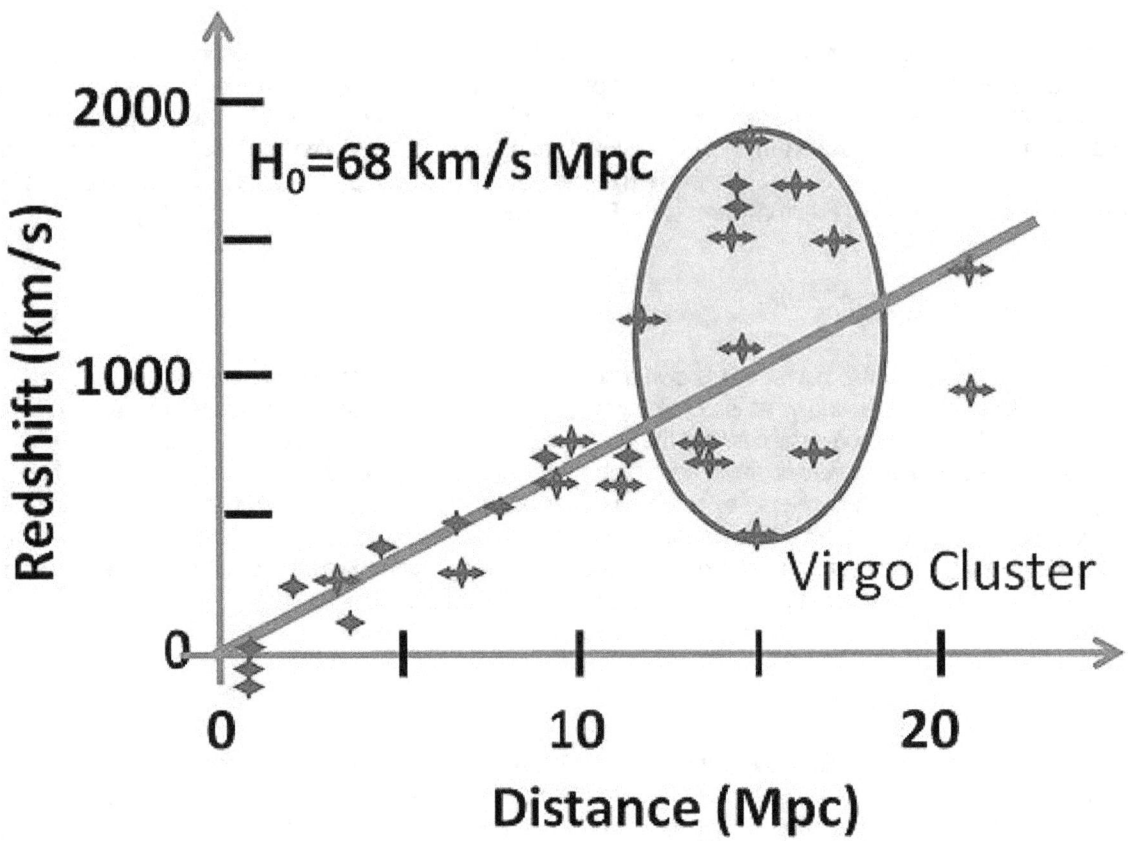

Fit of redshift velocities to Hubble's law.[30] Various estimates for the Hubble constant exist. The HST Key H_0 Group fitted type Ia supernovae for redshifts between 0.01 and 0.1 to find that $H_0 = 71 \pm 2$ (statistical) ± 6 (systematic) $km\,s^{-1}Mpc^{-1}$,[24] while Sandage et al. find $H_0 = 62.3 \pm 1.3$ (statistical) ± 5 (systematic) $km\,s^{-1}Mpc^{-1}$.[31]

Universe. Even for small z the velocity entering the Hubble law is no longer interpreted as a Doppler effect, although at small z the velocity-redshift relation for both interpretations is the same.

Hubble Diagram

Hubble's law can be easily depicted in a "Hubble Diagram" in which the velocity (assumed approximately proportional to the redshift) of an object is plotted with respect to its distance from the observer.[32] A straight line of positive slope on this diagram is the visual depiction of Hubble's law.

10.2.6   Cosmological constant abandoned

Main article: Cosmological constant

After Hubble's discovery was published, Albert Einstein abandoned his work on the cosmological constant, which he had designed to modify his equations of general relativity, to allow them to produce a static solution which, in their simplest form, model either an expanding or contracting universe.[33] After Hubble's discovery that the Universe was, in fact, expanding, Einstein called his faulty assumption that the Universe is static his "biggest mistake".[33] On its own, general relativity could predict the expansion of the Universe, which (through observations such as the bending of light by large masses, or the precession of the orbit of Mercury) could be experimentally observed and compared to his theoretical calculations using particular solutions of the equations he had originally formulated.

In 1931, Einstein made a trip to Mount Wilson to thank Hubble for providing the observational basis for modern cosmology.[34]

The cosmological constant has regained attention in recent decades as a hypothesis for dark energy.[35]

10.3   Interpretation

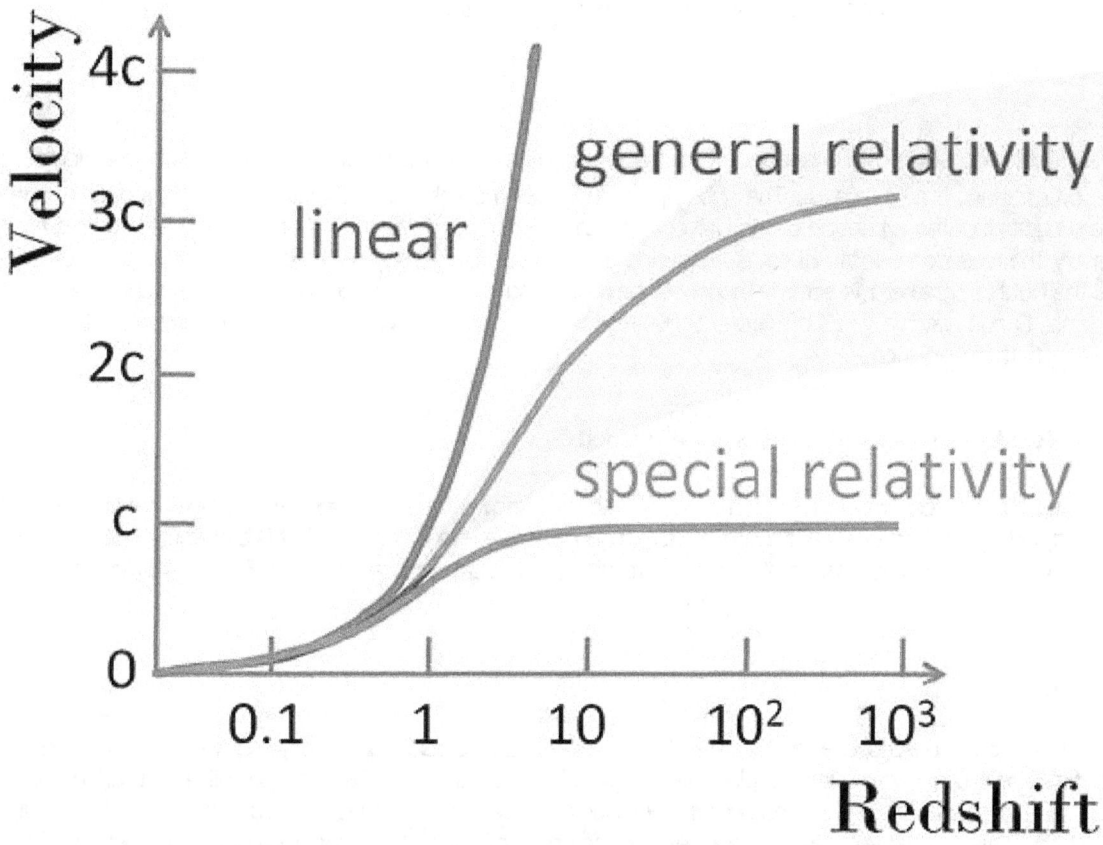

A variety of possible recessional velocity vs. redshift functions including the simple linear relation v = cz; a variety of possible shapes from theories related to general relativity; and a curve that does not permit speeds faster than light in accordance with special relativity. All curves are linear at low redshifts. See Davis and Lineweaver.[36]

The discovery of the linear relationship between redshift and distance, coupled with a supposed linear relation between recessional velocity and redshift, yields a straightforward mathematical expression for Hubble's Law as follows:

$v = H_0 D$

where

- v is the recessional velocity, typically expressed in km/s.

- H_0 is Hubble's constant and corresponds to the value of H (often termed the **Hubble parameter** which is a value that is time dependent and which can be expressed in terms of the scale factor) in the Friedmann equations taken at the time of observation denoted by the subscript 0. This value is the same throughout the Universe for a given comoving time.

- D is the proper distance (which can change over time, unlike the comoving distance, which is constant) from the galaxy to the observer, measured in mega parsecs (Mpc), in the 3-space defined by given cosmological time. (Recession velocity is just $v = dD/dt$).

Hubble's law is considered a fundamental relation between recessional velocity and distance. However, the relation between recessional velocity and redshift depends on the cosmological model adopted, and is not established except for small redshifts.

For distances D larger than the radius of the Hubble sphere r_{HS}, objects recede at a rate faster than the speed of light (*See* Uses of the proper distance for a discussion of the significance of this):

$$r_{HS} = \frac{c}{H_0} .$$

Since the Hubble "constant" is a constant only in space, not in time, the radius of the Hubble sphere may increase or decrease over various time intervals. The subscript '0' indicates the value of the Hubble constant today.[30] Current evidence suggests that the expansion of the Universe is accelerating (*see* Accelerating universe), meaning that, for any given galaxy, the recession velocity dD/dt is increasing over time as the galaxy moves to greater and greater distances; however, the Hubble parameter is actually thought to be decreasing with time, meaning that if we were to look at some *fixed* distance D and watch a series of different galaxies pass that distance, later galaxies would pass that distance at a smaller velocity than earlier ones.[37]

10.3.1   Redshift velocity and recessional velocity

Redshift can be measured by determining the wavelength of a known transition, such as hydrogen α-lines for distant quasars, and finding the fractional shift compared to a stationary reference. Thus redshift is a quantity unambiguous for experimental observation. The relation of redshift to recessional velocity is another matter. For an extensive discussion, see Harrison.[38]

Redshift velocity

The redshift z is often described as a *redshift velocity*, which is the recessional velocity that would produce the same redshift *if* it were caused by a linear Doppler effect (which, however, is not the case, as the shift is caused in part by a cosmological expansion of space, and because the velocities involved are too large to use a non-relativistic formula for Doppler shift). This redshift velocity can easily exceed the speed of light.[39] In other words, to determine the redshift velocity v_r, the relation:

$$v_{rs} \equiv cz,$$

is used.[40][41] That is, there is *no fundamental difference* between redshift velocity and redshift: they are rigidly proportional, and not related by any theoretical reasoning. The motivation behind the "redshift velocity" terminology is that the redshift velocity agrees with the velocity from a low-velocity simplification of the so-called Fizeau-Doppler formula[42]

$$z = \frac{\lambda_o}{\lambda_e} - 1 = \sqrt{\frac{1 + v/c}{1 - v/c}} - 1 \approx \frac{v}{c} .$$

Here, λ_o, λ_e are the observed and emitted wavelengths respectively. The "redshift velocity" v_r is not so simply related to real velocity at larger velocities, however, and this terminology leads to confusion if interpreted as a real velocity. Next, the connection between redshift or redshift velocity and recessional velocity is discussed. This discussion is based on Sartori.[43]

Recessional velocity

Suppose $R(t)$ is called the *scale factor* of the Universe, and increases as the Universe expands in a manner that depends upon the cosmological model selected. Its meaning is that all measured distances $D(t)$ between co-moving points increase proportionally to R. (The co-moving points are not moving relative to each other except as a result of the expansion of space.) In other words:

$$\frac{D(t)}{D(t_0)} = \frac{R(t)}{R(t_0)},$$

where t_0 is some reference time. If light is emitted from a galaxy at time t_e and received by us at t_0, it is red shifted due to the expansion of space, and this redshift z is simply:

$$z = \frac{R(t_0)}{1}$$

Suppose a galaxy is at distance D, and this distance changes with time at a rate $d_t D$. We call this rate of recession the "recession velocity" v_r:

$$v_r = d_t D = \frac{d_t R}{R} D.$$

We now define the Hubble constant as

$$H \equiv \frac{d_t R}{R},$$

and discover the Hubble law:

$$v_r = HD.$$

From this perspective, Hubble's law is a fundamental relation between (i) the recessional velocity contributed by the expansion of space and (ii) the distance to an object; the connection between redshift and distance is a crutch used to connect Hubble's law with observations. This law can be related to redshift z approximately by making a Taylor series expansion:

$$z = \frac{R(t_0)}{1} \qquad \frac{R(t_0)}{1 \approx (t_0 - t_e)H(t_0)}$$

If the distance is not too large, all other complications of the model become small corrections and the time interval is simply the distance divided by the speed of light:

$$z \approx (t_0 - t_e)H(t_0) \approx \frac{D}{}$$

According to this approach, the relation $cz = v_r$ is an approximation valid at low redshifts, to be replaced by a relation at large redshifts that is model-dependent. See velocity-redshift figure.

10.3.2 Observability of parameters

Strictly speaking, neither v nor D in the formula are directly observable, because they are properties *now* of a galaxy, whereas our observations refer to the galaxy in the past, at the time that the light we currently see left it.

For relatively nearby galaxies (redshift z much less than unity), v and D will not have changed much, and v can be estimated using the formula $v = zc$ where c is the speed of light. This gives the empirical relation found by Hubble.

For distant galaxies, v (or D) cannot be calculated from z without specifying a detailed model for how H changes with time. The redshift is not even directly related to the recession velocity at the time the light set out, but it does have a simple interpretation: *(1+z)* is the factor by which the Universe has expanded while the photon was travelling towards the observer.

10.3.3 Expansion velocity vs relative velocity

In using Hubble's law to determine distances, only the velocity due to the expansion of the Universe can be used. Since gravitationally interacting galaxies move relative to each other independent of the expansion of the Universe, these relative velocities, called peculiar velocities, need to be accounted for in the application of Hubble's law.

The Finger of God effect is one result of this phenomenon. In systems that are gravitationally bound, such as galaxies or our planetary system, the expansion of space is a much weaker effect than the attractive force of gravity.

10.3.4 Idealized Hubble's Law

The mathematical derivation of an idealized Hubble's Law for a uniformly expanding universe is a fairly elementary theorem of geometry in 3-dimensional Cartesian/Newtonian coordinate space, which, considered as a metric space, is entirely homogeneous and isotropic (properties do not vary with location or direction). Simply stated the theorem is this:

> *Any two points which are moving away from the origin, each along straight lines and with speed proportional to distance from the origin, will be moving away from each other with a speed proportional to their distance apart.*

In fact this applies to non-Cartesian spaces as long as they are locally homogeneous and isotropic; specifically to the negatively and positively curved spaces frequently considered as cosmological models (see shape of the universe).

An observation stemming from this theorem is that seeing objects recede from us on Earth is not an indication that Earth is near to a center from which the expansion is occurring, but rather that *every* observer in an expanding universe will see objects receding from them.

10.3.5 Ultimate fate and age of the universe

The value of the Hubble parameter changes over time, either increasing or decreasing depending on the value of the so-called deceleration parameter q, which is defined by

$$q = -\left(1 + \frac{\dot{H}}{H^2}\right).$$

In a universe with a deceleration parameter equal to zero, it follows that $H = 1/t$, where t is the time since the Big Bang. A non-zero, time-dependent value of q simply requires integration of the Friedmann equations backwards from the present time to the time when the comoving horizon size was zero.

It was long thought that q was positive, indicating that the expansion is slowing down due to gravitational attraction. This would imply an age of the Universe less than $1/H$ (which is about 14 billion years). For instance, a value for q of 1/2

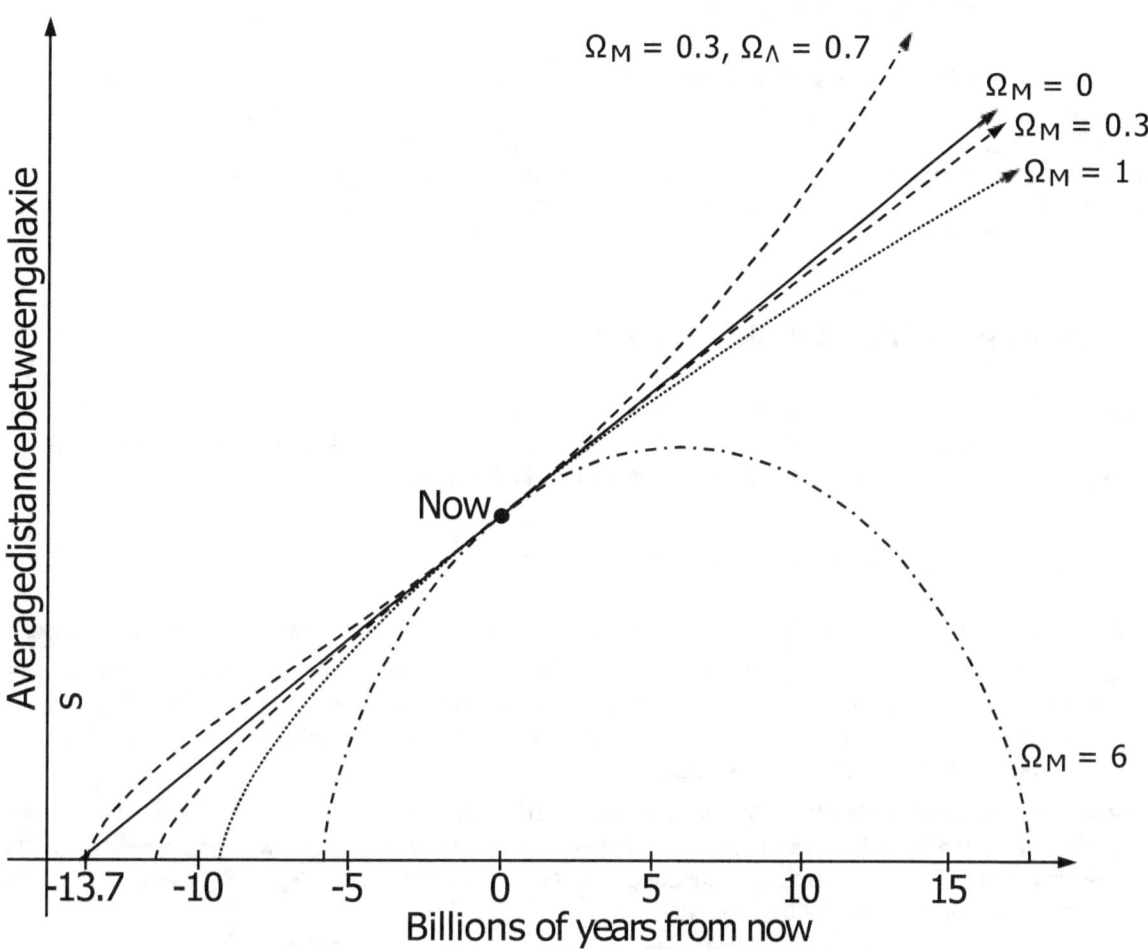

The age and ultimate fate of the universe can be determined by measuring the Hubble constant today and extrapolating with the observed value of the deceleration parameter, uniquely characterized by values of density parameters (ΩM for matter and $\Omega \Lambda$ for dark energy). A "closed universe" with $\Omega M > 1$ and $\Omega \Lambda = 0$ comes to an end in a Big Crunch and is considerably younger than its Hubble age. An "open universe" with $\Omega M \leq 1$ and $\Omega \Lambda = 0$ expands forever and has an age that is closer to its Hubble age. For the accelerating universe with nonzero $\Omega \Lambda$ that we inhabit, the age of the universe is coincidentally very close to the Hubble age.

(once favoured by most theorists) would give the age of the Universe as $2/(3H)$. The discovery in 1998 that q is apparently negative means that the Universe could actually be older than $1/H$. However, estimates of the age of the universe are very close to $1/H$.

10.3.6   Olbers' paradox

Main article: Olbers' paradox

The expansion of space summarized by the Big Bang interpretation of Hubble's Law is relevant to the old conundrum known as Olbers' paradox: if the Universe were infinite, static, and filled with a uniform distribution of stars, then every line of sight in the sky would end on a star, and the sky would be as bright as the surface of a star. However, the night sky is largely dark. Since the 17th century, astronomers and other thinkers have proposed many possible ways to resolve this paradox, but the currently accepted resolution depends in part on the Big Bang theory and in part on the Hubble expansion. In a universe that exists for a finite amount of time, only the light of a finite number of stars has had a chance to reach us yet, and the paradox is resolved. Additionally, in an expanding universe, distant objects recede from us, which causes the light emanating from them to be redshifted and diminished in brightness.[44]

10.3.7 Dimensionless Hubble parameter

Instead of working with Hubble's constant, a common practice is to introduce the **dimensionless Hubble parameter**, usually denoted by h, and to write the Hubble's parameter H_0 as $h \times 100$ km s^{-1} Mpc^{-1}, all the uncertainty relative of the value of H_0 being then relegated on h.[45] If a subscript is presented after h, it refers to the value of h used in that text's preceding calculation, and is equal to $H_0 / 100$. Currently $h = 0.678$, which can be represented as $h_{0.678}$. This should not be confused with the dimensionless value of Hubble's constant, usually expressed in terms of Planck units, with current value of $H_0 \times t_P = 1.18 \times 10^{-61}$.

10.4 Determining the Hubble constant

The value of the Hubble constant is estimated by measuring the redshift of distant galaxies and then determining the distances to the same galaxies (by some other method than Hubble's law). Uncertainties in the physical assumptions used to determine these distances have caused varying estimates of the Hubble constant.

10.4.1 Earlier measurement and discussion approaches

For most of the second half of the 20th century the value of H_0 was estimated to be between 50 and 90 (km/s)/Mpc.

The value of the Hubble constant was the topic of a long and rather bitter controversy between Gérard de Vaucouleurs, who claimed the value was around 100, and Allan Sandage, who claimed the value was near 50.[25] In 1996, a debate moderated by John Bahcall between Gustav Tammann and Sidney van den Bergh was held in similar fashion to the earlier Shapley-Curtis debate over these two competing values.

This previously wide variance in estimates was partially resolved with the introduction of the ΛCDM model of the Universe in the late 1990s. With the ΛCDM model observations of high-redshift clusters at X-ray and microwave wavelengths using the Sunyaev-Zel'dovich effect, measurements of anisotropies in the cosmic microwave background radiation, and optical surveys all gave a value of around 70 for the constant.

More recent measurements from the Planck mission indicate a lower value of around 67.[13]

See table of measurements above for many recent and older measurements.

10.4.2 Acceleration of the expansion

Main article: Accelerating universe

A value for q measured from standard candle observations of Type Ia supernovae, which was determined in 1998 to be negative, surprised many astronomers with the implication that the expansion of the Universe is currently "accelerating"[46] (although the Hubble factor is still decreasing with time, as mentioned above in the Interpretation section; see the articles on dark energy and the ΛCDM model).

10.5 Derivation of the Hubble parameter

Start with the Friedmann equation:

$$H^2 \equiv \left(\frac{\dot{a}}{a}\right)^2 = \frac{8\pi G}{3} - \frac{kc^2}{a^2}$$

where H is the Hubble parameter, a is the scale factor, G is the gravitational constant, k is the normalised spatial curvature of the Universe and equal to -1, 0, or $+1$, and Λ is the cosmological constant.

10.5.1   Matter-dominated universe (with a cosmological constant)

If the Universe is matter-dominated, then the mass density of the Universe ρ can just be taken to include matter so

$$\rho = \rho_m(a) = \frac{\rho_{mo}}{a^3},$$

where ρ_{mo} is the density of matter today. We know for nonrelativistic particles that their mass density decreases proportional to the inverse volume of the Universe, so the equation above must be true. We can also define (see density parameter for Ω_m)

$$\rho_c = \frac{3H^2}{8\pi G};$$

$$\Omega_m \equiv \frac{\rho_{mo}}{\rho_c} = \frac{8\pi G}{3H_0^2}\rho_{mo};$$

so $\rho = \rho_c \Omega_m / a^3$. Also, by definition,

$$\overline{(a_0 / \rho_0)^2}$$

and

$$\Omega_\Lambda \equiv \frac{\Lambda c^2}{3H_0^2},$$

where the subscript nought refers to the values today, and $a_0 = 1$. Substituting all of this into the Friedmann equation at the start of this section and replacing a with $a = 1/(1 + z)$ gives

$$H^2(z) = H_0^2 \left(\Omega_M(1 + z)^3 + \Omega_k(1 + z)^2 + \Omega_\Lambda \right).$$

10.5.2   Matter- and dark energy-dominated universe

If the Universe is both matter-dominated and dark energy- dominated, then the above equation for the Hubble parameter will also be a function of the equation of state of dark energy. So now:

$$\rho = \rho_m(a) + \rho_{de}(a),$$

where ρ_{de} is the mass density of the dark energy. By definition, an equation of state in cosmology is $P = w\rho c^2$, and if we substitute this into the fluid equation, which describes how the mass density of the Universe evolves with time,

$$\rho + 3\frac{\dot{a}}{a}\left(\rho + \frac{P}{c^2} \right) = 0;$$

$$\frac{d\rho}{\rho} \qquad \frac{da}{a}(1 + w).$$

If w is constant,

$\ln \rho = -3(1 + w) \ln a;$

$\rho = a^{-3(1+w)}.$

Therefore, for dark energy with a constant equation of state w, $\rho_{de}(a) = \rho_{de0}a^{-3(1+w)}$. If we substitute this into the Friedman equation in a similar way as before, but this time set $k = 0$, which is assuming we live in a spatially flat universe, (see Shape of the Universe)

$$H^2(z) = H_0^2 \left(\Omega_M(1 + z)^3 + \Omega_{de}(1 + z)^{3(1+w)} \right).$$

If dark energy does not have a constant equation-of-state w, then

$$\rho_{de}(a) = \rho_{de0}e^{-3\int \frac{da}{a}(1 + w(a))},$$

and to solve this we must parametrize $w(a)$, for example if $w(a) = w_0 + w_a(1 - a)$, giving

$$H^2(z) = H_0^2 \left(\Omega_M a^{-3} + \Omega_{de}a^{-3(1+w_0+w_a)}e^{-3w_a(1-a)} \right).$$

Other ingredients have been formulated recently.[47][48][49] In certain era, where the high energy experiments seem to have a reliable access in analyzing the property of the matter dominating the background geometry, with this era we mean the quark-gluon plasma, the transport properties have been taken into consideration. Therefore, the evolution of the Hubble parameter and of other essential cosmological parameters, in such a background are found to be considerably (non-negligibly) different than their evolution in an ideal, gaseous, non-viscous background.

10.6 Units derived from the Hubble constant

10.6.1 Hubble time

The Hubble constant H_0 has units of inverse time, i.e. $H_0^{-1} = \frac{1}{67.8 \text{km/(sMpc)}} = 4.55 \cdot 10^{17} \text{s} = 14.4$ billion years. This is somewhat longer than the age of the universe of 13.8 billion years. The Hubble time is the age it would have had if the expansion had been linear, and it is different from the real age of the universe because the expansion isn't linear.

We currently appear to be approaching a period where the expansion is exponential due to the increasing dominance of vacuum energy. In this regime, the Hubble parameter is constant, and the universe grows by a factor e each Hubble time:

$$H \equiv \frac{\dot{a}}{a}$$

Over long periods of time, the dynamics are complicated by general relativity, dark energy, inflation, etc., as explained above.

10.6.2 Hubble length

The Hubble length or Hubble distance is a unit of distance in cosmology, defined as cH_0^{-1} — the speed of light multiplied by the Hubble time. It is equivalent to 4,228 million parsecs or 13.8 billion light years. (The numerical value of the Hubble length in light years is, by definition, equal to that of the Hubble time in years.) The Hubble distance would be the distance between the Earth and the galaxies which are *currently* receding from us at the speed of light, as can be seen by substituting $D = c/H_0$ into the equation for Hubble's law, $v = H_0 D$.

10.6.3   Hubble volume

Main article: Hubble volume

The Hubble volume is sometimes defined as a volume of the Universe with a comoving size of c/H_0. The exact definition varies: it is sometimes defined as the volume of a sphere with radius c/H_0, or alternatively, a cube of side c/H_0. Some cosmologists even use the term Hubble volume to refer to the volume of the observable universe, although this has a radius approximately three times larger.

10.7   See also

- Cosmology

- Dark energy

- Dark matter

- Tests of general relativity

10.8   Notes

[1] Riess, A. et al. (September 1998). "Observational Evidence from Supernovae for an Accelerating Universe and a Cosmological Constant". *The Astronomical Journal* 116 (3): 1009–1038. arXiv:astro-ph/9805201. Bibcode:1998AJ....116.1009R. doi:10.1086/300499.

[2] Perlmutter, S. et al. (June 1999). "Measurements of Omega and Lambda from 42 High-Redshift Supernovae". *The Astrophysical Journal* 517 (2): 565–586. arXiv:astro-ph/9812133. Bibcode:1999ApJ...517..565P. doi:10.1086/307221.

[3] Coles, P., ed. (2001). *Routledge Critical Dictionary of the New Cosmology*. Routledge. p. 202. ISBN 0-203-16457-1.

[4] "Hubble Flow". *The Swinburne Astronomy Online Encyclopedia of Astronomy*. Swinburne University of Technology. Retrieved 2013-05-14.

[5] Lemaître, G. (1927). "Un univers homogène de masse constante et de rayon croissant rendant compte de la vitesse radiale des nébuleuses extra-galactiques". *Annales de la Société Scientifique de Bruxelles A* 47: 49–56. Bibcode:1927ASSB...47...49L. Partially translated in Lemaître, G. (1931). "Expansion of the universe, A homogeneous universe of constant mass and increasing radius accounting for the radial velocity of extra-galactic nebulae". *Monthly Notices of the Royal Astronomical Society* 91: 483–490. Bibcode:1931MNRAS..91..483L. doi:10.1093/mnras/91.5.483.

[6] van den Bergh, S. (2011). "The Curious Case of Lemaitre's Equation No. 24". *Journal of the Royal Astronomical Society of Canada* 105 (4): 151. arXiv:1106.1195. Bibcode:2011JRASC.105..151V.

[7] Block, D. L. (2012). "Georges Lemaitre and Stiglers Law of Eponymy". In Holder, R. D.; Mitton, S. *Georges Lemaître: Life, Science and Legacy*. Astrophysics and Space Science Library 395. pp. 89–96. arXiv:1106.3928. Bibcode:2012ASSL..395...89B. doi:10.1007/978-3-642-32254-9_8. ISBN 978-3-642-32253-2.

[8] Reich, E. S. (27 June 2011). "Edwin Hubble in translation trouble". *Nature News*. doi:10.1038/news.2011.385.

[9] Livio, M. (2011). "Lost in translation: Mystery of the missing text solved". *Nature* 479 (7372): 171. Bibcode:2011Natur.479..171L. doi:10.1038/479171a.

[10] Livio, M.; Riess, A. (2013). "Measuring the Hubble constant". *Physics Today* 66 (10): 41. Bibcode:2013PhT....66j..41L. doi:10.1063/PT.3.2148.

[11] Hubble, E. (1929). "A relation between distance and radial velocity among extra-galactic nebulae". *Proceedings of the National Academy of Sciences* 15 (3): 168–73. Bibcode:1929PNAS...15..168H. doi:10.1073/pnas.15.3.168. PMC 522427. PMID 16557160.

[12] Longair, M. S. (2006). *The Cosmic Century.* Cambridge University Press. p. 109. ISBN 0-521-47436-1.

[13] Bucher, P. A. R.; *et al.* (Planck Collaboration) (2013). "Planck 2013 results. I. Overview of products and scientific Results". arXiv:1303.5062 [astro-ph.CO].

[14] "Planck reveals an almost perfect universe". ESA. 21 March 2013. Retrieved 2013-03-21.

[15] "Planck Mission Brings Universe Into Sharp Focus". JPL. 21 March 2013. Retrieved 2013-03-21.

[16] Overbye, D. (21 March 2013). "An infant universe, born before we knew". *New York Times.* Retrieved 2013-03-21.

[17] Boyle, A. (21 March 2013). "Planck probe's cosmic 'baby picture' revises universe's vital statistics". *NBC News.* Retrieved 2013-03-21.

[18] Bennett, C. L. et al. (2013). "Nine-year Wilkinson Microwave Anisotropy Probe (WMAP) observations: Final maps and results". *The Astrophysical Journal Supplement Series* 208 (2): 20. arXiv:1212.5225. Bibcode:2013ApJS..208...20B. doi:10.1088/0067-0049/208/2/20.

[19] Jarosik, N. et al. (2011). "Seven-year Wilkinson Microwave Anisotropy Probe (WMAP) observations: Sky maps, systematic errors, and basic results". *The Astrophysical Journal Supplement Series* 192 (2): 14. arXiv:1001.4744. Bibcode:2011ApJS..192...14J. doi:10.1088/0067-0049/192/2/14.

[20] Results for H_0 and other cosmological parameters obtained by fitting a variety of models to several combinations of WMAP and other data are available at the NASA's LAMBDA website.

[21] Hinshaw, G.; *et al.* (WMAP Collaboration) (2009). "Five-year Wilkinson Microwave Anisotropy Probe observations: Data processing, sky maps, and basic results". *The Astrophysical Journal Supplement* 180 (2): 225–245. arXiv:0803.0732. Bibcode:2009ApJS..180..225H. doi:10.1088/0067-0049/180/2/225.

[22] Spergel, D. N.; *et al.* (WMAP Collaboration) (2007). "Three-year Wilkinson Microwave Anisotropy Probe (WMAP) Observations: Implications for cosmology". *The Astrophysical Journal Supplement Series* 170 (2): 377–408. arXiv:astro-ph/0603449. Bibcode:2007ApJS..170..377S. doi:10.1086/513700.

[23] Bonamente, M.; Joy, M. K.; Laroque, S. J.; Carlstrom, J. E.; Reese, E. D.; Dawson, K. S. (2006). "Determination of the cosmic distance scale from Sunyaev–Zel'dovich effect and Chandra X- ray measurements of high- redshift galaxy clusters". *The Astrophysical Journal* 647: 25. arXiv:astro-ph/0512349. Bibcode:2006ApJ...647...25B. doi:10.1086/505291.

[24] Freedman, W. L. et al. (2001). "Final results from the Hubble Space Telescope Key Project to measure the Hubble constant". *The Astrophysical Journal* 553 (1): 47–72. arXiv:astro-ph/0012376. Bibcode:2001ApJ...553...47F. doi:10.1086/320638.

[25] Overbye, D. (1999). "Prologue". *Lonely Hearts of the Cosmos* (2nd ed.). HarperCollins. p. 1*ff*. ISBN 978-0-316-64896-7.

[26] Sandage, A. R. (1958). "Current problems in the extragalactic distance scale". *The Astrophysical Journal* 127 (3): 513–526. Bibcode:1958ApJ...127..513S. doi:10.1086/146483.

[27] Friedman, A. (1922). "Über die Krümmung des Raumes". *Zeitschrift für Physik* 10 (1): 377–386. Bibcode:1922ZPhy...10..377F. doi:10.1007/BF01332580. Translated in Friedmann, A. (1999). "On the Curvature of Space". *General Relativity and Gravitation* 31 (12): 1991–2000. Bibcode:1999GReGr..31.1991F. doi:10.1023/A:1026751225741.

[28] Block, David. "A Hubble Eclipse: Lemaitre and Censorship". *Cornell University Library.* Retrieved 12 December 2014.

[29] van den Bergh, Sydney. "The Curious Case of Lemaitre's Equation No. 24". *Cornell University Library.* Retrieved 12 December 2014.

[30] Keel, W. C. (2007). *The Road to Galaxy Formation* (2nd ed.). Springer. pp. 7–8. ISBN 3-540-72534-2.

[31] Weinberg, S. (2008). *Cosmology.* Oxford University Press. p. 28. ISBN 0-19-852682-2.

[32] Kirshner, R. P. (2003). "Hubble's diagram and cosmic expansion". *Proceedings of the National Academy of Sciences* 101 (1): 8–13. Bibcode:2003PNAS..101....8K. doi:10.1073/pnas.2536799100.

[33] "What is a Cosmological Constant?". Goddard Space Flight Center. Retrieved 2013-10-17.

[34] Isaacson, W. (2007). *Einstein: His Life and Universe.* Simon & Schuster. p. 354. ISBN 0-7432-6473-8.

[35] "Einstein's Biggest Blunder? Dark Energy May Be Consistent With Cosmological Constant". Science Daily. 28 November 2007. Retrieved 2013-06-02.

[36] Davis, T. M.; Lineweaver, C. H. (2001). "Superluminal Recessional Velocities". *AIP Conference Proceedings* **555**: 348–351. arXiv:astro-ph/0011070. Bibcode:2001AIPC..555..348D. doi:10.1063/1.1363540.

[37] "Is the universe expanding faster than the speed of light?". *Ask an Astronomer at Cornell University.* Archived from the original on 23 November 2003. Retrieved 5 June 2015.

[38] Harrison, E. (1992). "The redshift-distance and velocity-distance laws". *The Astrophysical Journal* **403**: 28–31. Bibcode:1993ApJ...403...28H. doi:10.1086/172179.

[39] Madsen, M. S. (1995). *The Dynamic Cosmos.* CRC Press. p. 35. ISBN 0-412-62300-5.

[40] Dekel, A.; Ostriker, J. P. (1999). *Formation of Structure in the Universe.* Cambridge University Press. p. 164. ISBN 0-521-58632-1.

[41] Padmanabhan, T. (1993). *Structure formation in the universe.* Cambridge University Press. p. 58. ISBN 0-521-42486-0.

[42] Sartori, L. (1996). *Understanding Relativity.* University of California Press. p. 163, Appendix 5B. ISBN 0-520-20029-2.

[43] Sartori, L. (1996). *Understanding Relativity.* University of California Press. pp. 304–305. ISBN 0-520-20029-2.

[44] Chase, S. I.; Baez, J. C. (2004). "Olbers' Paradox". *The Original Usenet Physics FAQ.* Retrieved 2013-10-17. See also Asimov, I. (1974). "The Black of Night". *Asimov on Astronomy.* Doubleday. ISBN 0-385-04111-X.

[45] Peebles, P. J. E. (1993). *Principles of Physical Cosmology.* Princeton University Press.

[46] Perlmutter, S. (2003). "Supernovae, Dark Energy, and the Accelerating Universe" (PDF). *Physics Today* **56** (4): 53–60. Bibcode:2003PhT....56d..53P. doi:10.1063/1.1580050.

[47] Tawfik, A.; Harko, T. (2012). "Quark-hadron phase transitions in the viscous early universe". *Physical Review D* **85** (8): 084032. arXiv:1108.5697. Bibcode:2012PhRvD..85h4032T. doi:10.1103/PhysRevD.85.084032.

[48] Tawfik, A. (2011). "The Hubble parameter in the early universe with viscous QCD matter and finite cosmological constant". *Annalen der Physik* **523** (5): 423. arXiv:1102.2626. Bibcode:2011AnP...523..423T. doi:10.1002/andp.201100038.

[49] Tawfik, A.; Wahba, M.; Mansour, H.; Harko, T. (2011). "Viscous quark-gluon plasma in the early universe". *Annalen der Physik* **523** (3): 194. arXiv:1001.2814. Bibcode:2011AnP...523..194T. doi:10.1002/andp.201000052.

10.9   References

- Hubble, E. P. (1937). *The Observational Approach to Cosmology.* Clarendon Press. LCCN 38011865.

- Kutner, M. (2003). *Astronomy: A Physical Perspective.* Cambridge University Press. ISBN 0-521-52927-1.

- Liddle, A. R. (2003). *An Introduction to Modern Cosmology* (2nd ed.). John Wiley & Sons. ISBN 0-470-84835-9.

10.10   Further reading

- Freedman, W. L.; Madore, B. F. (2010). "The Hubble Constant". *Annual Review of Astronomy and Astrophysics* **48**: 673. arXiv:1004.1856. Bibcode:2010ARA&A..48..673F. doi:10.1146/annurev-astro-082708-101829.

10.11 External links

- NASA's WAMP - Big Bang Expansion: the Hubble Constant

- The Hubble Key Project

- The Hubble Diagram Project

- Merrifield, Michael (2009). "Hubble Constant". *Sixty Symbols*. Brady Haran for the University of Nottingham.

- Hubble's quantum law.

Hubble Constant
calculated using different survey methods

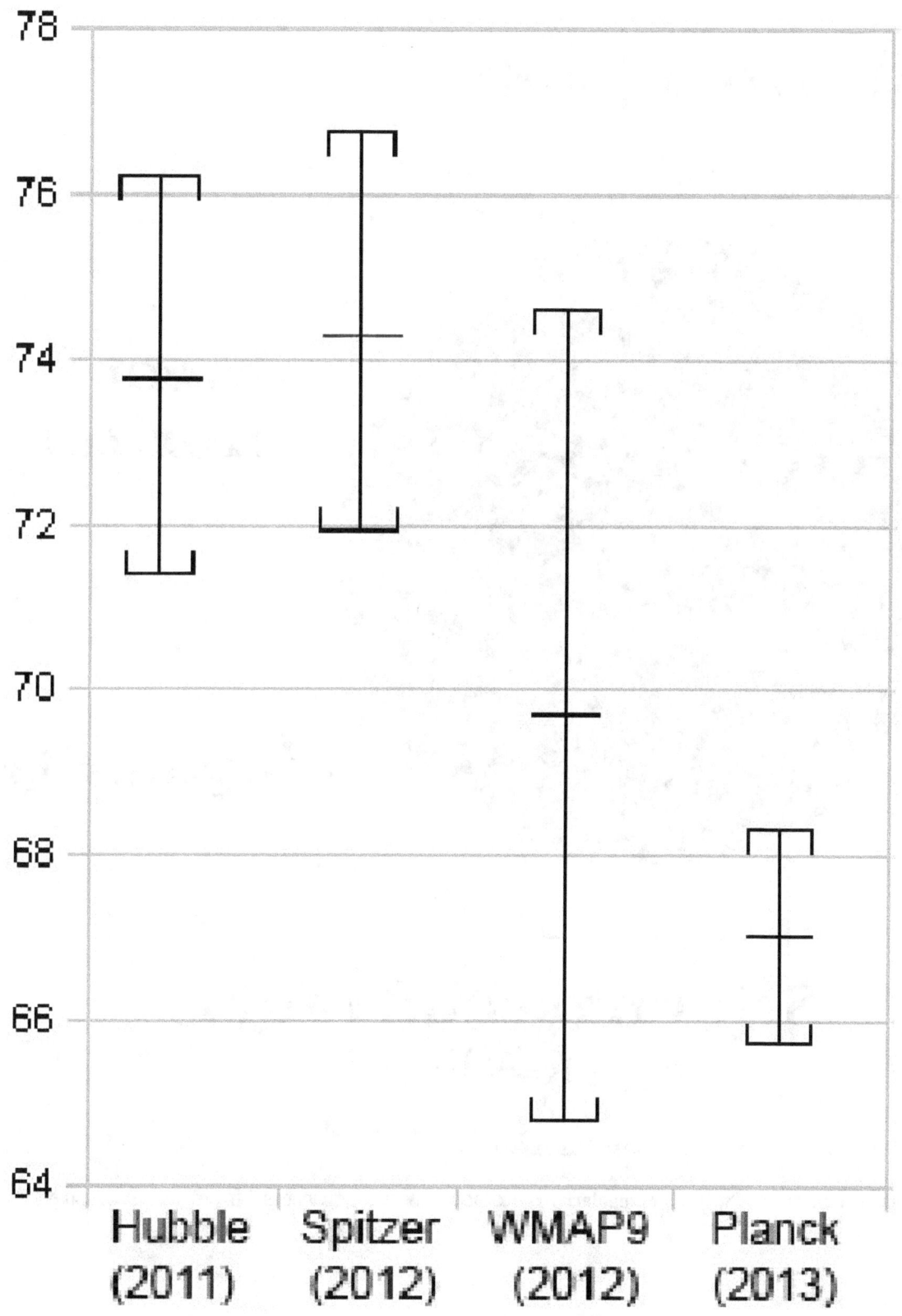

Chapter 11

Gravitational singularity

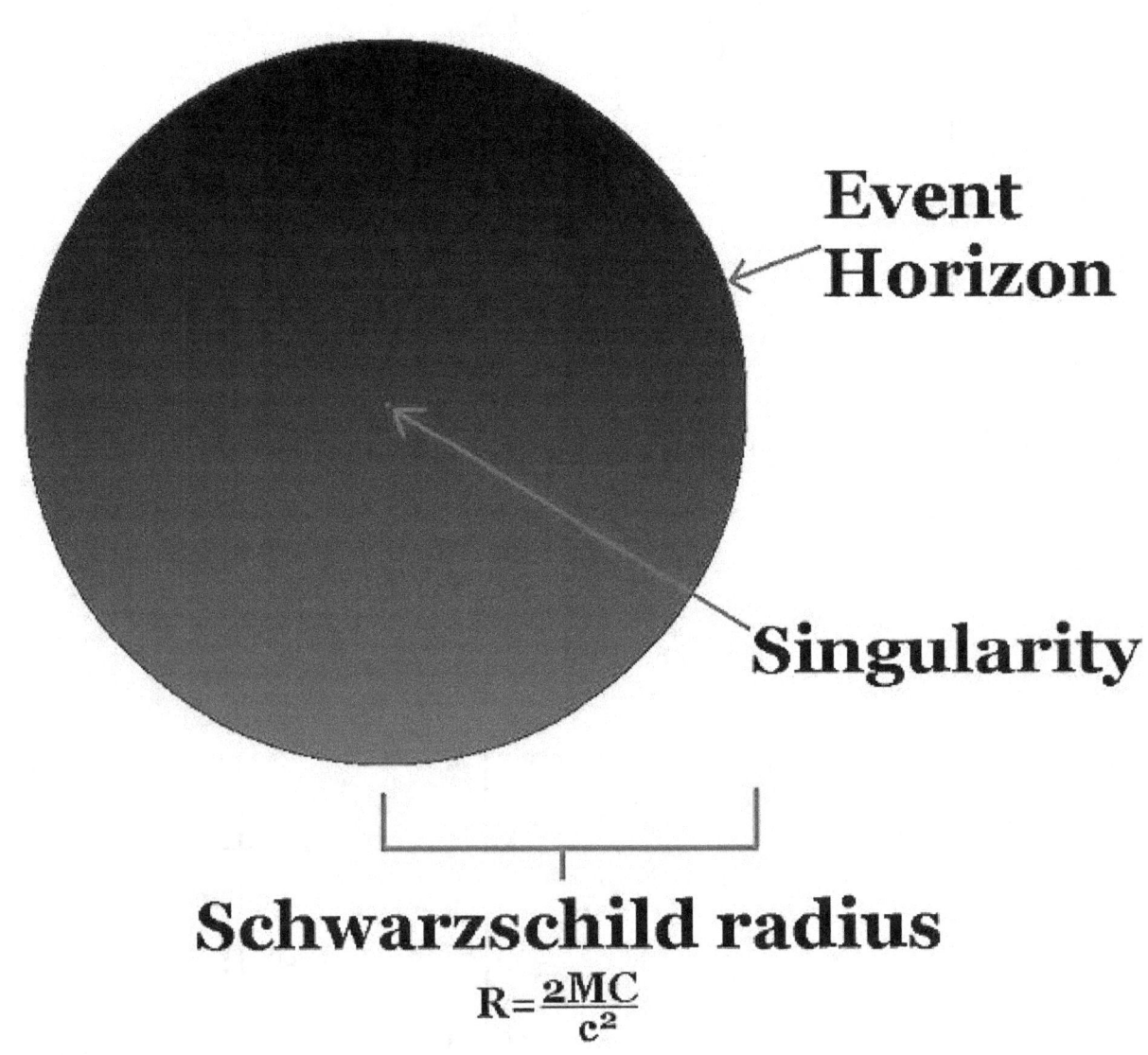

A simple illustration of a non-spinning Black Hole and its singularity

A **gravitational singularity** or **spacetime singularity** is a location where the quantities that are used to measure the gravitational field become infinite in a way that does not depend on the coordinate system. These quantities are the scalar invariant curvatures of spacetime, which includes a measure of the density of matter.

For the purposes of proving the Penrose–Hawking singularity theorems, a spacetime with a singularity is defined to be one that contains geodesics that cannot be extended in a smooth manner.[1] The end of such a geodesic is considered to be the singularity. This is a different definition, useful for proving theorems.

The two most important types of spacetime singularities are ***curvature singularities*** and ***conical singularities***.[2] Singularities can also be divided according to whether or not they are covered by an event horizon (naked singularities are not covered).[3] According to general relativity, the initial state of the universe, at the beginning of the Big Bang, was a singularity.[4] Both general relativity and quantum mechanics break down in describing the Big Bang,[5] but in general, quantum mechanics does not permit particles to inhabit a space smaller than their wavelengths.[6] Another type of singularity predicted by general relativity is inside a black hole: any star collapsing beyond a certain point (the Schwarzschild radius) would form a black hole, inside which a singularity (covered by an event horizon) would be formed, as all the matter would flow into a certain point (or a circular line, if the black hole is rotating).[7] This is again according to general relativity without quantum mechanics, which forbids wavelike particles entering a space smaller than their wavelength. These hypothetical singularities are also known as curvature singularities.

11.1   Interpretation

Many theories in physics have mathematical singularities of one kind or another. Equations for these physical theories predict that the ball of mass of some quantity becomes infinite or increases without limit. This is generally a sign for a missing piece in the theory, as in the ultraviolet catastrophe, renormalization, and instability of a hydrogen atom predicted by the Larmor formula.

In supersymmetry, a singularity in the moduli space happens usually when there are additional massless degrees of freedom in that certain point. Similarly, it is thought that singularities in spacetime often mean that there are additional degrees of freedom that exist only within the vicinity of the singularity. The same fields related to the whole spacetime also exist; for example, the electromagnetic field. In known examples of string theory, the latter degrees of freedom are related to closed strings, while the degrees of freedom are "stuck" to the singularity and related either to open strings or to the twisted sector of an orbifold.

Some theories, such as the theory of loop quantum gravity suggest that singularities may not exist. The idea is that due to quantum gravity effects, there is a minimum distance beyond which the force of gravity no longer continues to increase as the distance between the masses becomes shorter.

The Einstein-Cartan-Sciama-Kibble theory of gravity naturally averts the gravitational singularity at the Big Bang.[8] This theory extends general relativity to matter with intrinsic angular momentum (spin) by removing a constraint of the symmetry of the affine connection and regarding its antisymmetric part, the torsion tensor, as a variable in varying the action. The minimal coupling between torsion and Dirac spinors generates a spin–spin interaction in fermionic matter, which becomes dominant at extremely high densities and prevents the scale factor of the Universe from reaching zero. The Big Bang is replaced by a cusp-like Big Bounce at which the matter has an enormous but finite density and before which the Universe was contracting.

11.2   Types

11.2.1   Curvature

Solutions to the equations of general relativity or another theory of gravity (such as supergravity) often result in encountering points where the metric blows up to infinity. However, many of these points are completely regular, and the infinities are merely a result of using an inappropriate coordinate system at this point. In order to test whether there is a singularity at a certain point, one must check whether at this point diffeomorphism invariant quantities (i.e. scalars) become infinite. Such quantities are the same in every coordinate system, so these infinities will not "go away" by a change of coordinates.

An example is the Schwarzschild solution that describes a non-rotating, uncharged black hole. In coordinate systems convenient for working in regions far away from the black hole, a part of the metric becomes infinite at the event horizon. However, spacetime at the event horizon is regular. The regularity becomes evident when changing to another coordinate system (such as the Kruskal coordinates), where the metric is perfectly smooth. On the other hand, in the center of the black hole, where the metric becomes infinite as well, the solutions suggest a singularity exists. The existence of the singularity can be verified by noting that the Kretschmann scalar, being the square of the Riemann tensor i.e. $R_{\mu\nu\rho\sigma}R^{\mu\nu\rho\sigma}$, which is diffeomorphism invariant, is infinite. While in a non-rotating black hole the singularity occurs at a single point in the model coordinates, called a "point singularity". In a rotating black hole, also known as a Kerr black hole, the singularity occurs on a ring (a circular line), known as a "ring singularity". Such a singularity may also theoretically become a wormhole.[9]

More generally, a spacetime is considered singular if it is geodesically incomplete, meaning that there are freely-falling particles whose motion cannot be determined beyond a finite time, being after the point of reaching the singularity. For example, any observer inside the event horizon of a non-rotating black hole would fall into its center within a finite period of time. The classical version of the Big Bang cosmological model of the universe contains a causal singularity at the start of time (t=0), where all time-like geodesics have no extensions into the past. Extrapolating backward to this hypothetical time 0 results in a universe with all spatial dimensions of size zero, infinite density, infinite temperature, and infinite space-time curvature.

11.2.2 Conical

A conical singularity occurs when there is a point where the limit of every diffeomorphism invariant quantity is finite, in which case spacetime is not smooth at the point of the limit itself. Thus, spacetime looks like a cone around this point, where the singularity is located at the tip of the cone. The metric can be finite everywhere if a suitable coordinate system is used.

An example of such a conical singularity is a cosmic string.

11.2.3 Naked

Main article: Naked singularity

Until the early 1990s, it was widely believed that general relativity hides every singularity behind an event horizon, making naked singularities impossible. This is referred to as the cosmic censorship hypothesis. However, in 1991, physicists Stuart Shapiro and Saul Teukolsky performed computer simulations of a rotating plane of dust that indicated that general relativity might allow for "naked" singularities. What these objects would actually look like in such a model is unknown. Nor is it known whether singularities would still arise if the simplifying assumptions used to make the simulation were removed.

11.3 Entropy

Further information: Black hole, Hawking radiation and Entropy

Before Stephen Hawking came up with the concept of Hawking radiation, the question of black holes having entropy was avoided. However, this concept demonstrates that black holes can radiate energy, which conserves entropy and solves the incompatibility problems with the second law of thermodynamics. Entropy, however, implies heat and therefore temperature. The loss of energy also suggests that black holes do not last forever, but rather "evaporate" slowly. Small black holes tend to be hotter whereas larger ones tend to be colder. All known black hole candidates are so large that their temperature is far below that of the cosmic background radiation, so they are all gaining energy. They will not begin to lose energy until a cosmological redshift of more than one million is reached, rather than the thousand or so since the background radiation formed.

11.4   See also

- 0-dimensional singularity: magnetic monopole
- 1-dimensional singularity: cosmic string
- 2-dimensional singularity: domain wall
- Fuzzball (string theory)
- Penrose-Hawking singularity theorems

11.5   Notes

[1] Moulay, Emmanuel. "The universe and photons" (PDF). FQXi Foundational Questions Institute. Retrieved 26 December 2012.

[2] Uggla, Claes. "Spacetime singularities". *Einstein Online*. Max Planck Institute for Gravitational Physics.

[3] Patrick Di Justo; Kevin Grazier; Patrick Grazier & Kevin Grazier (2010). *The Science of Battlestar Galactica*. New York: John Wiley & Sons. p. 181. ISBN 978-0470399095.

[4] Wald, p. 99

[5] Hawking, Stephen. "The Beginning of Time". *Stephen Hawking: The Official Website*. Cambridge University. Retrieved 26 December 2012.

[6] Zebrowski, Ernest (2000). *A History of the Circle: Mathematical Reasoning and the Physical Universe*. Piscataway NJ: Rutgers University Press. p. 180. ISBN 978-0813528984.

[7] Curiel, Erik & Peter Bokulich. "Singularities and Black Holes". *Stanford Encyclopedia of Philosophy*. Center for the Study of Language and Information, Stanford University. Retrieved 26 December 2012.

[8] Poplawski, N. J. (2012). "Nonsingular, big-bounce cosmology from spinor-torsion coupling". *Physical Review D* 85: 107502. arXiv:1111.4595. Bibcode:2012PhRvD..85j7502P. doi:10.1103/PhysRevD.85.107502.

[9] If a rotating singularity is given a uniform electrical charge, a repellent force results, causing a ring singularity to form. The effect may be a stable wormhole, a non-point-like puncture in spacetime that may be connected to a second ring singularity on the other end. Although such wormholes are often suggested as routes for faster-than-light travel, such suggestions ignore the problem of escaping the black hole at the other end, or even of surviving the immense tidal forces in the tightly curved interior of the wormhole.

11.6   References

- Hawking, S. W.; Penrose, R. (1970), "The Singularities of Gravitational Collapse and Cosmology", *Proc R. Soc. A* 314 (1519): 529–548, Bibcode:1970RSPSA.314..529H, doi:10.1098/rspa.1970.0021 (Free access.)

- Shapiro, Stuart L.; Teukolsky, Saul A. (1991). "Formation of naked singularities: The violation of cosmic censorship". *Physical Review Letters* 66 (8): 994–997. Bibcode:1991PhRvL..66..994S. doi:10.1103/PhysRevLett.66.994. PMID 10043968.

- Robert M. Wald (1984). *General Relativity*. University of Chicago Press. ISBN 0-226-87033-2.

- Misner, Charles W.; Thorne, Kip; Wheeler, John Archibald (1973). *Gravitation*. W. H. Freeman. ISBN 0-7167-0344-0. §31.2 The nonsingularity of the gravitational radius, and following sections; §34 Global Techniques, Horizons, and Singularity Theorems

- Roger Penrose(1996)"Chandrasekhar, Black Holes, and Singularities"

- Roger Penrose(1999)"The Question of Cosmic Censorship"

- T. P. Singh"Gravitational Collapse, Black Holes and Naked Singularities"

11.7 Further reading

- *The Elegant Universe* by Brian Greene. This book provides a layman's introduction to string theory, although some of the views expressed are already becoming outdated. His use of common terms and his providing of examples throughout the text help the layperson understand the basics of string theory.

Chapter 12

Subatomic particle

In the physical sciences, **subatomic particles** are particles much smaller than atoms.[1] There are two types of subatomic particles: elementary particles, which according to current theories are not made of other particles; and *composite* particles.[2] Particle physics and nuclear physics study these particles and how they interact.[3]

In particle physics, the concept of a particle is one of several concepts inherited from classical physics. But it also reflects the modern understanding that at the quantum scale matter and energy behave very differently from what much of everyday experience would lead us to expect.

The idea of a particle underwent serious rethinking when experiments showed that light could behave like a stream of particles (called photons) as well as exhibit wave-like properties. This led to the new concept of wave–particle duality to reflect that quantum-scale "particles" behave like both particles and waves (also known as wavicles). Another new concept, the uncertainty principle, states that some of their properties taken together, such as their simultaneous position and momentum, cannot be measured exactly.[4] In more recent times, wave–particle duality has been shown to apply not only to photons but to increasingly massive particles as well.[5]

Interactions of particles in the framework of quantum field theory are understood as creation and annihilation of *quanta* of corresponding fundamental interactions. This blends particle physics with field theory.

12.1   Classification

12.1.1   By statistics

Main article: Spin–statistics theorem
Any subatomic particle, like any particle in the 3-dimensional space that obeys laws of quantum mechanics, can be either a boson (an integer spin) or a fermion (a half-integer spin).

12.1.2   By composition

The elementary particles of the Standard Model include:[6]

- Six "flavors" of quarks: up, down, bottom, top, strange, and charm;

- Six types of leptons: electron, electron neutrino, muon, muon neutrino, tau, tau neutrino;

- Twelve gauge bosons (force carriers): the photon of electromagnetism, the three W and Z bosons of the weak force, and the eight gluons of the strong force;

- The Higgs boson.

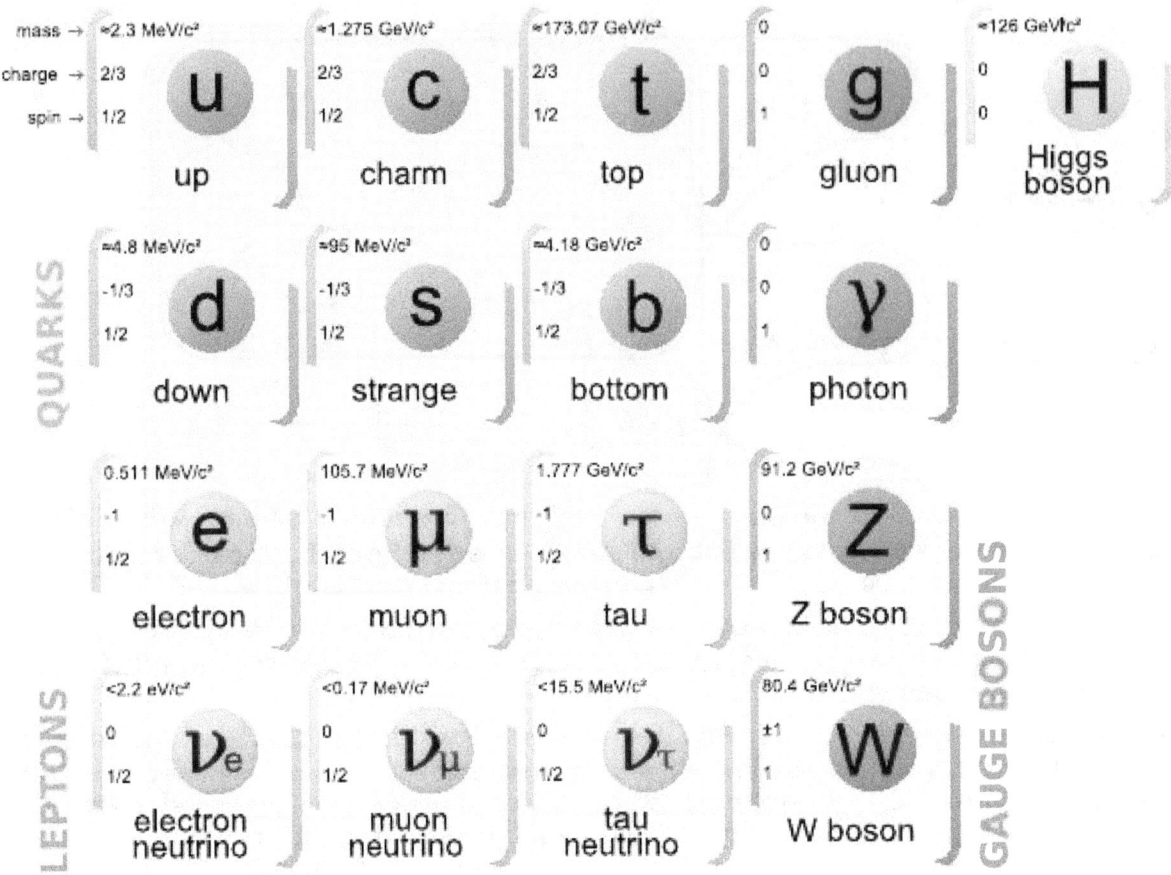

The Standard Model classification of particles

Various extensions of the Standard Model predict the existence of an elementary graviton particle and many other elementary particles.

Composite subatomic particles (such as protons or atomic nuclei) are bound states of two or more elementary particles. For example, a proton is made of two up quarks and one down quark, while the atomic nucleus of helium-4 is composed of two protons and two neutrons. Composite particles include all hadrons: these include baryons (such as protons and neutrons) and mesons (such as pions and kaons).

12.1.3 By mass

In special relativity, the energy of a particle at rest equals its mass times the speed of light squared ($E = mc^2$). That is, mass can be expressed in terms of energy and vice versa. If a particle has a frame of reference where it lies at rest, then it has a positive rest mass and is referred to as *massive*.

All composite particles are massive. Baryons (meaning "heavy") tend to have greater mass than mesons (meaning "intermediate"), which in turn tend to be heavier than leptons (meaning "lightweight"), but the heaviest lepton (the tau particle) is heavier than the two lightest flavours of baryons (nucleons). It is also certain that any particle with an electric charge is massive.

All massless particles (particles whose invariant mass is zero) are elementary. These include the photon and gluon, although the latter cannot be isolated.

The question of the masses of neutrinos is uncertain.

12.2   Other properties

Through the work of Albert Einstein, Louis de Broglie, and many others, current scientific theory holds that *all* particles also have a wave nature.[7] This has been verified not only for elementary particles but also for compound particles like atoms and even molecules. In fact, according to traditional formulations of non-relativistic quantum mechanics, wave–particle duality applies to all objects, even macroscopic ones; although the wave properties of macroscopic objects cannot be detected due to their small wavelengths.[8]

Interactions between particles have been scrutinized for many centuries, and a few simple laws underpin how particles behave in collisions and interactions. The most fundamental of these are the laws of conservation of energy and conservation of momentum, which let us make calculations of particle interactions on scales of magnitude that range from stars to quarks.[9] These are the prerequisite basics of Newtonian mechanics, a series of statements and equations in *Philosophiae Naturalis Principia Mathematica*, originally published in 1687.

12.3   Dividing an atom

The negatively charged electron has a mass equal to $1/1836$ of that of a hydrogen atom. The remainder of the hydrogen atom's mass comes from the positively charged proton. The atomic number of an element is the number of protons in its nucleus. Neutrons are neutral particles having a mass slightly greater than that of the proton. Different isotopes of the same element contain the same number of protons but differing numbers of neutrons. The mass number of an isotope is the total number of nucleons (neutrons and protons collectively).

Chemistry concerns itself with how electron sharing binds atoms into structures such as crystals and molecules. Nuclear physics deals with how protons and neutrons arrange themselves in nuclei. The study of subatomic particles, atoms and molecules, and their structure and interactions, requires quantum mechanics. Analyzing processes that change the numbers and types of particles requires quantum field theory. The study of subatomic particles *per se* is called particle physics. The term *high-energy physics* is nearly synonymous to "particle physics" since creation of particles requires high energies: it occurs only as a result of cosmic rays, or in particle accelerators. Particle phenomenology systematizes the knowledge about subatomic particles obtained from these experiments.

12.4   History

Main articles: History of subatomic physics and Timeline of particle discoveries

The term "*subatomic* particle" is largely a retronym of 1960s made to distinguish a big number of baryons and mesons (that comprise hadrons) from particles that are now thought to be truly elementary. Before that hadrons were usually classified as "elementary" because their composition was unknown.

A list of important discoveries follows:

12.5   See also

- *Atom: Journey Across the Subatomic Cosmos* (book)
- CPT invariance
- Dark Matter
- Hot spot effect in subatomic physics
- List of fictional elements, materials, isotopes and atomic particles
- List of particles

- Poincaré symmetry

- Ylem

12.6 References

[1] "Subatomic particles". NTD. Retrieved 5 June 2012.

[2] Bolonkin, Alexander (2011). *Universe, Human Immortality and Future Human Evaluation*. Elsevier. p. 25. ISBN 9780124158016.

[3] Fritzsch, Harald (2005). *Elementary Particles*. World Scientific. pp. 11–20. ISBN 978-981-256-141-1.

[4] Heisenberg, W. (1927), "Über den anschaulichen Inhalt der quantentheoretischen Kinematik und Mechanik", *Zeitschrift für Physik* (in German) **43** (3–4): 172–198, Bibcode:1927ZPhy...43..172H, doi:10.1007/BF01397280.

[5] Arndt, Markus; Nairz, Olaf; Vos-Andreae, Julian; Keller, Claudia; Van Der Zouw, Gerbrand; Zeilinger, Anton (2000). "Wave-particle duality of C60 molecules". *Nature* **401** (6754): 680–682. Bibcode:1999Natur.401..680A. doi:10.1038/44348. PMID 18494170.

[6] Cottingham, W. N.; Greenwood, D. A. (2007). *An introduction to the standard model of particle physics*. Cambridge University Press. p. 1. ISBN 978-0-521-85249-4.

[7] Walter Greiner (2001). *Quantum Mechanics: An Introduction*. Springer. p. 29. ISBN 3-540-67458-6.

[8] R. Eisberg & R. Resnick (1985). *Quantum Physics of Atoms, Molecules, Solids, Nuclei, and Particles* (2nd ed.). John Wiley & Sons. pp. 59–60. ISBN 0-471-87373-X. For both large and small wavelengths, both matter and radiation have both particle and wave aspects. [...] But the wave aspects of their motion become more difficult to observe as their wavelengths become shorter. [...] For ordinary macroscopic particles the mass is so large that the momentum is always sufficiently large to make the de Broglie wavelength small enough to be beyond the range of experimental detection, and classical mechanics reigns supreme.

[9] Isaac Newton (1687). Newton's Laws of Motion (*Philosophiae Naturalis Principia Mathematica*)

[10] Klemperer, Otto (1959). *Electron Physics: The Physics of the Free Electron*. Academic Press.

[11] Some sources such as The Strange Quark indicate 1947.

[12] http://press.web.cern.ch/press-releases/2014/06/cern-experiments-report-new-higgs-boson-measurements

12.7 Further reading

General readers

- Feynman, R.P. & Weinberg, S. (1987). *Elementary Particles and the Laws of Physics: The 1986 Dirac Memorial Lectures*. Cambridge Univ. Press.

- Brian Greene (1999). *The Elegant Universe*. W.W. Norton & Company. ISBN 0-393-05858-1.

- Oerter, Robert (2006). *The Theory of Almost Everything: The Standard Model, the Unsung Triumph of Modern Physics*. Plume.

- Schumm, Bruce A. (2004). *Deep Down Things: The Breathtaking Beauty of Particle Physics*. Johns Hopkins University Press. ISBN 0-8018-7971-X.

- Martinus Veltman (2003). *Facts and Mysteries in Elementary Particle Physics*. World Scientific. ISBN 981-238-149-X.

Textbooks

- Coughlan, G. D., J. E. Dodd, and B. M. Gripaios (2006). *The Ideas of Particle Physics: An Introduction for Scientists*, 3rd ed. Cambridge Univ. Press. An undergraduate text for those not majoring in physics.

- Griffiths, David J. (1987). *Introduction to Elementary Particles*. Wiley, John & Sons, Inc. ISBN 0-471-60386-4.

- Kane, Gordon L. (1987). *Modern Elementary Particle Physics*. Perseus Books. ISBN 0-201-11749-5.

12.8   External links

- particleadventure.org: The Standard Model.

- cpepweb.org: Particle chart.

- University of California: Particle Data Group.

- Annotated Physics Encyclopædia: Quantum Field Theory.

- Jose Galvez: Chapter 1 Electrodynamics (pdf).

Chapter 13

Atom

For other uses, see Atom (disambiguation).

An **atom** is the smallest constituent unit of ordinary matter that has the properties of a chemical element.[1] Every solid, liquid, gas, and plasma is made up of neutral or ionized atoms. Atoms are very small; typical sizes are around 100 pm (a ten-billionth of a meter, in the short scale).[2] However, atoms do not have well defined boundaries, and there are different ways to define their size which give different but close values.

Atoms are small enough that classical physics give noticeably incorrect results. Through the development of physics, atomic models have incorporated quantum principles to better explain and predict the behavior.

Every atom is composed of a nucleus and one or more electrons bound to the nucleus. The nucleus is made of one or more protons and typically a similar number of neutrons (none in hydrogen-1). Protons and neutrons are called nucleons. Over 99.94% of the atom's mass is in the nucleus. The protons have a positive electric charge, the electrons have a negative electric charge, and the neutrons have no electric charge. If the number of protons and electrons are equal, that atom is electrically neutral. If an atom has more or fewer electrons than protons, then it has an overall negative or positive charge, respectively, and it is called an ion.

Electrons of an atom are attracted to the protons in an atomic nucleus by this electromagnetic force. The protons and neutrons in the nucleus are attracted to each other by a different force, the nuclear force, which is usually stronger than the electromagnetic force repelling the positively charged protons from one another. Under certain circumstances the repelling electromagnetic force becomes stronger than the nuclear force, and nucleons can be ejected from the nucleus, leaving behind a different element: nuclear decay resulting in nuclear transmutation.

The number of protons in the nucleus defines to what chemical element the atom belongs: for example, all copper atoms contain 29 protons. The number of neutrons defines the isotope of the element.[3] The number of electrons influences the magnetic properties of an atom. Atoms can attach to one or more other atoms by chemical bonds to form chemical compounds such as molecules. The ability of atoms to associate and dissociate is responsible for most of the physical changes observed in nature, and is the subject of the discipline of chemistry.

Not all the matter of the universe is composed of atoms. Dark matter comprises more of the Universe than matter, and is composed not of atoms, but of particles of a currently unknown type.

13.1   History of atomic theory

Main article: Atomic theory

13.1.1   Atoms in philosophy

Main article: Atomism

The idea that matter is made up of discrete units is a very old one, appearing in many ancient cultures such as Greece and India. The word "atom", in fact, was coined by ancient Greek philosophers. However, these ideas were founded in philosophical and theological reasoning rather than evidence and experimentation. As a result, their views on what atoms look like and how they behave were incorrect. They also could not convince everybody, so atomism was but one of a number of competing theories on the nature of matter. It was not until the 19th century that the idea was embraced and refined by scientists, when the blossoming science of chemistry produced discoveries that only the concept of atoms could explain.

13.1.2   First evidence-based theory

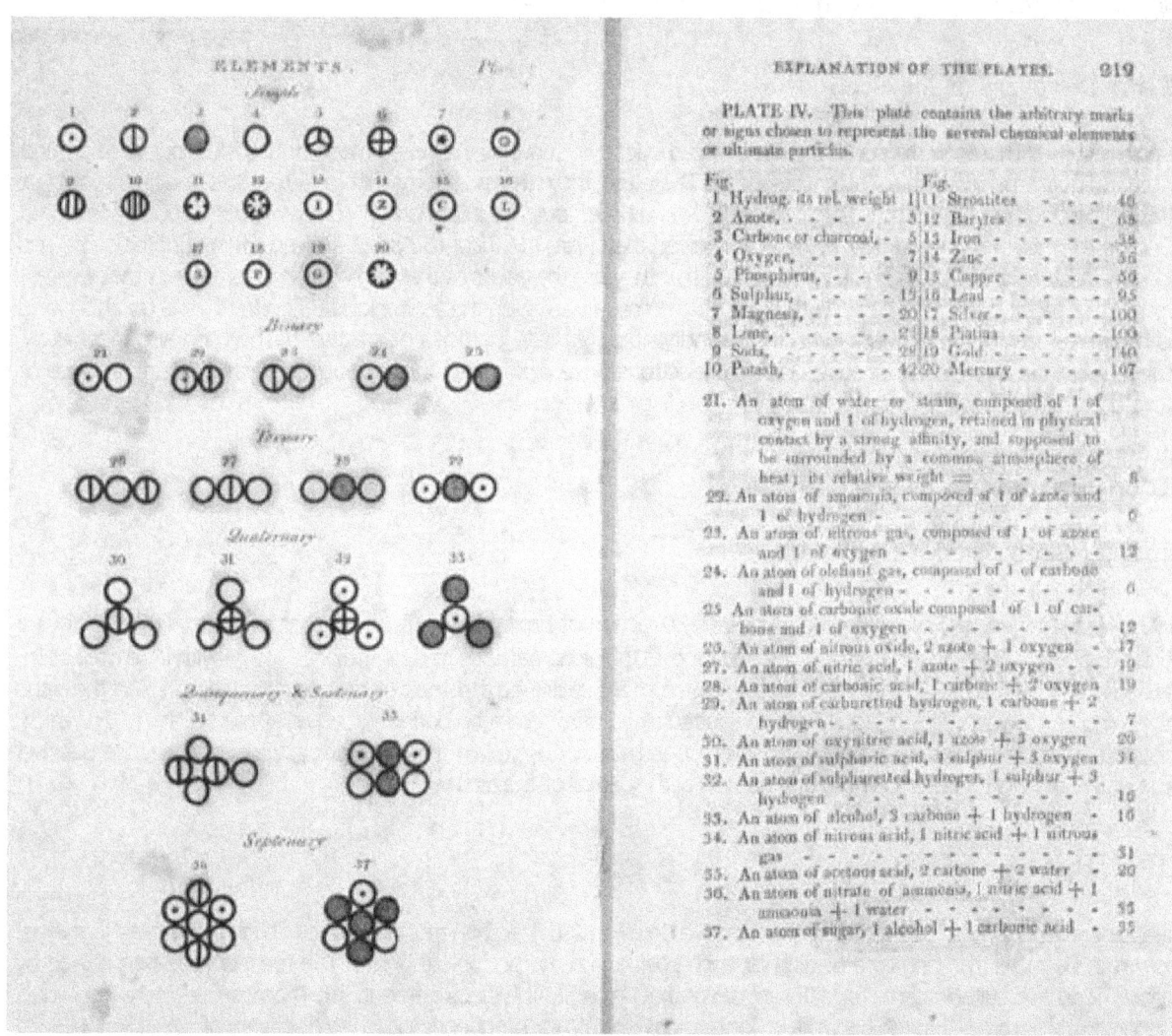

Various atoms and molecules as depicted in John Dalton's A New System of Chemical Philosophy (1808).

In the early 1800s, John Dalton used the concept of atoms to explain why elements always react in ratios of small whole numbers (the law of multiple proportions). For instance, there are two types of tin oxide: one is 88.1% tin and 11.9% oxygen and the other is 78.7% tin and 21.3% oxygen (tin(II) oxide and tin dioxide respectively). This means that 100g of tin will combine either with 13.5g or 27g of oxygen. 13.5 and 27 form a ratio of 1:2, a ratio of small whole numbers. This common pattern in chemistry suggested to Dalton that elements react in whole number multiples of discrete units—in other words, atoms. In the case of tin oxides, one tin atom will combine with either one or two oxygen atoms.[4]

Dalton also believed atomic theory could explain why water absorbs different gases in different proportions. For example, he found that water absorbs carbon dioxide far better than it absorbs nitrogen.[5] Dalton hypothesized this was due to the differences in mass and complexity of the gases' respective particles. Indeed, carbon dioxide molecules (CO_2) are heavier and larger than nitrogen molecules (N_2).

13.1.3   Brownian motion

In 1827, botanist Robert Brown used a microscope to look at dust grains floating in water and discovered that they moved about erratically, a phenomenon that became known as "Brownian motion". This was thought to be caused by water molecules knocking the grains about. In 1905 Albert Einstein produced the first mathematical analysis of the motion.[6][7][8] French physicist Jean Perrin used Einstein's work to experimentally determine the mass and dimensions of atoms, thereby conclusively verifying Dalton's atomic theory.[9]

13.1.4   Discovery of the electron

The physicist J. J. Thomson measured the mass of cathode rays, showing they were made of particles, but were around 1800 times lighter than the lightest atom, hydrogen. Therefore, they were not atoms, but a new particle, the first *subatomic* particle to be discovered, which he originally called "*corpuscle*" but was later named *electron*, after particles postulated by George Johnstone Stoney in 1874. He also showed they were identical to particles given off by photoelectric and radioactive materials.[10] It was quickly recognized that they are the particles that carry electric currents in metal wires, and carry the negative electric charge within atoms. Thomson was given the 1906 Nobel Prize in Physics for this work. Thus he overturned the belief that atoms are the indivisible, ultimate particles of matter.[11] Thomson also incorrectly postulated that the low mass, negatively charged electrons were distributed throughout the atom in a uniform sea of positive charge. This became known as the plum pudding model.

13.1.5   Discovery of the nucleus

Main article: Geiger-Marsden experiment

In 1909, Hans Geiger and Ernest Marsden, under the direction of Ernest Rutherford, bombarded a metal foil with alpha particles to observe how they scattered. They expected all the alpha particles to pass straight through with little deflection, because Thomson's model said that the charges in the atom are so diffuse that their electric fields could not affect the alpha particles much. However, Geiger and Marsden spotted alpha particles being deflected by angles greater than 90°, which was supposed to be impossible according to Thomson's model. To explain this, Rutherford proposed that the positive charge of the atom is concentrated in a tiny nucleus at the center of the atom.[12]

13.1.6   Discovery of isotopes

While experimenting with the products of radioactive decay, in 1913 radiochemist Frederick Soddy discovered that there appeared to be more than one type of atom at each position on the periodic table.[13] The term isotope was coined by Margaret Todd as a suitable name for different atoms that belong to the same element. J.J. Thomson created a technique for separating atom types through his work on ionized gases, which subsequently led to the discovery of stable isotopes.[14]

13.1.7   Bohr model

Main article: Bohr model

In 1913 the physicist Niels Bohr proposed a model in which the electrons of an atom were assumed to orbit the nucleus but could only do so in a finite set of orbits, and could jump between these orbits only in discrete changes of energy corresponding to absorption or radiation of a photon.[15] This quantization was used to explain why the electrons orbits are stable (given that normally, charges in acceleration, including circular motion, lose kinetic energy which is emitted as electromagnetic radiation, see *synchrotron radiation*) and why elements absorb and emit electromagnetic radiation in discrete spectra.[16]

Later in the same year Henry Moseley provided additional experimental evidence in favor of Niels Bohr's theory. These results refined Ernest Rutherford's and Antonius Van den Broek's model, which proposed that the atom contains in its nucleus a number of positive nuclear charges that is equal to its (atomic) number in the periodic table. Until these experiments, atomic number was not known to be a physical and experimental quantity. That it is equal to the atomic nuclear charge remains the accepted atomic model today.[17]

13.1.8   Chemical bonding explained

Chemical bonds between atoms were now explained, by Gilbert Newton Lewis in 1916, as the interactions between their constituent electrons.[18] As the chemical properties of the elements were known to largely repeat themselves according to the periodic law,[19] in 1919 the American chemist Irving Langmuir suggested that this could be explained if the electrons in an atom were connected or clustered in some manner. Groups of electrons were thought to occupy a set of electron shells about the nucleus.[20]

13.1.9   Further developments in quantum physics

The Stern–Gerlach experiment of 1922 provided further evidence of the quantum nature of the atom. When a beam of silver atoms was passed through a specially shaped magnetic field, the beam was split based on the direction of an atom's angular momentum, or spin. As this direction is random, the beam could be expected to spread into a line. Instead, the beam was split into two parts, depending on whether the atomic spin was oriented up or down.[21]

In 1924, Louis de Broglie proposed that all particles behave to an extent like waves. In 1926, Erwin Schrödinger used this idea to develop a mathematical model of the atom that described the electrons as three-dimensional waveforms rather than point particles. A consequence of using waveforms to describe particles is that it is physically impossible to obtain precise values for both the position and momentum of a particle at the same time; this became known as the uncertainty principle, formulated by Werner Heisenberg in 1926. In this concept, for a given accuracy in measuring a position one could only obtain a range of probable values for momentum, and vice versa. This model was able to explain observations of atomic behavior that previous models could not, such as certain structural and spectral patterns of atoms larger than hydrogen. Thus, the planetary model of the atom was discarded in favor of one that described atomic orbital zones around the nucleus where a given electron is most likely to be observed.[22][23]

13.1.10   Discovery of the neutron

The development of the mass spectrometer allowed the mass of atoms to be measured with increased accuracy. The device uses a magnet to bend the trajectory of a beam of ions, and the amount of deflection is determined by the ratio of an atom's mass to its charge. The chemist Francis William Aston used this instrument to show that isotopes had different masses. The atomic mass of these isotopes varied by integer amounts, called the whole number rule.[24] The explanation for these different isotopes awaited the discovery of the neutron, an uncharged particle with a mass similar to the proton, by the physicist James Chadwick in 1932. Isotopes were then explained as elements with the same number of protons, but different numbers of neutrons within the nucleus.[25]

13.1.11   Fission, high-energy physics and condensed matter

In 1938, the German chemist Otto Hahn, a student of Rutherford, directed neutrons onto uranium atoms expecting to get transuranium elements. Instead, his chemical experiments showed barium as a product.[26] A year later, Lise Meitner and her nephew Otto Frisch verified that Hahn's result were the first experimental *nuclear fission*.[27][28] In 1944, Hahn received the Nobel prize in chemistry. Despite Hahn's efforts, the contributions of Meitner and Frisch were not recognized.[29]

In the 1950s, the development of improved particle accelerators and particle detectors allowed scientists to study the impacts of atoms moving at high energies.[30] Neutrons and protons were found to be hadrons, or composites of smaller particles called quarks. The standard model of particle physics was developed that so far has successfully explained the properties of the nucleus in terms of these sub-atomic particles and the forces that govern their interactions.[31]

13.2 Structure

13.2.1 Subatomic particles

Main article: Subatomic particle

Though the word *atom* originally denoted a particle that cannot be cut into smaller particles, in modern scientific usage the atom is composed of various subatomic particles. The constituent particles of an atom are the electron, the proton and the neutron; all three are fermions. However, the hydrogen-1 atom has no neutrons and the hydron ion has no electrons.

The electron is by far the least massive of these particles at 9.11×10^{-31} kg, with a negative electrical charge and a size that is too small to be measured using available techniques.[32] It is the lightest particle with a positive rest mass measured. Under ordinary conditions, electrons are bound to the positively charged nucleus by the attraction created from opposite electric charges. If an atom has more or fewer electrons than its atomic number, then it becomes respectively negatively or positively charged as a whole; a charged atom is called an ion. Electrons have been known since the late 19th century, mostly thanks to J.J. Thomson; see history of subatomic physics for details.

Protons have a positive charge and a mass 1,836 times that of the electron, at 1.6726×10^{-27} kg. The number of protons in an atom is called its atomic number. Ernest Rutherford (1919) observed that nitrogen under alpha-particle bombardment ejects what appeared to be hydrogen nuclei. By 1920 he had accepted that the hydrogen nucleus is a distinct particle within the atom and named it proton.

Neutrons have no electrical charge and have a free mass of 1,839 times the mass of the electron,[33] or 1.6929×10^{-27} kg, the heaviest of the three constituent particles, but it can be reduced by the nuclear binding energy. Neutrons and protons (collectively known as nucleons) have comparable dimensions—on the order of 2.5×10^{-15} m—although the 'surface' of these particles is not sharply defined.[34] The neutron was discovered in 1932 by the English physicist James Chadwick.

In the Standard Model of physics, electrons are truly elementary particles with no internal structure. However, both protons and neutrons are composite particles composed of elementary particles called quarks. There are two types of quarks in atoms, each having a fractional electric charge. Protons are composed of two up quarks (each with charge $+^2/_3$) and one down quark (with a charge of $-^1/_3$). Neutrons consist of one up quark and two down quarks. This distinction accounts for the difference in mass and charge between the two particles.[35][36]

The quarks are held together by the strong interaction (or strong force), which is mediated by gluons. The protons and neutrons, in turn, are held to each other in the nucleus by the nuclear force, which is a residuum of the strong force that has somewhat different range-properties (see the article on the nuclear force for more). The gluon is a member of the family of gauge bosons, which are elementary particles that mediate physical forces.[35][36]

13.2.2 Nucleus

Main article: Atomic nucleus

All the bound protons and neutrons in an atom make up a tiny atomic nucleus, and are collectively called nucleons. The radius of a nucleus is approximately equal to $1.07\sqrt[3]{A}$ fm, where A is the total number of nucleons.[37] This is much smaller than the radius of the atom, which is on the order of 10^5 fm. The nucleons are bound together by a short-ranged attractive potential called the residual strong force. At distances smaller than 2.5 fm this force is much more powerful than the electrostatic force that causes positively charged protons to repel each other.[38]

Atoms of the same element have the same number of protons, called the atomic number. Within a single element, the number of neutrons may vary, determining the isotope of that element. The total number of protons and neutrons determine the nuclide. The number of neutrons relative to the protons determines the stability of the nucleus, with certain isotopes undergoing radioactive decay.[39]

The proton, the electron, and the neutron are classified as fermions. Fermions obey the Pauli exclusion principle which prohibits *identical* fermions, such as multiple protons, from occupying the same quantum state at the same time. Thus, every proton in the nucleus must occupy a quantum state different from all other protons, and the same applies to all neutrons of the nucleus and to all electrons of the electron cloud. However, a proton and a neutron are allowed to occupy the same quantum state.[40]

For atoms with low atomic numbers, a nucleus that has more neutrons than protons tends to drop to a lower energy state through radioactive decay so that the neutron–proton ratio is closer to one. However, as the atomic number increases, a higher proportion of neutrons is required to offset the mutual repulsion of the protons. Thus, there are no stable nuclei with equal proton and neutron numbers above atomic number $Z = 20$ (calcium) and as Z increases, the neutron–proton ratio of stable isotopes increases.[40] The stable isotope with the highest proton–neutron ratio is lead-208 (about 1.5).

The number of protons and neutrons in the atomic nucleus can be modified, although this can require very high energies because of the strong force. Nuclear fusion occurs when multiple atomic particles join to form a heavier nucleus, such as through the energetic collision of two nuclei. For example, at the core of the Sun protons require energies of 3–10 keV to overcome their mutual repulsion—the coulomb barrier—and fuse together into a single nucleus.[41] Nuclear fission is the opposite process, causing a nucleus to split into two smaller nuclei—usually through radioactive decay. The nucleus can also be modified through bombardment by high energy subatomic particles or photons. If this modifies the number of protons in a nucleus, the atom changes to a different chemical element.[42][43]

If the mass of the nucleus following a fusion reaction is less than the sum of the masses of the separate particles, then the difference between these two values can be emitted as a type of usable energy (such as a gamma ray, or the kinetic energy of a beta particle), as described by Albert Einstein's mass–energy equivalence formula, $E = mc^2$, where m is the mass loss and c is the speed of light. This deficit is part of the binding energy of the new nucleus, and it is the non-recoverable loss of the energy that causes the fused particles to remain together in a state that requires this energy to separate.[44]

The fusion of two nuclei that create larger nuclei with lower atomic numbers than iron and nickel—a total nucleon number of about 60—is usually an exothermic process that releases more energy than is required to bring them together.[45] It is this energy-releasing process that makes nuclear fusion in stars a self-sustaining reaction. For heavier nuclei, the binding energy per nucleon in the nucleus begins to decrease. That means fusion processes producing nuclei that have atomic numbers higher than about 26, and atomic masses higher than about 60, is an endothermic process. These more massive nuclei can not undergo an energy-producing fusion reaction that can sustain the hydrostatic equilibrium of a star.[40]

13.2.3   Electron cloud

Main articles: Atomic orbital and Electron configuration

The electrons in an atom are attracted to the protons in the nucleus by the electromagnetic force. This force binds the electrons inside an electrostatic potential well surrounding the smaller nucleus, which means that an external source of energy is needed for the electron to escape. The closer an electron is to the nucleus, the greater the attractive force. Hence electrons bound near the center of the potential well require more energy to escape than those at greater separations.

Electrons, like other particles, have properties of both a particle and a wave. The electron cloud is a region inside the potential well where each electron forms a type of three-dimensional standing wave—a wave form that does not move relative to the nucleus. This behavior is defined by an atomic orbital, a mathematical function that characterises the probability that an electron appears to be at a particular location when its position is measured.[46] Only a discrete (or quantized) set of these orbitals exist around the nucleus, as other possible wave patterns rapidly decay into a more stable form.[47] Orbitals can have one or more ring or node structures, and they differ from each other in size, shape and orientation.[48]

Each atomic orbital corresponds to a particular energy level of the electron. The electron can change its state to a higher energy level by absorbing a photon with sufficient energy to boost it into the new quantum state. Likewise, through spontaneous emission, an electron in a higher energy state can drop to a lower energy state while radiating the excess energy as a photon. These characteristic energy values, defined by the differences in the energies of the quantum states, are responsible for atomic spectral lines.[47]

The amount of energy needed to remove or add an electron—the electron binding energy—is far less than the binding energy of nucleons. For example, it requires only 13.6 eV to strip a ground-state electron from a hydrogen atom,[49] compared to 2.23 *million* eV for splitting a deuterium nucleus.[50] Atoms are electrically neutral if they have an equal number of protons and electrons. Atoms that have either a deficit or a surplus of electrons are called ions. Electrons that are farthest from the nucleus may be transferred to other nearby atoms or shared between atoms. By this mechanism, atoms are able to bond into molecules and other types of chemical compounds like ionic and covalent network crystals.[51]

13.3   Properties

13.3.1   Nuclear properties

Main articles: Isotope, Stable isotope, List of nuclides and List of elements by stability of isotopes

By definition, any two atoms with an identical number of *protons* in their nuclei belong to the same chemical element. Atoms with equal numbers of protons but a different number of *neutrons* are different isotopes of the same element. For example, all hydrogen atoms admit exactly one proton, but isotopes exist with no neutrons (hydrogen-1, by far the most common form,[52] also called protium), one neutron (deuterium), two neutrons (tritium) and more than two neutrons. The known elements form a set of atomic numbers, from the single proton element hydrogen up to the 118-proton element ununoctium.[53] All known isotopes of elements with atomic numbers greater than 82 are radioactive.[54][55]

About 339 nuclides occur naturally on Earth,[56] of which 254 (about 75%) have not been observed to decay, and are referred to as "stable isotopes". However, only 90 of these nuclides are stable to all decay, even in theory. Another 164 (bringing the total to 254) have not been observed to decay, even though in theory it is energetically possible. These are also formally classified as "stable". An additional 34 radioactive nuclides have half-lives longer than 80 million years, and are long-lived enough to be present from the birth of the solar system. This collection of 288 nuclides are known as primordial nuclides. Finally, an additional 51 short-lived nuclides are known to occur naturally, as daughter products of primordial nuclide decay (such as radium from uranium), or else as products of natural energetic processes on Earth, such as cosmic ray bombardment (for example, carbon-14).[57][note 1]

For 80 of the chemical elements, at least one stable isotope exists. As a rule, there is only a handful of stable isotopes for each of these elements, the average being 3.2 stable isotopes per element. Twenty-six elements have only a single stable isotope, while the largest number of stable isotopes observed for any element is ten, for the element tin. Elements 43, 61, and all elements numbered 83 or higher have no stable isotopes.[58]

Stability of isotopes is affected by the ratio of protons to neutrons, and also by the presence of certain "magic numbers" of neutrons or protons that represent closed and filled quantum shells. These quantum shells correspond to a set of energy levels within the shell model of the nucleus; filled shells, such as the filled shell of 50 protons for tin, confers unusual stability on the nuclide. Of the 254 known stable nuclides, only four have both an odd number of protons *and* odd number of neutrons: hydrogen-2 (deuterium), lithium-6, boron-10 and nitrogen-14. Also, only four naturally occurring, radioactive odd–odd nuclides have a half-life over a billion years: potassium-40, vanadium-50, lanthanum-138 and tantalum-180m. Most odd–odd nuclei are highly unstable with respect to beta decay, because the decay products are even–even, and are therefore more strongly bound, due to nuclear pairing effects.[58]

13.3.2   Mass

Main articles: Atomic mass and mass number

The large majority of an atom's mass comes from the protons and neutrons that make it up. The total number of these particles (called "nucleons") in a given atom is called the mass number. It is a positive integer and dimensionless (instead of having dimension of mass), because it expresses a count. An example of use of a mass number is "carbon-12," which has 12 nucleons (six protons and six neutrons).

The actual mass of an atom at rest is often expressed using the unified atomic mass unit (u), also called dalton (Da). This unit is defined as a twelfth of the mass of a free neutral atom of carbon-12, which is approximately 1.66×10^{-27} kg.[59] Hydrogen-1 (the lightest isotope of hydrogen which is also the nuclide with the lowest mass) has an atomic weight of 1.007825 u.[60] The value of this number is called the atomic mass. A given atom has an atomic mass approximately equal (within 1%) to its mass number times the atomic mass unit (for example the mass of a nitrogen-14 is roughly 14 u). However, this number will not be exactly an integer except in the case of carbon-12 (see below).[61] The heaviest stable atom is lead-208,[54] with a mass of 207.9766521 u.[62]

As even the most massive atoms are far too light to work with directly, chemists instead use the unit of moles. One mole of atoms of any element always has the same number of atoms (about 6.022×10^{23}). This number was chosen so that if an element has an atomic mass of 1 u, a mole of atoms of that element has a mass close to one gram. Because of the definition of the unified atomic mass unit, each carbon-12 atom has an atomic mass of exactly 12 u, and so a mole of carbon-12 atoms weighs exactly 0.012 kg.[59]

13.3.3   Shape and size

Main article: Atomic radius

Atoms lack a well-defined outer boundary, so their dimensions are usually described in terms of an atomic radius. This is a measure of the distance out to which the electron cloud extends from the nucleus.[2] However, this assumes the atom to exhibit a spherical shape, which is only obeyed for atoms in vacuum or free space. Atomic radii may be derived from the distances between two nuclei when the two atoms are joined in a chemical bond. The radius varies with the location of an atom on the atomic chart, the type of chemical bond, the number of neighboring atoms (coordination number) and a quantum mechanical property known as spin.[63] On the periodic table of the elements, atom size tends to increase when moving down columns, but decrease when moving across rows (left to right).[64] Consequently, the smallest atom is helium with a radius of 32 pm, while one of the largest is caesium at 225 pm.[65]

When subjected to external forces, like electrical fields, the shape of an atom may deviate from spherical symmetry. The deformation depends on the field magnitude and the orbital type of outer shell electrons, as shown by group-theoretical considerations. Aspherical deviations might be elicited for instance in crystals, where large crystal-electrical fields may occur at low-symmetry lattice sites. Significant ellipsoidal deformations have recently been shown to occur for sulfur ions[66] and chalcogen ions[67] in pyrite-type compounds.

Atomic dimensions are thousands of times smaller than the wavelengths of light (400–700 nm) so they cannot be viewed using an optical microscope. However, individual atoms can be observed using a scanning tunneling microscope. To visualize the minuteness of the atom, consider that a typical human hair is about 1 million carbon atoms in width.[68] A single drop of water contains about 2 sextillion (2×10^{21}) atoms of oxygen, and twice the number of hydrogen atoms.[69] A single carat diamond with a mass of 2×10^{-4} kg contains about 10 sextillion (10^{22}) atoms of carbon.[note 2] If an apple were magnified to the size of the Earth, then the atoms in the apple would be approximately the size of the original apple.[70]

13.3.4   Radioactive decay

Main article: Radioactive decay
Every element has one or more isotopes that have unstable nuclei that are subject to radioactive decay, causing the nucleus to emit particles or electromagnetic radiation. Radioactivity can occur when the radius of a nucleus is large compared with the radius of the strong force, which only acts over distances on the order of 1 fm.[71]

The most common forms of radioactive decay are:[72][73]

- Alpha decay: this process is caused when the nucleus emits an alpha particle, which is a helium nucleus consisting of two protons and two neutrons. The result of the emission is a new element with a lower atomic number.

- Beta decay (and electron capture): these processes are regulated by the weak force, and result from a transformation of a neutron into a proton, or a proton into a neutron. The neutron to proton transition is accompanied by the emission of an electron and an antineutrino, while proton to neutron transition (except in electron capture) causes the emission of a positron and a neutrino. The electron or positron emissions are called beta particles. Beta decay either increases or decreases the atomic number of the nucleus by one. Electron capture is more common than positron emission, because it requires less energy. In this type of decay, an electron is absorbed by the nucleus, rather than a positron emitted from the nucleus. A neutrino is still emitted in this process, and a proton changes to a neutron.

- Gamma decay: this process results from a change in the energy level of the nucleus to a lower state, resulting in the emission of electromagnetic radiation. The excited state of a nucleus which results in gamma emission usually occurs following the emission of an alpha or a beta particle. Thus, gamma decay usually follows alpha or beta decay.

Other more rare types of radioactive decay include ejection of neutrons or protons or clusters of nucleons from a nucleus, or more than one beta particle. An analog of gamma emission which allows excited nuclei to lose energy in a different way, is internal conversion— a process that produces high-speed electrons that are not beta rays, followed by production of high-energy photons that are not gamma rays. A few large nuclei explode into two or more charged fragments of varying masses plus several neutrons, in a decay called spontaneous nuclear fission.

Each radioactive isotope has a characteristic decay time period—the *half-life*—that is determined by the amount of time needed for half of a sample to decay. This is an exponential decay process that steadily decreases the proportion of the remaining isotope by 50% every half-life. Hence after two half-lives have passed only 25% of the isotope is present, and so forth.[71]

13.3.5 Magnetic moment

Main articles: Electron magnetic moment and Nuclear magnetic moment

Elementary particles possess an intrinsic quantum mechanical property known as spin. This is analogous to the angular momentum of an object that is spinning around its center of mass, although strictly speaking these particles are believed to be point-like and cannot be said to be rotating. Spin is measured in units of the reduced Planck constant ($\hbar$), with electrons, protons and neutrons all having spin ½ $\hbar$, or "spin-½". In an atom, electrons in motion around the nucleus possess orbital angular momentum in addition to their spin, while the nucleus itself possesses angular momentum due to its nuclear spin.[74]

The magnetic field produced by an atom—its magnetic moment—is determined by these various forms of angular momentum, just as a rotating charged object classically produces a magnetic field. However, the most dominant contribution comes from electron spin. Due to the nature of electrons to obey the Pauli exclusion principle, in which no two electrons may be found in the same quantum state, bound electrons pair up with each other, with one member of each pair in a spin up state and the other in the opposite, spin down state. Thus these spins cancel each other out, reducing the total magnetic dipole moment to zero in some atoms with even number of electrons.[75]

In ferromagnetic elements such as iron, cobalt and nickel, an odd number of electrons leads to an unpaired electron and a net overall magnetic moment. The orbitals of neighboring atoms overlap and a lower energy state is achieved when the spins of unpaired electrons are aligned with each other, a spontaneous process known as an exchange interaction. When the magnetic moments of ferromagnetic atoms are lined up, the material can produce a measurable macroscopic field. Paramagnetic materials have atoms with magnetic moments that line up in random directions when no magnetic field is present, but the magnetic moments of the individual atoms line up in the presence of a field.[75][76]

The nucleus of an atom will have no spin when it has even numbers of both neutrons and protons, but for other cases of odd numbers, the nucleus may have a spin. Normally nuclei with spin are aligned in random directions because of thermal equilibrium. However, for certain elements (such as xenon-129) it is possible to polarize a significant proportion of the nuclear spin states so that they are aligned in the same direction—a condition called hyperpolarization. This has important applications in magnetic resonance imaging.[77][78]

13.3.6   Energy levels

The potential energy of an electron in an atom is negative, its dependence of its position reaches the minimum (the most absolute value) inside the nucleus, and vanishes when the distance from the nucleus goes to infinity, roughly in an inverse proportion to the distance. In the quantum-mechanical model, a bound electron can only occupy a set of states centered on the nucleus, and each state corresponds to a specific energy level; see time-independent Schrödinger equation for theoretical explanation. An energy level can be measured by the amount of energy needed to unbind the electron from the atom, and is usually given in units of electronvolts (eV). The lowest energy state of a bound electron is called the ground state, i.e. stationary state, while an electron transition to a higher level results in an excited state.[79] The electron's energy raises when *n* increases because the (average) distance to the nucleus increases. Dependence of the energy on ℓ is caused not by electrostatic potential of the nucleus, but by interaction between electrons.

For an electron to transition between two different states, e.g. grounded state to first excited level (ionization), it must absorb or emit a photon at an energy matching the difference in the potential energy of those levels, according to Niels Bohr model, what can be precisely calculated by the Schrödinger equation. Electrons jump between orbitals in a particle-like fashion. For example, if a single photon strikes the electrons, only a single electron changes states in response to the photon; see Electron properties.

The energy of an emitted photon is proportional to its frequency, so these specific energy levels appear as distinct bands in the electromagnetic spectrum.[80] Each element has a characteristic spectrum that can depend on the nuclear charge, subshells filled by electrons, the electromagnetic interactions between the electrons and other factors.[81]

When a continuous spectrum of energy is passed through a gas or plasma, some of the photons are absorbed by atoms, causing electrons to change their energy level. Those excited electrons that remain bound to their atom spontaneously emit this energy as a photon, traveling in a random direction, and so drop back to lower energy levels. Thus the atoms behave like a filter that forms a series of dark absorption bands in the energy output. (An observer viewing the atoms from a view that does not include the continuous spectrum in the background, instead sees a series of emission lines from the photons emitted by the atoms.) Spectroscopic measurements of the strength and width of atomic spectral lines allow the composition and physical properties of a substance to be determined.[82]

Close examination of the spectral lines reveals that some display a fine structure splitting. This occurs because of spin–orbit coupling, which is an interaction between the spin and motion of the outermost electron.[83] When an atom is in an external magnetic field, spectral lines become split into three or more components; a phenomenon called the Zeeman effect. This is caused by the interaction of the magnetic field with the magnetic moment of the atom and its electrons. Some atoms can have multiple electron configurations with the same energy level, which thus appear as a single spectral line. The interaction of the magnetic field with the atom shifts these electron configurations to slightly different energy levels, resulting in multiple spectral lines.[84] The presence of an external electric field can cause a comparable splitting and shifting of spectral lines by modifying the electron energy levels, a phenomenon called the Stark effect.[85]

If a bound electron is in an excited state, an interacting photon with the proper energy can cause stimulated emission of a photon with a matching energy level. For this to occur, the electron must drop to a lower energy state that has an energy difference matching the energy of the interacting photon. The emitted photon and the interacting photon then move off in parallel and with matching phases. That is, the wave patterns of the two photons are synchronized. This physical property is used to make lasers, which can emit a coherent beam of light energy in a narrow frequency band.[86]

13.3.7   Valence and bonding behavior

Main articles: Valence (chemistry) and Chemical bond

Valency is the combining power of an element. It is equal to number of hydrogen atoms that atom can combine or displace in forming compounds.[87] The outermost electron shell of an atom in its uncombined state is known as the valence shell, and the electrons in that shell are called valence electrons. The number of valence electrons determines the bonding behavior with other atoms. Atoms tend to chemically react with each other in a manner that fills (or empties) their outer valence shells.[88] For example, a transfer of a single electron between atoms is a useful approximation for bonds that form between atoms with one-electron more than a filled shell, and others that are one-electron short of a full shell, such as occurs in the compound sodium chloride and other chemical ionic salts. However, many elements display multiple valences, or tendencies to share differing numbers of electrons in different compounds. Thus, chemical bonding between these elements takes many forms of electron-sharing that are more than simple electron transfers. Examples include the element carbon and the organic compounds.[89]

The chemical elements are often displayed in a periodic table that is laid out to display recurring chemical properties, and elements with the same number of valence electrons form a group that is aligned in the same column of the table. (The horizontal rows correspond to the filling of a quantum shell of electrons.) The elements at the far right of the table have their outer shell completely filled with electrons, which results in chemically inert elements known as the noble gases.[90][91]

13.3.8 States

Main articles: State of matter and Phase (matter)

Quantities of atoms are found in different states of matter that depend on the physical conditions, such as temperature and pressure. By varying the conditions, materials can transition between solids, liquids, gases and plasmas.[92] Within a state, a material can also exist in different allotropes. An example of this is solid carbon, which can exist as graphite or diamond.[93] Gaseous allotropes exist as well, such as dioxygen and ozone.

At temperatures close to absolute zero, atoms can form a Bose–Einstein condensate, at which point quantum mechanical effects, which are normally only observed at the atomic scale, become apparent on a macroscopic scale.[94][95] This super-cooled collection of atoms then behaves as a single super atom, which may allow fundamental checks of quantum mechanical behavior.[96]

13.4 Identification

The scanning tunneling microscope is a device for viewing surfaces at the atomic level. It uses the quantum tunneling phenomenon, which allows particles to pass through a barrier that would normally be insurmountable. Electrons tunnel through the vacuum between two planar metal electrodes, on each of which is an adsorbed atom, providing a tunneling-current density that can be measured. Scanning one atom (taken as the tip) as it moves past the other (the sample) permits plotting of tip displacement versus lateral separation for a constant current. The calculation shows the extent to which scanning-tunneling-microscope images of an individual atom are visible. It confirms that for low bias, the microscope images the space-averaged dimensions of the electron orbitals across closely packed energy levels—the Fermi level local density of states.[97][98]

An atom can be ionized by removing one of its electrons. The electric charge causes the trajectory of an atom to bend when it passes through a magnetic field. The radius by which the trajectory of a moving ion is turned by the magnetic field is determined by the mass of the atom. The mass spectrometer uses this principle to measure the mass-to-charge ratio of ions. If a sample contains multiple isotopes, the mass spectrometer can determine the proportion of each isotope in the sample by measuring the intensity of the different beams of ions. Techniques to vaporize atoms include inductively coupled plasma atomic emission spectroscopy and inductively coupled plasma mass spectrometry, both of which use a plasma to vaporize samples for analysis.[99]

A more area-selective method is electron energy loss spectroscopy, which measures the energy loss of an electron beam within a transmission electron microscope when it interacts with a portion of a sample. The atom-probe tomograph has sub-nanometer resolution in 3-D and can chemically identify individual atoms using time-of-flight mass spectrometry.[100]

Spectra of excited states can be used to analyze the atomic composition of distant stars. Specific light wavelengths contained in the observed light from stars can be separated out and related to the quantized transitions in free gas atoms. These colors can be replicated using a gas-discharge lamp containing the same element.[101] Helium was discovered in this way in the spectrum of the Sun 23 years before it was found on Earth.[102]

13.5   Origin and current state

Atoms form about 4% of the total energy density of the observable Universe, with an average density of about 0.25 atoms/m^3.[103] Within a galaxy such as the Milky Way, atoms have a much higher concentration, with the density of matter in the interstellar medium (ISM) ranging from 10^5 to 10^9 atoms/m^3.[104] The Sun is believed to be inside the Local Bubble, a region of highly ionized gas, so the density in the solar neighborhood is only about 10^3 atoms/m^3.[105] Stars form from dense clouds in the ISM, and the evolutionary processes of stars result in the steady enrichment of the ISM with elements more massive than hydrogen and helium. Up to 95% of the Milky Way's atoms are concentrated inside stars and the total mass of atoms forms about 10% of the mass of the galaxy.[106] (The remainder of the mass is an unknown dark matter.)[107]

13.5.1   Formation

Electrons are thought to exist in the Universe since early stages of the Big Bang. Atomic nuclei forms in nucleosynthesis reactions. In about three minutes Big Bang nucleosynthesis produced most of the helium, lithium, and deuterium in the Universe, and perhaps some of the beryllium and boron.[108][109][110]

Ubiquitousness and stability of atoms relies on their binding energy, which means that an atom has a lower energy than an unbound system of the nucleus and electrons. Where the temperature is much higher than ionization potential, the matter exists in the form of plasma—a gas of positively charged ions (possibly, bare nuclei) and electrons. When the temperature drops below the ionization potential, atoms become statistically favorable. Atoms (complete with bound electrons) became to dominate over charged particles 380,000 years after the Big Bang—an epoch called recombination, when the expanding Universe cooled enough to allow electrons to become attached to nuclei.[111]

Since the Big Bang, which produced no carbon or heavier elements, atomic nuclei have been combined in stars through the process of nuclear fusion to produce more of the element helium, and (via the triple alpha process) the sequence of elements from carbon up to iron;[112] see stellar nucleosynthesis for details.

Isotopes such as lithium-6, as well as some beryllium and boron are generated in space through cosmic ray spallation.[113] This occurs when a high-energy proton strikes an atomic nucleus, causing large numbers of nucleons to be ejected.

Elements heavier than iron were produced in supernovae through the r-process and in AGB stars through the s-process, both of which involve the capture of neutrons by atomic nuclei.[114] Elements such as lead formed largely through the radioactive decay of heavier elements.[115]

13.5.2   Earth

Most of the atoms that make up the Earth and its inhabitants were present in their current form in the nebula that collapsed out of a molecular cloud to form the Solar System. The rest are the result of radioactive decay, and their relative proportion can be used to determine the age of the Earth through radiometric dating.[116][117] Most of the helium in the crust of the Earth (about 99% of the helium from gas wells, as shown by its lower abundance of helium-3) is a product of alpha decay.[118]

There are a few trace atoms on Earth that were not present at the beginning (i.e., not "primordial"), nor are results of radioactive decay. Carbon-14 is continuously generated by cosmic rays in the atmosphere.[119] Some atoms on Earth have been artificially generated either deliberately or as by-products of nuclear reactors or explosions.[120][121] Of the transuranic elements—those with atomic numbers greater than 92—only plutonium and neptunium occur naturally on Earth.[122][123] Transuranic elements have radioactive lifetimes shorter than the current age of the Earth[124] and thus identifiable quantities of these elements have long since decayed, with the exception of traces of plutonium-244 possibly deposited by cosmic dust.[125] Natural deposits of plutonium and neptunium are produced by neutron capture in uranium ore.[126]

The Earth contains approximately 1.33×10^{50} atoms.[127] Although small numbers of independent atoms of noble gases exist, such as argon, neon, and helium, 99% of the atmosphere is bound in the form of molecules, including carbon dioxide and diatomic oxygen and nitrogen. At the surface of the Earth, an overwhelming majority of atoms combine to form various compounds, including water, salt, silicates and oxides. Atoms can also combine to create materials that do not consist of discrete molecules, including crystals and liquid or solid metals.[128][129] This atomic matter forms networked arrangements that lack the particular type of small-scale interrupted order associated with molecular matter.[130]

13.5.3 Rare and theoretical forms

Superheavy elements

Main article: Transuranium element

While isotopes with atomic numbers higher than lead (82) are known to be radioactive, an "island of stability" has been proposed for some elements with atomic numbers above 103. These superheavy elements may have a nucleus that is relatively stable against radioactive decay.[131] The most likely candidate for a stable superheavy atom, unbihexium, has 126 protons and 184 neutrons.[132]

Exotic matter

Main article: Exotic matter

Each particle of matter has a corresponding antimatter particle with the opposite electrical charge. Thus, the positron is a positively charged antielectron and the antiproton is a negatively charged equivalent of a proton. When a matter and corresponding antimatter particle meet, they annihilate each other. Because of this, along with an imbalance between the number of matter and antimatter particles, the latter are rare in the universe. The first causes of this imbalance are not yet fully understood, although theories of baryogenesis may offer an explanation. As a result, no antimatter atoms have been discovered in nature.[133][134] However, in 1996 the antimatter counterpart of the hydrogen atom (antihydrogen) was synthesized at the CERN laboratory in Geneva.[135][136]

Other exotic atoms have been created by replacing one of the protons, neutrons or electrons with other particles that have the same charge. For example, an electron can be replaced by a more massive muon, forming a muonic atom. These types of atoms can be used to test the fundamental predictions of physics.[137][138][139]

13.6 See also

- History of quantum mechanics
- Infinite divisibility
- List of basic chemistry topics
- Timeline of atomic and subatomic physics
- Vector model of the atom
- Nuclear model
- Radioactive isotope

13.7 Notes

[1] For more recent updates see Interactive Chart of Nuclides (Brookhaven National Laboratory).

[2] A carat is 200 milligrams. By definition, carbon-12 has 0.012 kg per mole. The Avogadro constant defines 6×10^{23} atoms per mole.

13.8   References

[1] "Atom". *Compendium of Chemical Terminology (IUPAC Gold Book)* (2nd ed.). IUPAC. Retrieved 2015-04-25.

[2] Ghosh, D. C.; Biswas, R. (2002). "Theoretical calculation of Absolute Radii of Atoms and Ions. Part 1. The Atomic Radii". *Int. J. Mol. Sci.* **3**: 87–113. doi:10.3390/i3020087.

[3] Leigh, G. J., ed. (1990). *International Union of Pure and Applied Chemistry, Commission on the Nomenclature of Inorganic Chemistry, Nomenclature of Organic Chemistry – Recommendations 1990.* Oxford: Blackwell Scientific Publications. p. 35. ISBN 0-08-022369-9. An atom is the smallest unit quantity of an element that is capable of existence whether alone or in chemical combination with other atoms of the same or other elements.

[4] Andrew G. van Melsen (1952). *From Atomos to Atom.* Mineola, N.Y.: Dover Publications. ISBN 0-486-49584-1.

[5] Dalton, John. "On the Absorption of Gases by Water and Other Liquids", in *Memoirs of the Literary and Philosophical Society of Manchester.* 1803. Retrieved on August 29, 2007.

[6] Einstein, Albert (1905). "Über die von der molekularkinetischen Theorie der Wärme geforderte Bewegung von in ruhenden Flüssigkeiten suspendierten Teilchen" (PDF). *Annalen der Physik* (in German) **322** (8): 549–560. Bibcode:1905AnP...322..549E. doi:10.1002/andp.19053220806. Retrieved 4 February 2007.

[7] Mazo, Robert M. (2002). *Brownian Motion: Fluctuations, Dynamics, and Applications.* Oxford University Press. pp. 1–7. ISBN 0-19-851567-7. OCLC 48753074.

[8] Lee, Y.K.; Hoon, K. (1995). "Brownian Motion". Imperial College. Archived from the original on 18 December 2007. Retrieved 18 December 2007.

[9] Patterson, G. (2007). "Jean Perrin and the triumph of the atomic doctrine". *Endeavour* **31** (2): 50–53. doi:10.1016/j.endeavour.2007.05.003. PMID 17602746.

[10] Thomson, J. J. (August 1901). "On bodies smaller than atoms". *The Popular Science Monthly* (Bonnier Corp.): 323–335. Retrieved 2009-06-21.

[11] "J.J. Thomson". Nobel Foundation. 1906. Retrieved 20 December 2007.

[12] Rutherford, E. (1911). "The Scattering of α and β Particles by Matter and the Structure of the Atom" (PDF). *Philosophical Magazine* **21** (125): 669–88. doi:10.1080/14786440508637080.

[13] "Frederick Soddy, The Nobel Prize in Chemistry 1921". Nobel Foundation. Retrieved 18 January 2008.

[14] Thomson, Joseph John (1913). "Rays of positive electricity". *Proceedings of the Royal Society.* A **89** (607): 1–20. Bibcode:1913RSPSA..89....1T. doi:10.1098/rspa.1913.0057.

[15] Stern, David P. (16 May 2005). "The Atomic Nucleus and Bohr's Early Model of the Atom". NASA/Goddard Space Flight Center. Retrieved 20 December 2007.

[16] Bohr, Niels (11 December 1922). "Niels Bohr, The Nobel Prize in Physics 1922, Nobel Lecture". Nobel Foundation. Retrieved 16 February 2008.

[17] Pais, Abraham (1986). *Inward Bound: Of Matter and Forces in the Physical World.* New York: Oxford University Press. pp. 228–230. ISBN 0-19-851971-0.

[18] Lewis, Gilbert N. (1916). "The Atom and the Molecule". *Journal of the American Chemical Society* **38** (4): 762–786. doi:10.1021/ja02261a002.

[19] Scerri, Eric R. (2007). *The periodic table: its story and its significance.* Oxford University Press US. pp. 205–226. ISBN 0-19-530573-6.

[20] Langmuir, Irving (1919). "The Arrangement of Electrons in Atoms and Molecules". *Journal of the American Chemical Society* **41** (6): 868–934. doi:10.1021/ja02227a002.

[21] Scully, Marlan O.; Lamb, Willis E.; Barut, Asim (1987). "On the theory of the Stern-Gerlach apparatus". *Foundations of Physics* **17** (6): 575–583. Bibcode:1987FoPh...17..575S. doi:10.1007/BF01882788.

[22] Brown, Kevin (2007). "The Hydrogen Atom". MathPages. Retrieved 21 December 2007.

[23] Harrison, David M. (2000). "The Development of Quantum Mechanics". University of Toronto. Archived from the original on 25 December 2007. Retrieved 21 December 2007.

[24] Aston, Francis W. (1920). "The constitution of atmospheric neon". *Philosophical Magazine* **39** (6): 449–55. doi:10.1080/14786440408636058.

[25] Chadwick, James (12 December 1935). "Nobel Lecture: The Neutron and Its Properties". Nobel Foundation. Retrieved 21 December 2007.

[26] "Otto Hahn, Lise Meitner and Fritz Strassmann". *Chemical Achievers: The Human Face of the Chemical Sciences.* Chemical Heritage Foundation. Archived from the original on 24 October 2009. Retrieved 15 September 2009.

[27] Meitner, Lise; Frisch, Otto Robert (1939). "Disintegration of uranium by neutrons: a new type of nuclear reaction". *Nature* **143** (3615): 239–240. Bibcode:1939Natur.143..239M. doi:10.1038/143239a0.

[28] Schroeder, M. "Lise Meitner – Zur 125. Wiederkehr Ihres Geburtstages" (in German). Retrieved 4 June 2009.

[29] Crawford, E.; Sime, Ruth Lewin; Walker, Mark (1997). "A Nobel tale of postwar injustice". *Physics Today* **50** (9): 26–32. Bibcode:1997PhT....50i..26C. doi:10.1063/1.881933.

[30] Kullander, Sven (28 August 2001). "Accelerators and Nobel Laureates". Nobel Foundation. Retrieved 31 January 2008.

[31] "The Nobel Prize in Physics 1990". Nobel Foundation. 17 October 1990. Retrieved 31 January 2008.

[32] Demtröder, Wolfgang (2002). *Atoms, Molecules and Photons: An Introduction to Atomic- Molecular- and Quantum Physics* (1st ed.). Springer. pp. 39–42. ISBN 3-540-20631-0. OCLC 181435713.

[33] Woan, Graham (2000). *The Cambridge Handbook of Physics.* Cambridge University Press. p. 8. ISBN 0-521-57507-9. OCLC 224032426.

[34] MacGregor, Malcolm H. (1992). *The Enigmatic Electron.* Oxford University Press. pp. 33–37. ISBN 0-19-521833-7. OCLC 223372888.

[35] Particle Data Group (2002). "The Particle Adventure". Lawrence Berkeley Laboratory. Archived from the original on 4 January 2007. Retrieved 3 January 2007.

[36] Schombert, James (18 April 2006). "Elementary Particles". University of Oregon. Retrieved 3 January 2007.

[37] Jevremovic, Tatjana (2005). *Nuclear Principles in Engineering.* Springer. p. 63. ISBN 0-387-23284-2. OCLC 228384008.

[38] Pfeffer, Jeremy I.; Nir, Shlomo (2000). *Modern Physics: An Introductory Text.* Imperial College Press. pp. 330–336. ISBN 1-86094-250-4. OCLC 45900880.

[39] Wenner, Jennifer M. (10 October 2007). "How Does Radioactive Decay Work?". Carleton College. Retrieved 9 January 2008.

[40] Raymond, David (7 April 2006). "Nuclear Binding Energies". New Mexico Tech. Archived from the original on 11 December 2006. Retrieved 3 January 2007.

[41] Mihos, Chris (23 July 2002). "Overcoming the Coulomb Barrier". Case Western Reserve University. Retrieved 13 February 2008.

[42] Staff (30 March 2007). "ABC's of Nuclear Science". Lawrence Berkeley National Laboratory. Archived from the original on 5 December 2006. Retrieved 3 January 2007.

[43] Makhijani, Arjun; Saleska, Scott (2 March 2001). "Basics of Nuclear Physics and Fission". Institute for Energy and Environmental Research. Archived from the original on 16 January 2007. Retrieved 3 January 2007.

[44] Shultis, J. Kenneth; Faw, Richard E. (2002). *Fundamentals of Nuclear Science and Engineering.* CRC Press. pp. 10–17. ISBN 0-8247-0834-2. OCLC 123346507.

[45] Fewell, M. P. (1995). "The atomic nuclide with the highest mean binding energy". *American Journal of Physics* **63** (7): 653–658. Bibcode:1995AmJPh..63..653F. doi:10.1119/1.17828.

[46] Mulliken, Robert S. (1967). "Spectroscopy, Molecular Orbitals, and Chemical Bonding". *Science* **157** (3784): 13–24. Bibcode:1967Sci...157...13M. doi:10.1126/science.157.3784.13. PMID 5338306.

[47] Brucat, Philip J. (2008). "The Quantum Atom". University of Florida. Archived from the original on 7 December 2006. Retrieved 4 January 2007.

[48] Manthey, David (2001). "Atomic Orbitals". Orbital Central. Archived from the original on 10 January 2008. Retrieved 21 January 2008.

[49] Herter, Terry (2006). "Lecture 8: The Hydrogen Atom". Cornell University. Retrieved 14 February 2008.

[50] Bell, R. E.; Elliott, L. G. (1950). "Gamma-Rays from the Reaction $H^1(n,\gamma)D^2$ and the Binding Energy of the Deuteron". *Physical Review* 79 (2): 282–285. Bibcode:1950PhRv...79..282B. doi:10.1103/PhysRev.79.282.

[51] Smirnov, Boris M. (2003). *Physics of Atoms and Ions*. Springer. pp. 249–272. ISBN 0-387-95550-X.

[52] Matis, Howard S. (9 August 2000). "The Isotopes of Hydrogen". *Guide to the Nuclear Wall Chart*. Lawrence Berkeley National Lab. Archived from the original on 18 December 2007. Retrieved 21 December 2007.

[53] Weiss, Rick (17 October 2006). "Scientists Announce Creation of Atomic Element, the Heaviest Yet". Washington Post. Retrieved 21 December 2007.

[54] Sills, Alan D. (2003). *Earth Science the Easy Way*. Barron's Educational Series. pp. 131–134. ISBN 0-7641-2146-4. OCLC 51543743.

[55] Dumé, Belle (23 April 2003). "Bismuth breaks half-life record for alpha decay". Physics World. Archived from the original on 14 December 2007. Retrieved 21 December 2007.

[56] Lindsay, Don (30 July 2000). "Radioactives Missing From The Earth". Don Lindsay Archive. Archived from the original on 28 April 2007. Retrieved 23 May 2007.

[57] Tuli, Jagdish K. (April 2005). "Nuclear Wallet Cards". National Nuclear Data Center, Brookhaven National Laboratory. Retrieved 16 April 2011.

[58] CRC Handbook (2002).

[59] Mills, Ian; Cvitaš, Tomislav; Homann, Klaus; Kallay, Nikola; Kuchitsu, Kozo (1993). *Quantities, Units and Symbols in Physical Chemistry* (PDF) (2nd ed.). Oxford: International Union of Pure and Applied Chemistry, Commission on Physiochemical Symbols Terminology and Units, Blackwell Scientific Publications. p. 70. ISBN 0-632-03583-8. OCLC 27011505.

[60] Chieh, Chung (22 January 2001). "Nuclide Stability". University of Waterloo. Retrieved 4 January 2007.

[61] "Atomic Weights and Isotopic Compositions for All Elements". National Institute of Standards and Technology. Archived from the original on 31 December 2006. Retrieved 4 January 2007.

[62] Audi, G.; Wapstra, A.H.; Thibault, C. (2003). "The Ame2003 atomic mass evaluation (II)" (PDF). *Nuclear Physics A* 729 (1): 337–676. Bibcode:2003NuPhA.729..337A. doi:10.1016/j.nuclphysa.2003.11.003.

[63] Shannon, R. D. (1976). "Revised effective ionic radii and systematic studies of interatomic distances in halides and chalcogenides". *Acta Crystallographica A* 32 (5): 751–767. Bibcode:1976AcCrA..32..751S. doi:10.1107/S0567739476001551.

[64] Dong, Judy (1998). "Diameter of an Atom". The Physics Factbook. Archived from the original on 4 November 2007. Retrieved 19 November 2007.

[65] Zumdahl, Steven S. (2002). *Introductory Chemistry: A Foundation* (5th ed.). Houghton Mifflin. ISBN 0-618-34342-3. OCLC 173081482. Archived from the original on 4 March 2008. Retrieved 5 February 2008.

[66] Birkholz, M.; Rudert, R. (2008). "Interatomic distances in pyrite-structure disulfides – a case for ellipsoidal modeling of sulfur ions]" (PDF). *phys. stat. sol. b* 245: 1858–1864. Bibcode:2008PSSBR.245.1858B. doi:10.1002/pssb.200879532.

[67] Birkholz, M. (2014). "Modeling the Shape of Ions in Pyrite-Type Crystals". *Crystals* 4: 390–403. doi:10.3390/cryst4030390.

[68] Staff (2007). "Small Miracles: Harnessing nanotechnology". Oregon State University. Retrieved 7 January 2007.—describes the width of a human hair as 10^5 nm and 10 carbon atoms as spanning 1 nm.

[69] Padilla, Michael J.; Miaoulis, Ioannis; Cyr, Martha (2002). *Prentice Hall Science Explorer: Chemical Building Blocks*. Upper Saddle River, New Jersey USA: Prentice-Hall, Inc. p. 32. ISBN 0-13-054091-9. OCLC 47925884. There are 2,000,000,000,000,000,000,000 (that's 2 sextillion) atoms of oxygen in one drop of water—and twice as many atoms of hydrogen.

[70] Feynman, Richard (1995). *Six Easy Pieces*. The Penguin Group. p. 5. ISBN 978-0-14-027666-4. OCLC 40499574.

[71] "Radioactivity". Splung.com. Archived from the original on 4 December 2007. Retrieved 19 December 2007.

[72] L'Annunziata, Michael F. (2003). *Handbook of Radioactivity Analysis*. Academic Press. pp. 3–56. ISBN 0-12-436603-1. OCLC 16212955.

[73] Firestone, Richard B. (22 May 2000). "Radioactive Decay Modes". Berkeley Laboratory. Retrieved 7 January 2007.

[74] Hornak, J. P. (2006). "Chapter 3: Spin Physics". *The Basics of NMR*. Rochester Institute of Technology. Archived from the original on 3 February 2007. Retrieved 7 January 2007.

[75] Schroeder, Paul A. (25 February 2000). "Magnetic Properties". University of Georgia. Archived from the original on 29 April 2007. Retrieved 7 January 2007.

[76] Goebel, Greg (1 September 2007). "[4.3] Magnetic Properties of the Atom". *Elementary Quantum Physics*. In The Public Domain website. Retrieved 7 January 2007.

[77] Yarris, Lynn (Spring 1997). "Talking Pictures". *Berkeley Lab Research Review*. Archived from the original on 13 January 2008. Retrieved 9 January 2008.

[78] Liang, Z.-P.; Haacke, E. M. (1999). Webster, J. G., ed. *Encyclopedia of Electrical and Electronics Engineering: Magnetic Resonance Imaging*. vol. 2. John Wiley & Sons. pp. 412–426. ISBN 0-471-13946-7.

[79] Zeghbroeck, Bart J. Van (1998). "Energy levels". Shippensburg University. Archived from the original on 15 January 2005. Retrieved 23 December 2007.

[80] Fowles, Grant R. (1989). *Introduction to Modern Optics*. Courier Dover Publications. pp. 227–233. ISBN 0-486-65957-7. OCLC 18834711.

[81] Martin, W. C.; Wiese, W. L. (May 2007). "Atomic Spectroscopy: A Compendium of Basic Ideas, Notation, Data, and Formulas". National Institute of Standards and Technology. Archived from the original on 8 February 2007. Retrieved 8 January 2007.

[82] "Atomic Emission Spectra — Origin of Spectral Lines". Avogadro Web Site. Retrieved 10 August 2006.

[83] Fitzpatrick, Richard (16 February 2007). "Fine structure". University of Texas at Austin. Retrieved 14 February 2008.

[84] Weiss, Michael (2001). "The Zeeman Effect". University of California-Riverside. Archived from the original on 2 February 2008. Retrieved 6 February 2008.

[85] Beyer, H. F.; Shevelko, V. P. (2003). *Introduction to the Physics of Highly Charged Ions*. CRC Press. pp. 232–236. ISBN 0-7503-0481-2. OCLC 47150433.

[86] Watkins, Thayer. "Coherence in Stimulated Emission". San José State University. Archived from the original on 12 January 2008. Retrieved 23 December 2007.

[87] oxford dictionary – valency

[88] Reusch, William (16 July 2007). "Virtual Textbook of Organic Chemistry". Michigan State University. Retrieved 11 January 2008.

[89] "Covalent bonding – Single bonds". chemguide. 2000.

[90] Husted, Robert et al. (11 December 2003). "Periodic Table of the Elements". Los Alamos National Laboratory. Archived from the original on 10 January 2008. Retrieved 11 January 2008.

[91] Baum, Rudy (2003). "It's Elemental: The Periodic Table". Chemical & Engineering News. Retrieved 11 January 2008.

[92] Goodstein, David L. (2002). *States of Matter*. Courier Dover Publications. pp. 436–438. ISBN 0-13-843557-X.

[93] Brazhkin, Vadim V. (2006). "Metastable phases, phase transformations, and phase diagrams in physics and chemistry". *Physics-Uspekhi* 49 (7): 719–24. Bibcode:2006PhyU...49..719B. doi:10.1070/PU2006v049n07ABEH006013.

[94] Myers, Richard (2003). *The Basics of Chemistry*. Greenwood Press. p. 85. ISBN 0-313-31664-3. OCLC 50164580.

[95] Staff (9 October 2001). "Bose-Einstein Condensate: A New Form of Matter". National Institute of Standards and Technology. Archived from the original on 3 January 2008. Retrieved 16 January 2008.

[96] Colton, Imogen; Fyffe, Jeanette (3 February 1999). "Super Atoms from Bose-Einstein Condensation". The University of Melbourne. Archived from the original on 29 August 2007. Retrieved 6 February 2008.

[97] Jacox, Marilyn; Gadzuk, J. William (November 1997). "Scanning Tunneling Microscope". National Institute of Standards and Technology. Archived from the original on 7 January 2008. Retrieved 11 January 2008.

[98] "The Nobel Prize in Physics 1986". The Nobel Foundation. Retrieved 11 January 2008.—in particular, see the Nobel lecture by G. Binnig and H. Rohrer.

[99] Jakubowski, N.; Moens, Luc; Vanhaecke, Frank (1998). "Sector field mass spectrometers in ICP-MS". *Spectrochimica Acta Part B: Atomic Spectroscopy* **53** (13): 1739–63. Bibcode:1998AcSpe..53.1739J. doi:10.1016/S0584-8547(98)00222-5.

[100] Müller, Erwin W.; Panitz, John A.; McLane, S. Brooks (1968). "The Atom-Probe Field Ion Microscope". *Review of Scientific Instruments* **39** (1): 83–86. Bibcode:1968RScI...39...83M. doi:10.1063/1.1683116.

[101] Lochner, Jim; Gibb, Meredith; Newman, Phil (30 April 2007). "What Do Spectra Tell Us?". NASA/Goddard Space Flight Center. Archived from the original on 16 January 2008. Retrieved 3 January 2008.

[102] Winter, Mark (2007). "Helium". WebElements. Archived from the original on 30 December 2007. Retrieved 3 January 2008.

[103] Hinshaw, Gary (10 February 2006). "What is the Universe Made Of?". NASA/WMAP. Archived from the original on 31 December 2007. Retrieved 7 January 2008.

[104] Choppin, Gregory R.; Liljenzin, Jan-Olov; Rydberg, Jan (2001). *Radiochemistry and Nuclear Chemistry*. Elsevier. p. 441. ISBN 0-7506-7463-6. OCLC 162592180.

[105] Davidsen, Arthur F. (1993). "Far-Ultraviolet Astronomy on the Astro-1 Space Shuttle Mission". *Science* **259** (5093): 327–34. Bibcode:1993Sci...259..327D. doi:10.1126/science.259.5093.327. PMID 17832344.

[106] Lequeux, James (2005). *The Interstellar Medium*. Springer. p. 4. ISBN 3-540-21326-0. OCLC 133157789.

[107] Smith, Nigel (6 January 2000). "The search for dark matter". Physics World. Archived from the original on 16 February 2008. Retrieved 14 February 2008.

[108] Croswell, Ken (1991). "Boron, bumps and the Big Bang: Was matter spread evenly when the Universe began? Perhaps not; the clues lie in the creation of the lighter elements such as boron and beryllium". *New Scientist* (1794): 42. Archived from the original on 7 February 2008. Retrieved 14 January 2008.

[109] Copi, Craig J.; Schramm, DN; Turner, MS (1995). "Big-Bang Nucleosynthesis and the Baryon Density of the Universe". *Science* **267** (5195): 192–99. arXiv:astro-ph/9407006. Bibcode:1995Sci...267..192C. doi:10.1126/science.7809624. PMID 7809624.

[110] Hinshaw, Gary (15 December 2005). "Tests of the Big Bang: The Light Elements". NASA/WMAP. Archived from the original on 17 January 2008. Retrieved 13 January 2008.

[111] Abbott, Brian (30 May 2007). "Microwave (WMAP) All-Sky Survey". Hayden Planetarium. Retrieved 13 January 2008.

[112] Hoyle, F. (1946). "The synthesis of the elements from hydrogen". *Monthly Notices of the Royal Astronomical Society* **106**: 343–83. Bibcode:1946MNRAS.106..343H. doi:10.1093/mnras/106.5.343.

[113] Knauth, D. C.; Knauth, D. C.; Lambert, David L.; Crane, P. (2000). "Newly synthesized lithium in the interstellar medium". *Nature* **405** (6787): 656–58. doi:10.1038/35015028. PMID 10864316.

[114] Mashnik, Stepan G. (2000). "On Solar System and Cosmic Rays Nucleosynthesis and Spallation Processes". arXiv:astro-ph/0008382 [astro-ph].

[115] Kansas Geological Survey (4 May 2005). "Age of the Earth". University of Kansas. Retrieved 14 January 2008.

[116] Manuel 2001, pp. 407–430, 511–519.

[117] Dalrymple, G. Brent (2001). "The age of the Earth in the twentieth century: a problem (mostly) solved". *Geological Society, London, Special Publications* **190** (1): 205–21. Bibcode:2001GSLSP.190..205D. doi:10.1144/GSL.SP.2001.190.01.14. Retrieved 14 January 2008.

[118] Anderson, Don L.; Foulger, G. R.; Meibom, Anders (2 September 2006). "Helium: Fundamental models". MantlePlumes.org. Archived from the original on 8 February 2007. Retrieved 14 January 2007.

[119] Pennicott, Katie (10 May 2001). "Carbon clock could show the wrong time". PhysicsWeb. Archived from the original on 15 December 2007. Retrieved 14 January 2008.

[120] Yarris, Lynn (27 July 2001). "New Superheavy Elements 118 and 116 Discovered at Berkeley Lab". Berkeley Lab. Archived from the original on 9 January 2008. Retrieved 14 January 2008.

[121] Diamond, H et al. (1960). "Heavy Isotope Abundances in Mike Thermonuclear Device". *Physical Review* 119 (6): 2000–04. Bibcode:1960PhRv..119.2000D. doi:10.1103/PhysRev.119.2000.

[122] Poston Sr., John W. (23 March 1998). "Do transuranic elements such as plutonium ever occur naturally?". Scientific American.

[123] Keller, C. (1973). "Natural occurrence of lanthanides, actinides, and superheavy elements". *Chemiker Zeitung* 97 (10): 522–30. OSTI 4353086.

[124] Zaider, Marco; Rossi, Harald H. (2001). *Radiation Science for Physicians and Public Health Workers*. Springer. p. 17. ISBN 0-306-46403-9. OCLC 44110319.

[125] Manuel 2001, pp. 407–430,511–519.

[126] "Oklo Fossil Reactors". Curtin University of Technology. Archived from the original on 18 December 2007. Retrieved 15 January 2008.

[127] Weisenberger, Drew. "How many atoms are there in the world?". Jefferson Lab. Retrieved 16 January 2008.

[128] Pidwirny, Michael. "Fundamentals of Physical Geography". University of British Columbia Okanagan. Archived from the original on 21 January 2008. Retrieved 16 January 2008.

[129] Anderson, Don L. (2002). "The inner inner core of Earth". *Proceedings of the National Academy of Sciences* 99 (22): 13966–68. Bibcode:2002PNAS...9913966A. doi:10.1073/pnas.232565899. PMC 137819. PMID 12391308.

[130] Pauling, Linus (1960). *The Nature of the Chemical Bond*. Cornell University Press. pp. 5–10. ISBN 0-8014-0333-2. OCLC 17518275.

[131] Anonymous (2 October 2001). "Second postcard from the island of stability". *CERN Courier*. Archived from the original on 3 February 2008. Retrieved 14 January 2008.

[132] Jacoby, Mitch (2006). "As-yet-unsynthesized superheavy atom should form a stable diatomic molecule with fluorine". *Chemical & Engineering News* 84 (10): 19. doi:10.1021/cen-v084n010.p019a.

[133] Koppes, Steve (1 March 1999). "Fermilab Physicists Find New Matter-Antimatter Asymmetry". University of Chicago. Retrieved 14 January 2008.

[134] Cromie, William J. (16 August 2001). "A lifetime of trillionths of a second: Scientists explore antimatter". Harvard University Gazette. Retrieved 14 January 2008.

[135] Hijmans, Tom W. (2002). "Particle physics: Cold antihydrogen". *Nature* 419 (6906): 439–40. Bibcode:2002Natur.419..439H. doi:10.1038/419439a. PMID 12368837.

[136] Staff (30 October 2002). "Researchers 'look inside' antimatter". BBC News. Retrieved 14 January 2008.

[137] Barrett, Roger (1990). "The Strange World of the Exotic Atom". *New Scientist* (1728): 77–115. Archived from the original on 21 December 2007. Retrieved 4 January 2008.

[138] Indelicato, Paul (2004). "Exotic Atoms". *Physica Scripta* T112 (1): 20–26. arXiv:physics/0409058. Bibcode:2004PhST..112...20I. doi:10.1238/Physica.Topical.112a00020.

[139] Ripin, Barrett H. (July 1998). "Recent Experiments on Exotic Atoms". American Physical Society. Retrieved 15 February 2008.

13.9 Sources

- Manuel, Oliver (2001). *Origin of Elements in the Solar System: Implications of Post-1957 Observations.* Springer. ISBN 0-306-46562-0. OCLC 228374906.

13.10 Further reading

- Dalton, J. (1808). *A New System of Chemical Philosophy, Part 1.* London and Manchester: S. Russell.

- Gangopadhyaya, Mrinalkanti (1981). *Indian Atomism: History and Sources.* Atlantic Highlands, New Jersey: Humanities Press. ISBN 0-391-02177-X. OCLC 10916778.

- Harrison, Edward Robert (2003). *Masks of the Universe: Changing Ideas on the Nature of the Cosmos.* Cambridge University Press. ISBN 0-521-77351-2. OCLC 50441595.

- Iannone, A. Pablo (2001). *Dictionary of World Philosophy.* Routledge. ISBN 0-415-17995-5. OCLC 44541769.

- King, Richard (1999). *Indian philosophy: an introduction to Hindu and Buddhist thought.* Edinburgh University Press. ISBN 0-7486-0954-7.

- Levere, Trevor, H. (2001). *Transforming Matter – A History of Chemistry for Alchemy to the Buckyball.* The Johns Hopkins University Press. ISBN 0-8018-6610-3.

- Liddell, Henry George; Scott, Robert. "A Greek-English Lexicon". Perseus Digital Library.

- Liddell, Henry George; Scott, Robert. "ἄτομος". *A Greek-English Lexicon.* Perseus Digital Library. Retrieved 21 June 2010.

- McEvilley, Thomas (2002). *The shape of ancient thought: comparative studies in Greek and Indian philosophies.* Allworth Press. ISBN 1-58115-203-5.

- Moran, Bruce T. (2005). *Distilling Knowledge: Alchemy, Chemistry, and the Scientific Revolution.* Harvard University Press. ISBN 0-674-01495-2.

- Ponomarev, Leonid Ivanovich (1993). *The Quantum Dice.* CRC Press. ISBN 0-7503-0251-8. OCLC 26853108.

- Roscoe, Henry Enfield (1895). *John Dalton and the Rise of Modern Chemistry.* Century science series. New York: Macmillan. Retrieved 3 April 2011.

- Siegfried, Robert (2002). *From Elements to Atoms: A History of Chemical Composition.* DIANE. ISBN 0-87169-924-9. OCLC 186607849.

- Teresi, Dick (2003). *Lost Discoveries: The Ancient Roots of Modern Science.* Simon & Schuster. pp. 213–214. ISBN 0-7432-4379-X.

- Various (2002). Lide, David R., ed. *Handbook of Chemistry & Physics* (88th ed.). CRC. ISBN 0-8493-0486-5. OCLC 179976746. Archived from the original on 23 May 2008. Retrieved 23 May 2008.

- Wurtz, Charles Adolphe (1881). *The Atomic Theory.* New York: D. Appleton and company. ISBN 0-559-43636-X.

13.11 External links

- "Quantum Mechanics and the Structure of Atoms" on YouTube

- Freudenrich, Craig C. "How Atoms Work". How Stuff Works. Archived from the original on 8 January 2007. Retrieved 9 January 2007.

- "The Atom". *Free High School Science Texts: Physics*. Wikibooks. Retrieved 10 July 2010.

- Anonymous (2007). "The atom". Science aid+. Retrieved 10 July 2010.—a guide to the atom for teens.

- Anonymous (3 January 2006). "Atoms and Atomic Structure". BBC. Archived from the original on 2 January 2007. Retrieved 11 January 2007.

- Various (3 January 2006). "Physics 2000, Table of Contents". University of Colorado. Archived from the original on 14 January 2008. Retrieved 11 January 2008.

- Various (3 February 2006). "What does an atom look like?". University of Karlsruhe. Retrieved 12 May 2008.

THOMSON

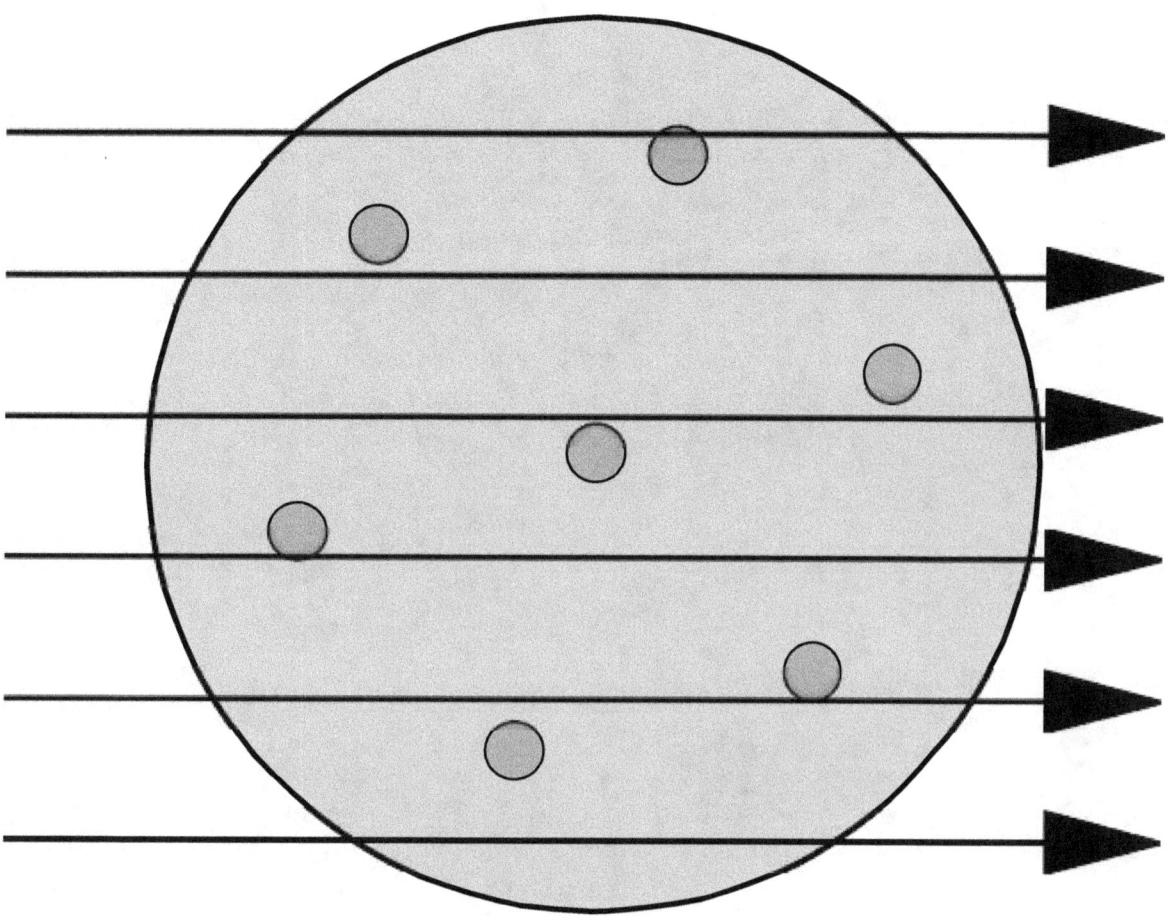

RUTHERFORD

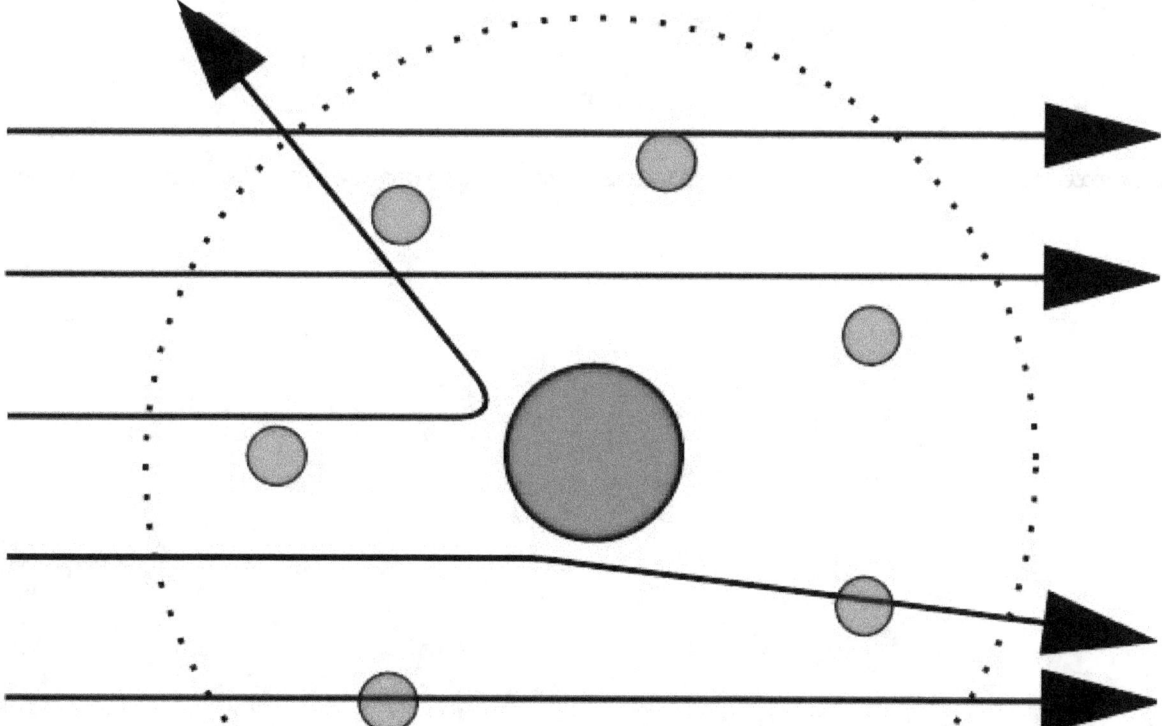

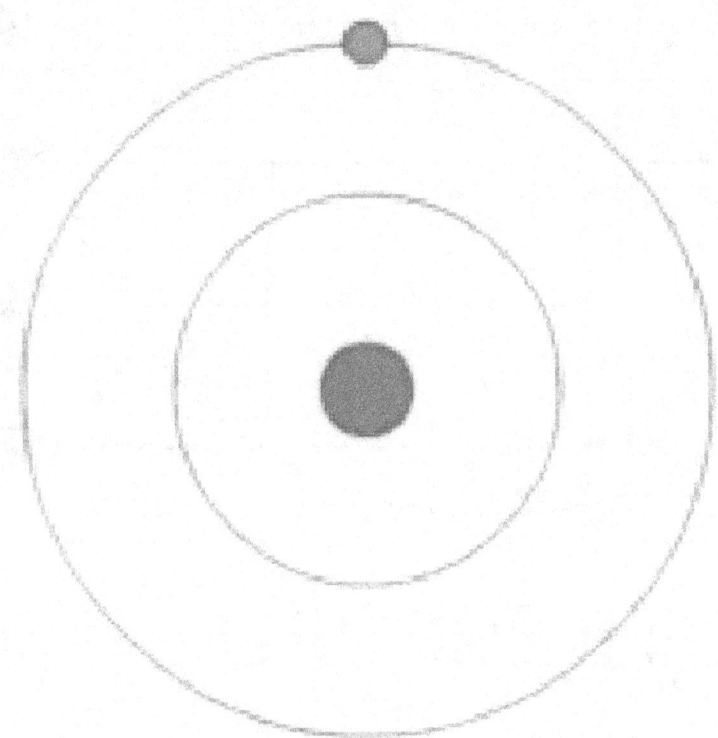

The Bohr model of the atom, with an electron making instantaneous "quantum leaps" from one orbit to another. This model is obsolete.

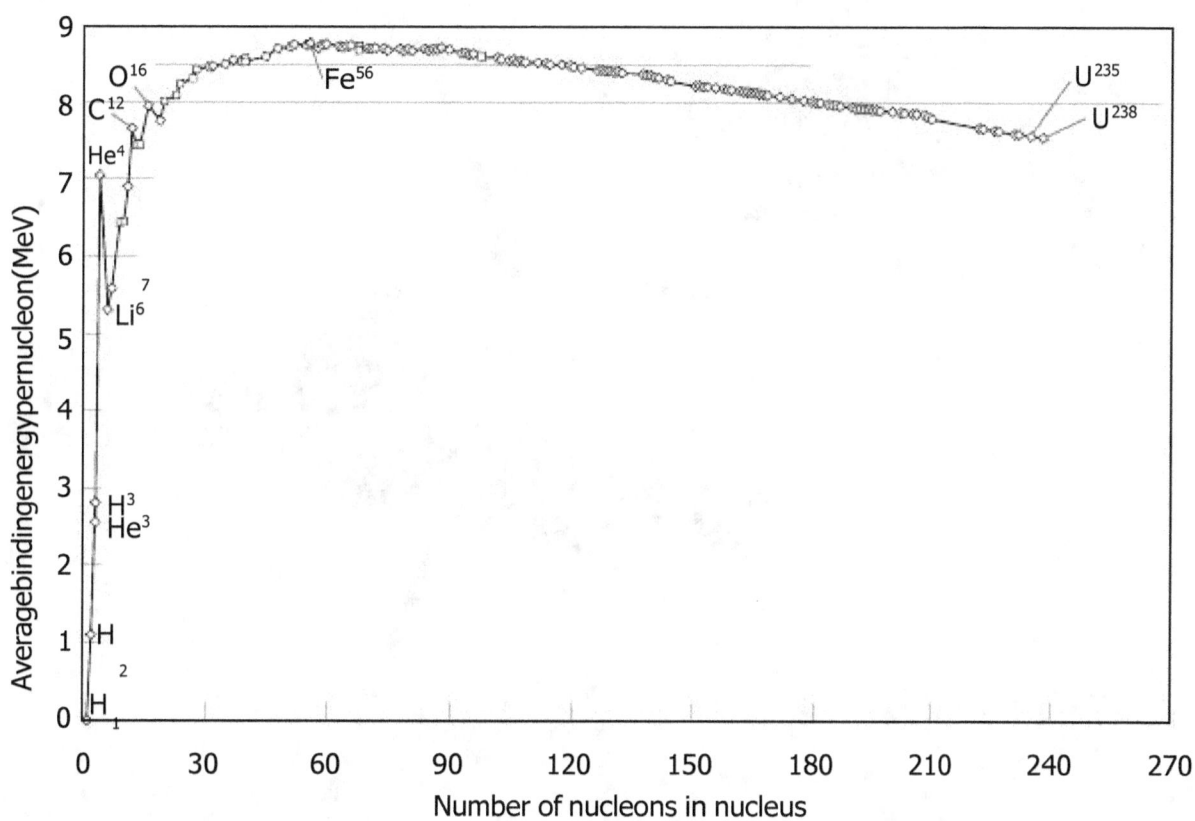

The binding energy needed for a nucleon to escape the nucleus, for various isotopes

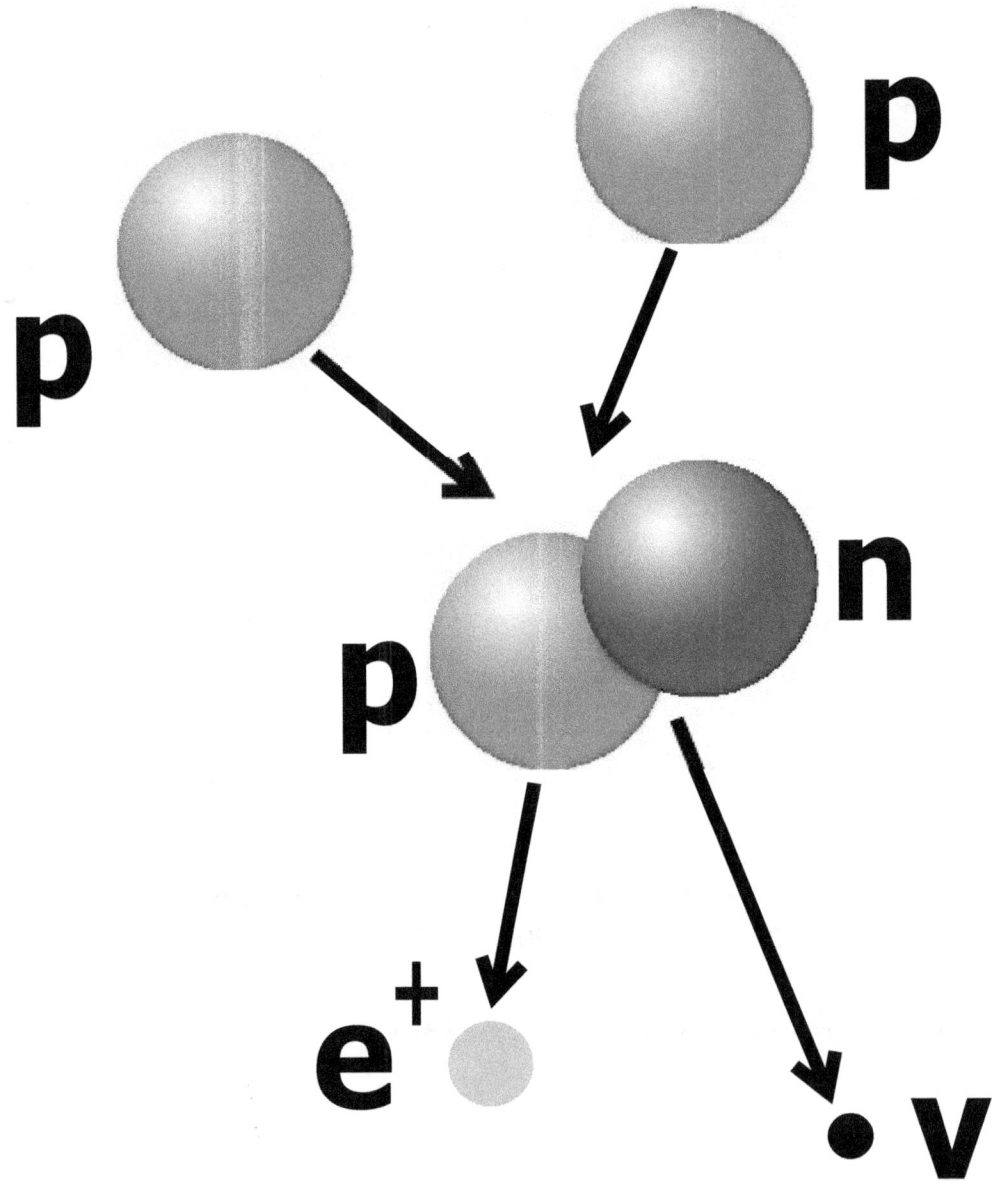

Illustration of a nuclear fusion process that forms a deuterium nucleus, consisting of a proton and a neutron, from two protons. A positron (e⁺)—an antimatter electron—is emitted along with an electron neutrino.

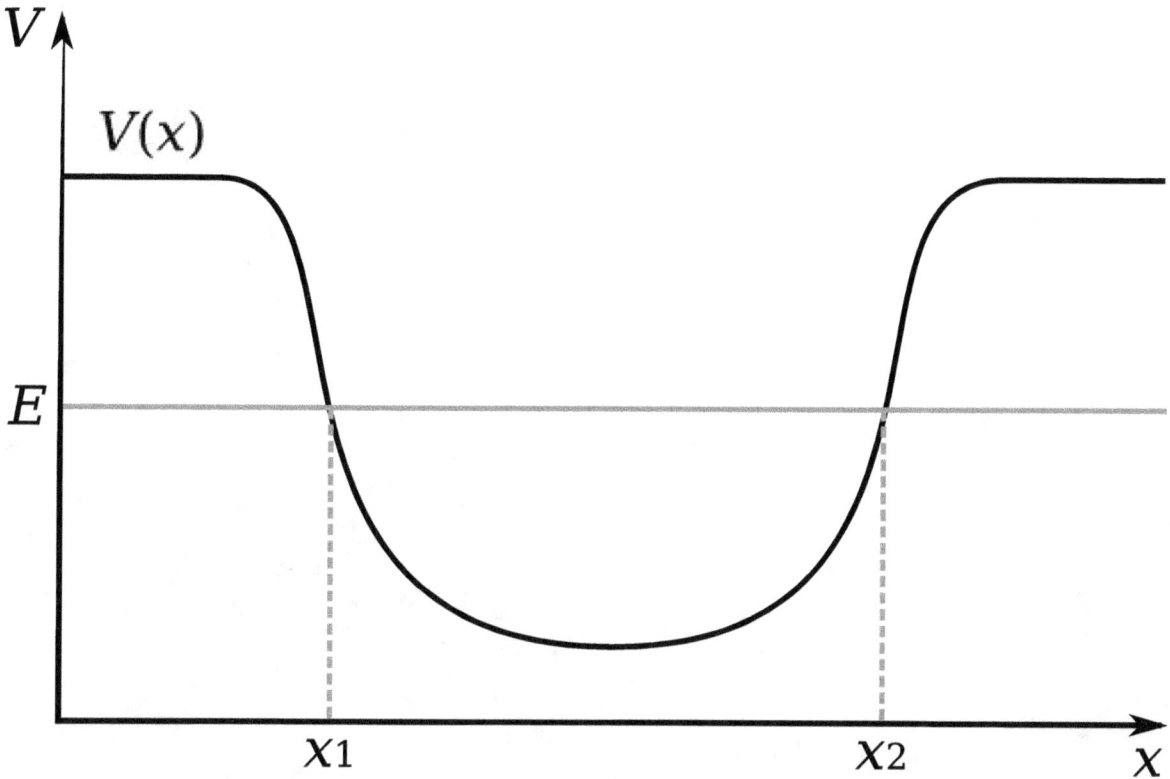

A potential well, showing, according to classical mechanics, the minimum energy V (x) needed to reach each position x. Classically, a particle with energy E is constrained to a range of positions between x₁ and x₂.

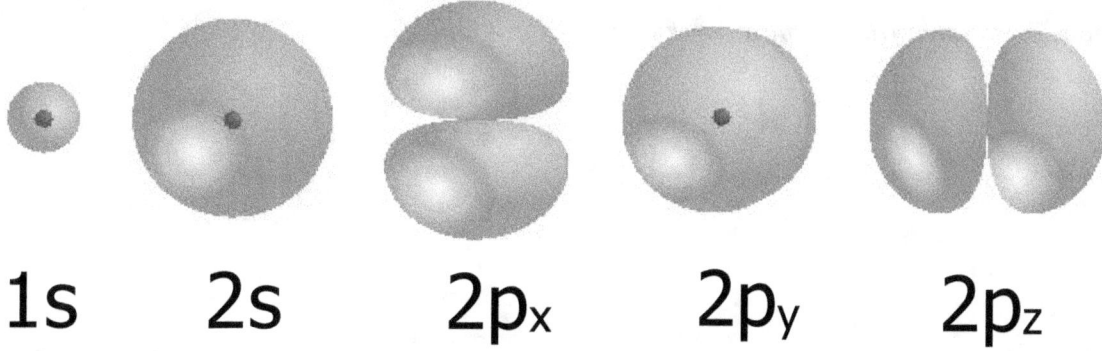

Wave functions of the first five atomic orbitals. The three 2p orbitals each display a single angular node that has an orientation and a minimum at the center.

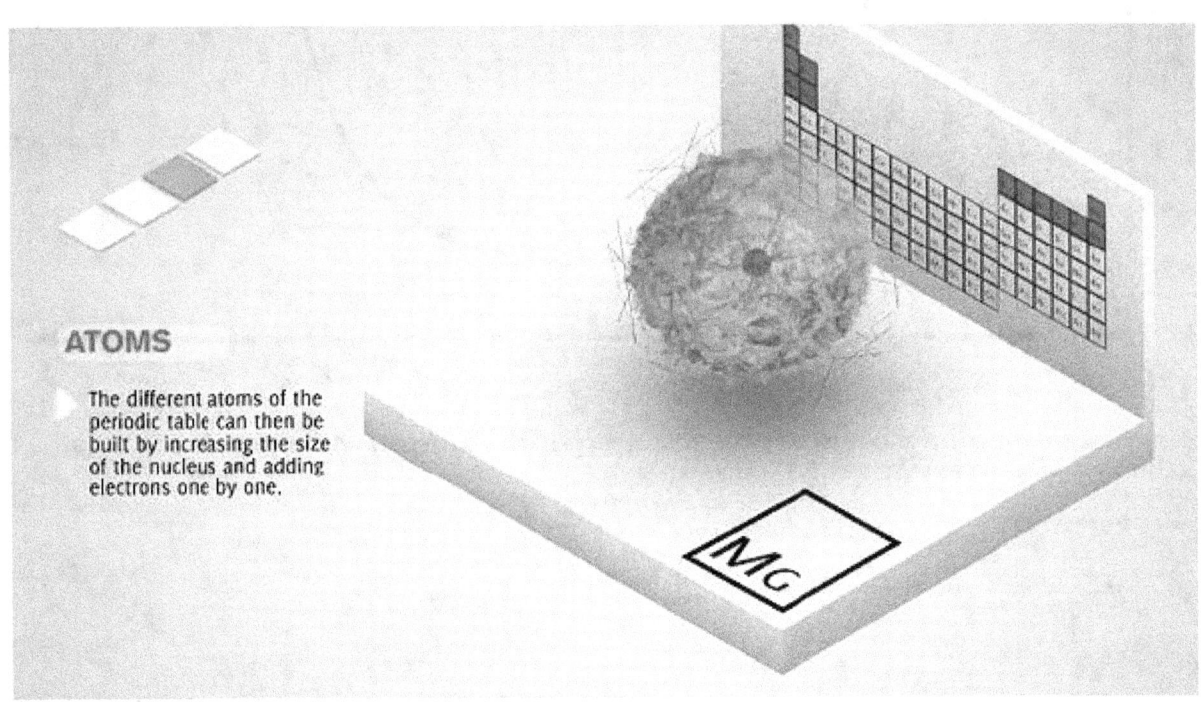

ATOMS

The different atoms of the periodic table can then be built by increasing the size of the nucleus and adding electrons one by one.

How atoms are constructed from electron orbitals and link to the periodic table

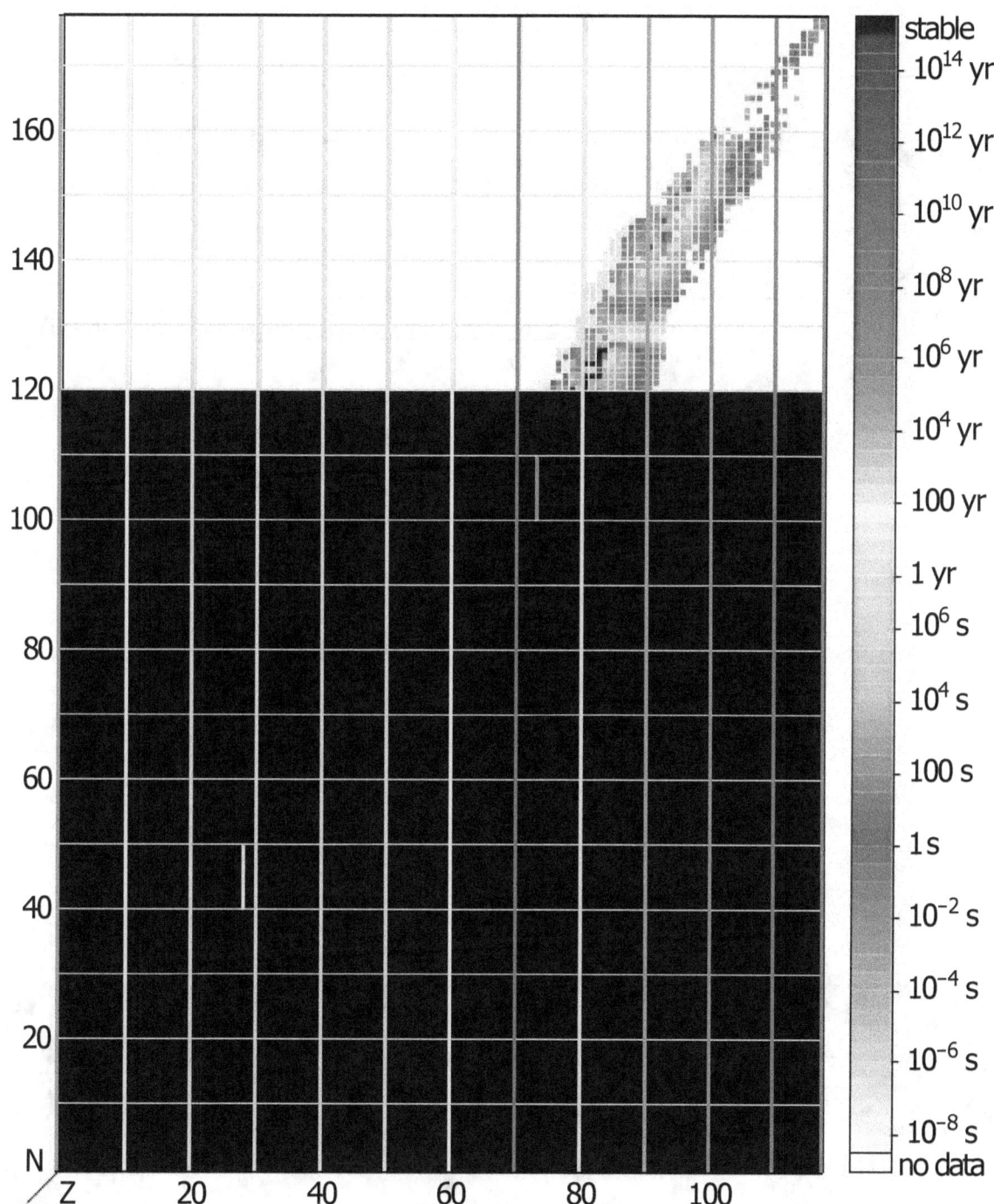

This diagram shows the half-life (T½) of various isotopes with Z protons and N neutrons.

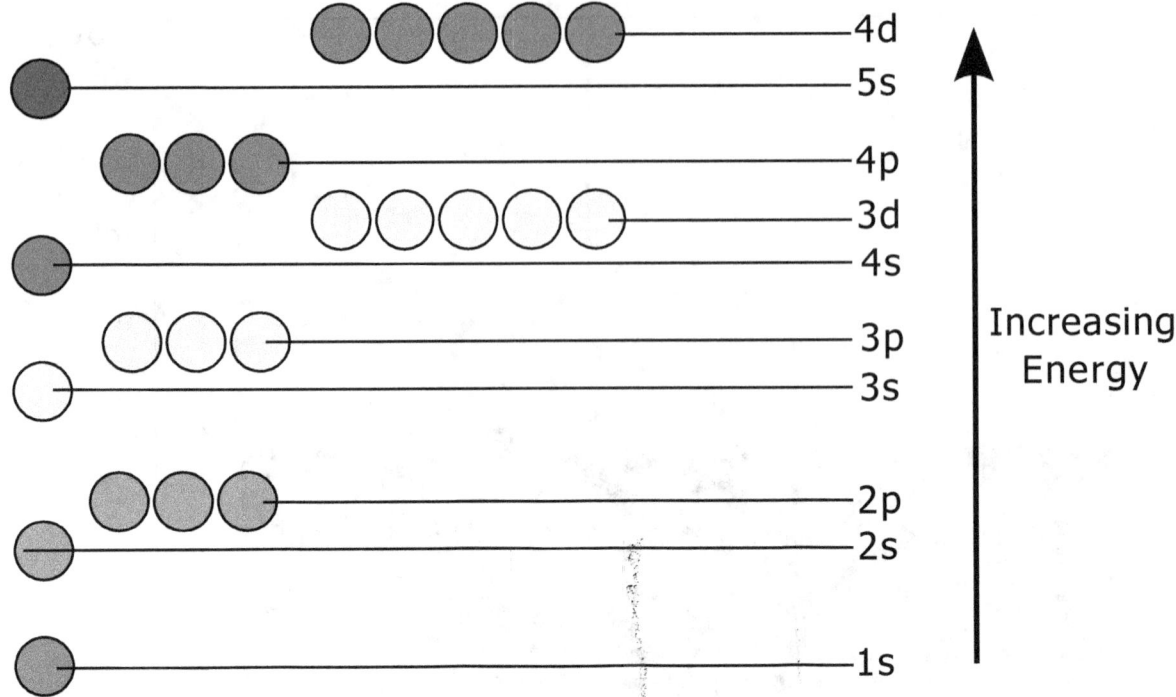

These electron's energy levels (not to scale) are sufficient for ground states of atoms up to cadmium ($5s^2\,4d^{10}$) inclusively. Do not forget that even the top of the diagram is lower than an unbound electron state.

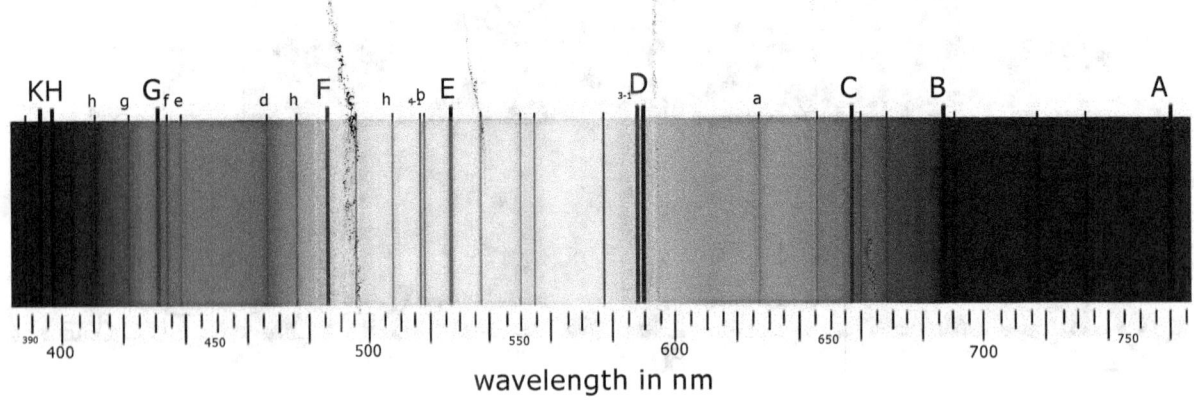

An example of absorption lines in a spectrum

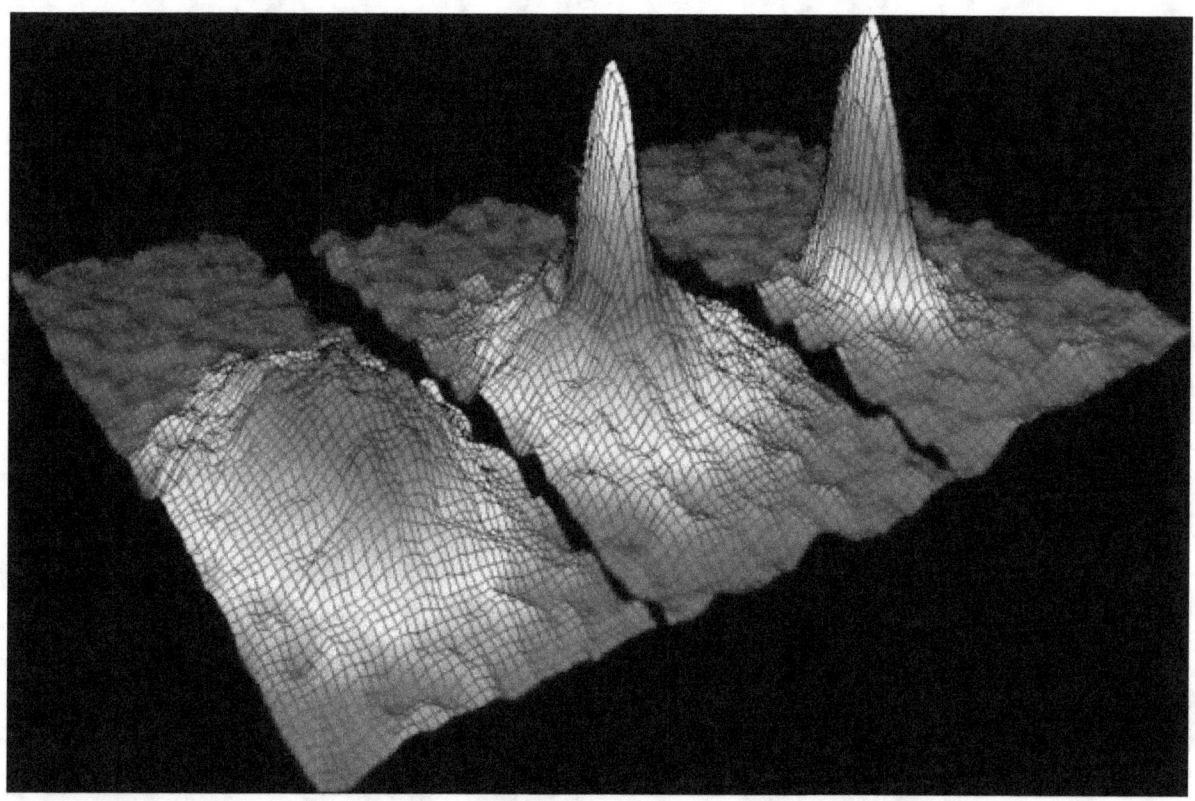

Snapshots illustrating the formation of a Bose–Einstein condensate

Scanning tunneling microscope image showing the individual atoms making up this gold (100) surface. The surface atoms deviate from the bulk crystal structure and arrange in columns several atoms wide with pits between them (See surface reconstruction).

Chapter 14

Gravity

See also Gravity (disambiguation), Gravitation (disambiguation) or Law of Gravity (disambiguation).

Gravity or **gravitation** is a natural phenomenon by which all things attract one another including stars, planets, galaxies

Hammer and feather drop: Apollo 15 astronaut David Scott on the Moon enacting the legend of Galileo's gravity experiment. (1.38 MB, ogg/Theora format).

and even light and sub-atomic particles. Gravity is responsible for the formation of the universe (e.g. creating spheres of hydrogen, igniting them under pressure to form stars and grouping them in to galaxies). Gravity is a cause of time dilation (time lapses more slowly in strong gravitation). Without gravity, the universe would be without thermal energy and composed only of equally spaced particles. On Earth, gravity gives weight to physical objects and causes the tides. Gravity has an infinite range, and it cannot be absorbed, transformed, or shielded against.

Gravity is most accurately described by the general theory of relativity (proposed by Albert Einstein in 1915) which describes gravity, not as a force, but as a consequence of the curvature of spacetime caused by the uneven distribution of mass/energy. For most applications, gravity is well approximated by Newton's law of universal gravitation, which postulates that the gravitational force of two bodies of mass is directly proportional to the product of their masses and inversely proportional to the square of the distance between them.

Gravity is the weakest of the four fundamental interactions of nature. The gravitational attraction is approximately 10^{-38} times the strength of the strong force (i.e. gravity is 38 orders of magnitude weaker), 10^{-36} times the strength of the electromagnetic force, and 10^{-29} times the strength of the weak force. As a consequence, gravity has a negligible influence on the behavior of sub-atomic particles, and plays no role in determining the internal properties of everyday matter. On the other hand, gravity is the dominant force at the macroscopic scale, that is the cause of the formation, shape, and trajectory (orbit) of astronomical bodies, including those of asteroids, comets, planets, stars, and galaxies. It is responsible for causing the Earth and the other planets to orbit the Sun; for causing the Moon to orbit the Earth; for the formation of tides; for natural convection, by which fluid flow occurs under the influence of a density gradient and gravity; for heating the interiors of forming stars and planets to very high temperatures; for solar system, galaxy, stellar formation and evolution; and for various other phenomena observed on Earth and throughout the universe.

In pursuit of a theory of everything, the merging of general relativity and quantum mechanics (or quantum field theory) into a more general theory of quantum gravity has become an area of research.

14.1 History of gravitational theory

Main article: History of gravitational theory

14.1.1 Scientific revolution

Modern work on gravitational theory began with the work of Galileo Galilei in the late 16th and early 17th centuries. In his famous (though possibly apocryphal[1]) experiment dropping balls from the Tower of Pisa, and later with careful measurements of balls rolling down inclines, Galileo showed that gravity accelerates all objects at the same rate. This was a major departure from Aristotle's belief that heavier objects accelerate faster.[2] Galileo postulated air resistance as the reason that lighter objects may fall more slowly in an atmosphere. Galileo's work set the stage for the formulation of Newton's theory of gravity.

14.1.2 Newton's theory of gravitation

Main article: Newton's law of universal gravitation

In 1687, English mathematician Sir Isaac Newton published *Principia*, which hypothesizes the inverse-square law of universal gravitation. In his own words, "I deduced that the forces which keep the planets in their orbs must [be] reciprocally as the squares of their distances from the centers about which they revolve: and thereby compared the force requisite to keep the Moon in her Orb with the force of gravity at the surface of the Earth; and found them answer pretty nearly."[3] The equation is the following:

$$F = G \frac{m_1 m_2}{r}$$

Where F is the force, m_1 and m_2 are the masses of the objects interacting, r is the distance between the centers of the masses and G is the gravitational constant.

Newton's theory enjoyed its greatest success when it was used to predict the existence of Neptune based on motions of Uranus that could not be accounted for by the actions of the other planets. Calculations by both John Couch Adams and Urbain Le Verrier predicted the general position of the planet, and Le Verrier's calculations are what led Johann Gottfried Galle to the discovery of Neptune.

Sir Isaac Newton, an English physicist who lived from 1642 to 1727

A discrepancy in Mercury's orbit pointed out flaws in Newton's theory. By the end of the 19th century, it was known that its orbit showed slight perturbations that could not be accounted for entirely under Newton's theory, but all searches for another perturbing body (such as a planet orbiting the Sun even closer than Mercury) had been fruitless. The issue was resolved in 1915 by Albert Einstein's new theory of general relativity, which accounted for the small discrepancy in Mercury's orbit.

Although Newton's theory has been superseded by the Einstein's general relativity, most modern non-relativistic gravitational calculations are still made using the Newton's theory because it is simpler to work with and it gives sufficiently accurate results for most applications involving sufficiently small masses, speeds and energies.

14.1.3 Equivalence principle

The equivalence principle, explored by a succession of researchers including Galileo, Loránd Eötvös, and Einstein, expresses the idea that all objects fall in the same way. The simplest way to test the weak equivalence principle is to drop two objects of different masses or compositions in a vacuum and see whether they hit the ground at the same time. Such experiments demonstrate that all objects fall at the same rate when other forces (such as air resistance and electromagnetic effects) are negligible. More sophisticated tests use a torsion balance of a type invented by Eötvös. Satellite experiments, for example STEP, are planned for more accurate experiments in space.[4]

Formulations of the equivalence principle include:

- The weak equivalence principle: *The trajectory of a point mass in a gravitational field depends only on its initial position and velocity, and is independent of its composition.*[5]

- The Einsteinian equivalence principle: *The outcome of any local non-gravitational experiment in a freely falling laboratory is independent of the velocity of the laboratory and its location in spacetime.*[6]

- The strong equivalence principle requiring both of the above.

14.1.4 General relativity

See also: Introduction to general relativity

In general relativity, the effects of gravitation are ascribed to spacetime curvature instead of a force. The starting point for general relativity is the equivalence principle, which equates free fall with inertial motion and describes free-falling inertial objects as being accelerated relative to non-inertial observers on the ground.[7][8] In Newtonian physics, however, no such acceleration can occur unless at least one of the objects is being operated on by a force.

Einstein proposed that spacetime is curved by matter, and that free-falling objects are moving along locally straight paths in curved spacetime. These straight paths are called geodesics. Like Newton's first law of motion, Einstein's theory states that if a force is applied on an object, it would deviate from a geodesic. For instance, we are no longer following geodesics while standing because the mechanical resistance of the Earth exerts an upward force on us, and we are non-inertial on the ground as a result. This explains why moving along the geodesics in spacetime is considered inertial.

Einstein discovered the field equations of general relativity, which relate the presence of matter and the curvature of spacetime and are named after him. The Einstein field equations are a set of 10 simultaneous, non-linear, differential equations. The solutions of the field equations are the components of the metric tensor of spacetime. A metric tensor describes a geometry of spacetime. The geodesic paths for a spacetime are calculated from the metric tensor.

Notable solutions of the Einstein field equations include:

- The Schwarzschild solution, which describes spacetime surrounding a spherically symmetric non-rotating uncharged massive object. For compact enough objects, this solution generated a black hole with a central singularity. For radial distances from the center which are much greater than the Schwarzschild radius, the accelerations predicted by the Schwarzschild solution are practically identical to those predicted by Newton's theory of gravity.

- The Reissner-Nordström solution, in which the central object has an electrical charge. For charges with a geometrized length which are less than the geometrized length of the mass of the object, this solution produces black holes with two event horizons.

- The Kerr solution for rotating massive objects. This solution also produces black holes with multiple event horizons.

- The Kerr-Newman solution for charged, rotating massive objects. This solution also produces black holes with multiple event horizons.

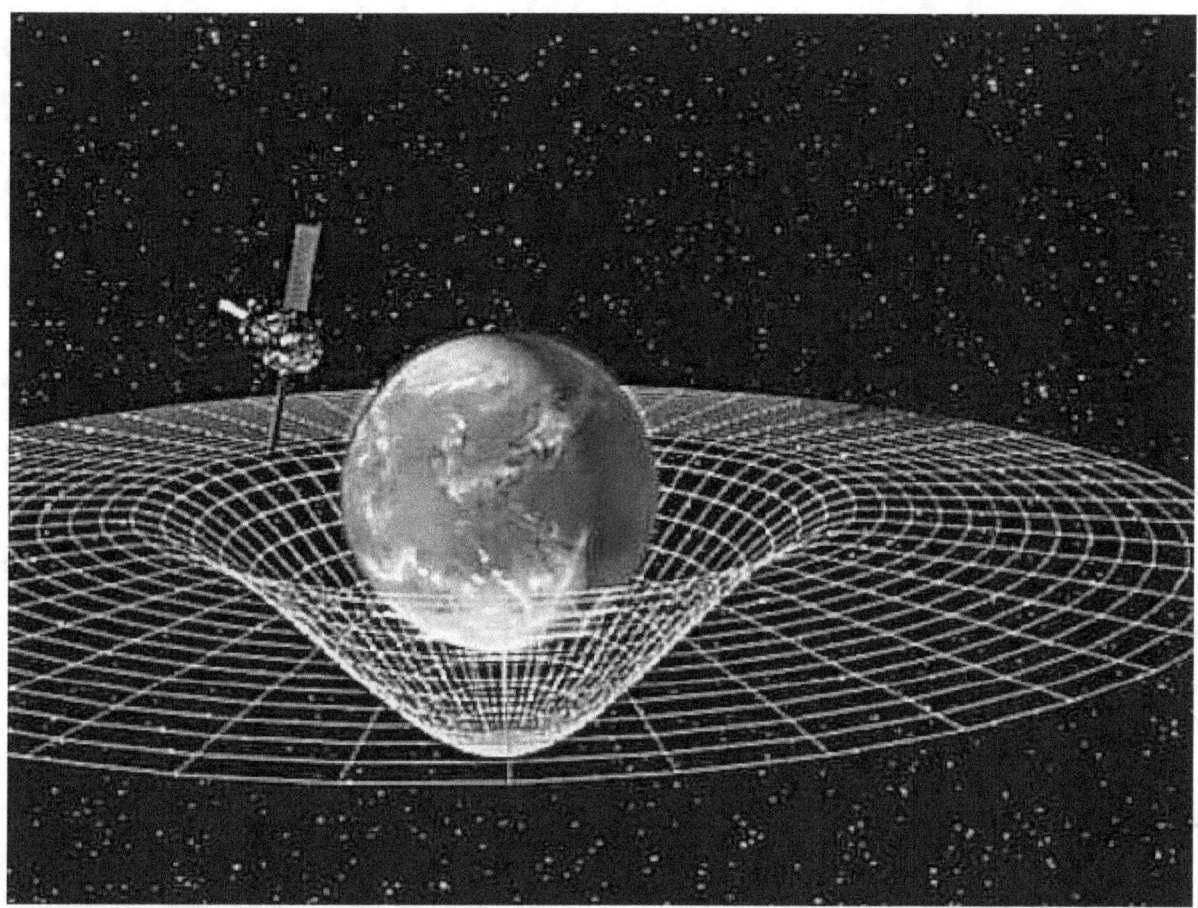

Two-dimensional analogy of spacetime distortion generated by the mass of an object. Matter changes the geometry of spacetime, this (curved) geometry being interpreted as gravity. White lines do not represent the curvature of space but instead represent the coordinate system imposed on the curved spacetime, which would be rectilinear in a flat spacetime.

- The cosmological Friedmann-Lemaître-Robertson-Walker solution, which predicts the expansion of the universe.

The tests of general relativity included the following:[9]

- General relativity accounts for the anomalous perihelion precession of Mercury.[10]

- The prediction that time runs slower at lower potentials has been confirmed by the Pound–Rebka experiment, the Hafele–Keating experiment, and the GPS.

- The prediction of the deflection of light was first confirmed by Arthur Stanley Eddington from his observations during the Solar eclipse of May 29, 1919.[11][12] Eddington measured starlight deflections twice those predicted by Newtonian corpuscular theory, in accordance with the predictions of general relativity. However, his interpretation of the results was later disputed.[13] More recent tests using radio interferometric measurements of quasars passing behind the Sun have more accurately and consistently confirmed the deflection of light to the degree predicted by general relativity.[14] See also gravitational lens.

- The time delay of light passing close to a massive object was first identified by Irwin I. Shapiro in 1964 in inter-planetary spacecraft signals.

- Gravitational radiation has been indirectly confirmed through studies of binary pulsars.

- Alexander Friedmann in 1922 found that Einstein equations have non-stationary solutions (even in the presence of the cosmological constant). In 1927 Georges Lemaître showed that static solutions of the Einstein equations, which are possible in the presence of the cosmological constant, are unstable, and therefore the static universe envisioned by Einstein could not exist. Later, in 1931, Einstein himself agreed with the results of Friedmann and Lemaître. Thus general relativity predicted that the Universe had to be non-static—it had to either expand or contract. The expansion of the universe discovered by Edwin Hubble in 1929 confirmed this prediction.[15]

- The theory's prediction of frame dragging was consistent with the recent Gravity Probe B results.[16]

- General relativity predicts that light should lose its energy when travelling away from the massive bodies. The group of Radek Wojtak of the Niels Bohr Institute at the University of Copenhagen collected data from 8000 galaxy clusters and found that the light coming from the cluster centers tended to be red-shifted compared to the cluster edges, confirming the energy loss due to gravity.[17]

14.1.5 Gravity and quantum mechanics

Main articles: Graviton and Quantum gravity

In the decades after the discovery of general relativity, it was realized that general relativity is incompatible with quantum mechanics.[18] It is possible to describe gravity in the framework of quantum field theory like the other fundamental forces, such that the attractive force of gravity arises due to exchange of virtual gravitons, in the same way as the electromagnetic force arises from exchange of virtual photons.[19][20] This reproduces general relativity in the classical limit. However, this approach fails at short distances of the order of the Planck length,[18] where a more complete theory of quantum gravity (or a new approach to quantum mechanics) is required.

14.2 Specifics

14.2.1 Earth's gravity

Main article: Earth's gravity

Every planetary body (including the Earth) is surrounded by its own gravitational field, which exerts an attractive force on all objects. Assuming a spherically symmetrical planet, the strength of this field at any given point above the surface is proportional to the planetary body's mass and inversely proportional to the square of the distance from the center of the body.

The strength of the gravitational field is numerically equal to the acceleration of objects under its influence. The rate of acceleration of falling objects near the Earth's surface varies very slightly depending on elevation, latitude, and other factors. For purposes of weights and measures, a standard gravity value is defined by the International Bureau of Weights and Measures, under the International System of Units (SI).

That value, denoted g, is $g = 9.80665$ m/s^2 (32.1740 ft/s^2).[21][22]

The standard value of 9.80665 m/s^2 is the one originally adopted by the International Committee on Weights and Measures in 1901 for 45° latitude, even though it has been shown to be too high by about five parts in ten thousand.[23] This value has persisted in meteorology and in some standard atmospheres as the value for 45° latitude even though it applies more precisely to latitude of 45°32'33".[24]

Assuming the standardized value for g and ignoring air resistance, this means that an object falling freely near the Earth's surface increases its velocity by 9.80665 m/s (32.1740 ft/s or 22 mph) for each second of its descent. Thus, an object starting from rest will attain a velocity of 9.80665 m/s (32.1740 ft/s) after one second, approximately 19.62 m/s (64.4 ft/s) after two seconds, and so on, adding 9.80665 m/s (32.1740 ft/s) to each resulting velocity. Also, again ignoring air resistance, any and all objects, when dropped from the same height, will hit the ground at the same time. It is relevant to note that Earth's gravity doesn't have exactly the same value in all regions. There are slight variations in different parts of the globe due to latitude, surface features such as mountains and ridges, and perhaps unusually high or low sub-surface densities.[25]

  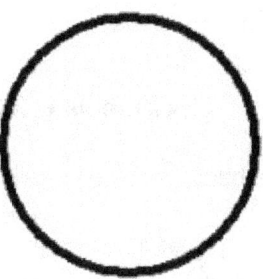

If an object with comparable mass to that of the Earth were to fall towards it, then the corresponding acceleration of the Earth would be observable.

According to Newton's 3rd Law, the Earth itself experiences a force equal in magnitude and opposite in direction to that which it exerts on a falling object. This means that the Earth also accelerates towards the object until they collide. Because the mass of the Earth is huge, however, the acceleration imparted to the Earth by this opposite force is negligible in comparison to the object's. If the object doesn't bounce after it has collided with the Earth, each of them then exerts a repulsive contact force on the other which effectively balances the attractive force of gravity and prevents further acceleration.

The force of gravity on Earth is the resultant (vector sum) of two forces: (a) The gravitational attraction in accordance with Newton's universal law of gravitation, and (b) the centrifugal force, which results from the choice of an earthbound, rotating frame of reference. At the equator, the force of gravity is the weakest due to the centrifugal force caused by the Earth's rotation. The force of gravity varies with latitude and increases from about 9.780 m/s^2 at the Equator to about 9.832 m/s^2 at the poles.

14.2.2   Equations for a falling body near the surface of the Earth

Main article: Equations for a falling body

Under an assumption of constant gravitational attraction, Newton's law of universal gravitation simplifies to $F = mg$, where m is the mass of the body and g is a constant vector with an average magnitude of 9.81 m/s^2 on Earth. This resulting force is the object's weight. The acceleration due to gravity is equal to this g. An initially stationary object which is allowed to fall freely under gravity drops a distance which is proportional to the square of the elapsed time. The image on the right, spanning half a second, was captured with a stroboscopic flash at 20 flashes per second. During the first $^1/20$ of a second the ball drops one unit of distance (here, a unit is about 12 mm); by $^2/20$ it has dropped at total of 4 units; by $^3/20$, 9 units and so on.

Under the same constant gravity assumptions, the potential energy, Ep, of a body at height h is given by $Ep = mgh$ (or $Ep = Wh$, with W meaning weight). This expression is valid only over small distances h from the surface of the Earth. Similarly, the expression $h = \frac{v^2}{2g}$ for the maximum height reached by a vertically projected body with initial velocity v is useful for small heights and small initial velocities only.

14.2.3   Gravity and astronomy

The application of Newton's law of gravity has enabled the acquisition of much of the detailed information we have about the planets in our solar system, the mass of the Sun, and details of quasars; even the existence of dark matter is inferred using Newton's law of gravity. Although we have not traveled to all the planets nor to the Sun, we know their masses. These masses are obtained by applying the laws of gravity to the measured characteristics of the orbit. In space an object maintains its orbit because of the force of gravity acting upon it. Planets orbit stars, stars orbit Galactic Centers, galaxies orbit a center of mass in clusters, and clusters orbit in superclusters. The force of gravity exerted on one object by another is directly proportional to the product of those objects' masses and inversely proportional to the square of the distance between them.

14.2.4 Gravitational radiation

Main article: Gravitational wave

In general relativity, gravitational radiation is generated in situations where the curvature of spacetime is oscillating, such as is the case with co-orbiting objects. The gravitational radiation emitted by the Solar System is far too small to measure. However, gravitational radiation has been indirectly observed as an energy loss over time in binary pulsar systems such as PSR B1913+16. It is believed that neutron star mergers and black hole formation may create detectable amounts of gravitational radiation. Gravitational radiation observatories such as the Laser Interferometer Gravitational Wave Observatory (LIGO) have been created to study the problem. No confirmed detections have been made of this hypothetical radiation.

14.2.5 Speed of gravity

Main article: Speed of gravity

In December 2012, a research team in China announced that it had produced measurements of the phase lag of Earth tides during full and new moons which seem to prove that the speed of gravity is equal to the speed of light.[27] This means that if the Sun suddenly disappeared, the Earth would keep orbiting it normally for 8 minutes, which is the time light takes to travel that distance. The team's findings were released in the Chinese Science Bulletin in February 2013.[28]

14.3 Anomalies and discrepancies

There are some observations that are not adequately accounted for, which may point to the need for better theories of gravity or perhaps be explained in other ways.

- **Extra-fast stars:** Stars in galaxies follow a distribution of velocities where stars on the outskirts are moving faster than they should according to the observed distributions of normal matter. Galaxies within galaxy clusters show a similar pattern. Dark matter, which would interact gravitationally but not electromagnetically, would account for the discrepancy. Various modifications to Newtonian dynamics have also been proposed.

- **Flyby anomaly:** Various spacecraft have experienced greater acceleration than expected during gravity assist maneuvers.

- **Accelerating expansion:** The metric expansion of space seems to be speeding up. Dark energy has been proposed to explain this. A recent alternative explanation is that the geometry of space is not homogeneous (due to clusters of galaxies) and that when the data are reinterpreted to take this into account, the expansion is not speeding up after all,[29] however this conclusion is disputed.[30]

- **Anomalous increase of the astronomical unit:** Recent measurements indicate that planetary orbits are widening faster than if this were solely through the sun losing mass by radiating energy.

- **Extra energetic photons:** Photons travelling through galaxy clusters should gain energy and then lose it again on the way out. The accelerating expansion of the universe should stop the photons returning all the energy, but even taking this into account photons from the cosmic microwave background radiation gain twice as much energy as expected. This may indicate that gravity falls off *faster* than inverse-squared at certain distance scales.[31]

- **Extra massive hydrogen clouds:** The spectral lines of the Lyman-alpha forest suggest that hydrogen clouds are more clumped together at certain scales than expected and, like dark flow, may indicate that gravity falls off *slower* than inverse-squared at certain distance scales.[31]

- **Power:** Proposed extra dimensions could explain why the gravity force is so weak.[32]

14.4 Alternative theories

Main article: Alternatives to general relativity

14.4.1 Historical alternative theories

- Aristotelian theory of gravity

- Le Sage's theory of gravitation (1784) also called LeSage gravity, proposed by Georges-Louis Le Sage, based on a fluid-based explanation where a light gas fills the entire universe.

- Ritz's theory of gravitation, *Ann. Chem. Phys.* 13, 145, (1908) pp. 267–271, Weber-Gauss electrodynamics applied to gravitation. Classical advancement of perihelia.

- Nordström's theory of gravitation (1912, 1913), an early competitor of general relativity.

- Kaluza Klein theory (1921)

- Whitehead's theory of gravitation (1922), another early competitor of general relativity.

14.4.2 Recent alternative theories

- Brans–Dicke theory of gravity (1961) [33]

- Induced gravity (1967), a proposal by Andrei Sakharov according to which general relativity might arise from quantum field theories of matter

- $f(R)$ gravity (1970)

- Horndeski theory (1974) [34]

- Supergravity (1976)

- String theory

- In the modified Newtonian dynamics (MOND) (1981), Mordehai Milgrom proposes a modification of Newton's Second Law of motion for small accelerations [35]

- The self-creation cosmology theory of gravity (1982) by G.A. Barber in which the Brans-Dicke theory is modified to allow mass creation

- Loop quantum gravity (1988) by Carlo Rovelli, Lee Smolin, and Abhay Ashtekar

- Nonsymmetric gravitational theory (NGT) (1994) by John Moffat

- Tensor–vector–scalar gravity (TeVeS) (2004), a relativistic modification of MOND by Jacob Bekenstein

- Gravity as an entropic force, gravity arising as an emergent phenomenon from the thermodynamic concept of entropy.

- In the superfluid vacuum theory the gravity and curved space-time arise as a collective excitation mode of non-relativistic background superfluid.
- Chameleon theory (2004) by Justin Khoury and Amanda Weltman.
- Pressuron theory (2013) by Olivier Minazzoli and Aurélien Hees.

14.5   See also

- Angular momentum
- Anti-gravity, the idea of neutralizing or repelling gravity
- Artificial gravity
- Birkeland current
- Gravitational wave
- Gravitational wave background
- Cosmic gravitational wave background
- Einstein–Infeld–Hoffmann equations
- Escape velocity, the minimum velocity needed to escape from a gravity well
- g-force, a measure of acceleration
- Gauge gravitation theory
- Gauss's law for gravity
- Gravitational binding energy
- Gravity assist
- Gravity gradiometry
- Gravity Recovery and Climate Experiment
- Gravity Research Foundation
- Jovian–Plutonian gravitational effect
- Kepler's third law of planetary motion
- Lagrangian point
- Micro-g environment, also called microgravity
- Mixmaster dynamics
- *n*-body problem
- Newton's laws of motion
- Pioneer anomaly
- Scalar theories of gravitation
- Speed of gravity
- Standard gravitational parameter
- Standard gravity
- Weightlessness

14.6 Footnotes

[1] Ball, Phil (June 2005). "Tall Tales". *Nature News*. doi:10.1038/news050613-10.

[2] Galileo (1638), *Two New Sciences*; First Day Salviati speaks: "If this were what Aristotle meant you would burden him with another error which would amount to a falsehood; because, since there is no such sheer height available on earth, it is clear that Aristotle could not have made the experiment; yet he wishes to give us the impression of his having performed it when he speaks of such an effect as one which we see."

[3]   • Chandrasekhar, Subrahmanyan (2003). *Newton's Principia for the common reader*. Oxford: Oxford University Press. (pp.1–2). The quotation comes from a memorandum thought to have been written about 1714. As early as 1645 Ismaël Bullialdus had argued that any force exerted by the Sun on distant objects would have to follow an inverse-square law. However, he also dismissed the idea that any such force did exist. See, for example,
Linton, Christopher M. (2004). *From Eudoxus to Einstein—A History of Mathematical Astronomy*. Cambridge: Cambridge University Press. p. 225. ISBN 978-0-521-82750-8.

[4] M.C.W.Sandford (2008). "STEP: Satellite Test of the Equivalence Principle". Rutherford Appleton Laboratory. Retrieved 2011-10-14.

[5] Paul S Wesson (2006). *Five-dimensional Physics*. World Scientific. p. 82. ISBN 981-256-661-9.

[6] Haugen, Mark P.; C. Lämmerzahl (2001). *Principles of Equivalence: Their Role in Gravitation Physics and Experiments that Test Them*. Springer. arXiv:gr-qc/0103067. ISBN 978-3-540-41236-6.

[7] "Gravity and Warped Spacetime". black-holes.org. Retrieved 2010-10-16.

[8] Dmitri Pogosyan. "Lecture 20: Black Holes—The Einstein Equivalence Principle". University of Alberta. Retrieved 2011-10-14.

[9] Pauli, Wolfgang Ernst (1958). "Part IV. General Theory of Relativity". *Theory of Relativity*. Courier Dover Publications. ISBN 978-0-486-64152-2.

[10] Max Born (1924), *Einstein's Theory of Relativity* (The 1962 Dover edition, page 348 lists a table documenting the observed and calculated values for the precession of the perihelion of Mercury, Venus, and Earth.)

[11] Dyson, F.W.; Eddington, A.S.; Davidson, C.R. (1920). "A Determination of the Deflection of Light by the Sun's Gravitational Field, from Observations Made at the Total Eclipse of May 29, 1919". *Phil. Trans. Roy. Soc. A* **220** (571–581): 291–333. Bibcode:1920RSPTA.220..291D. doi:10.1098/rsta.1920.0009.. Quote, p. 332: "Thus the results of the expeditions to Sobral and Principe can leave little doubt that a deflection of light takes place in the neighbourhood of the sun and that it is of the amount demanded by Einstein's generalised theory of relativity, as attributable to the sun's gravitational field."

[12] Weinberg, Steven (1972). *Gravitation and cosmology*. John Wiley & Sons.. Quote, p. 192: "About a dozen stars in all were studied, and yielded values 1.98 ± 0.11" and 1.61 ± 0.31", in substantial agreement with Einstein's prediction $\theta\odot$ = 1.75"."

[13] Earman, John; Glymour, Clark (1980). "Relativity and Eclipses: The British eclipse expeditions of 1919 and their predecessors". *Historical Studies in the Physical Sciences* **11**: 49–85. doi:10.2307/27757471.

[14] Weinberg, Steven (1972). *Gravitation and cosmology*. John Wiley & Sons. p. 194.

[15] See W.Pauli, 1958, pp.219–220

[16] "NASA's Gravity Probe B Confirms Two Einstein Space-Time Theories". Nasa.gov. Retrieved 2013-07-23.

[17] Bhattacharjee, Yudhijit. "Galaxy Clusters Validate Einstein's Theory". News.sciencemag.org. Retrieved 2013-07-23.

[18] Randall, Lisa (2005). *Warped Passages: Unraveling the Universe's Hidden Dimensions*. Ecco. ISBN 0-06-053108-8.

[19] Feynman, R. P.; Morinigo, F. B.; Wagner, W. G.; Hatfield, B. (1995). *Feynman lectures on gravitation*. Addison-Wesley. ISBN 0-201-62734-5.

[20] Zee, A. (2003). *Quantum Field Theory in a Nutshell*. Princeton University Press. ISBN 0-691-01019-6.

[21] Bureau International des Poids et Mesures (2006). "The International System of Units (SI)" (PDF) (8th ed.). p. 131. Retrieved 2009-11-25. Unit names are normally printed in Roman (upright) type ...   Symbols for quantities are generally single letters set in an italic font, although they may be qualified by further information in subscripts or superscripts or in brackets.

[22] "SI Unit rules and style conventions". National Institute For Standards and Technology (USA). September 2004. Retrieved 2009-11-25. Variables and quantity symbols are in italic type. Unit symbols are in Roman type.

[23] List, R. J. editor, 1968, Acceleration of Gravity, *Smithsonian Meteorological Tables*, Sixth Ed. Smithsonian Institution, Washington, D.C., p. 68.

[24] U.S. Standard Atmosphere, 1976, U.S. Government Printing Office, Washington, D.C., 1976. (Linked file is very large.)

[25] "Astronomy Picture of the Day".

[26] "Milky Way Emerges as Sun Sets over Paranal". *www.eso.org*. European Southern Obseevatory. Retrieved 29 April 2015.

[27] Chinese scientists find evidence for speed of gravity, astrowatch.com, 12/28/12.

[28] TANG, Ke Yun; HUA ChangCai; WEN Wu; CHI ShunLiang; YOU QingYu; YU Dan (February 2013). "Observational evidences for the speed of the gravity based on the Earth tide" (PDF). *Chinese Science Bulletin* 58 (4-5): 474–477. doi:10.1007/s11434-012-5603-3. Retrieved 12 June 2013.

[29] Dark energy may just be a cosmic illusion, *New Scientist*, issue 2646, 7 March 2008.

[30] Swiss-cheese model of the cosmos is full of holes, *New Scientist*, issue 2678, 18 October 2008.

[31] Chown, Marcus (16 March 2009). "Gravity may venture where matter fears to tread". *New Scientist* (2699). Retrieved 4 August 2013.

[32] CERN (20 January 2012). "Extra dimensions, gravitons, and tiny black holes".

[33] Brans, C.H. (Mar 2014). "Jordan-Brans-Dicke Theory". *Scholarpedia* 9: 31358. Bibcode:2014Schpj...931358B. doi:10.4249/scholarpedia.31358.

[34] Horndeski, G.W. (Sep 1974). "Second-Order Scalar-Tensor Field Equations in a Four-Dimensional Space". *International Journal of Theoretical Physics* 88 (10): 363–384. Bibcode:1974IJTP...10..363H. doi:10.1007/BF01807638.

[35] Milgrom, M. (Jun 2014). "The MOND paradigm of modified dynamics". *Scholarpedia* 9: 31410. Bibcode:2014SchpJ...931410M. doi:10.4249/scholarpedia.31410.

14.7 References

- Halliday, David; Robert Resnick; Kenneth S. Krane (2001). *Physics v. 1*. New York: John Wiley & Sons. ISBN 0-471-32057-9.

- Serway, Raymond A.; Jewett, John W. (2004). *Physics for Scientists and Engineers* (6th ed.). Brooks/Cole. ISBN 0-534-40842-7.

- Tipler, Paul (2004). *Physics for Scientists and Engineers: Mechanics, Oscillations and Waves, Thermodynamics* (5th ed.). W. H. Freeman. ISBN 0-7167-0809-4.

14.8 Further reading

- Thorne, Kip S.; Misner, Charles W.; Wheeler, John Archibald (1973). *Gravitation*. W.H. Freeman. ISBN 0-7167-0344-0.

14.9 External links

- Hazewinkel, Michiel, ed. (2001), "Gravitation", *Encyclopedia of Mathematics*, Springer, ISBN 978-1-55608-010-4

- Hazewinkel, Michiel, ed. (2001), "Gravitation, theory of", *Encyclopedia of Mathematics*, Springer, ISBN 978-1-55608-010-4

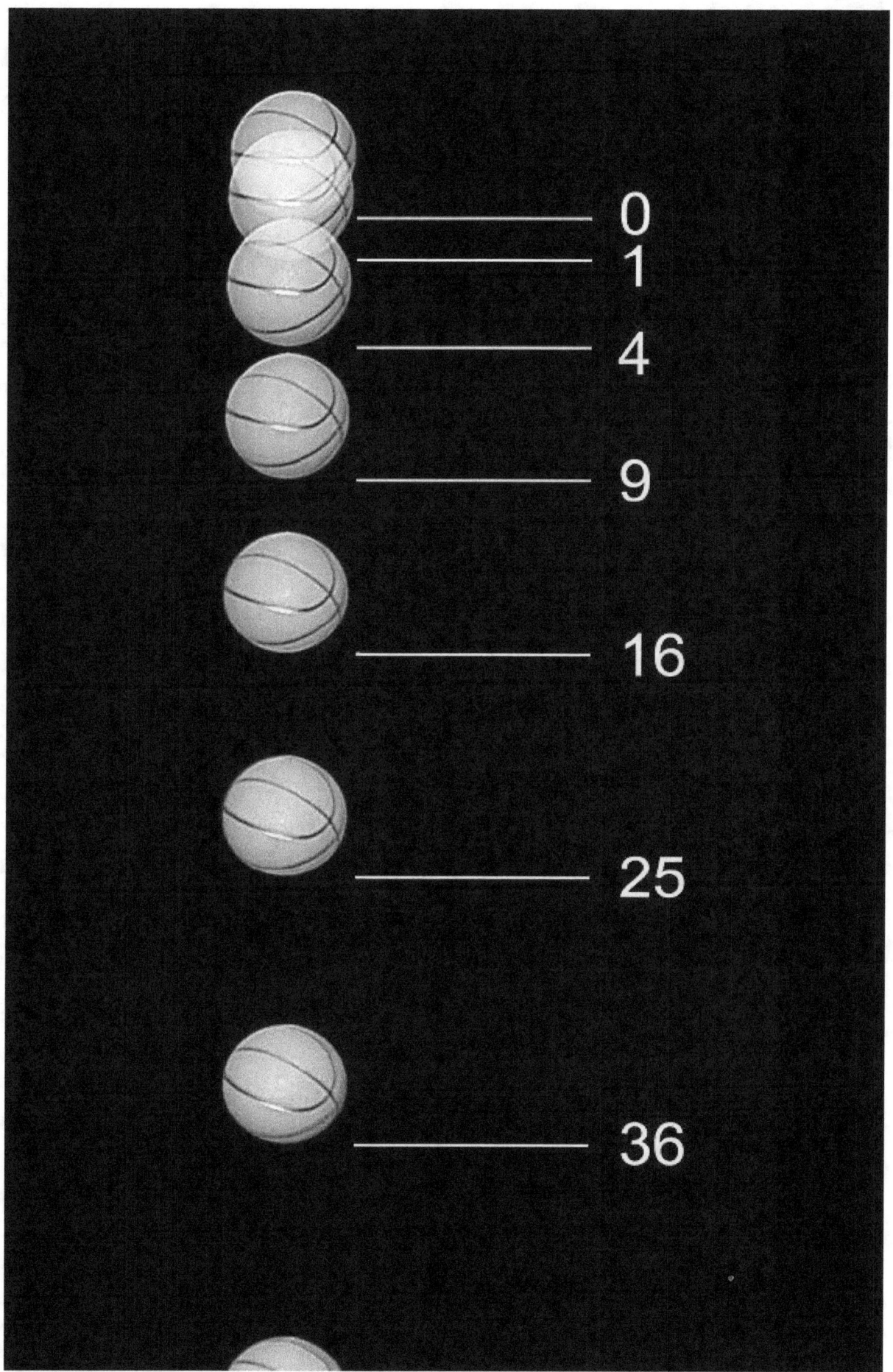

Gravity acts on stars that conform our Milky Way.[26]

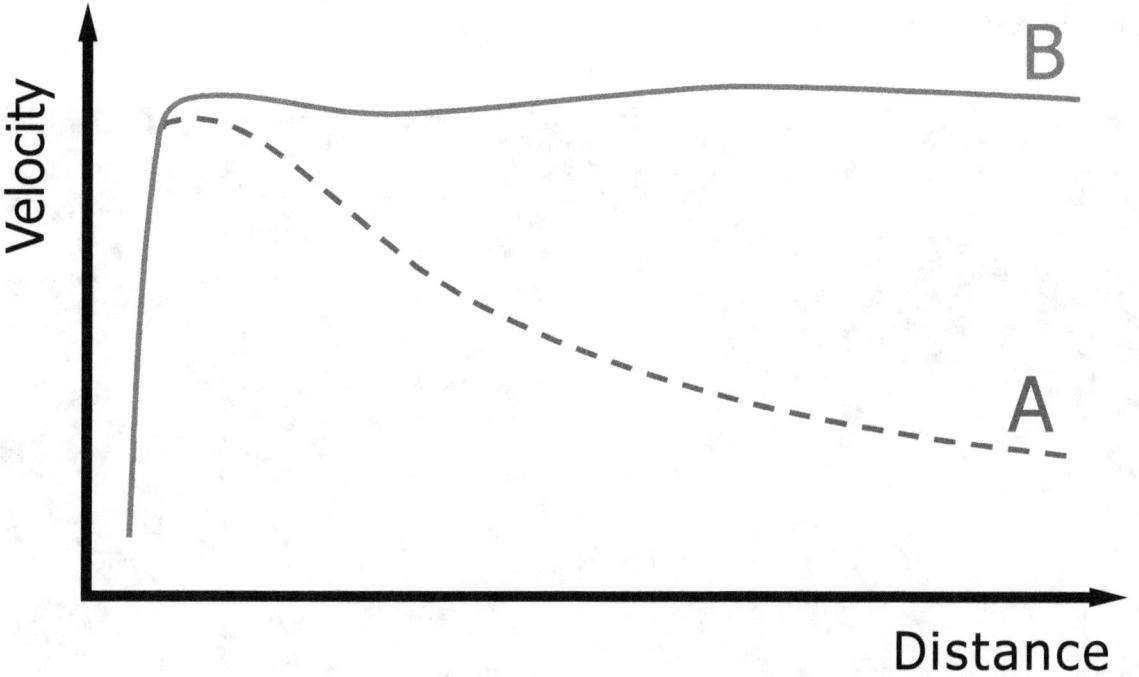

*Rotation curve of a typical spiral galaxy: predicted (**A**) and observed (**B**). The discrepancy between the curves is attributed to dark matter.*

Chapter 15

Star

For other uses, see Star (disambiguation).

A star-forming region in the Large Magellanic Cloud.

A **star** is a luminous sphere of plasma held together by its own gravity. The nearest star to Earth is the Sun. Other stars, mostly in the Milky Way, are visible from Earth during the night, appearing as a multitude of fixed luminous points in the sky due to their immense distance from Earth. Historically, the most prominent stars were grouped into constellations and asterisms, and the brightest stars gained proper names. Extensive catalogues of stars have been assembled by astronomers, which provide standardized star designations.

For at least a portion of its life, a star shines due to thermonuclear fusion of hydrogen into helium in its core, releasing energy that traverses the star's interior and then radiates into outer space. Once the hydrogen in the core of a star is nearly exhausted, almost all naturally occurring elements heavier than helium are created by stellar nucleosynthesis during the

False-color imagery of the Sun, a G-type main-sequence star, the closest to Earth

star's lifetime and, for some stars, by supernova nucleosynthesis when it explodes. Near the end of its life, a star can also contain degenerate matter. Astronomers can determine the mass, age, metallicity (chemical composition), and many other properties of a star by observing its motion through space, luminosity, and spectrum respectively. The total mass of a star is the principal determinant of its evolution and eventual fate. Other characteristics of a star, including diameter and temperature, change over its life, while the star's environment affects its rotation and movement. A plot of the temperature of many stars against their luminosities, known as a Hertzsprung–Russell diagram (H–R diagram), allows the age and evolutionary state of a star to be determined.

A star's life begins with the gravitational collapse of a gaseous nebula of material composed primarily of hydrogen, along with helium and trace amounts of heavier elements. Once the stellar core is sufficiently dense, hydrogen becomes steadily converted into helium through nuclear fusion, releasing energy in the process.[1] The remainder of the star's interior carries energy away from the core through a combination of radiative and convective processes. The star's internal pressure prevents it from collapsing further under its own gravity. Once the hydrogen fuel at the core is exhausted, a star with at least 0.4 times the mass of the Sun[2] expands to become a red giant, in some cases fusing heavier elements at the core or in shells around the core. The star then evolves into a degenerate form, recycling a portion of its matter into the interstellar environment, where it will contribute to the formation of a new generation of stars with a higher proportion of heavy elements.[3] Meanwhile, the core becomes a stellar remnant: a white dwarf, a neutron star, or (if it is sufficiently massive) a black hole.

Binary and multi-star systems consist of two or more stars that are gravitationally bound, and generally move around each other in stable orbits. When two such stars have a relatively close orbit, their gravitational interaction can have a significant impact on their evolution.[4] Stars can form part of a much larger gravitationally bound structure, such as a star cluster or a galaxy.

15.1   Observation history

People have seen patterns in the stars since ancient times.[5] This 1690 depiction of the constellation of Leo, the lion, is by Johannes Hevelius.[6]

Historically, stars have been important to civilizations throughout the world. They have been part of religious practices and used for celestial navigation and orientation. Many ancient astronomers believed that stars were permanently affixed to a heavenly sphere, and that they were immutable. By convention, astronomers grouped stars into constellations and used them to track the motions of the planets and the inferred position of the Sun.[5] The motion of the Sun against the background stars (and the horizon) was used to create calendars, which could be used to regulate agricultural practices.[7] The Gregorian calendar, currently used nearly everywhere in the world, is a solar calendar based on the angle of the Earth's rotational axis relative to its local star, the Sun.

The oldest accurately dated star chart appeared in ancient Egyptian astronomy in 1534 BC.[8] The earliest known star catalogues were compiled by the ancient Babylonian astronomers of Mesopotamia in the late 2nd millennium BC, during the Kassite Period (*ca.* 1531–1155 BC).[9]

The first star catalogue in Greek astronomy was created by Aristillus in approximately 300 BC, with the help of Timocharis.[10] The star catalog of Hipparchus (2nd century BC) included 1020 stars and was used to assemble Ptolemy's star catalogue.[11] Hipparchus is known for the discovery of the first recorded *nova* (new star).[12] Many of the constellations and star names in use today derive from Greek astronomy.

In spite of the apparent immutability of the heavens, Chinese astronomers were aware that new stars could appear.[13] In 185 AD, they were the first to observe and write about a supernova, now known as the SN 185.[14] The brightest stellar event in recorded history was the SN 1006 supernova, which was observed in 1006 and written about by the Egyptian astronomer Ali ibn Ridwan and several Chinese astronomers.[15] The SN 1054 supernova, which gave birth to the Crab Nebula, was also observed by Chinese and Islamic astronomers.[16][17][18]

Medieval Islamic astronomers gave Arabic names to many stars that are still used today, and they invented numerous astronomical instruments that could compute the positions of the stars. They built the first large observatory research institutes, mainly for the purpose of producing *Zij* star catalogues.[19] Among these, the *Book of Fixed Stars* (964) was written by the Persian astronomer Abd al-Rahman al-Sufi, who observed a number of stars, star clusters (including the Omicron Velorum and Brocchi's Clusters) and galaxies (including the Andromeda Galaxy).[20] According to A. Zahoor, in the 11th century, the Persian polymath scholar Abu Rayhan Biruni described the Milky Way galaxy as a multitude of fragments having the properties of nebulous stars, and also gave the latitudes of various stars during a lunar eclipse in 1019.[21]

According to Josep Puig, the Andalusian astronomer Ibn Bajjah proposed that the Milky Way was made up of many stars which almost touched one another and appeared to be a continuous image due to the effect of refraction from sublunary material, citing his observation of the conjunction of Jupiter and Mars on 500 AH (1106/1107 AD) as evidence.[22] Early European astronomers such as Tycho Brahe identified new stars in the night sky (later termed *novae*), suggesting that the heavens were not immutable. In 1584 Giordano Bruno suggested that the stars were like the Sun, and may have other planets, possibly even Earth-like, in orbit around them,[23] an idea that had been suggested earlier by the ancient Greek philosophers, Democritus and Epicurus,[24] and by medieval Islamic cosmologists[25] such as Fakhr al-Din al-Razi.[26] By the following century, the idea of the stars being the same as the Sun was reaching a consensus among astronomers. To explain why these stars exerted no net gravitational pull on the Solar System, Isaac Newton suggested that the stars were equally distributed in every direction, an idea prompted by the theologian Richard Bentley.[27]

The Italian astronomer Geminiano Montanari recorded observing variations in luminosity of the star Algol in 1667. Edmond Halley published the first measurements of the proper motion of a pair of nearby "fixed" stars, demonstrating that they had changed positions from the time of the ancient Greek astronomers Ptolemy and Hipparchus.[23]

William Herschel was the first astronomer to attempt to determine the distribution of stars in the sky. During the 1780s, he performed a series of gauges in 600 directions, and counted the stars observed along each line of sight. From this he deduced that the number of stars steadily increased toward one side of the sky, in the direction of the Milky Way core. His son John Herschel repeated this study in the southern hemisphere and found a corresponding increase in the same direction.[28] In addition to his other accomplishments, William Herschel is also noted for his discovery that some stars do not merely lie along the same line of sight, but are also physical companions that form binary star systems.

The science of stellar spectroscopy was pioneered by Joseph von Fraunhofer and Angelo Secchi. By comparing the spectra of stars such as Sirius to the Sun, they found differences in the strength and number of their absorption lines—the dark lines in a stellar spectra due to the absorption of specific frequencies by the atmosphere. In 1865 Secchi began classifying stars into spectral types.[29] However, the modern version of the stellar classification scheme was developed by Annie J. Cannon during the 1900s.

The first direct measurement of the distance to a star (61 Cygni at 11.4 light-years) was made in 1838 by Friedrich Bessel using the parallax technique. Parallax measurements demonstrated the vast separation of the stars in the heavens.[23] Observation of double stars gained increasing importance during the 19th century. In 1834, Friedrich Bessel observed changes in the proper motion of the star Sirius, and inferred a hidden companion. Edward Pickering discovered the first spectroscopic binary in 1899 when he observed the periodic splitting of the spectral lines of the star Mizar in a 104-day period. Detailed observations of many binary star systems were collected by astronomers such as William Struve and S. W. Burnham, allowing the masses of stars to be determined from computation of the orbital elements. The first solution to the problem of deriving an orbit of binary stars from telescope observations was made by Felix Savary in 1827.[30] The twentieth century saw increasingly rapid advances in the scientific study of stars. The photograph became a valuable astronomical tool. Karl Schwarzschild discovered that the color of a star, and hence its temperature, could be determined by comparing the visual magnitude against the photographic magnitude. The development of the photoelectric photometer allowed very precise measurements of magnitude at multiple wavelength intervals. In 1921 Albert A. Michelson made the first measurements of a stellar diameter using an interferometer on the Hooker telescope.[31]

Important theoretical work on the physical structure of stars occurred during the first decades of the twentieth century. In 1913, the Hertzsprung-Russell diagram was developed, propelling the astrophysical study of stars. Successful models were developed to explain the interiors of stars and stellar evolution. Cecilia Payne-Gaposchkin first proposed that stars were made primarily of hydrogen and helium in her 1925 PhD thesis.[32] The spectra of stars were further understood through advances in quantum physics. This allowed the chemical composition of the stellar atmosphere to be determined.[33]

With the exception of supernovae, individual stars have primarily been observed in our Local Group of galaxies,[34] and especially in the visible part of the Milky Way (as demonstrated by the detailed star catalogues available for our galaxy).[35] But some stars have been observed in the M100 galaxy of the Virgo Cluster, about 100 million light years from the Earth.[36] In the Local Supercluster it is possible to see star clusters, and current telescopes could in principle observe faint individual stars in the Local Cluster[37] (see Cepheids). However, outside the Local Supercluster of galaxies, neither individual stars nor clusters of stars have been observed. The only exception is a faint image of a large star cluster containing hundreds of thousands of stars located at a distance of one billion light years[38]—ten times further than the most distant star cluster previously observed.

15.2 Designations

Main articles: Stellar designation, Astronomical naming conventions and Star catalogue

The concept of the constellation was known to exist during the Babylonian period. Ancient sky watchers imagined that prominent arrangements of stars formed patterns, and they associated these with particular aspects of nature or their myths. Twelve of these formations lay along the band of the ecliptic and these became the basis of astrology.[39] Many of the more prominent individual stars were also given names, particularly with Arabic or Latin designations.

As well as certain constellations and the Sun itself, individual stars have their own myths.[40] To the Ancient Greeks, some "stars", known as planets (Greek πλανήτης (planētēs), meaning "wanderer"), represented various important deities, from which the names of the planets Mercury, Venus, Mars, Jupiter and Saturn were taken.[40] (Uranus and Neptune were also Greek and Roman gods, but neither planet was known in Antiquity because of their low brightness. Their names were assigned by later astronomers.)

Circa 1600, the names of the constellations were used to name the stars in the corresponding regions of the sky. The German astronomer Johann Bayer created a series of star maps and applied Greek letters as designations to the stars in each constellation. Later a numbering system based on the star's right ascension was invented and added to John Flamsteed's star catalogue in his book *"Historia coelestis Britannica"* (the 1712 edition), whereby this numbering system came to be called *Flamsteed designation* or *Flamsteed numbering*.[41][42]

The only internationally recognized authority for naming celestial bodies is the International Astronomical Union (IAU).[43] A number of private companies sell names of stars, which the British Library calls an unregulated commercial enterprise.[44][45] However, the IAU has disassociated itself from this commercial practice, and these names are neither recognized by the IAU nor used by them.[46] One such star naming company is the International Star Registry, which, during the 1980s, was accused of deceptive practice for making it appear that the assigned name was official. This now-discontinued ISR practice was informally labeled a scam and a fraud,[47][48][49][50] and the New York City Department of Consumer Affairs issued a violation against ISR for engaging in a deceptive trade practice.[51][52]

15.3 Units of measurement

Although stellar parameters can be expressed in SI units or CGS units, it is often most convenient to express mass, luminosity, and radii in solar units, based on the characteristics of the Sun:

Large lengths, such as the radius of a giant star or the semi-major axis of a binary star system, are often expressed in terms of the astronomical unit —approximately equal to the mean distance between the Earth and the Sun (150 million km or 93 million miles).

15.4   Formation and evolution

Main article: Stellar evolution

Stars form within extended regions of higher density in the interstellar medium, although the density is still lower than the inside of a vacuum chamber. These regions - known as *molecular clouds* - consist mostly of hydrogen, with about 23 to 28 percent helium and a few percent heavier elements. One example of such a star-forming region is the Orion Nebula.[55] As massive stars form from molecular clouds, they powerfully illuminate those clouds. They also ionize the hydrogen, creating an H II region.

All stars spend the majority of their existence as *main sequence stars*, fueled primarily by the nuclear fusion of hydrogen into helium within their cores. However, stars of different masses have markedly different properties at various stages of their development. The ultimate fate of more massive stars differs from that of less massive stars, as do their luminosity and the impact they have on their environment. Accordingly, astronomers often group stars by their mass:[56]

- *Very low mass stars* with masses below 0.5 $M\odot$ are fully convective and distribute helium evenly throughout the whole star while on the main sequence. Therefore they never undergo shell burning, never become red giants, and are theorized to become helium white dwarfs which simply cool off after exhausting their hydrogen.[57] However, as the lifetime of 0.5 $M\odot$ stars is longer than the age of the universe, no such star has yet reached the white dwarf stage.

- *Low mass stars* (including the Sun), with a main sequence mass above about 0.5 $M\odot$ and below 1.8–2.5 $M\odot$ depending on composition, do become red giants when their core hydrogen is depleted, then ignite a degenerate helium core in a helium flash, develop a degenerate carbon-oxygen core on the asymptotic giant branch, and finally produce a planetary nebula to become a white dwarf.

- *Intermediate-mass stars*, between 1.8–2.5 $M\odot$ and 5–10 $M\odot$, pass through similar evolutionary stages to the low mass stars, but after a relatively short period on the RGB they ignite helium without a flash and spend an extended period in the red clump before forming a degenerate carbon-oxygen core.

- *Massive stars* generally have a minimum mass of 7–10 $M\odot$, but this may be as low as 5–6 $M\odot$. After exhausting the hydrogen at the core these stars become supergiants and go on to fuse elements heavier than helium. They end their lives when their cores collapse and they explode as supernovae.

15.4.1   Star formation

Main article: Star formation

The formation of a star begins with gravitational instability within a molecular cloud, caused by regions of higher density - often triggered by shock-waves from nearby supernovae (massive stellar explosions), the collision of different molecular clouds, or the collision of galaxies (as in a starburst galaxy). Once a region reaches a sufficient density of matter to satisfy the criteria for Jeans instability, it begins to collapse under its own gravitational force.[58]

As the cloud collapses, individual conglomerations of dense dust and gas form "Bok globules". As a globule collapses and the density increases, the gravitational energy converts into heat and the temperature rises. When the protostellar cloud has approximately reached the stable condition of hydrostatic equilibrium, a protostar forms at the core.[59] These pre–main sequence stars are often surrounded by a protoplanetary disk and powered mainly by the release of gravitational energy. The period of gravitational contraction lasts about 10 to 15 million years.

Early stars of less than 2 $M_\odot$ are called T Tauri stars, while those with greater mass are Herbig Ae/Be stars. These newly formed stars emit jets of gas along their axis of rotation, which may reduce the angular momentum of the collapsing star and result in small patches of nebulosity known as Herbig–Haro objects.[60][61] These jets, in combination with radiation from nearby massive stars, may help to drive away the surrounding cloud from which the star was formed.[62]

Early in their development, T Tauri stars follow the Hayashi track—they contract and decrease in luminosity while remaining at roughly the same temperature. Less massive T Tauri stars follow this track to the main sequence, while more massive stars turn onto the Henyey track.

15.4.2 Main sequence

Main article: Main sequence

Stars spend about 90% of their existence fusing hydrogen into helium in high-temperature and high-pressure reactions near the core. Such stars are said to be on the main sequence and are called dwarf stars. Starting at zero-age main sequence, the proportion of helium in a star's core will steadily increase, the rate of nuclear fusion at the core will slowly increase, as will the star's temperature and luminosity.[63] The Sun, for example, is estimated to have increased in luminosity by about 40% since it reached the main sequence 4.6 billion (4.6×10^9) years ago.[64]

Every star generates a stellar wind of particles that causes a continual outflow of gas into space. For most stars, the mass lost is negligible. The Sun loses 10^{-14} $M_\odot$ every year,[65] or about 0.01% of its total mass over its entire lifespan. However, very massive stars can lose 10^{-7} to 10^{-5} $M_\odot$ each year, significantly affecting their evolution.[66] Stars that begin with more than 50 $M_\odot$ can lose over half their total mass while on the main sequence.[67]

The duration that a star spends on the main sequence depends primarily on the amount of fuel it has to fuse and the rate at which it fuses that fuel, i.e. its initial mass and its luminosity. For the Sun, its life is estimated to be about 10 billion (10^{10}) years. Massive stars consume their fuel very rapidly and are short-lived. Low mass stars consume their fuel very slowly. Stars less massive than 0.25 $M_\odot$, called red dwarfs, are able to fuse nearly all of their mass as fuel while stars of about 1 $M_\odot$ can only use about 10% of their mass as fuel. The combination of their slow fuel-consumption and relatively large usable fuel supply allows about 0.25 $M_\odot$ stars to last for about one trillion (10^{12}) years according to stellar-evolution calculations, while the least-massive hydrogen-fusing stars (0.08 $M_\odot$) will last for about 12 trillion years. Red dwarfs become hotter and more luminous as they accumulate helium. When they eventually run out of hydrogen, they contract into a white dwarf and start to cool.[57] However, since the lifespan of such stars is greater than the current age of the universe (13.8 billion years), no stars under about 0.85 $M_\odot$[68] are expected to have moved off the main sequence.

Besides mass, the elements heavier than helium can play a significant role in the evolution of stars. Astronomers consider all elements heavier than helium "metals", and call the chemical concentration of these elements the metallicity. The metallicity can influence the duration that a star will burn its fuel, control the formation of magnetic fields[69] and modify the strength of the stellar wind.[70] Older, population II stars have substantially less metallicity than the younger, population I stars due to the composition of the molecular clouds from which they formed. Over time these clouds become increasingly enriched in heavier elements as older stars die and shed portions of their atmospheres.

15.4.3 Post–main sequence

Main article: Red giant

As stars of at least 0.4 $M_\odot$[2] exhaust their supply of hydrogen at their core, they start to fuse hydrogen in a shell outside the helium core. Their outer layers expand and cool greatly to form a red giant. In about 5 billion years, when the Sun enters this phase, it will expand to a maximum radius of roughly 1 astronomical unit (150 million kilometres), 250 times its present size. As a giant, the Sun will lose roughly 30% of its current mass.[64][71]

As hydrogen shell burning produces more helium, the core increases in mass and temperature. In a red giant of up to 2.25 $M_\odot$, the helium core becomes degenerate before it is compressed enough to start helium fusion. When the temperature increases sufficiently helium fusion begins explosively in the helium flash and the star rapidly shrinks in radius, increases its surface temperature, and moves to the horizontal branch. For more massive stars, the helium core fusion starts before the core becomes degenerate and the star spends a period in the red clump before the outer convective envelope collapses and the star moves to the horizontal branch.[4]

After the star has consumed the helium at the core, fusion continues in a shell around a hot core of carbon and oxygen. The star then follows an evolutionary path (the asymptotic giant branch or AGB) that parallels the original red giant phase at a higher luminosity. The more massive AGB stars may undergo a brief period of carbon fusion before the core becomes degenerate.

Massive stars

Main article: Red supergiant

During their helium-burning phase, very high-mass stars with more than nine solar masses expand to form red supergiants. Once this fuel is exhausted at the core, they continue to fuse elements heavier than helium.

The core contracts until the temperature and pressure suffice to fuse carbon (see carbon burning process). This process continues, with the successive stages being fueled by neon (see neon burning process), oxygen (see oxygen burning process), and silicon (see silicon burning process). Near the end of the star's life, fusion continues along a series of onion-layer shells within the star. Each shell fuses a different element, with the outermost shell fusing hydrogen; the next shell fusing helium, and so forth.[72]

The final stage occurs when a massive star begins producing iron. Since iron nuclei are more tightly bound than any heavier nuclei, any fusion beyond iron does not produce a net release of energy—the process would, on the contrary, consume energy. Likewise, since they are more tightly bound than all lighter nuclei, energy cannot be released by fission.[73] In relatively old, very massive stars, a large core of inert iron will accumulate in the center of the star. The heavier elements in these stars can work their way to the surface, forming evolved objects known as Wolf-Rayet stars that have a dense stellar wind which sheds the outer atmosphere.

Collapse

As a star's core shrinks, the intensity of radiation from that surface increases, creating such radiation pressure on the outer shell of gas that it will push those layers away, forming a planetary nebula. If what remains after the outer atmosphere has been shed is less than 1.4 $M_\odot$, it shrinks to a relatively tiny object about the size of Earth, known as a white dwarf. White dwarfs lack the mass for further gravitational compression to take place.[74] The electron-degenerate matter inside a white dwarf is no longer a plasma, even though stars are generally referred to as being spheres of plasma. Eventually, white dwarfs fade into black dwarfs over a very long period of time.

In larger stars, fusion continues until the iron core has grown so large (more than 1.4 $M_\odot$) that it can no longer support its own mass. This core will suddenly collapse as its electrons are driven into its protons, forming neutrons, neutrinos and gamma rays in a burst of electron capture and inverse beta decay. The shockwave formed by this sudden collapse causes the rest of the star to explode in a supernova. Supernovae become so bright that they may briefly outshine the star's entire home galaxy. When they occur within the Milky Way, supernovae have historically been observed by naked-eye observers as "new stars" where none seemingly existed before.[75]

Supernova explosions blow away the star's outer layers, leaving remnants such as the Crab Nebula.[75] There remains a neutron star (which sometimes manifests itself as a pulsar or X-ray burster) or, in the case of the largest stars (large enough to leave a remnant greater than roughly 4 $M_\odot$), a black hole.[76] In a neutron star the matter is in a state known as neutron-degenerate matter, with a more exotic form of degenerate matter, QCD matter, possibly present in the core. Within a black hole the matter is in a state that is not currently understood.

The blown-off outer layers of dying stars include heavy elements, which may be recycled during the formation of new stars. These heavy elements allow the formation of rocky planets. The outflow from supernovae and the stellar wind of large stars play an important part in shaping the interstellar medium.[75]

15.5   Distribution

In addition to isolated stars, a multi-star system can consist of two or more gravitationally bound stars that orbit each other. The simplest and most common multi-star system is a binary star, but systems of three or more stars are also found. For reasons of orbital stability, such multi-star systems are often organized into hierarchical sets of binary stars.[77] Larger groups called star clusters also exist. These range from loose stellar associations with only a few stars, up to enormous globular clusters with hundreds of thousands of stars.

It has been a long-held assumption that the majority of stars occur in gravitationally bound, multiple-star systems. This is particularly true for very massive O and B class stars, where 80% of the stars are believed to be part of multiple-star systems. However the proportion of single star systems increases for smaller stars, so that only 25% of red dwarfs are known to have stellar companions. As 85% of all stars are red dwarfs, most stars in the Milky Way are likely single from birth.[78]

Stars are not spread uniformly across the universe, but are normally grouped into galaxies along with interstellar gas and dust. A typical galaxy contains hundreds of billions of stars, and there are more than 100 billion (10^{11}) galaxies in the observable universe.[79] A 2010 star count estimate was 300 sextillion (3×10^{23}) in the observable universe.[80] While it is often believed that stars only exist within galaxies, intergalactic stars have been discovered.[81]

The nearest star to the Earth, apart from the Sun, is Proxima Centauri, which is 39.9 trillion kilometres, or 4.2 light-years away. Travelling at the orbital speed of the Space Shuttle (8 kilometres per second—almost 30,000 kilometres per hour), it would take about 150,000 years to get there.[82] Distances like this are typical inside galactic discs, including in the vicinity of the solar system.[83] Stars can be much closer to each other in the centres of galaxies and in globular clusters, or much farther apart in galactic halos.

Due to the relatively vast distances between stars outside the galactic nucleus, collisions between stars are thought to be rare. In denser regions such as the core of globular clusters or the galactic center, collisions can be more common.[84] Such collisions can produce what are known as blue stragglers. These abnormal stars have a higher surface temperature than the other main sequence stars with the same luminosity in the cluster.[85]

15.6   Characteristics

Almost everything about a star is determined by its initial mass, including essential characteristics such as luminosity and size, as well as its evolution, lifespan, and eventual fate.

15.6.1   Age

Most stars are between 1 billion and 10 billion years old. Some stars may even be close to 13.8 billion years old—the observed age of the universe. The oldest star yet discovered, HD 140283, nicknamed Methuselah star, is an estimated 14.46 ± 0.8 billion years old.[86] (Due to the uncertainty in the value, this age for the star does not conflict with the age of the Universe, determined by the Planck satellite as 13.798 ± 0.037.[86])

The more massive the star, the shorter its lifespan, primarily because massive stars have greater pressure on their cores, causing them to burn hydrogen more rapidly. The most massive stars last an average of a few million years, while stars of minimum mass (red dwarfs) burn their fuel very slowly and can last tens to hundreds of billions of years.[87][88]

15.6.2   Chemical composition

See also: Metallicity and Molecules in stars

> "From a chemist's point of view, the surface or interior of a star...is boring—there are no molecules there."--Roald Hoffmann[89]

When stars form in the present Milky Way galaxy they are composed of about 71% hydrogen and 27% helium,[90] as measured by mass, with a small fraction of heavier elements. Typically the portion of heavy elements is measured in terms of the iron content of the stellar atmosphere, as iron is a common element and its absorption lines are relatively easy to measure. The portion of heavier elements may be an indicator of the likelihood that the star has a planetary system.[91]

The star with the lowest iron content ever measured is the dwarf HE1327-2326, with only 1/200,000th the iron content of the Sun.[92] By contrast, the super-metal-rich star μ Leonis has nearly double the abundance of iron as the Sun, while the planet-bearing star 14 Herculis has nearly triple the iron.[93] There also exist chemically peculiar stars that show unusual abundances of certain elements in their spectrum; especially chromium and rare earth elements.[94]

15.6.3 Diameter

Due to their great distance from the Earth, all stars except the Sun appear to the unaided eye as shining points in the night sky that twinkle because of the effect of the Earth's atmosphere. The Sun is also a star, but it is close enough to the Earth to appear as a disk instead, and to provide daylight. Other than the Sun, the star with the largest apparent size is R Doradus, with an angular diameter of only 0.057 arcseconds.[95]

The disks of most stars are much too small in angular size to be observed with current ground-based optical telescopes, and so interferometer telescopes are required to produce images of these objects. Another technique for measuring the angular size of stars is through occultation. By precisely measuring the drop in brightness of a star as it is occulted by the Moon (or the rise in brightness when it reappears), the star's angular diameter can be computed.[96]

Stars range in size from neutron stars, which vary anywhere from 20 to 40 km (25 mi) in diameter, to supergiants like Betelgeuse in the Orion constellation, which has a diameter approximately 1,070 times that of the Sun—about 1,490,171,880 km (925,949,878 mi). Betelgeuse, however, has a much lower density than the Sun.[97]

15.6.4 Kinematics

Main article: Stellar kinematics

The motion of a star relative to the Sun can provide useful information about the origin and age of a star, as well as the structure and evolution of the surrounding galaxy. The components of motion of a star consist of the radial velocity toward or away from the Sun, and the traverse angular movement, which is called its proper motion.

Radial velocity is measured by the doppler shift of the star's spectral lines, and is given in units of km/s. The proper motion of a star is determined by precise astrometric measurements in units of milli-arc seconds (mas) per year. By determining the parallax of a star, the proper motion can then be converted into units of velocity. Stars with high rates of proper motion are likely to be relatively close to the Sun, making them good candidates for parallax measurements.[99]

Once both rates of movement are known, the space velocity of the star relative to the Sun or the galaxy can be computed. Among nearby stars, it has been found that younger population I stars have generally lower velocities than older, population II stars. The latter have elliptical orbits that are inclined to the plane of the galaxy.[100] A comparison of the kinematics of nearby stars has also led to the identification of stellar associations. These are most likely groups of stars that share a common point of origin in giant molecular clouds.[101]

15.6.5 Magnetic field

Main article: Stellar magnetic field

The magnetic field of a star is generated within regions of the interior where convective circulation occurs. This movement of conductive plasma functions like a dynamo, generating magnetic fields that extend throughout the star. The strength of the magnetic field varies with the mass and composition of the star, and the amount of magnetic surface activity depends upon the star's rate of rotation. This surface activity produces starspots, which are regions of strong magnetic fields and lower than normal surface temperatures. Coronal loops are arching magnetic fields that reach out into the corona from active regions. Stellar flares are bursts of high-energy particles that are emitted due to the same magnetic activity.[102]

Young, rapidly rotating stars tend to have high levels of surface activity because of their magnetic field. The magnetic field can act upon a star's stellar wind, functioning as a brake to gradually slow the rate of rotation with time. Thus, older stars such as the Sun have a much slower rate of rotation and a lower level of surface activity. The activity levels of slowly rotating stars tend to vary in a cyclical manner and can shut down altogether for periods of time.[103] During the Maunder minimum, for example, the Sun underwent a 70-year period with almost no sunspot activity.

15.6.6 Mass

Main article: Stellar mass

One of the most massive stars known is Eta Carinae,[104] which, with 100–150 times as much mass as the Sun, will have a lifespan of only several million years. Sudies of the most massive open clusters suggests 150 $M_\odot$ as an upper limit for stars in the current era of the universe.[105] This represents an empirical value for the theoretical limit on the formation of massive stars due to increasing radiation pressure on the accreting gas cloud. Several stars in the R136 cluster in the Large Magellanic Cloud have been measured with larger masses,[106] but it has been determined that they could have been created through the collision and merger of massive stars in close binary systems, sidestepping the 150 $M_\odot$ limit on massive star formation.[107]

The first stars to form after the Big Bang may have been larger, up to 300 $M_\odot$ or more,[108] due to the complete absence of elements heavier than lithium in their composition. This generation of supermassive population III stars are likely to have existed in the very early universe (i.e., at high redshift) and may have started the production of chemical elements heavier than hydrogen that are needed for the later formation of planets and life. In June 2015, astronomers reported evidence for Population III stars in the Cosmos Redshift 7 galaxy at z = 6.60. [109][110]

With a mass only 80 times that of Jupiter (M_J), 2MASS J0523-1403 is the smallest known star undergoing nuclear fusion in its core.[111] For stars with similar metallicity to the Sun, the theoretical minimum mass the star can have, and still undergo fusion at the core, is estimated to be about 75 M_J.[112][113] When the metallicity is very low, however, a recent study of the faintest stars found that the minimum star size seems to be about 8.3% of the solar mass, or about 87 M_J.[113][114] Smaller bodies are called brown dwarfs, which occupy a poorly defined grey area between stars and gas giants.

The combination of the radius and the mass of a star determines the surface gravity. Giant stars have a much lower surface gravity than main sequence stars, while the opposite is the case for degenerate, compact stars such as white dwarfs. The surface gravity can influence the appearance of a star's spectrum, with higher gravity causing a broadening of the absorption lines.[33]

15.6.7 Rotation

Main article: Stellar rotation

The rotation rate of stars can be determined through spectroscopic measurement, or more exactly determined by tracking the rotation rate of starspots. Young stars can have a rapid rate of rotation greater than 100 km/s at the equator. The B-class star Achernar, for example, has an equatorial rotation velocity of about 225 km/s or greater, causing its equator to be slung outward and giving it an equatorial diameter that is more than 50% larger than the distance between the poles. This rate of rotation is just below the critical velocity of 300 km/s where the star would break apart.[115] By contrast, the Sun only rotates once every 25 – 35 days, with an equatorial velocity of 1.994 km/s. The star's magnetic field and the stellar wind serve to slow a main sequence star's rate of rotation by a significant amount as it evolves on the main sequence.[116]

Degenerate stars have contracted into a compact mass, resulting in a rapid rate of rotation. However they have relatively low rates of rotation compared to what would be expected by conservation of angular momentum—the tendency of a rotating body to compensate for a contraction in size by increasing its rate of spin. A large portion of the star's angular momentum is dissipated as a result of mass loss through the stellar wind.[117] In spite of this, the rate of rotation for a pulsar can be very rapid. The pulsar at the heart of the Crab nebula, for example, rotates 30 times per second.[118] The rotation rate of the pulsar will gradually slow due to the emission of radiation.

15.6.8   Temperature

The surface temperature of a main sequence star is determined by the rate of energy production at the core and by its radius, and is often estimated from the star's color index.[119] The temperature is normally given as the effective temperature, which is the temperature of an idealized black body that radiates its energy at the same luminosity per surface area as the star. Note that the effective temperature is only a representative value, as the temperature increases toward the core.[120] The temperature in the core region of a star is several million kelvins.[121]

The stellar temperature will determine the rate of ionization of various elements, resulting in characteristic absorption lines in the spectrum. The surface temperature of a star, along with its visual absolute magnitude and absorption features, is used to classify a star (see classification below).[33]

Massive main sequence stars can have surface temperatures of 50,000 K. Smaller stars such as the Sun have surface temperatures of a few thousand K. Red giants have relatively low surface temperatures of about 3,600 K; but they also have a high luminosity due to their large exterior surface area.[122]

15.7   Radiation

The energy produced by stars, as a product of nuclear fusion, radiates into space as both electromagnetic radiation and particle radiation. The particle radiation emitted by a star is manifested as the stellar wind,[123] which streams from the outer layers as free protons, and electrically charged alpha, and beta particles. Although almost massless there also exists a steady stream of neutrinos emanating from the star's core.

The production of energy at the core is the reason stars shine so brightly: every time two or more atomic nuclei fuse together to form a single atomic nucleus of a new heavier element, gamma ray photons are released from the nuclear fusion product. This energy is converted to other forms of electromagnetic energy of lower frequency, such as visible light, by the time it reaches the star's outer layers.

The color of a star, as determined by the most intense frequency of the visible light, depends on the temperature of the star's outer layers, including its photosphere.[124] Besides visible light, stars also emit forms of electromagnetic radiation that are invisible to the human eye. In fact, stellar electromagnetic radiation spans the entire electromagnetic spectrum, from the longest wavelengths of radio waves through infrared, visible light, ultraviolet, to the shortest of X-rays, and gamma rays. From the standpoint of total energy emitted by a star, not all components of stellar electromagnetic radiation are significant, but all frequencies provide insight into the star's physics.

Using the stellar spectrum, astronomers can also determine the surface temperature, surface gravity, metallicity and rotational velocity of a star. If the distance of the star is known, such as by measuring the parallax, then the luminosity of the star can be derived. The mass, radius, surface gravity, and rotation period can then be estimated based on stellar models. (Mass can be calculated for stars in binary systems by measuring their orbital velocities and distances. Gravitational microlensing has been used to measure the mass of a single star.[125]) With these parameters, astronomers can also estimate the age of the star.[126]

15.7.1   Luminosity

The luminosity of a star is the amount of light and other forms of radiant energy it radiates per unit of time. It has units of power. The luminosity of a star is determined by the radius and the surface temperature. However, many stars do not radiate a uniform flux (the amount of energy radiated per unit area) across their entire surface. The rapidly rotating star Vega, for example, has a higher energy flux at its poles than along its equator.[127]

Surface patches with a lower temperature and luminosity than average are known as starspots. Small, *dwarf* stars such as our Sun generally have essentially featureless disks with only small starspots. Larger, *giant* stars have much larger, more obvious starspots,[128] and they also exhibit strong stellar limb darkening. That is, the brightness decreases towards the edge of the stellar disk.[129] Red dwarf flare stars such as UV Ceti may also possess prominent starspot features.[130]

15.7.2 Magnitude

Main articles: Apparent magnitude and Absolute magnitude

The apparent brightness of a star is expressed in terms of its apparent magnitude, which is the brightness of a star and is a function of the star's luminosity, distance from Earth, and the altering of the star's light as it passes through Earth's atmosphere. Intrinsic or absolute magnitude is directly related to a star's luminosity and is what the apparent magnitude a star would be if the distance between the Earth and the star were 10 parsecs (32.6 light-years).

Both the apparent and absolute magnitude scales are logarithmic units: one whole number difference in magnitude is equal to a brightness variation of about 2.5 times[132] (the 5th root of 100 or approximately 2.512). This means that a first magnitude star (+1.00) is about 2.5 times brighter than a second magnitude (+2.00) star, and approximately 100 times brighter than a sixth magnitude star (+6.00). The faintest stars visible to the naked eye under good seeing conditions are about magnitude +6.

On both apparent and absolute magnitude scales, the smaller the magnitude number, the brighter the star; the larger the magnitude number, the fainter. The brightest stars, on either scale, have negative magnitude numbers. The variation in brightness (ΔL) between two stars is calculated by subtracting the magnitude number of the brighter star (m) from the magnitude number of the fainter star (m), then using the difference as an exponent for the base number 2.512; that is to say:

$$\Delta m = m_f - m_b$$

$$2.512^{\Delta m} = \Delta L$$

Relative to both luminosity and distance from Earth, a star's absolute magnitude (M) and apparent magnitude (m) are not equivalent;[132] for example, the bright star Sirius has an apparent magnitude of −1.44, but it has an absolute magnitude of +1.41.

The Sun has an apparent magnitude of −26.7, but its absolute magnitude is only +4.83. Sirius, the brightest star in the night sky as seen from Earth, is approximately 23 times more luminous than the Sun, while Canopus, the second brightest star in the night sky with an absolute magnitude of −5.53, is approximately 14,000 times more luminous than the Sun. Despite Canopus being vastly more luminous than Sirius, however, Sirius appears brighter than Canopus. This is because Sirius is merely 8.6 light-years from the Earth, while Canopus is much farther away at a distance of 310 light-years.

As of 2006, the star with the highest known absolute magnitude is LBV 1806-20, with a magnitude of −14.2. This star is at least 5,000,000 times more luminous than the Sun.[133] The least luminous stars that are currently known are located in the NGC 6397 cluster. The faintest red dwarfs in the cluster were magnitude 26, while a 28th magnitude white dwarf was also discovered. These faint stars are so dim that their light is as bright as a birthday candle on the Moon when viewed from the Earth.[134]

15.8 Classification

Main article: Stellar classification

The current stellar classification system originated in the early 20th century, when stars were classified from *A* to *Q* based on the strength of the hydrogen line.[136] It was not known at the time that the major influence on the line strength was temperature; the hydrogen line strength reaches a peak at over 9000 K, and is weaker at both hotter and cooler temperatures. When the classifications were reordered by temperature, it more closely resembled the modern scheme.[137]

Stars are given a single-letter classification according to their spectra, ranging from type *O*, which are very hot, to *M*, which are so cool that molecules may form in their atmospheres. The main classifications in order of decreasing surface temperature are: *O, B, A, F, G, K,* and *M*. A variety of rare spectral types have special classifications. The most common of these are types *L* and *T*, which classify the coldest low-mass stars and brown dwarfs. Each letter has 10 sub-divisions, numbered from 0 to 9, in order of decreasing temperature. However, this system breaks down at extreme high temperatures: class *O0* and *O1* stars may not exist.[138]

In addition, stars may be classified by the luminosity effects found in their spectral lines, which correspond to their spatial size and is determined by the surface gravity. These range from *0* (hypergiants) through *III* (giants) to *V* (main sequence dwarfs); some authors add *VII* (white dwarfs). Most stars belong to the main sequence, which consists of ordinary hydrogen-burning stars. These fall along a narrow, diagonal band when graphed according to their absolute magnitude and spectral type.[138] The Sun is a main sequence *G2V* yellow dwarf of intermediate temperature and ordinary size.

Additional nomenclature, in the form of lower-case letters, can follow the spectral type to indicate peculiar features of the spectrum. For example, an "*e*" can indicate the presence of emission lines; "*m*" represents unusually strong levels of metals, and "*var*" can mean variations in the spectral type.[138]

White dwarf stars have their own class that begins with the letter *D*. This is further sub-divided into the classes *DA*, *DB*, *DC*, *DO*, *DZ*, and *DQ*, depending on the types of prominent lines found in the spectrum. This is followed by a numerical value that indicates the temperature index.[139]

15.9   Variable stars

Main article: Variable star

Variable stars have periodic or random changes in luminosity because of intrinsic or extrinsic properties. Of the intrinsically variable stars, the primary types can be subdivided into three principal groups.

During their stellar evolution, some stars pass through phases where they can become pulsating variables. Pulsating variable stars vary in radius and luminosity over time, expanding and contracting with periods ranging from minutes to years, depending on the size of the star. This category includes Cepheid and cepheid-like stars, and long-period variables such as Mira.[140]

Eruptive variables are stars that experience sudden increases in luminosity because of flares or mass ejection events.[140] This group includes protostars, Wolf-Rayet stars, and Flare stars, as well as giant and supergiant stars.

Cataclysmic or explosive variable stars are those that undergo a dramatic change in their properties. This group includes novae and supernovae. A binary star system that includes a nearby white dwarf can produce certain types of these spectacular stellar explosions, including the nova and a Type 1a supernova.[4] The explosion is created when the white dwarf accretes hydrogen from the companion star, building up mass until the hydrogen undergoes fusion.[141] Some novae are also recurrent, having periodic outbursts of moderate amplitude.[140]

Stars can also vary in luminosity because of extrinsic factors, such as eclipsing binaries, as well as rotating stars that produce extreme starspots.[140] A notable example of an eclipsing binary is Algol, which regularly varies in magnitude from 2.3 to 3.5 over a period of 2.87 days.

15.10   Structure

Main article: Stellar structure

The interior of a stable star is in a state of hydrostatic equilibrium: the forces on any small volume almost exactly counterbalance each other. The balanced forces are inward gravitational force and an outward force due to the pressure gradient within the star. The pressure gradient is established by the temperature gradient of the plasma; the outer part of the star is cooler than the core. The temperature at the core of a main sequence or giant star is at least on the order of 10^7 K. The resulting temperature and pressure at the hydrogen-burning core of a main sequence star are sufficient for nuclear fusion to occur and for sufficient energy to be produced to prevent further collapse of the star.[142][143]

As atomic nuclei are fused in the core, they emit energy in the form of gamma rays. These photons interact with the surrounding plasma, adding to the thermal energy at the core. Stars on the main sequence convert hydrogen into helium, creating a slowly but steadily increasing proportion of helium in the core. Eventually the helium content becomes predominant and energy production ceases at the core. Instead, for stars of more than 0.4 $M\odot$, fusion occurs in a slowly expanding shell around the degenerate helium core.[144]

In addition to hydrostatic equilibrium, the interior of a stable star will also maintain an energy balance of thermal equilibrium. There is a radial temperature gradient throughout the interior that results in a flux of energy flowing toward the exterior. The outgoing flux of energy leaving any layer within the star will exactly match the incoming flux from below.

The radiation zone is the region within the stellar interior where radiative transfer is sufficiently efficient to maintain the flux of energy. In this region the plasma will not be perturbed and any mass motions will die out. If this is not the case, however, then the plasma becomes unstable and convection will occur, forming a convection zone. This can occur, for example, in regions where very high energy fluxes occur, such as near the core or in areas with high opacity as in the outer envelope.[143]

The occurrence of convection in the outer envelope of a main sequence star depends on the mass. Stars with several times the mass of the Sun have a convection zone deep within the interior and a radiative zone in the outer layers. Smaller stars such as the Sun are just the opposite, with the convective zone located in the outer layers.[145] Red dwarf stars with less than 0.4 $M_\odot$ are convective throughout, which prevents the accumulation of a helium core.[2] For most stars the convective zones will also vary over time as the star ages and the constitution of the interior is modified.[143]

The portion of a star that is visible to an observer is called the photosphere. This is the layer at which the plasma of the star becomes transparent to photons of light. From here, the energy generated at the core becomes free to propagate out into space. It is within the photosphere that sun spots, or regions of lower than average temperature, appear.

Above the level of the photosphere is the stellar atmosphere. In a main sequence star such as the Sun, the lowest level of the atmosphere is the thin chromosphere region, where spicules appear and stellar flares begin. This is surrounded by a transition region, where the temperature rapidly increases within a distance of only 100 km (62 mi). Beyond this is the corona, a volume of super-heated plasma that can extend outward to several million kilometres.[146] The existence of a corona appears to be dependent on a convective zone in the outer layers of the star.[145] Despite its high temperature, the corona emits very little light. The corona region of the Sun is normally only visible during a solar eclipse.

From the corona, a stellar wind of plasma particles expands outward from the star, propagating until it interacts with the interstellar medium. For the Sun, the influence of its solar wind extends throughout the bubble-shaped region of the heliosphere.[147]

15.11 Nuclear fusion reaction pathways

Main article: Stellar nucleosynthesis

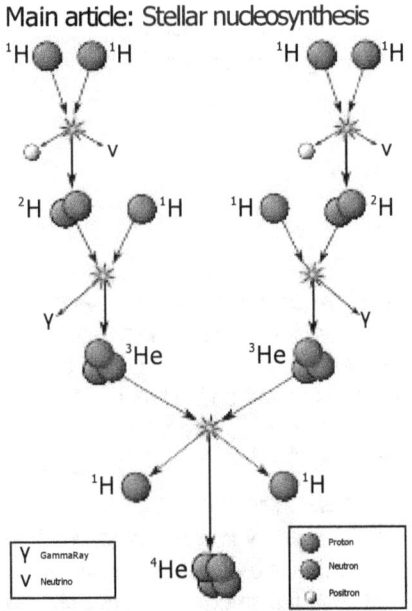

Overview of the proton-proton chain

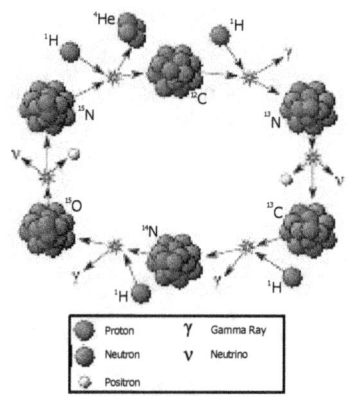

The carbon-nitrogen-oxygen cycle

A variety of different nuclear fusion reactions take place inside the cores of stars, depending upon their mass and composition, as part of stellar nucleosynthesis. The net mass of the fused atomic nuclei is smaller than the sum of the constituents. This lost mass is released as electromagnetic energy, according to the mass-energy equivalence relationship $E = mc^2$.[1]

The hydrogen fusion process is temperature-sensitive, so a moderate increase in the core temperature will result in a significant increase in the fusion rate. As a result the core temperature of main sequence stars only varies from 4 million kelvin for a small M-class star to 40 million kelvin for a massive O-class star.[121]

In the Sun, with a 10-million-kelvin core, hydrogen fuses to form helium in the proton-proton chain reaction:[148]

$$4\,^1H \rightarrow 2\,^2H + 2e^+ + 2v_e \ (4.0\,MeV + 1.0\,MeV)$$
$$2\,^1H + 2\,^2H \rightarrow 2\,^3He + 2\gamma \ (5.5\,MeV)$$
$$2\,^3He \rightarrow\ ^4He + 2\,^1H \ (12.9\,MeV)$$

These reactions result in the overall reaction:

$$4\,^1H \rightarrow\ ^4He + 2e^+ + 2\gamma + 2v_e \ (26.7\,MeV)$$

where e^+ is a positron, γ is a gamma ray photon, v_e is a neutrino, and H and He are isotopes of hydrogen and helium, respectively. The energy released by this reaction is in millions of electron volts, which is actually only a tiny amount of energy. However enormous numbers of these reactions occur constantly, producing all the energy necessary to sustain the star's radiation output.

In more massive stars, helium is produced in a cycle of reactions catalyzed by carbon—the carbon-nitrogen-oxygen cycle.[148]

In evolved stars with cores at 100 million kelvin and masses between 0.5 and 10 $M\odot$, helium can be transformed into carbon in the triple-alpha process that uses the intermediate element beryllium:[148]

$$^4He +\ ^4He + 92\,keV \rightarrow\ ^{8*}Be$$
$$^4He +\ ^{8*}Be + 67\,keV \rightarrow\ ^{12*}C$$
$$^{12*}C \rightarrow\ ^{12}C + \gamma + 7.4\,MeV$$

For an overall reaction of:

$$3\,^4He \rightarrow\ ^{12}C + \gamma + 7.2\,MeV$$

In massive stars, heavier elements can also be burned in a contracting core through the neon burning process and oxygen burning process. The final stage in the stellar nucleosynthesis process is the silicon burning process that results in the production of the stable isotope iron-56. Fusion can not proceed any further except through an endothermic process, and so further energy can only be produced through gravitational collapse.[148]

The example below shows the amount of time required for a star of 20 $M\odot$ to consume all of its nuclear fuel. As an O-class main sequence star, it would be 8 times the solar radius and 62,000 times the Sun's luminosity.[150]

15.12   See also

- Exoplanet host stars
- Lists of stars
- List of largest known stars
- Outline of astronomy
- Sidereal time
- Star clocks
- Star count
- Stars and planetary systems in fiction
- Stellar astronomy
- Stellar dynamics
- *Twinkle Twinkle Little Star* (children's nursery rhyme)

15.13   References

[1] Bahcall, John N. (June 29, 2000). "How the Sun Shines". Nobel Foundation. Retrieved 2006-08-30.

[2] Richmond, Michael. "Late stages of evolution for low-mass stars". Rochester Institute of Technology. Retrieved 2006-08-04.

[3] "Stellar Evolution & Death". NASA Observatorium. Archived from the original on 2008-02-10. Retrieved 2006-06-08.

[4] Iben, Icko, Jr. (1991). "Single and binary star evolution". *Astrophysical Journal Supplement Series* 76: 55–114. Bibcode:1991ApJS...76...55I. doi:10.1086/191565.

[5] Forbes, George (1909). *History of Astronomy*. London: Watts & Co. ISBN 1-153-62774-4.

[6] Hevelius, Johannis (1690). *Firmamentum Sobiescianum, sive Uranographia*. Gdansk.

[7] Tøndering, Claus. "Other ancient calendars". WebExhibits. Retrieved 2006-12-10.

[8] von Spaeth, Ove (2000). "Dating the Oldest Egyptian Star Map". *Centaurus International Magazine of the History of Mathematics, Science and Technology* 42 (3): 159–179. Bibcode:2000Cent...42..159V. doi:10.1034/j.1600-0498.2000.420301.x. Retrieved 2007-10-21.

[9] North, John (1995). *The Norton History of Astronomy and Cosmology*. New York and London: W.W. Norton & Company. pp. 30–31. ISBN 0-393-03656-1.

[10] Murdin, P. (November 2000). "Aristillus (c. 200 BC)". *Encyclopedia of Astronomy and Astrophysics*. Bibcode:2000eaa..bookE3440. doi:10.1888/0333750888/3440. ISBN 0-333-75088-8.

[11] Grasshoff, Gerd (1990). *The history of Ptolemy's star catalogue*. Springer. pp. 1–5. ISBN 0-387-97181-5.

[12] Pinotsis, Antonios D. "Astronomy in Ancient Rhodes". Section of Astrophysics, Astronomy and Mechanics, Department of Physics, University of Athens. Retrieved 2009-06-02.

[13] Clark, D. H.; Stephenson, F. R. (June 29, 1981). "The Historical Supernovae". *Supernovae: A survey of current research; Proceedings of the Advanced Study Institute*. Cambridge, England: Dordrecht, D. Reidel Publishing Co. pp. 355–370. Bibcode:1982sscr.conf..355C.

[14] Zhao, Fu-Yuan; Strom, R. G.; Jiang, Shi-Yang (2006). "The Guest Star of AD185 Must Have Been a Supernova". *Chinese Journal of Astronomy and Astrophysics* 6 (5): 635–640. Bibcode:2006ChJAA...6..635Z. doi:10.1088/1009-9271/6/5/17.

[15] "Astronomers Peg Brightness of History's Brightest Star". NAOA News. March 5, 2003. Retrieved 2006-06-08.

[16] Frommert, Hartmut; Kronberg, Christine (August 30, 2006). "Supernova 1054 – Creation of the Crab Nebula". *SEDS*. University of Arizona.

[17] Duyvendak, J. J. L. (April 1942). "Further Data Bearing on the Identification of the Crab Nebula with the Supernova of 1054 A.D. Part I. The Ancient Oriental Chronicles". *Publications of the Astronomical Society of the Pacific* 54 (318): 91–94. Bibcode:1942PASP...54...91D. doi:10.1086/125409.
Mayall, N. U.; Oort, Jan Hendrik (April 1942). "Further Data Bearing on the Identification of the Crab Nebula with the Supernova of 1054 A.D. Part II. The Astronomical Aspects". *Publications of the Astronomical Society of the Pacific* 54 (318): 95–104. Bibcode:1942PASP...54...95M. doi:10.1086/125410.

[18] Brecher, K. et al. (1983). "Ancient records and the Crab Nebula supernova". *The Observatory* 103: 106–113. Bibcode:1983Obs...103..106B.

[19] Kennedy, Edward S. (1962). "Review: *The Observatory in Islam and Its Place in the General History of the Observatory* by Aydin Sayili". *Isis* 53 (2): 237–239. doi:10.1086/349558.

[20] Jones, Kenneth Glyn (1991). *Messier's nebulae and star clusters*. Cambridge University Press. p. 1. ISBN 0-521-37079-5.

[21] Zahoor, A. (1997). "Al-Biruni". Hasanuddin University. Archived from the original on 2008-06-26. Retrieved 2007-10-21.

[22] Montada, Josep Puig (September 28, 2007). "Ibn Bajja". Stanford Encyclopedia of Philosophy. Retrieved 2008-07-11.

[23] Drake, Stephen A. (August 17, 2006). "A Brief History of High-Energy (X-ray & Gamma-Ray) Astronomy". NASA HEASARC. Retrieved 2006-08-24.

[24] Greskovic, Peter; Rudy, Peter (July 24, 2006). "Exoplanets". ESO. Retrieved 2012-06-15.

[25] Ahmad, I. A. (1995). "The impact of the Qur'anic conception of astronomical phenomena on Islamic civilization". *Vistas in Astronomy* 39 (4): 395–403 [402]. Bibcode:1995VA.....39..395A. doi:10.1016/0083-6656(95)00033-X.

[26] Setia, Adi (2004). "Fakhr Al-Din Al-Razi on Physics and the Nature of the Physical World: A Preliminary Survey". *Islam & Science* 2 (2) – via Questia.

[27] Hoskin, Michael (1998). "The Value of Archives in Writing the History of Astronomy". Space Telescope Science Institute. Retrieved 2006-08-24.

[28] Proctor, Richard A. (1870). "Are any of the nebulæ star-systems?". *Nature* 1 (13): 331–333. Bibcode:1870Natur...1..331P. doi:10.1038/001331a0.

[29] MacDonnell, Joseph. "Angelo Secchi, S.J. (1818–1878) the Father of Astrophysics". Fairfield University. Archived from the original on 2011-07-21. Retrieved 2006-10-02.

[30] Aitken, Robert G. (1964). *The Binary Stars*. New York: Dover Publications Inc. p. 66. ISBN 0-486-61102-7.

[31] Michelson, A. A.; Pease, F. G. (1921). "Measurement of the diameter of Alpha Orionis with the interferometer". *Astrophysical Journal* 53: 249–259. Bibcode:1921ApJ....53..249M. doi:10.1086/142603.

[32] "" Payne-Gaposchkin, Cecilia Helena." CWP". University of California. Retrieved 2013-02-21.

[33] Unsöld, Albrecht (2001). *The New Cosmos* (5th ed.). New York: Springer. pp. 180–185, 215–216. ISBN 3-540-67877-8.

[34] e. g. Battinelli, Paolo; Demers, Serge; Letarte, Bruno (2003). "Carbon Star Survey in the Local Group. V. The Outer Disk of M31". *The Astronomical Journal* 125 (3): 1298–1308. Bibcode:2003AJ....125.1298B. doi:10.1086/346274.

[35] "Millennium Star Atlas marks the completion of ESA's Hipparcos Mission". ESA. December 8, 1997. Retrieved 2007-08-05.

[36] Villard, Ray; Freedman, Wendy L. (October 26, 1994). "Hubble Space Telescope Measures Precise Distance to the Most Remote Galaxy Yet". Hubble Site. Retrieved 2007-08-05.

[37] "Hubble Completes Eight-Year Effort to Measure Expanding Universe". Hubble Site. May 25, 1999. Retrieved 2007-08-02.

[38] "UBC Prof., alumnus discover most distant star clusters: a billion light-years away.". UBC Public Affairs. January 8, 2007. Retrieved 2015-06-28.

[39] Koch-Westenholz, Ulla; Koch, Ulla Susanne (1995). *Mesopotamian astrology: an introduction to Babylonian and Assyrian celestial divination*. Carsten Niebuhr Institute Publications **19**. Museum Tusculanum Press. p. 163. ISBN 87-7289-287-0.

[40] Coleman, Leslie S. "Myths, Legends and Lore". Frosty Drew Observatory. Retrieved 2012-06-15.

[41] "Naming Astronomical Objects". International Astronomical Union (IAU). Retrieved 2009-01-30.

[42] "Naming Stars". Students for the Exploration and Development of Space (SEDS). Retrieved 2009-01-30.

[43] Lyall, Francis; Larsen, Paul B. (2009). "Chapter 7: The Moon and Other Celestial Bodies". *Space Law: A Treatise*. Ashgate Publishing, Ltd. p. 176. ISBN 0-7546-4390-5.

[44] "Star naming". Scientia Astrophysical Organization. 2005. Retrieved 2010-06-29.

[45] "Disclaimer: Name a star, name a rose and other, similar enterprises". *British Library*. The British Library Board. Archived from the original on 2010-01-19. Retrieved 2010-06-29.

[46] Andersen, Johannes. "Buying Stars and Star Names". International Astronomical Union. Retrieved 2010-06-24.

[47] Pliat, Phil (September–October 2006). "Name Dropping: Want to Be a Star?". *Skeptical Inquirer* **30.5**. Retrieved 2010-06-29.

[48] Adams, Cecil (April 1, 1998). "Can you pay $35 to get a star named after you?". The Straight Dope. Retrieved 2006-08-13.

[49] Golden, Frederick; Faflick, Philip (January 11, 1982). "Science: Stellar Idea or Cosmic Scam?". *Times Magazine* (Time Inc.). Retrieved 2010-06-24.

[50] Di Justo, Patrick (December 26, 2001). "Buy a Star, But It's Not Yours". *Wired* (Condé Nast Digital). Retrieved 2010-06-29.

[51] Plait, Philip C. (2002). *Bad astronomy: misconceptions and misuses revealed, from astrology to the moon landing "hoax"*. John Wiley and Sons. pp. 237–240. ISBN 0-471-40976-6.

[52] Sclafani, Tom (May 8, 1998). "Consumer Affairs Commissioner Polonetsky Warns Consumers: "Buying A Star Won't Make You One"". National Astronomy and Ionosphere Center, Aricebo Observatory. Retrieved 2010-06-24.

[53] Sackmann, I.-J.; Boothroyd, A. I. (2003). "Our Sun. V. A Bright Young Sun Consistent with Helioseismology and Warm Temperatures on Ancient Earth and Mars". *The Astrophysical Journal* **583** (2): 1024–1039. arXiv:astro-ph/0210128. Bibcode:2003ApJ...583.1024S. doi:10.1086/345408.

[54] Tripathy, S. C.; Antia, H. M. (1999). "Influence of surface layers on the seismic estimate of the solar radius". *Solar Physics* **186** (1/2): 1–11. Bibcode:1999SoPh..186....1T. doi:10.1023/A:1005116830445.

[55] Woodward, P. R. (1978). "Theoretical models of star formation". *Annual review of astronomy and astrophysics* **16** (1): 555–584. Bibcode:1978ARA&A..16..555W. doi:10.1146/annurev.aa.16.090178.003011.

[56] Kwok, Sun (2000). *The origin and evolution of planetary nebulae*. Cambridge astrophysics series **33**. Cambridge University Press. pp. 103–104. ISBN 0-521-62313-8.

[57] Adams, Fred C.; Laughlin, Gregory; Graves, Genevieve J. M. "Red Dwarfs and the End of the Main Sequence" (PDF). *Gravitational Collapse: From Massive Stars to Planets*. Revista Mexicana de Astronomía y Astrofísica. pp. 46–49. Bibcode:2004RMxAC..22...46A. Retrieved 2008-06-24.

[58] Smith, Michael David (2004). *The Origin of Stars*. Imperial College Press. pp. 57–68. ISBN 1-86094-501-5.

[59] Seligman, Courtney. "Slow Contraction of Protostellar Cloud". *Self-published*. Archived from the original on 2008-06-23. Retrieved 2006-09-05.

[60] Bally, J.; Morse, J.; Reipurth, B. (1996). "The Birth of Stars: Herbig-Haro Jets, Accretion and Proto-Planetary Disks". In Benvenuti, Piero; Macchetto, F. D.; Schreier, Ethan J. *Science with the Hubble Space Telescope – II. Proceedings of a workshop held in Paris, France, December 4–8, 1995*. Space Telescope Science Institute. p. 491. Bibcode:1996swhs.conf..491B.

[61] Smith, Michael David (2004). *The origin of stars*. Imperial College Press. p. 176. ISBN 1-86094-501-5.

[62] Megeath, Tom (May 11, 2010). "Herschel finds a hole in space". ESA. Retrieved 2010-05-17.

[63] Mengel, J. G. et al. (1979). "Stellar evolution from the zero-age main sequence". *Astrophysical Journal Supplement Series* 40: 733–791. Bibcode:1979ApJS...40..733M. doi:10.1086/190603.

[64] Sackmann, I. J.; Boothroyd, A. I.; Kraemer, K. E. (1993). "Our Sun. III. Present and Future". *Astrophysical Journal* 418: 457. Bibcode:1993ApJ...418..457S. doi:10.1086/173407.

[65] Wood, B. E. et al. (2002). "Measured Mass-Loss Rates of Solar-like Stars as a Function of Age and Activity". *The Astrophysical Journal* 574 (1): 412–425. arXiv:astro-ph/0203437. Bibcode:2002ApJ...574..412W. doi:10.1086/340797.

[66] de Loore, C.; de Greve, J. P.; Lamers, H. J. G. L. M. (1977). "Evolution of massive stars with mass loss by stellar wind". *Astronomy and Astrophysics* 61 (2): 251–259. Bibcode:1977A&A....61..251D.

[67] "The evolution of stars between 50 and 100 times the mass of the Sun". Royal Greenwich Observatory. Retrieved 2006-09-07.

[68] "Main Sequence Lifetime". *Swinburne Astronomy Online Encyclopedia of Astronomy.* Swinburne University of Technology.

[69] Pizzolato, N. et al. (2001). "Subphotospheric convection and magnetic activity dependence on metallicity and age: Models and tests". *Astronomy & Astrophysics* 373 (2): 597–607. Bibcode:2001A&A...373..597P. doi:10.1051/0004-6361:20010626.

[70] "Mass loss and Evolution". UCL Astrophysics Group. June 18, 2004. Archived from the original on 2004-11-22. Retrieved 2006-08-26.

[71] Schröder, K.-P.; Smith, Robert Connon (2008). "Distant future of the Sun and Earth revisited". *Monthly Notices of the Royal Astronomical Society* 386 (1): 155. arXiv:0801.4031. Bibcode:2008MNRAS.386..155S. doi:10.1111/j.1365-2966.2008.13022.x. See also Palmer, Jason (February 22, 2008). "Hope dims that Earth will survive Sun's death". *NewScientist.com news service.* Retrieved 2008-03-24.

[72] "What is a star?". Royal Greenwich Observatory. Retrieved 2006-09-07.

[73] Sneden, Christopher (February 8, 2001). "Astronomy: The age of the Universe". *Nature* 409 (6821): 673–675. doi:10.1038/35055646. ISSN 0028-0836.

[74] Liebert, J. (1980). "White dwarf stars". *Annual review of astronomy and astrophysics* 18 (2): 363–398. Bibcode:1980ARA&A..18..363L. doi:10.1146/annurev.aa.18.090180.002051.

[75] "Introduction to Supernova Remnants". Goddard Space Flight Center. April 6, 2006. Retrieved 2006-07-16.

[76] Fryer, C. L. (2003). "Black-hole formation from stellar collapse". *Classical and Quantum Gravity* 20 (10): S73–S80. Bibcode:2003CQGra..20S..73F doi:10.1088/0264-9381/20/10/309.

[77] Szebehely, Victor G.; Curran, Richard B. (1985). *Stability of the Solar System and Its Minor Natural and Artificial Bodies.* Springer. ISBN 90-277-2046-0.

[78] "Most Milky Way Stars Are Single" (Press release). Harvard-Smithsonian Center for Astrophysics. January 30, 2006. Retrieved 2006-07-16.

[79] "What is a galaxy? How many stars in a galaxy / the Universe?". Royal Greenwich Observatory. Retrieved 2006-07-18.

[80] Borenstein, Seth (December 1, 2010). "Universe's Star Count Could Triple". *CBS News.* Retrieved 2011-07-14.

[81] "Hubble Finds Intergalactic Stars". Hubble News Desk. January 14, 1997. Retrieved 2006-11-06.

[82] 3.99×10^{13} km / (3×10^4 km/h × 24 × 365.25) = 1.5×10^5 years.

[83] Holmberg, J.; Flynn, C. (2000). "The local density of matter mapped by Hipparcos". *Monthly Notices of the Royal Astronomical Society* 313 (2): 209–216. arXiv:astro-ph/9812404. Bibcode:2000MNRAS.313..209H. doi:10.1046/j.1365-8711.2000.02905.x.

[84] "Astronomers: Star collisions are rampant, catastrophic". CNN News. June 2, 2000. Archived from the original on 2007-01-07. Retrieved 2014-01-21.

[85] Lombardi, Jr., J. C. et al. (2002). "Stellar Collisions and the Interior Structure of Blue Stragglers". *The Astrophysical Journal* 568 (2): 939–953. arXiv:astro-ph/0107388. Bibcode:2002ApJ...568..939L. doi:10.1086/339060.

[86] H. E. Bond; E. P. Nelan; D. A. VandenBerg; G. H. Schaefer; D. Harmer (2013). "HD 140283: A Star in the Solar Neighborhood that Formed Shortly After the Big Bang". *The Astrophysical Journal Letters* **765** (1): L12. arXiv:1302.3180. Bibcode:2013ApJ...765L..12B. doi:10.1088/2041-8205/765/1/L12.

[87] Naftilan, S. A.; Stetson, P. B. (July 13, 2006). "How do scientists determine the ages of stars? Is the technique really accurate enough to use it to verify the age of the universe?". Scientific American. Retrieved 2007-05-11.

[88] Laughlin, G.; Bodenheimer, P.; Adams, F. C. (1997). "The End of the Main Sequence". *The Astrophysical Journal* **482** (1): 420–432. Bibcode:1997ApJ...482..420L. doi:10.1086/304125.

[89] Stellar Molecules » American Scientist

[90] Irwin, Judith A. (2007). *Astrophysics: Decoding the Cosmos.* John Wiley and Sons. p. 78. ISBN 0-470-01306-0.

[91] Fischer, D. A.; Valenti, J. (2005). "The Planet-Metallicity Correlation". *The Astrophysical Journal* **622** (2): 1102–1117. Bibcode:2005ApJ...622.1102F. doi:10.1086/428383.

[92] "Signatures Of The First Stars". ScienceDaily. April 17, 2005. Retrieved 2006-10-10.

[93] Feltzing, S.; Gonzalez, G. (2000). "The nature of super-metal-rich stars: Detailed abundance analysis of 8 super-metal-rich star candidates". *Astronomy & Astrophysics* **367** (1): 253–265. Bibcode:2001A&A...367..253F. doi:10.1051/0004-6361:20000477.

[94] Gray, David F. (1992). *The Observation and Analysis of Stellar Photospheres.* Cambridge University Press. pp. 413–414. ISBN 0-521-40868-7.

[95] "The Biggest Star in the Sky". ESO. March 11, 1997. Retrieved 2006-07-10.

[96] Ragland, S.; Chandrasekhar, T.; Ashok, N. M. (1995). "Angular Diameter of Carbon Star Tx-Piscium from Lunar Occultation Observations in the Near Infrared". *Journal of Astrophysics and Astronomy* **16**: 332. Bibcode:1995JApAS..16..332R.

[97] Davis, Kate (December 1, 2000). "Variable Star of the Month—December, 2000: Alpha Orionis". AAVSO. Archived from the original on 2006-07-12. Retrieved 2006-08-13.

[98] Loktin, A. V. (September 2006). "Kinematics of stars in the Pleiades open cluster". *Astronomy Reports* **50** (9): 714–721. Bibcode:2006ARep...50..714L. doi:10.1134/S1063772906090058.

[99] "Hipparcos: High Proper Motion Stars". ESA. September 10, 1999. Retrieved 2006-10-10.

[100] Johnson, Hugh M. (1957). "The Kinematics and Evolution of Population I Stars". *Publications of the Astronomical Society of the Pacific* **69** (406): 54. Bibcode:1957PASP...69...54J. doi:10.1086/127012.

[101] Elmegreen, B.; Efremov, Y. N. (1999). "The Formation of Star Clusters". *American Scientist* **86** (3): 264. Bibcode:1998AmSci..86..264E. doi:10.1511/1998.3.264. Archived from the original on March 23, 2005. Retrieved 2006-08-23.

[102] Brainerd, Jerome James (July 6, 2005). "X-rays from Stellar Coronas". The Astrophysics Spectator. Retrieved 2007-06-21.

[103] Berdyugina, Svetlana V. (2005). "Starspots: A Key to the Stellar Dynamo". Living Reviews. Retrieved 2007-06-21.

[104] Smith, Nathan (1998). "The Behemoth Eta Carinae: A Repeat Off ender". *Mercury Magazine* (Astronomical Society of the Pacific) **27**: 20. Retrieved 2006-08-13.

[105] Weidner, C.; Kroupa, P. (February 11, 2004). "Evidence for a fundamental stellar upper mass limit from clustered star formation" (PDF). *Monthly Notices of the Royal Astronomical Society* **348** (1): 187–191. doi:10.1111/j.1365-2966.2004.07340.x. ISSN 0035-8711.

[106] Hainich, R.; Rühling, U.; Todt, H.; Oskinova, L. M.; Liermann, A.; Gräfener, G.; Foellmi, C.; Schnurr, O.; Hamann, W. -R. (2014). "The Wolf-Rayet stars in the Large Magellanic Cloud". *Astronomy & Astrophysics* **565**: A27. arXiv:1401.5474. Bibcode:2014A&A...565A..27H. doi:10.1051/0004-6361/201322696.

[107] Banerjee, Sambaran; Kroupa, Pavel; Oh, Seungkyung (October 21, 2012). "The emergence of super-canonical stars in R136-type starburst clusters". *Monthly Notices of the Royal Astronomical Society* **426** (2): 1416–1426. doi:10.1111/j.1365-2966.2012.21672.x. ISSN 0035-8711.

[108] "Ferreting Out The First Stars". Harvard-Smithsonian Center for Astrophysics. September 22, 2005. Retrieved 2006-09-05.

[109] Sobral, David; Matthee, Jorryt; Darvish, Behnam; Schaerer, Daniel; Mobasher, Bahram; Röttgering, Huub J. A.; Santos, Sérgio; Hemmati, Shoubaneh (4 June 2015). "Evidence For POPIII-Like Stellar Populations In The Most Luminous LYMAN-α Emitters At The Epoch Of Re-Ionisation: Spectroscopic Confirmation" (PDF). *The Astrophysical Journal*. Retrieved 17 June 2015.

[110] Overbye, Dennis (17 June 2015). "Astronomers Report Finding Earliest Stars That Enriched Cosmos". *New York Times*. Retrieved 17 June 2015.

[111] "2MASS J05233822-1403022". SIMBAD - Centre de Données astronomiques de Strasbourg. Retrieved 14 December 2013.

[112] Boss, Alan (April 3, 2001). "Are They Planets or What?". Carnegie Institution of Washington. Archived from the original on 2006-09-28. Retrieved 2006-06-08.

[113] Shiga, David (August 17, 2006). "Mass cut-off between stars and brown dwarfs revealed". New Scientist. Archived from the original on 2006-11-14. Retrieved 2006-08-23.

[114] Leadbeater, Elli (August 18, 2006). "Hubble glimpses faintest stars". BBC. Retrieved 2006-08-22.

[115] "Flattest Star Ever Seen". ESO. June 11, 2003. Retrieved 2006-10-03.

[116] Fitzpatrick, Richard (February 13, 2006). "Introduction to Plasma Physics: A graduate course". The University of Texas at Austin. Archived from the original on 2010-01-04. Retrieved 2006-10-04.

[117] Villata, Massimo (1992). "Angular momentum loss by a stellar wind and rotational velocities of white dwarfs". *Monthly Notices of the Royal Astronomical Society* 257 (3): 450–454. Bibcode:1992MNRAS.257..450V.

[118] "A History of the Crab Nebula". ESO. May 30, 1996. Retrieved 2006-10-03.

[119] Strobel, Nick (August 20, 2007). "Properties of Stars: Color and Temperature". *Astronomy Notes*. Primis/McGraw-Hill, Inc. Archived from the original on 2007-06-26. Retrieved 2007-10-09.

[120] Seligman, Courtney. "Review of Heat Flow Inside Stars". *Self-published*. Retrieved 2007-07-05.

[121] "Main Sequence Stars". The Astrophysics Spectator. February 16, 2005. Retrieved 2006-10-10.

[122] Zeilik, Michael A.; Gregory, Stephan A. (1998). *Introductory Astronomy & Astrophysics* (4th ed.). Saunders College Publishing. p. 321. ISBN 0-03-006228-4.

[123] Koppes, Steve (June 20, 2003). "University of Chicago physicist receives Kyoto Prize for lifetime achievements in science". The University of Chicago News Office. Retrieved 2012-06-15.

[124] "The Colour of Stars". Australian Telescope Outreach and Education. Retrieved 2006-08-13.

[125] "Astronomers Measure Mass of a Single Star—First Since the Sun". Hubble News Desk. July 15, 2004. Retrieved 2006-05-24.

[126] Garnett, D. R.; Kobulnicky, H. A. (2000). "Distance Dependence in the Solar Neighborhood Age-Metallicity Relation". *The Astrophysical Journal* 532 (2): 1192–1196. arXiv:astro-ph/9912031. Bibcode:2000ApJ...532.1192G. doi:10.1086/308617.

[127] Staff (January 10, 2006). "Rapidly Spinning Star Vega has Cool Dark Equator". National Optical Astronomy Observatory. Retrieved 2007-11-18.

[128] Michelson, A. A.; Pease, F. G. (2005). "Starspots: A Key to the Stellar Dynamo". *Living Reviews in Solar Physics* (Max Planck Society).

[129] Manduca, A.; Bell, R. A.; Gustafsson, B. (1977). "Limb darkening coefficients for late-type giant model atmospheres". *Astronomy and Astrophysics* 61 (6): 809–813. Bibcode:1977A&A....61..809M.

[130] Chugainov, P. F. (1971). "On the Cause of Periodic Light Variations of Some Red Dwarf Stars". *Information Bulletin on Variable Stars* 520: 1–3. Bibcode:1971IBVS..520....1C.

[131] "Magnitude". National Solar Observatory—Sacramento Peak. Archived from the original on 2008-02-06. Retrieved 2006-08-23.

[132] "Luminosity of Stars". Australian Telescope Outreach and Education. Retrieved 2006-08-13.

[133] Hoover, Aaron (January 15, 2004). "Star may be biggest, brightest yet observed". HubbleSite. Archived from the original on 2007-08-07. Retrieved 2006-06-08.

[134] "Faintest Stars in Globular Cluster NGC 6397". HubbleSite. August 17, 2006. Retrieved 2006-06-08.

[135] Smith, Gene (April 16, 1999). "Stellar Spectra". University of California, San Diego. Retrieved 2006-10-12.

[136] Fowler, A. (February 1891). "The Draper Catalogue of Stellar Spectra". *Nature* **45**: 427–8. Bibcode:1892Natur..45..427F. doi:10.1038/045427a0.

[137] Jaschek, Carlos; Jaschek, Mercedes (1990). *The Classification of Stars*. Cambridge University Press. pp. 31–48. ISBN 0-521-38996-8.

[138] MacRobert, Alan M. "The Spectral Types of Stars". Sky and Telescope. Retrieved 2006-07-19.

[139] "White Dwarf (wd) Stars". White Dwarf Research Corporation. Archived from the original on 2009-10-08. Retrieved 2006-07-19.

[140] "Types of Variable". AAVSO. May 11, 2010. Retrieved 2010-08-20.

[141] "Cataclysmic Variables". NASA Goddard Space Flight Center. 2004-11-01. Retrieved 2006-06-08.

[142] Hansen, Carl J.; Kawaler, Steven D.; Trimble, Virginia (2004). *Stellar Interiors*. Springer. pp. 32–33. ISBN 0-387-20089-4.

[143] Schwarzschild, Martin (1958). *Structure and Evolution of the Stars*. Princeton University Press. ISBN 0-691-08044-5.

[144] "Formation of the High Mass Elements". Smoot Group. Retrieved 2006-07-11.

[145] "What is a Star?". NASA. 2006-09-01. Retrieved 2006-07-11.

[146] "The Glory of a Nearby Star: Optical Light from a Hot Stellar Corona Detected with the VLT" (Press release). ESO. August 1, 2001. Retrieved 2006-07-10.

[147] Burlaga, L. F. et al. (2005). "Crossing the Termination Shock into the Heliosheath: Magnetic Fields". *Science* **309** (5743): 2027–2029. Bibcode:2005Sci...309.2027B. doi:10.1126/science.1117542. PMID 16179471.

[148] Wallerstein, G. et al. (1999). "Synthesis of the elements in stars: forty years of progress" (PDF). *Reviews of Modern Physics* **69** (4): 995–1084. Bibcode:1997RvMP...69..995W. doi:10.1103/RevModPhys.69.995. Retrieved 2006-08-04.

[149] Girardi, L.; Bressan, A.; Bertelli, G.; Chiosi, C. (2000). "Evolutionary tracks and isochrones for low- and intermediate-mass stars: From 0.15 to 7 M $_u$, and from Z=0.0004 to 0.03". *Astronomy and Astrophysics Supplement* **141** (3): 371–383. arXiv:astro-ph/9910164. Bibcode:2000A&AS..141..371G. doi:10.1051/aas:2000126.

[150] Woosley, S. E.; Heger, A.; Weaver, T. A. (2002). "The evolution and explosion of massive stars". *Reviews of Modern Physics* **74** (4): 1015–1071. Bibcode:2002RvMP...74.1015W. doi:10.1103/RevModPhys.74.1015.

[151] 11.5 days is 0.0315 years.

15.14   Further reading

- Pickover, Cliff (2001). *The Stars of Heaven*. Oxford University Press. ISBN 0-19-514874-6.

- Gribbin, John; Gribbin, Mary (2001). *Stardust: Supernovae and Life—The Cosmic Connection*. Yale University Press. ISBN 0-300-09097-8.

- Hawking, Stephen (1988). *A Brief History of Time*. Bantam Books. ISBN 0-553-17521-1.

15.15   External links

- Kaler, James. "Portraits of Stars and their Constellations". University of Illinois. Retrieved 2010-08-20.

- "Query star by identifier, coordinates or reference code". *SIMBAD*. Centre de Données astronomiques de Strasbourg. Retrieved 2010-08-20.

- "How To Decipher Classification Codes". Astronomical Society of South Australia. Retrieved 2010-08-20.

- "Live Star Chart". Dobsonian Telescope Community. Retrieved 2010-08-20. View the stars above your location

- Prialnick, Dina et al. (2001). "Stars: Stellar Atmospheres, Structure, & Evolution". University of St. Andrews. Retrieved 2010-08-20.

The constellation of Leo as it can be seen by the naked eye. Lines have been added.

Alpha Centauri A and B over limb of Saturn

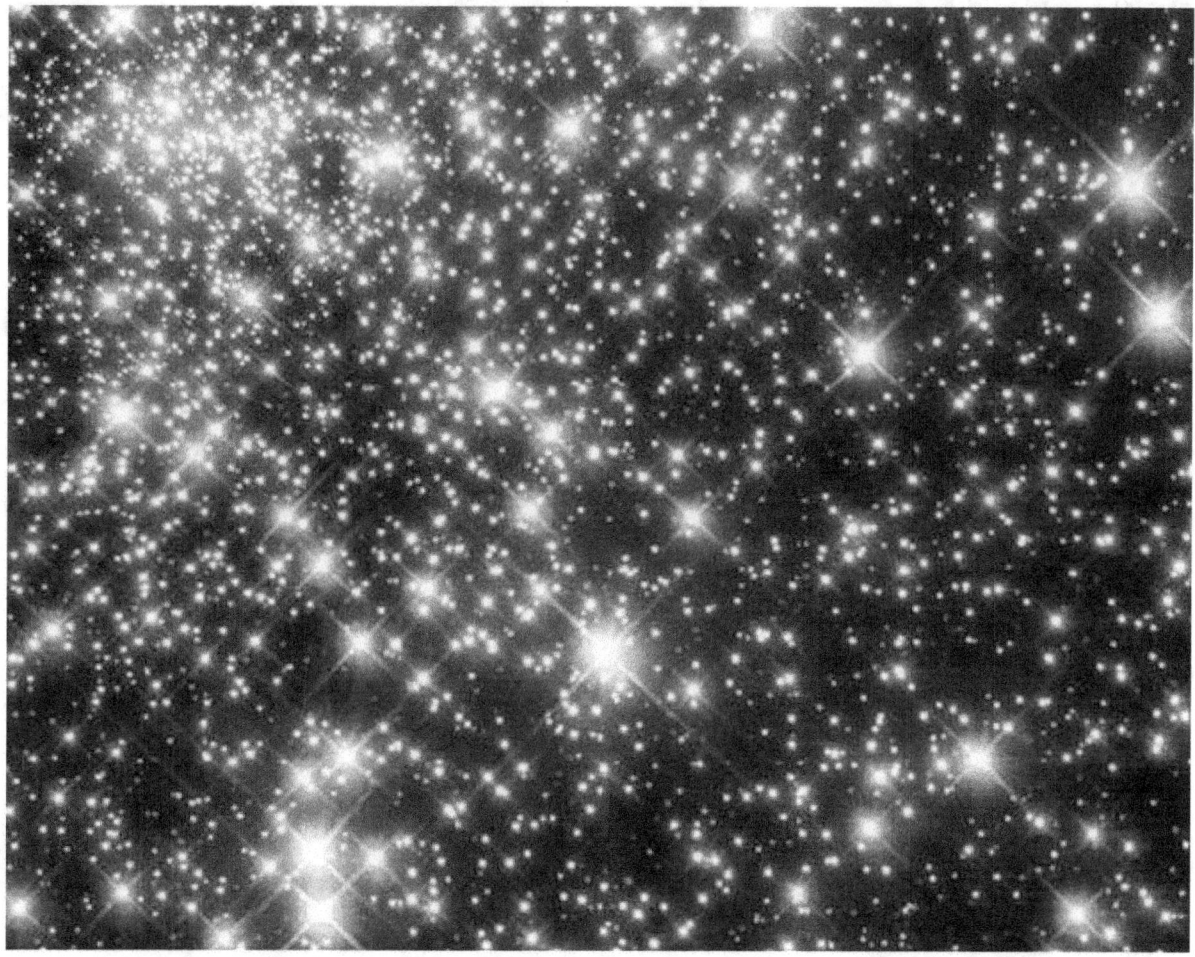

This view contains blue stars known as "Blue stragglers", for their apparent location on the Hertzsprung–Russell diagram

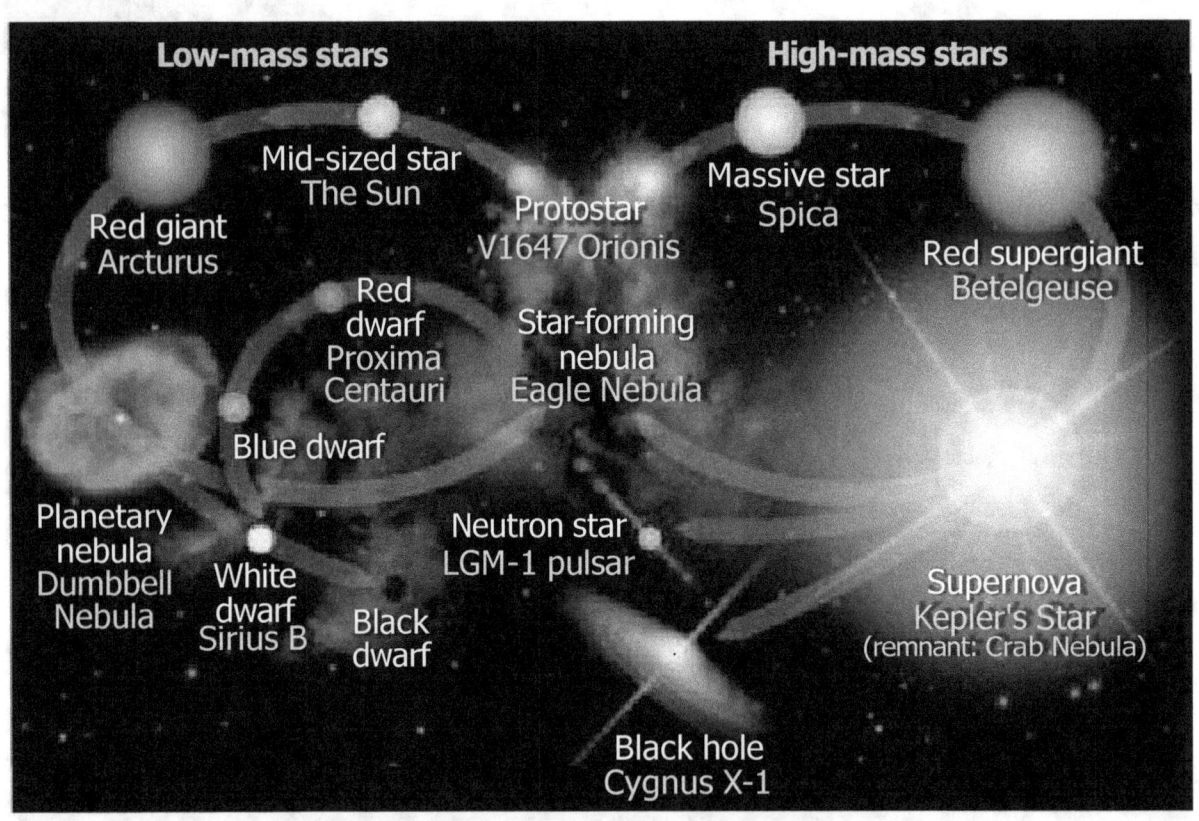

Stellar evolution of low-mass (left cycle) and high-mass (right cycle) stars, with examples in italics

Artist's conception of the birth of a star within a dense molecular cloud.

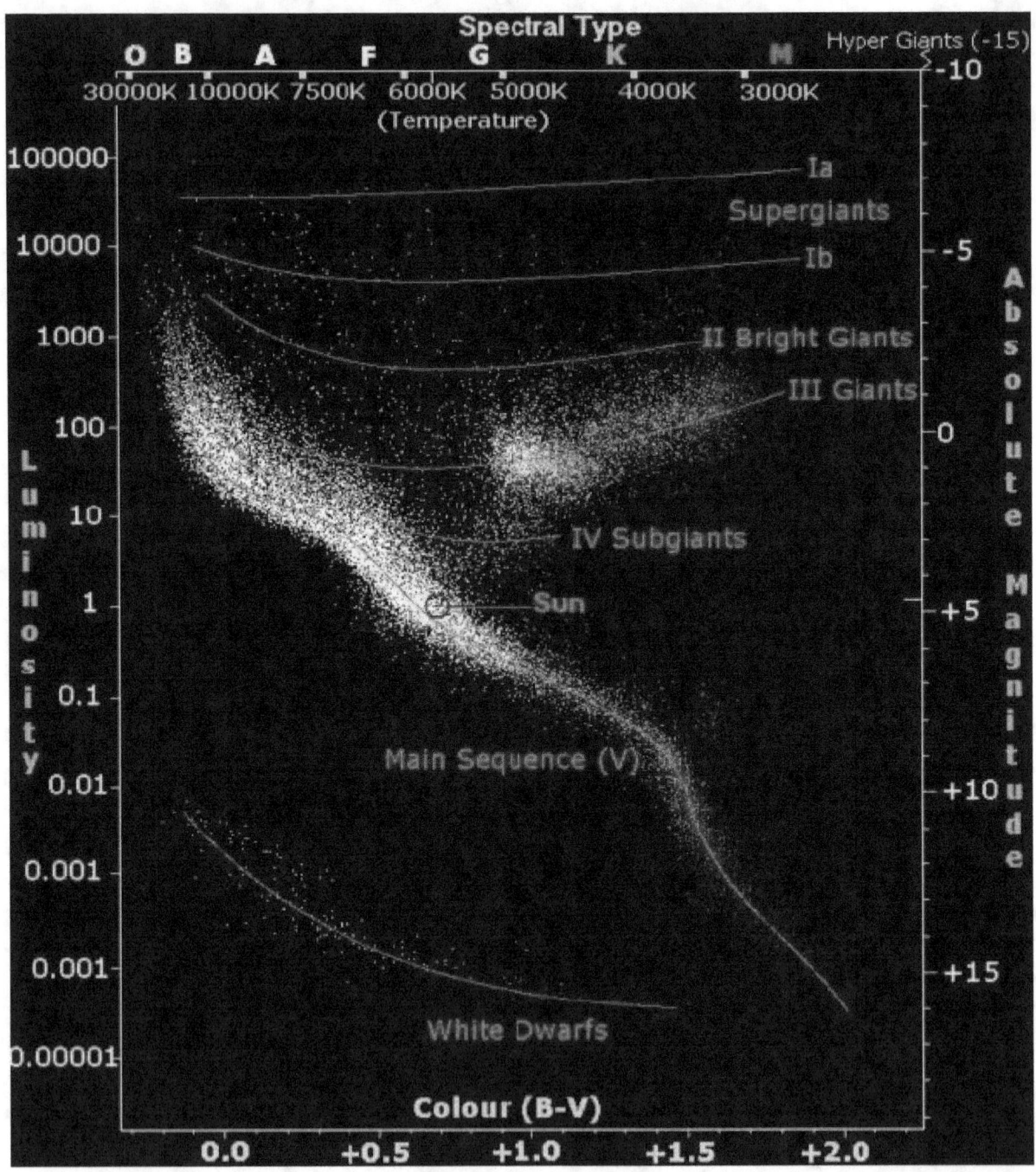

An example of a Hertzsprung–Russell diagram for a set of stars that includes the Sun (center). (See "Classification" below.)

The Crab Nebula, remnants of a supernova that was first observed around 1050 AD

A white dwarf star in orbit around Sirius (artist's impression).

The Sun is the nearest star to Earth.

Stars vary widely in size. In each image in the sequence, the right-most object appears as the left-most object in the next panel. The Earth appears at right in panel 1 and the Sun is second from the right in panel 3. The rightmost star at panel 6 is UY Scuti, the largest known star.

The Pleiades, an open cluster of stars in the constellation of Taurus. These stars share a common motion through space.[98]

Surface magnetic field of SU Aur (a young star of T Tauri type), reconstructed by means of Zeeman-Doppler imaging

The reflection nebula NGC 1999 is brilliantly illuminated by V380 Orionis (center), a variable star with about 3.5 times the mass of the Sun. The black patch of sky is a vast hole of empty space and not a dark nebula as previously thought.

The asymmetrical appearance of Mira, an oscillating variable star.

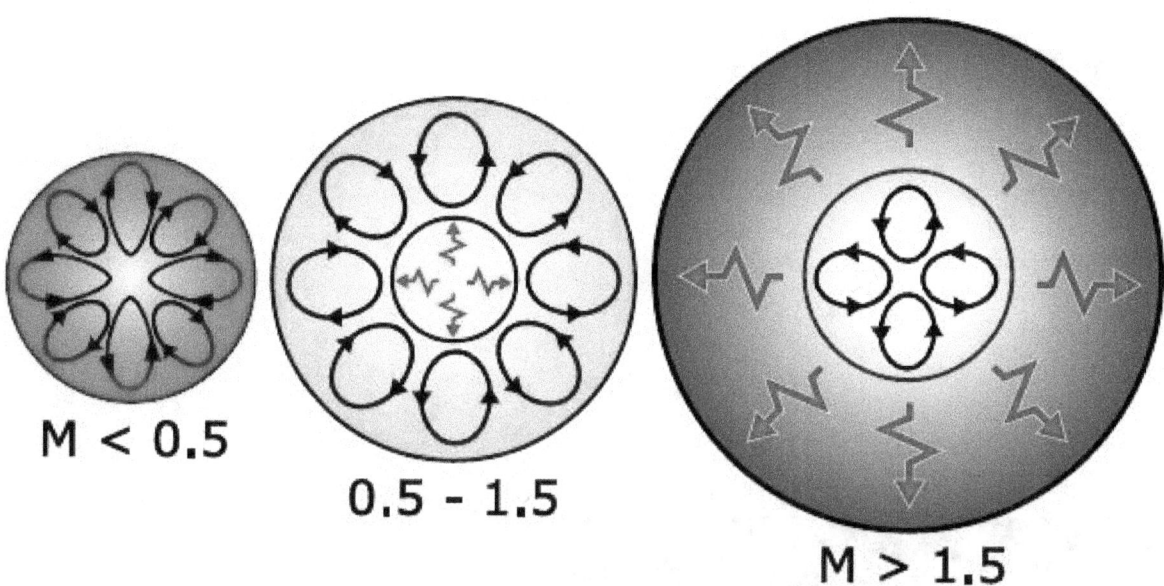

Internal structures of main sequence stars, convection zones with arrowed cycles and radiative zones with red flashes. To the left a **low-mass** *red dwarf, in the center a* **mid-sized** *yellow dwarf and at the right a* **massive** *blue-white main sequence star.*

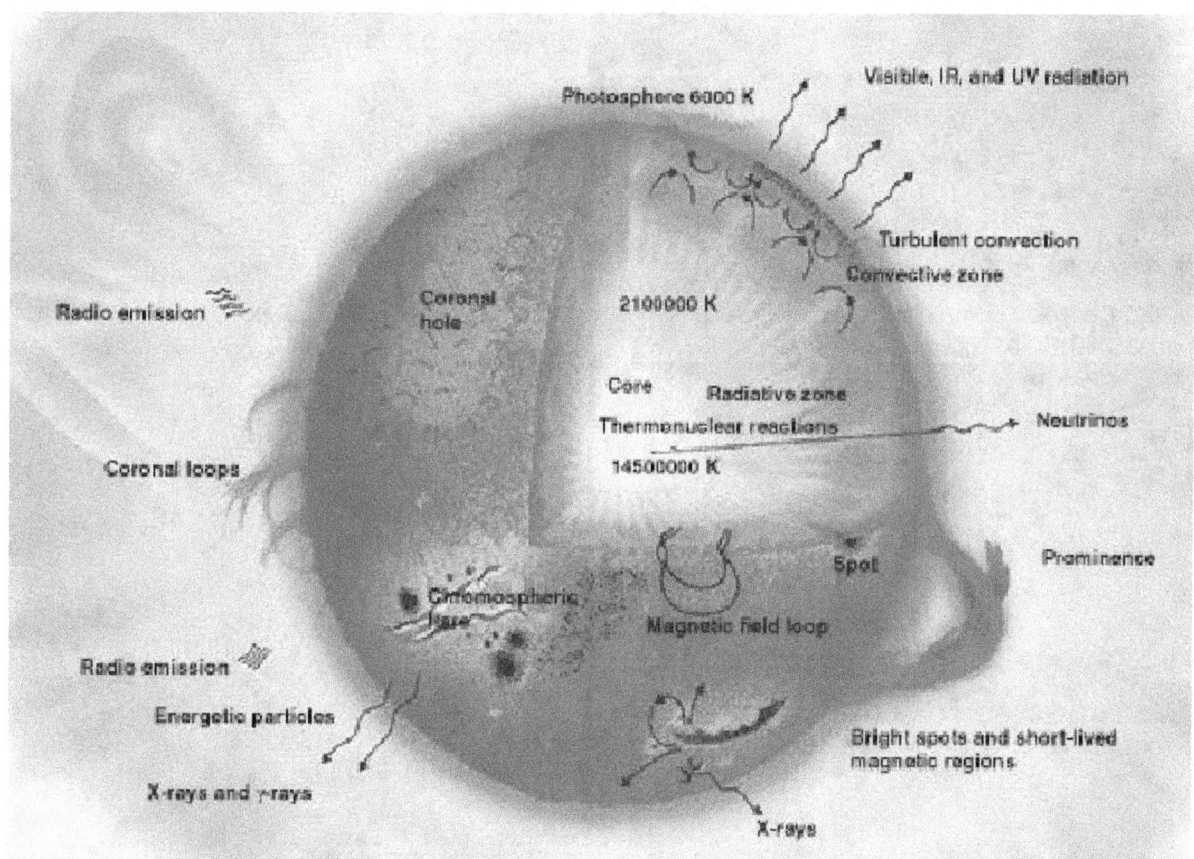

This diagram shows a cross-section of the Sun.

Chapter 16

Galaxy

This article is about the astronomical structure. For other uses, see Galaxy (disambiguation).

NGC 4414, a typical spiral galaxy in the constellation Coma Berenices, is about 55,000 light-years in diameter and approximately 60 million light-years away from Earth

The Sombrero galaxy (M104), a bright nearby spiral galaxy.

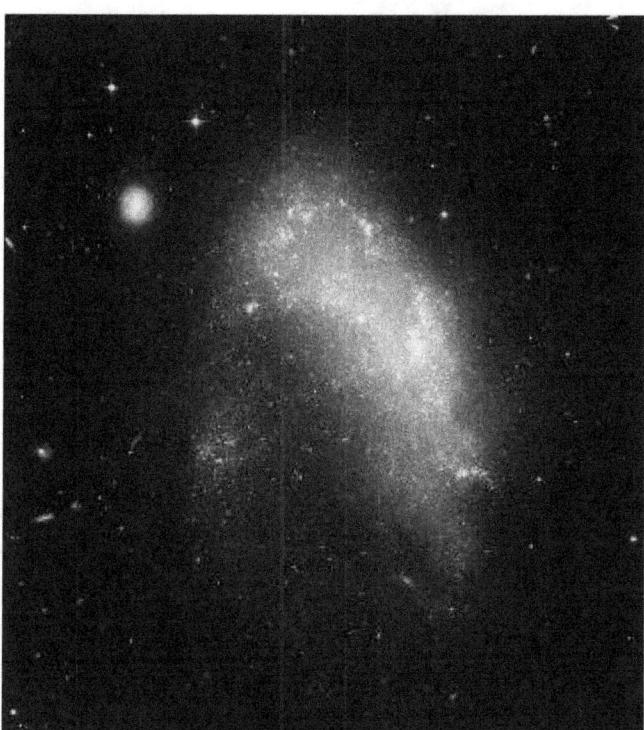

NGC 1427A, an example of an irregular galaxy, 52 million light-years away

A **galaxy** is a gravitationally bound system of stars, stellar remnants, interstellar gas and dust, and dark matter.[1][2] The word galaxy is derived from the Greek *galaxias* (γαλαξίας), literally "milky", a reference to the Milky Way. Examples of galaxies range from dwarfs with just a few thousand (10^3) stars to giants with one hundred trillion (10^{14}) stars,[3] each orbiting their galaxy's own center of mass. Galaxies can be categorized according to their visual morphology, including elliptical,[4] spiral, and irregular.[5] Many galaxies are believed to have supermassive black holes at an active center. The Milky Way's central black hole, known as Sagittarius A*, has a mass four million times that of our Sun.[6] As of May 2015, EGS-zs8-1 is the most distant known galaxy, estimated to be 13.1 billion light-years away and to have 15% of the mass of the Milky Way.[7][8][9][10]

There are approximately 170 billion (1.7×10^{11}) galaxies in the observable universe.[11] Most of the galaxies are 1,000 to 100,000 parsecs in diameter and usually separated by distances on the order of millions of parsecs (or megaparsecs). The space between galaxies is filled with a tenuous gas with an average density less than one atom per cubic meter. The majority of galaxies are gravitationally organized into associations known as galaxy groups, clusters, and superclusters. At the largest scale, these associations are generally arranged into sheets and filaments that are surrounded by immense voids.[12]

16.1   Etymology

The word *galaxy* derives from the Greek term for our own galaxy, *galaxias* (γαλαξίας, "milky one"), or *kyklos galaktikos* ("milky circle")[13] due to its appearance as a "milky" band of light in the sky. In Greek mythology, Zeus places his son born by a mortal woman, the infant Heracles, on Hera's breast while she is asleep so that the baby will drink her divine milk and will thus become immortal. Hera wakes up while breastfeeding and then realizes she is nursing an unknown baby: she pushes the baby away and a jet of her milk sprays the night sky, producing the faint band of light known as the Milky Way.[14][15]

In the astronomical literature, the capitalized word "Galaxy" is often used to refer to our galaxy, the Milky Way, to distinguish it from the other galaxies in our universe. The English term *Milky Way* can be traced back to a story by Chaucer c. 1380:

> "See yonder, lo, the Galaxyë
> Which men clepeth *the Milky Wey*,
> For hit is whyt."
> —Geoffrey Chaucer, The House of Fame[13]

When William Herschel assembled his catalog of deep sky objects in 1786, he used the term *spiral nebula* for certain objects such as M31. These would later be recognized as conglomerations of stars when the true distance to these objects began to be appreciated, and they would later be termed *island universes*. However, the word *Universe* was understood to mean the entirety of existence, so this expression fell into disuse and the objects instead became known as galaxies.[16]

16.2   Nomenclature

Tens of thousands of galaxies have been catalogued, but only a few have well-established names, such as the Andromeda Galaxy, the Magellanic clouds, the Whirlpool Galaxy and the Sombrero Galaxy. Astronomers work with numbers from certain catalogues, such as the Messier catalogue, the NGC (New General Catalogue), the IC (Index Catalogue), the CGCG (Catalogue of Galaxies and of Clusters of Galaxies), the MCG (Morphological Catalogue of Galaxies) and UGC (Uppsala General Catalogue of Galaxies). All of the well-known galaxies appear in one or more of these catalogues but each time under a different number. For example, the Messier 109, a spiral system which has the number 109 in the catalogue of Messier also codes NCG3992, UGC6937, CGCG 269-023, MCG +09-20-044 and PGC 37617.

Because it is customary in science to assign names to most of the studied objects, even to the smallest ones, the Belgian astrophysicist Gerard Bodifee and the classicist Michel Berger started a new catalogue (CNG-Catalogue of Named Galaxies)[17] in which a thousand well-known galaxies are given meaningful, descriptive names in Latin (or Latinized Greek)[18] in accordance with the binomial nomenclature that one uses in other sciences such as biology, anatomy, paleontology and in other fields of astronomy such as the geography of Mars. One of the arguments to do so is that these impressive objects deserve better than uninspired codes. For instance, Bodifee and Berger propose the informal, descriptive name *Callimorphus Ursae Majoris* for the well-formed barred galaxy Messier 109 in Ursa Major.

16.3   Observation history

The realization that we live in a galaxy, and that ours is one among many, parallels major discoveries that were made about the Milky Way and other nebulae in the night sky.

16.3.1   Milky Way

Main article: Milky Way

The Greek philosopher Democritus (450–370 BC) proposed that the bright band on the night sky known as the Milky Way might consist of distant stars.[19] Aristotle (384–322 BC), however, believed the Milky Way to be caused by "the ignition of the fiery exhalation of some stars that were large, numerous and close together" and that the "ignition takes place in the upper part of the atmosphere, in the region of the World that is continuous with the heavenly motions."[20] The Neoplatonist philosopher Olympiodorus the Younger (c. 495–570 AD) was critical of this view, arguing that if the Milky Way is sublunary (situated between Earth and the Moon) it should appear different at different times and places on Earth, and that it should have parallax, which it does not. In his view, the Milky Way is celestial.[21]

According to Mohani Mohamed, the Arabian astronomer Alhazen (965–1037) made the first attempt at observing and measuring the Milky Way's parallax,[22] and he thus "determined that because the Milky Way had no parallax, it must be remote from the Earth, not belonging to the atmosphere."[23] The Persian astronomer al-Bīrūnī (973–1048) proposed the Milky Way galaxy to be "a collection of countless fragments of the nature of nebulous stars."[24][25] The Andalusian astronomer Ibn Bajjah ("Avempace", d. 1138) proposed that the Milky Way is made up of many stars that almost touch one another and appear to be a continuous image due to the effect of refraction from sublunary material,[20][26] citing his observation of the conjunction of Jupiter and Mars as evidence of this occurring when two objects are near.[20] In the 14th century, the Syrian-born Ibn Qayyim proposed the Milky Way galaxy to be "a myriad of tiny stars packed together in the sphere of the fixed stars."[27]

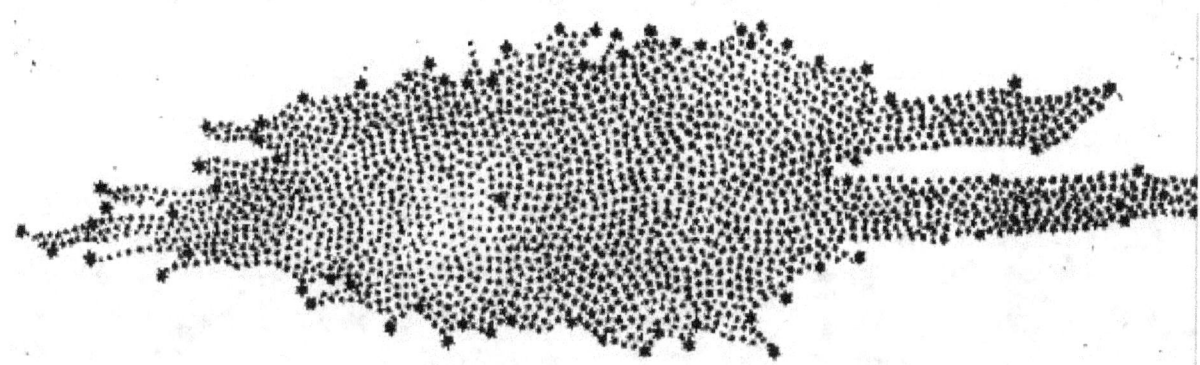

The shape of the Milky Way as estimated from star counts by William Herschel in 1785; the solar system was assumed to be near the center.

Actual proof of the Milky Way consisting of many stars came in 1610 when the Italian astronomer Galileo Galilei used a telescope to study the Milky Way and discovered that it is composed of a huge number of faint stars.[28][29] In 1750 the English astronomer Thomas Wright, in his *An original theory or new hypothesis of the Universe*, speculated (correctly) that the galaxy might be a rotating body of a huge number of stars held together by gravitational forces, akin to the solar system but on a much larger scale. The resulting disk of stars can be seen as a band on the sky from our perspective inside the disk.[30][31] In a treatise in 1755, Immanuel Kant elaborated on Wright's idea about the structure of the Milky Way.[32]

The first project to describe the shape of the Milky Way and the position of the Sun was undertaken by William Herschel in 1785 by counting the number of stars in different regions of the sky. He produced a diagram of the shape of the galaxy with the solar system close to the center.[33][34] Using a refined approach, Kapteyn in 1920 arrived at the picture of a small (diameter about 15 kiloparsecs) ellipsoid galaxy with the Sun close to the center. A different method by Harlow Shapley based on the cataloguing of globular clusters led to a radically different picture: a flat disk with diameter approximately 70 kiloparsecs and the Sun far from the center.[31] Both analyses failed to take into account the absorption of light by interstellar dust present in the galactic plane, but after Robert Julius Trumpler quantified this effect in 1930 by studying open clusters, the present picture of our host galaxy, the Milky Way, emerged.[35]

A fish-eye mosaic of the Milky Way arching at a high inclination across the night sky, shot from a dark-sky location in Chile

16.3.2 Distinction from other nebulae

A few galaxies outside the Milky Way are visible in the night sky to the unaided eye. In the 10th century, the Persian astronomer Al-Sufi made the earliest recorded identification of the Andromeda Galaxy, describing it as a "small cloud".[36] In 964, Al-Sufi identified the Large Magellanic Cloud in his *Book of Fixed Stars*; it was not seen by Europeans until Magellan's voyage in the 16th century.[37][38] The Andromeda Galaxy was independently noted by Simon Marius in 1612.[36]

In 1750, Thomas Wright speculated (correctly) that the Milky Way is a flattened disk of stars, and that some of the nebulae visible in the night sky might be separate Milky Ways.[31][39] In 1755, Immanuel Kant used the term "island Universe" to describe these distant nebulae.

Photograph of the "Great Andromeda Nebula" from 1899, later identified as the Andromeda Galaxy

Toward the end of the 18th century, Charles Messier compiled a catalog containing the 109 brightest celestial objects having nebulous appearance. Subsequently, William Herschel assembled a catalog of 5,000 nebulae.[31] In 1845, Lord Rosse constructed a new telescope and was able to distinguish between elliptical and spiral nebulae. He also managed to make out individual point sources in some of these nebulae, lending credence to Kant's earlier conjecture.[40]

In 1912, Vesto Slipher made spectrographic studies of the brightest spiral nebulae to determine their composition. Slipher discovered that the spiral nebulae have high Doppler shifts, indicating that they are moving at a rate exceeding the velocity of the stars he had measured. He found that the majority of these nebulae are moving away from us.[41][42]

In 1917, Heber Curtis observed nova S Andromedae within the "Great Andromeda Nebula" (as the Andromeda Galaxy, Messier object M31, was then known). Searching the photographic record, he found 11 more novae. Curtis noticed that these novae were, on average, 10 magnitudes fainter than those that occurred within our galaxy. As a result he was able to come up with a distance estimate of 150,000 parsecs. He became a proponent of the so-called "island universes" hypothesis, which holds that spiral nebulae are actually independent galaxies.[43]

In 1920 the so-called Great Debate took place between Harlow Shapley and Heber Curtis, concerning the nature of the Milky Way, spiral nebulae, and the dimensions of the Universe. To support his claim that the Great Andromeda Nebula is an external galaxy, Curtis noted the appearance of dark lanes resembling the dust clouds in the Milky Way, as well as the significant Doppler shift.[44]

In 1922, the Estonian astronomer Ernst Öpik gave a distance determination that supported the theory that the Andromeda Nebula is indeed a distant extra-galactic object.[45] Using the new 100 inch Mt. Wilson telescope, Edwin Hubble was able to resolve the outer parts of some spiral nebulae as collections of individual stars and identified some Cepheid variables, thus allowing him to estimate the distance to the nebulae: they were far too distant to be part of the Milky Way.[46] In 1936 Hubble produced a classification of galactic morphology that is used to this day.[47]

16.3.3   Modern research

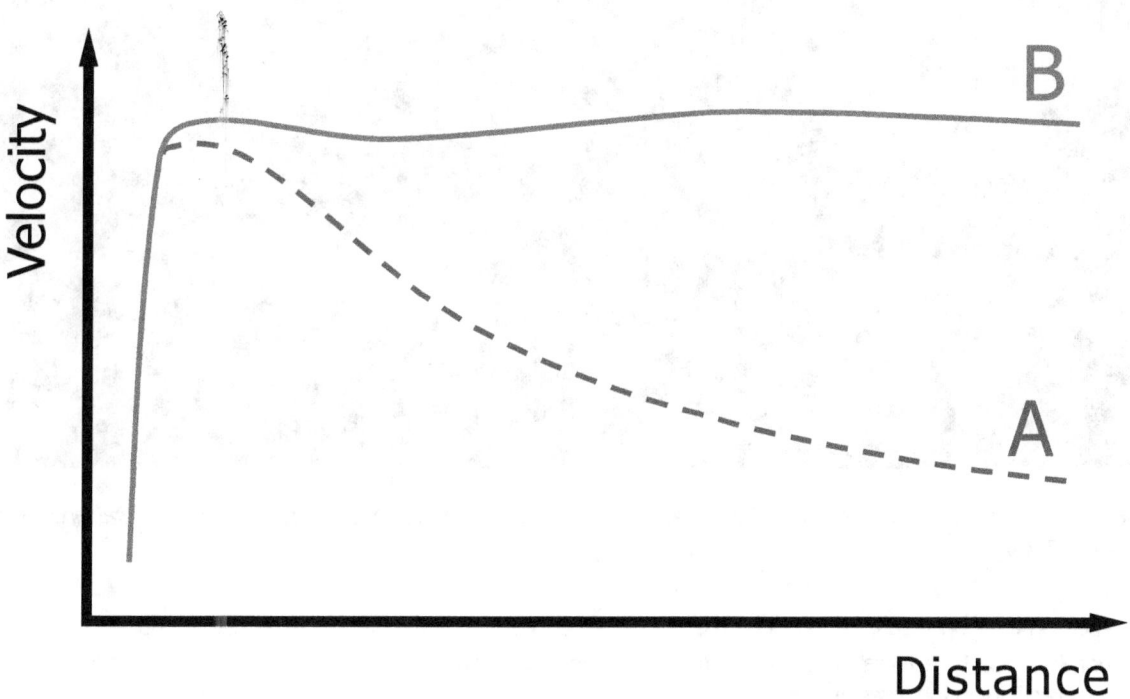

Rotation curve of a typical spiral galaxy: predicted based on the visible matter (A) and observed (B). The distance is from the galactic core.

In 1944, Hendrik van de Hulst predicted that microwave radiation with wavelength of 21 cm would be detectable from interstellar atomic hydrogen gas;[48] and in 1951 it was observed. This radiation is not affected by dust absorption, and so its Doppler shift can be used to map the motion of the gas in our galaxy. These observations led to the hypothesis of a rotating bar structure in the center of our galaxy.[49] With improved radio telescopes, hydrogen gas could also be traced in other galaxies. In the 1970s, Vera Rubin uncovered a discrepancy between observed galactic rotation speed and that predicted by the visible mass of stars and gas. Today, the galaxy rotation problem is thought to be explained by the presence of large quantities of unseen dark matter.[50][51] A concept known as the universal rotation curve of spirals, moreover, shows that the problem is ubiquitous in these objects.

Beginning in the 1990s, the Hubble Space Telescope yielded improved observations. Among other things, Hubble data helped establish that the missing dark matter in our galaxy cannot solely consist of inherently faint and small stars.[52] The Hubble Deep Field, an extremely long exposure of a relatively empty part of the sky, provided evidence that there are about 125 billion (1.25×10^{11}) galaxies in the Universe.[53] Improved technology in detecting the spectra invisible to humans (radio telescopes, infrared cameras, and x-ray telescopes) allow detection of other galaxies that are not detected by Hubble. Particularly, galaxy surveys in the Zone of Avoidance (the region of the sky blocked by the Milky Way) have revealed a number of new galaxies.[54]

16.4 Types and morphology

Main article: Galaxy morphological classification

Galaxies come in three main types: ellipticals, spirals, and irregulars. A slightly more extensive description of galaxy

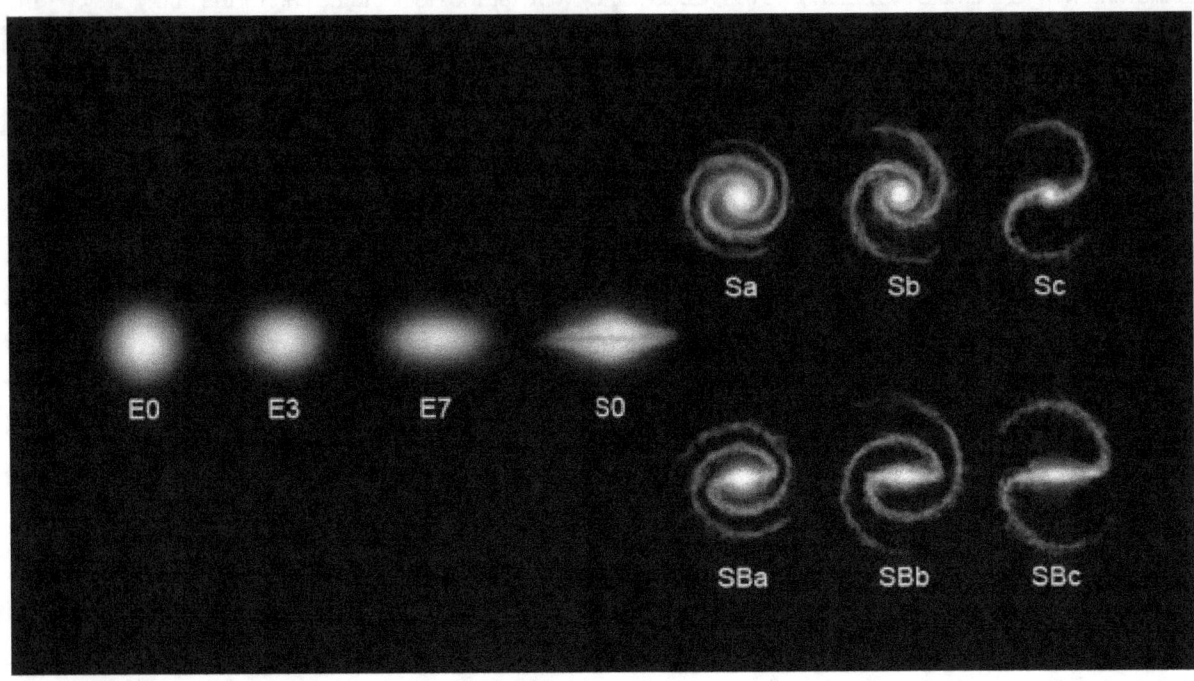

Types of galaxies according to the Hubble classification scheme: an E *indicates a type of elliptical galaxy; an* S *is a spiral; and* SB *is a barred-spiral galaxy.*[note 1]

types based on their appearance is given by the Hubble sequence. Since the Hubble sequence is entirely based upon visual morphological type, it may miss certain important characteristics of galaxies such as star formation rate in starburst galaxies and activity in the cores of active galaxies.[5]

16.4.1 Ellipticals

Main article: Elliptical galaxy

The Hubble classification system rates elliptical galaxies on the basis of their ellipticity, ranging from E0, being nearly spherical, up to E7, which is highly elongated. These galaxies have an ellipsoidal profile, giving them an elliptical appearance regardless of the viewing angle. Their appearance shows little structure and they typically have relatively little interstellar matter. Consequently these galaxies also have a low portion of open clusters and a reduced rate of new star formation. Instead they are dominated by generally older, more evolved stars that are orbiting the common center of gravity in random directions. The stars contain low abundances of heavy elements because star formation ceases after the initial burst. In this sense they have some similarity to the much smaller globular clusters.[55]

The largest galaxies are giant ellipticals. Many elliptical galaxies are believed to form due to the interaction of galaxies, resulting in a collision and merger. They can grow to enormous sizes (compared to spiral galaxies, for example), and giant elliptical galaxies are often found near the core of large galaxy clusters.[56] Starburst galaxies are the result of such a galactic collision that can result in the formation of an elliptical galaxy.[55]

NGC 3923 Elliptical Shell Galaxy-Hubble Space Telescope photograph

Shell galaxy

A shell galaxy is a type of elliptical galaxy where the stars in the galaxy's halo are arranged in concentric shells. About 1/10 tenth of elliptical galaxies have a shell-like structure, which has never been observed in spiral galaxies. The shell-like structures are thought to develop when a larger galaxy absorbs a smaller companion galaxy. As the two galaxy centers approach, the centers start to oscillate around a center point, the oscillation creates gravitational ripples forming the shells of stars, similar to ripples spreading on water. For example, galaxy NGC 3923 has over twenty shells.[57]

16.4.2   Spirals

Main articles: Spiral galaxy and Barred spiral galaxy
Spiral galaxies resemble spiraling pinwheels. Though the stars and other visible material contained in such a galaxy lie

The Pinwheel Galaxy, NGC 5457.

mostly on a plane, the majority of mass in spiral galaxies exists in a roughly spherical halo of dark matter that extends beyond the visible component, as demonstrated by the universal rotation curve concept.[58]

Spiral galaxies consist of a rotating disk of stars and interstellar medium, along with a central bulge of generally older stars. Extending outward from the bulge are relatively bright arms. In the Hubble classification scheme, spiral galaxies are listed as type *S*, followed by a letter (*a*, *b*, or *c*) that indicates the degree of tightness of the spiral arms and the size of the central bulge. An *Sa* galaxy has tightly wound, poorly defined arms and possesses a relatively large core region. At the other extreme, an *Sc* galaxy has open, well-defined arms and a small core region.[59] A galaxy with poorly defined arms is sometimes referred to as a flocculent spiral galaxy; in contrast to the grand design spiral galaxy that has prominent and well-defined spiral arms.[60]

It appears the reason that some spiral galaxies are fat and bulging while some are flat discs is because of how fast they rotate.[61]

In spiral galaxies, the spiral arms do have the shape of approximate logarithmic spirals, a pattern that can be theoretically shown to result from a disturbance in a uniformly rotating mass of stars. Like the stars, the spiral arms rotate around the center, but they do so with constant angular velocity. The spiral arms are thought to be areas of high-density matter, or "density waves".[62] As stars move through an arm, the space velocity of each stellar system is modified by the gravitational force of the higher density. (The velocity returns to normal after the stars depart on the other side of the arm.) This effect is akin to a "wave" of slowdowns moving along a highway full of moving cars. The arms are visible because the high density facilitates star formation, and therefore they harbor many bright and young stars.[63]

NGC 1300, an example of a barred spiral galaxy.

Hoag's Object, an example of a ring galaxy

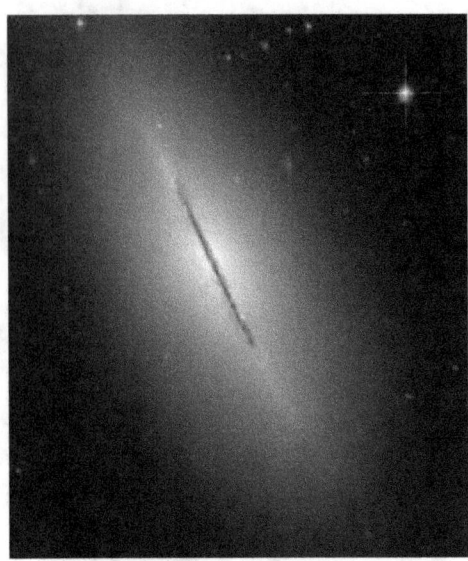

NGC 5866, an example of a lenticular galaxy

Barred Spiral Galaxy

A majority of spiral galaxies, including our own Milky Way galaxy, have a linear, bar-shaped band of stars that extends outward to either side of the core, then merges into the spiral arm structure.[64] In the Hubble classification scheme, these are designated by an *SB*, followed by a lower-case letter (*a*, *b* or *c*) that indicates the form of the spiral arms (in the same manner as the categorization of normal spiral galaxies). Bars are thought to be temporary structures that can occur as a result of a density wave radiating outward from the core, or else due to a tidal interaction with another galaxy.[65] Many barred spiral galaxies are active, possibly as a result of gas being channeled into the core along the arms.[66]

Our own galaxy, the Milky Way, is a large disk-shaped barred-spiral galaxy[67] about 30 kiloparsecs in diameter and a kiloparsec thick. It contains about two hundred billion (2×10^{11})[68] stars and has a total mass of about six hundred billion (6×10^{11}) times the mass of the Sun.[69]

16.4.3   Other morphologies

- Peculiar galaxies are galactic formations that develop unusual properties due to tidal interactions with other galaxies. An example of this is the ring galaxy, which possesses a ring-like structure of stars and interstellar medium surrounding a bare core. A ring galaxy is thought to occur when a smaller galaxy passes through the core of a spiral galaxy.[70] Such an event may have affected the Andromeda Galaxy, as it displays a multi-ring-like structure when viewed in infrared radiation.[71]

- A lenticular galaxy is an intermediate form that has properties of both elliptical and spiral galaxies. These are categorized as Hubble type S0, and they possess ill-defined spiral arms with an elliptical halo of stars[72] (barred lenticular galaxies receive Hubble classification SB0.)

- Irregular galaxies are galaxies that can not be readily classified into an elliptical or spiral morphology. An Irr-I galaxy has some structure but does not align cleanly with the Hubble classification scheme. Irr-II galaxies do not possess any structure that resembles a Hubble classification, and may have been disrupted.[73] Nearby examples of (dwarf) irregular galaxies include the Magellanic Clouds.

- An ultra diffuse galaxy (UDG) is an extremely-low-density galaxy. The galaxy may be the same size as the Milky Way but has a visible star count of only 1% of the Milky Way. The lack of luminosity is because their is a lack of star-forming gas in the galaxy which results in old stellar populations.

16.4.4   Dwarfs

Main article: Dwarf galaxy

Despite the prominence of large elliptical and spiral galaxies, most galaxies in the Universe are dwarf galaxies. These galaxies are relatively small when compared with other galactic formations, being about one hundredth the size of the Milky Way, containing only a few billion stars. Ultra-compact dwarf galaxies have recently been discovered that are only 100 parsecs across.[74]

Many dwarf galaxies may orbit a single larger galaxy; the Milky Way has at least a dozen such satellites, with an estimated 300–500 yet to be discovered.[75] Dwarf galaxies may also be classified as elliptical, spiral, or irregular. Since small dwarf ellipticals bear little resemblance to large ellipticals, they are often called dwarf spheroidal galaxies instead.

A study of 27 Milky Way neighbors found that in all dwarf galaxies, the central mass is approximately 10 million solar masses, regardless of whether the galaxy has thousands or millions of stars. This has led to the suggestion that galaxies are largely formed by dark matter, and that the minimum size may indicate a form of warm dark matter incapable of gravitational coalescence on a smaller scale.[76]

16.5   Unusual dynamics and activities

16.5.1   Interacting

Main article: Interacting galaxy

Interactions between galaxies are relatively frequent, and they can play an important role in galactic evolution. Near misses between galaxies result in warping distortions due to tidal interactions, and may cause some exchange of gas and dust.[77][78] Collisions occur when two galaxies pass directly through each other and have sufficient relative momentum not to merge. The stars of interacting galaxies will usually not collide, but the gas and dust within the two forms will interact, sometimes triggering star formation. A collision can severely distort the shape of the galaxies, forming bars, rings or tail-like structures.[77][78]

At the extreme of interactions are galactic mergers. In this case the relative momentum of the two galaxies is insufficient to allow the galaxies to pass through each other. Instead, they gradually merge to form a single, larger galaxy. Mergers can result in significant changes to morphology, as compared to the original galaxies. In the case where one of the galaxies is much more massive, however, the result is known as cannibalism. In this case the larger galaxy will remain relatively undisturbed by the merger, while the smaller galaxy is torn apart. The Milky Way galaxy is currently in the process of cannibalizing the Sagittarius Dwarf Elliptical Galaxy and the Canis Major Dwarf Galaxy.[77][78]

16.5.2   Starburst

Main article: Starburst galaxy

Stars are created within galaxies from a reserve of cold gas that forms into giant molecular clouds. Some galaxies have been observed to form stars at an exceptional rate, known as a starburst. Should they continue to do so, however, they would consume their reserve of gas in a time frame lower than the lifespan of the galaxy. Hence starburst activity usually lasts for only about ten million years, a relatively brief period in the history of a galaxy. Starburst galaxies were more common during the early history of the Universe,[80] and, at present, still contribute an estimated 15% to the total star production rate.[81]

Starburst galaxies are characterized by dusty concentrations of gas and the appearance of newly formed stars, including massive stars that ionize the surrounding clouds to create H II regions.[82] These massive stars produce supernova explosions, resulting in expanding remnants that interact powerfully with the surrounding gas. These outbursts trigger a chain reaction of star building that spreads throughout the gaseous region. Only when the available gas is nearly consumed or dispersed does the starburst activity come to an end.[80]

Starbursts are often associated with merging or interacting galaxies. The prototype example of such a starburst-forming

The Antennae Galaxies are undergoing a collision that will result in their eventual merger.

interaction is M82, which experienced a close encounter with the larger M81. Irregular galaxies often exhibit spaced knots of starburst activity.[83]

16.5.3   Active nucleus

Main article: Active galactic nucleus

A portion of the observable galaxies are classified as active. That is, a significant portion of the total energy output from the galaxy is emitted by a source other than the stars, dust and interstellar medium.

The standard model for an active galactic nucleus is based upon an accretion disc that forms around a supermassive black hole (SMBH) at the core region. The radiation from an active galactic nucleus results from the gravitational energy of matter as it falls toward the black hole from the disc.[84] In about 10% of these objects, a diametrically opposed pair of energetic jets ejects particles from the core at velocities close to the speed of light. The mechanism for producing these

M82, a starburst galaxy that has ten times the star formation of a "normal" galaxy.[79]

jets is still not well understood.[85]

Active galaxies that emit high-energy radiation in the form of x-rays are classified as Seyfert galaxies or quasars, depending on the luminosity.

Blazars

Main article: Blazars

Blazars are believed to be an active galaxy with a relativistic jet that is pointed in the direction of Earth. A radio galaxy emits radio frequencies from relativistic jets. A unified model of these types of active galaxies explains their differences based on the viewing angle of the observer.[85]

LINERS

Possibly related to active galactic nuclei (as well as starburst regions) are low-ionization nuclear emission-line regions (LINERs). The emission from LINER-type galaxies is dominated by weakly ionized elements.[86] Approximately one-third of nearby galaxies are classified as containing LINER nuclei.[84][86][87]

A jet of particles is being emitted from the core of the elliptical radio galaxy M87.

Seyfert Galaxy

Main article: Seyfert Galaxy

Seyfert galaxies are one of the two largest groups of active galaxies, along with quasars. They have quasar-like nuclei (very luminous, distant and bright sources of electromagnetic radiation) with very high surface brightnesses but unlike quasars, their host galaxies are clearly detectable. Seyfert galaxies account for about 10% of all galaxies. Seen in visible light, most Seyfert galaxies look like normal spiral galaxies, but when studied under other wavelengths, the luminosity of their cores is equivalent to the luminosity of whole galaxies the size of the Milky Way.

Quasar

Main article: Quasar

Quasars (/ˈkweɪzɑr/) or quasi-stellar radio sources are the most energetic and distant members of a class of objects called active galactic nuclei (AGN). Quasars are extremely luminous and were first identified as being high redshift sources of electromagnetic energy, including radio waves and visible light, that appeared to be similar to stars, rather than extended sources similar to galaxies. Their luminosity can be 100 times greater than that of the Milky Way.

16.5.4   Luminous infrared galaxy

Main article: Luminous infrared galaxy

Luminous Infrared Galaxies or (LIRG's) are galaxies with luminosities, the measurement of brightness, above 1011 L⊙. LIRG's are more abundant than starburst galaxies, Seyfert galaxies and quasi-stellar objects at comparable luminosity. Infrared galaxies emit more energy in the infrared than at all other wavelengths combined. An LIRG's luminosity is 100 billion times that of our sun.

16.6   Formation and evolution

Main article: Galaxy formation and evolution

Galactic formation and evolution is an active area of research in astrophysics.

16.6.1   Formation

Current cosmological models of the early Universe are based on the Big Bang theory. About 300,000 years after this event, atoms of hydrogen and helium began to form, in an event called recombination. Nearly all the hydrogen was neutral (non-ionized) and readily absorbed light, and no stars had yet formed. As a result this period has been called the "dark ages". It was from density fluctuations (or anisotropic irregularities) in this primordial matter that larger structures began to appear. As a result, masses of baryonic matter started to condense within cold dark matter halos.[89][90] These primordial structures would eventually become the galaxies we see today.

Early galaxies

Evidence for the early appearance of galaxies was found in 2006, when it was discovered that the galaxy IOK-1 has an unusually high redshift of 6.96, corresponding to just 750 million years after the Big Bang and making it the most distant and primordial galaxy yet seen.[91] While some scientists have claimed other objects (such as Abell 1835 IR1916) have higher redshifts (and therefore are seen in an earlier stage of the Universe's evolution), IOK-1's age and composition have been more reliably established. In December 2012, astronomers reported that the UDFj-39546284 is the most distant object known and has a redshift value of 11.9. The object, is estimated to have existed around "380 million years"[92] after the Big Bang (which was about 13.8 billion years ago),[93] is about 13.42 billion light years away. The existence of such early protogalaxies suggests that they must have grown in the so-called "dark ages".[89] As of May 5, 2015, the galaxy EGS-zs8-1 is the most distant and earliest galaxy measured, forming 670 million years after the Big Bang. The light from EGS-zs8-1 has taken 13 billion years to reach Earth, and is now 30 billion light-years away, because of the expansion of the universe during 13 billion years.[94][95][96][96][97][98]

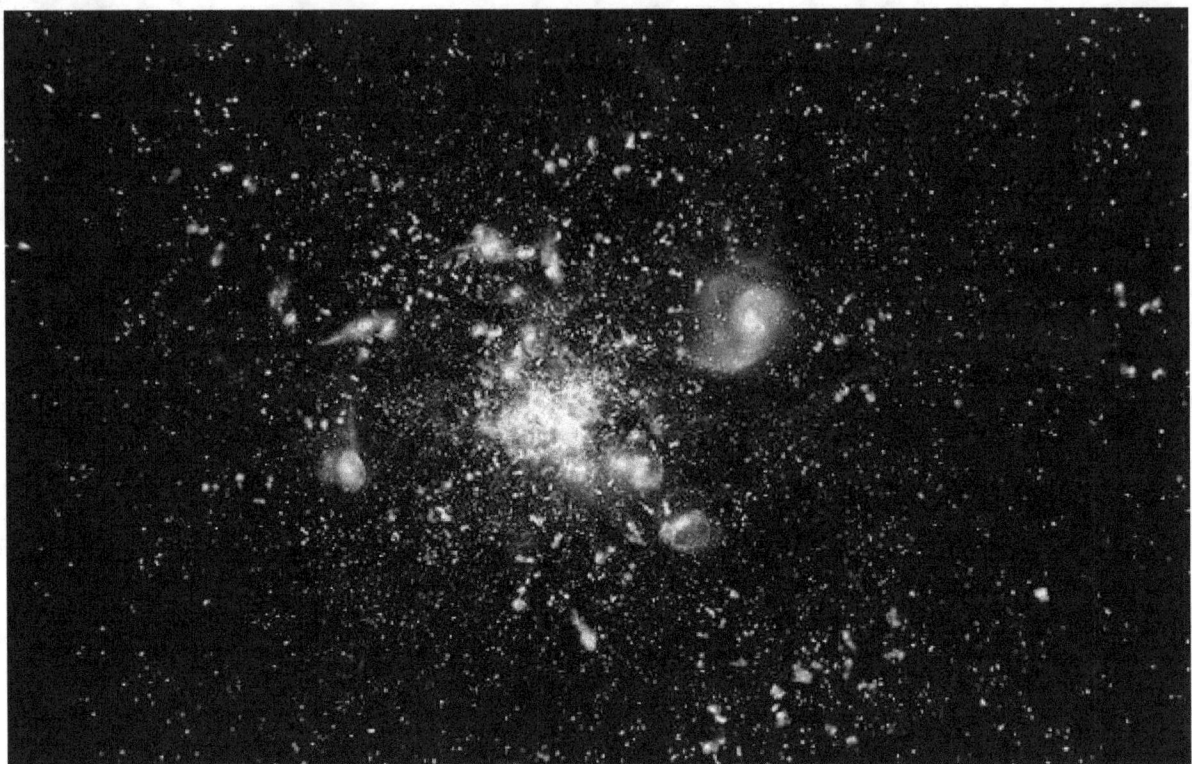

Artist's impression of a protocluster forming in the early Universe.[88]

Early galaxy formation

The detailed process by which early galaxies formed is an open question in astrophysics. Theories can be divided into two categories: top-down and bottom-up. In top-down theories (such as the Eggen–Lynden-Bell–Sandage [ELS] model), protogalaxies form in a large-scale simultaneous collapse lasting about one hundred million years.[99] In bottom-up theories (such as the Searle-Zinn [SZ] model), small structures such as globular clusters form first, and then a number of such bodies accrete to form a larger galaxy.[100]

Once protogalaxies began to form and contract, the first halo stars (called Population III stars) appeared within them. These were composed almost entirely of hydrogen and helium, and may have been massive. If so, these huge stars would have quickly consumed their supply of fuel and became supernovae, releasing heavy elements into the interstellar medium.[101] This first generation of stars re-ionized the surrounding neutral hydrogen, creating expanding bubbles of space through which light could readily travel.[102]

In June 2015, astronomers reported evidence for Population III stars in the Cosmos Redshift 7 galaxy at z = 6.60. Such stars are likely to have existed in the very early universe (i.e., at high redshift), and may have started the production of chemical elements heavier than hydrogen that are needed for the later formation of planets and life as we know it.[103][104]

16.6.2   Evolution

Within a billion years of a galaxy's formation, key structures begin to appear. Globular clusters, the central supermassive black hole, and a galactic bulge of metal-poor Population II stars form. The creation of a supermassive black hole appears to play a key role in actively regulating the growth of galaxies by limiting the total amount of additional matter added.[105] During this early epoch, galaxies undergo a major burst of star formation.[106]

During the following two billion years, the accumulated matter settles into a galactic disc.[107] A galaxy will continue to absorb infalling material from high-velocity clouds and dwarf galaxies throughout its life.[108] This matter is mostly hydrogen and helium. The cycle of stellar birth and death slowly increases the abundance of heavy elements, eventually

Artist's impression of a young galaxy accreting material.

allowing the formation of planets.[109]

Hubble eXtreme Deep Field (XDF)

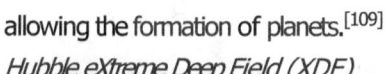

XDF view field compared to the angular size of the Moon. Several thousand galaxies, each consisting of billions of stars, are in this small view.

XDF (2012) view: Each light speck is a galaxy, some of which are as old as 13.2 billion years[110] – the observable universe is estimated to contain 200 billion galaxies.

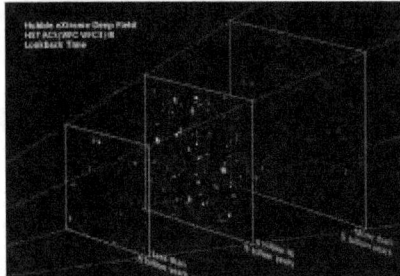

XDF image shows (from left) fully mature galaxies, nearly mature galaxies (from 5 to 9 billion years ago), protogalaxies, blazing with young stars (beyond 9 billion years).

The evolution of galaxies can be significantly affected by interactions and collisions. Mergers of galaxies were common during the early epoch, and the majority of galaxies were peculiar in morphology.[111] Given the distances between the stars, the great majority of stellar systems in colliding galaxies will be unaffected. However, gravitational stripping of the interstellar gas and dust that makes up the spiral arms produces a long train of stars known as tidal tails. Examples of these formations can be seen in NGC 4676[112] or the Antennae Galaxies.[113]

The Milky Way galaxy and the nearby Andromeda Galaxy are moving toward each other at about 130 km/s, and—depending upon the lateral movements—the two might collide in about five to six billion years. Although the Milky Way has never collided with a galaxy as large as Andromeda before, evidence of past collisions of the Milky Way with smaller dwarf galaxies is increasing.[114]

Such large-scale interactions are rare. As time passes, mergers of two systems of equal size become less common. Most bright galaxies have remained fundamentally unchanged for the last few billion years, and the net rate of star formation probably also peaked approximately ten billion years ago.[115]

16.6.3 Future trends

Spiral galaxies, like the Milky Way, produce new generations of stars as long as they have dense molecular clouds of interstellar hydrogen in their spiral arms.[116] Elliptical galaxies are largely devoid of this gas, and so form few new stars.[117] The supply of star-forming material is finite; once stars have converted the available supply of hydrogen into heavier elements, new star formation will come to an end.[118][119]

The current era of star formation is expected to continue for up to one hundred billion years, and then the "stellar age" will wind down after about ten trillion to one hundred trillion years (10^{13}–10^{14} years), as the smallest, longest-lived stars in our universe, tiny red dwarfs, begin to fade. At the end of the stellar age, galaxies will be composed of compact objects: brown dwarfs, white dwarfs that are cooling or cold ("black dwarfs"), neutron stars, and black holes. Eventually, as a result of gravitational relaxation, all stars will either fall into central supermassive black holes or be flung into intergalactic space as a result of collisions.[118][120]

16.7   Larger-scale structures

Main articles: Observable universe § Large-scale structure, Galaxy filament and Galaxy groups and clusters

Deep sky surveys show that galaxies are often found in groups and clusters. Solitary galaxies that have not significantly interacted with another galaxy of comparable mass during the past billion years are relatively scarce. Only about 5% of the galaxies surveyed have been found to be truly isolated; however, these isolated formations may have interacted and even merged with other galaxies in the past, and may still be orbited by smaller, satellite galaxies. Isolated galaxies[note 2] can produce stars at a higher rate than normal, as their gas is not being stripped by other nearby galaxies.[121]

On the largest scale, the Universe is continually expanding, resulting in an average increase in the separation between individual galaxies (see Hubble's law). Associations of galaxies can overcome this expansion on a local scale through their mutual gravitational attraction. These associations formed early in the Universe, as clumps of dark matter pulled their respective galaxies together. Nearby groups later merged to form larger-scale clusters. This on-going merger process (as well as an influx of infalling gas) heats the inter-galactic gas within a cluster to very high temperatures, reaching 30–100 megakelvins.[122] About 70–80% of the mass in a cluster is in the form of dark matter, with 10–30% consisting of this heated gas and the remaining few percent of the matter in the form of galaxies.[123]

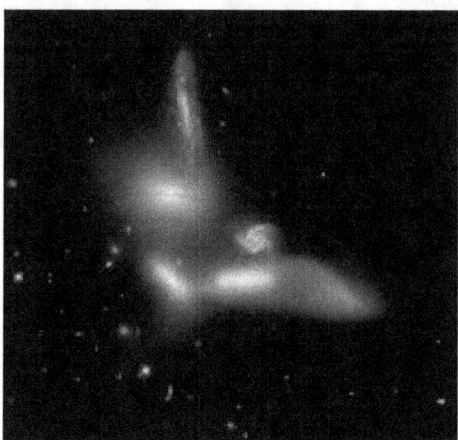

Seyfert's Sextet is an example of a compact galaxy group.

Most galaxies in the Universe are gravitationally bound to a number of other galaxies. These form a fractal-like hierarchical distribution of clustered structures, with the smallest such associations being termed groups. A group of galaxies is the most common type of galactic cluster, and these formations contain a majority of the galaxies (as well as most of the baryonic mass) in the Universe.[124][125] To remain gravitationally bound to such a group, each member galaxy must have a sufficiently low velocity to prevent it from escaping (see Virial theorem). If there is insufficient kinetic energy, however, the group may evolve into a smaller number of galaxies through mergers.[126]

Clusters of galaxies consist of hundreds to thousands of galaxies bound together by gravity.[127] Clusters of galaxies are often dominated by a single giant elliptical galaxy, known as the brightest cluster galaxy, which, over time, tidally destroys its satellite galaxies and adds their mass to its own.[128]

Superclusters contain tens of thousands of galaxies, which are found in clusters, groups and sometimes individually. At the supercluster scale, galaxies are arranged into sheets and filaments surrounding vast empty voids.[129] Above this scale, the Universe appears to be the same in all directions (isotropic and homogeneous).[130]

The Milky Way galaxy is a member of an association named the Local Group, a relatively small group of galaxies that has a diameter of approximately one megaparsec. The Milky Way and the Andromeda Galaxy are the two brightest galaxies within the group; many of the other member galaxies are dwarf companions of these two galaxies.[131] The Local Group itself is a part of a cloud-like structure within the Virgo Supercluster, a large, extended structure of groups and clusters of galaxies centered on the Virgo Cluster.[132] And the Virgo Supercluster itself is a part of the Pisces-Cetus Supercluster Complex, a giant galaxy filament.

16.8 Multi-wavelength observation

See also: Observational astronomy

This ultraviolet image of Andromeda shows blue regions containing young, massive stars.

The peak radiation of most stars lies in the visible spectrum, so the observation of the stars that form galaxies has been a major component of optical astronomy. It is also a favorable portion of the spectrum for observing ionized H II regions, and for examining the distribution of dusty arms.

The dust present in the interstellar medium is opaque to visual light. It is more transparent to far-infrared, which can be used to observe the interior regions of giant molecular clouds and galactic cores in great detail.[133] Infrared is also used to observe distant, red-shifted galaxies that were formed much earlier in the history of the Universe. Water vapor and carbon dioxide absorb a number of useful portions of the infrared spectrum, so high-altitude or space-based telescopes are used for infrared astronomy.

The first non-visual study of galaxies, particularly active galaxies, was made using radio frequencies. The atmosphere is nearly transparent to radio between 5 MHz and 30 GHz. (The ionosphere blocks signals below this range.)[134] Large radio interferometers have been used to map the active jets emitted from active nuclei. Radio telescopes can also be used to observe neutral hydrogen (via 21 cm radiation), including, potentially, the non-ionized matter in the early Universe that later collapsed to form galaxies.[135]

Ultraviolet and X-ray telescopes can observe highly energetic galactic phenomena. An ultraviolet flare was observed when a star in a distant galaxy was torn apart from the tidal forces of a black hole.[136] The distribution of hot gas in galactic clusters can be mapped by X-rays. The existence of supermassive black holes at the cores of galaxies was confirmed through X-ray astronomy.[137]

16.9 See also

- Dark galaxy

- Galactic orientation

- Galaxy formation and evolution

- Illustris project

- List of galaxies

- List of nearest galaxies

- Luminous infrared galaxy

- Supermassive black hole

- Timeline of knowledge about galaxies, clusters of galaxies, and large-scale structure

16.10   Notes

[1] Galaxies to the left side of the Hubble classification scheme are sometimes referred to as "early-type", while those to the right are "late-type".

[2] The term "field galaxy" is sometimes used to mean an isolated galaxy, although the same term is also used to describe galaxies that do not belong to a cluster but may be a member of a group of galaxies.

16.11   References

[1] Sparke & Gallagher III 2000, p. i

[2] Hupp, E.; Roy, S.; Watzke, M. (August 12, 2006). "NASA Finds Direct Proof of Dark Matter". NASA. Retrieved April 17, 2007.

[3] Uson, J. M.; Boughn, S. P.; Kuhn, J. R. (1990). "The central galaxy in Abell 2029 – An old supergiant". *Science* **250** (4980): 539–540. Bibcode:1990Sci...250..539U. doi:10.1126/science.250.4980.539.

[4] Hoover, A. (June 16, 2003). "UF Astronomers: Universe Slightly Simpler Than Expected". Hubble News Desk. Retrieved March 4, 2011. Based upon:

- Graham, A. W.; Guzman, R. (2003). "HST Photometry of Dwarf Elliptical Galaxies in Coma, and an Explanation for the Alleged Structural Dichotomy between Dwarf and Bright Elliptical Galaxies". *Astronomical Journal* **125** (6): 2936–2950. arXiv:astro-ph/0303391. Bibcode:2003AJ....125.2936G. doi:10.1086/374992.

[5] Jarrett, T. H. "Near-Infrared Galaxy Morphology Atlas". California Institute of Technology. Retrieved January 9, 2007.

[6] Finley, D.; Aguilar, D. (November 2, 2005). "Astronomers Get Closest Look Yet At Milky Way's Mysterious Core". National Radio Astronomy Observatory. Retrieved August 10, 2006.

[7] Oesch, P.A. et al. (May 3, 2015). "A Spectroscopic Redshift Measurement for a Luminous Lyman Break Galaxy at z=7.730 using Keck/MOSFIRE". *ArXiv*. arXiv:1502.05399. Bibcode:2015arXiv150205399O. Retrieved May 6, 2015.

[8] Staff (May 5, 2015). "Astronomers unveil the farthest galaxy". Phys.org. Retrieved May 6, 2015.

[9] Overbye, Dennis (May 5, 2015). "Astronomers Measure Distance to Farthest Galaxy Yet". *New York Times*. Retrieved May 6, 2015.

[10] Borenstein, Seth (May 5, 2015). "Astronomers find farthest galaxy: 13.1 billion light-years". *AP News*. Retrieved May 6, 2015.

[11] Gott III, J. R. et al. (2005). "A Map of the Universe". *Astrophysical Journal* **624** (2): 463–484. arXiv:astro-ph/0310571. Bibcode:2005ApJ...624..463G. doi:10.1086/428890.

[12] "Galaxy Clusters and Large-Scale Structure". University of Cambridge. Retrieved January 15, 2007.

[13] Harper, D. "galaxy". *Online Etymology Dictionary*. Retrieved November 11, 2011.

[14] Waller & Hodge 2003, p. 91

[15] Koneãn˘, Lubomír. "Emblematics, Agriculture, and Mythography in The Origin of the Milky Way" (PDF). Academy of Sciences of the Czech Republic. Archived from the original (PDF) on July 20, 2006. Retrieved January 5, 2007.

[16] Rao, J. (September 2, 2005). "Explore the Archer's Realm". Space.com. Retrieved January 3, 2007.

[17] Bodifée G. & Berger M. (2010). "CNG-Catalogue of Named Galaxies" (PDF). Retrieved January 17, 2014.

[18] "Contemporary Latin". Retrieved January 22, 2014.

[19] Plutarch (2006). *The Complete Works Volume 3: Essays and Miscellanies.* Chapter 3: Echo Library. p. 66. ISBN 978-1-4068-3224-2.

[20] Montada, J. P. (September 28, 2007). "Ibn Bajja". *Stanford Encyclopedia of Philosophy.* Retrieved July 11, 2008.

[21] Heidarzadeh 2008, pp. 23–25

[22] Mohamed 2000, pp. 49–50

[23] Bouali, H.-E.; Zghal, M.; Lakhdar, Z. B. (2005). "Popularisation of Optical Phenomena: Establishing the First Ibn Al-Haytham Workshop on Photography" (PDF). The Education and Training in Optics and Photonics Conference. Retrieved July 8, 2008.

[24] O'Connor, John J.; Robertson, Edmund F., "Abu Rayhan Muhammad ibn Ahmad al-Biruni", *MacTutor History of Mathematics archive,* University of St Andrews.

[25] Al-Biruni 2004, p. 87

[26] Heidarzadeh 2008, p. 25, Table 2.1

[27] Livingston, J. W. (1971). "Ibn Qayyim al-Jawziyyah: A Fourteenth Century Defense against Astrological Divination and Al-chemical Transmutation". *Journal of the American Oriental Society* 91 (1): 96–103 [99]. doi:10.2307/600445. JSTOR 600445.

[28] Galileo Galilei, *Sidereus Nuncius* (Venice, (Italy): Thomas Baglioni, 1610), pages 15 and 16.
English translation: Galileo Galilei with Edward Stafford Carlos, trans., *The Sidereal Messenger* (London, England: Rivingtons, 1880), pages 42 and 43.

[29] O'Connor, J. J.; Robertson, E. F. (November 2002). "Galileo Galilei". University of St. Andrews. Retrieved January 8, 2007.

[30] Thomas Wright, *An Original Theory or New Hypothesis of the Universe* ... (London, England: H. Chapelle, 1750). From p.48:
" ... the stars are not infinitely dispersed and distributed in a promiscuous manner throughout all the mundane space, without order or design, ... this phaenomenon [is] no other than a certain effect arising from the observer's situation, ... To a spectator placed in an indefinite space, ... it [i.e., the Milky Way (*Via Lactea*)] [is] a vast ring of stars ... "
On page 73, Wright called the Milky Way the *Vortex Magnus* (the great whirlpool) and estimated its diameter at 8.64×10^{12} miles (13.9×10^{12} km).

[31] Evans, J. C. (November 24, 1998). "Our Galaxy". George Mason University. Retrieved January 4, 2007.

[32] Immanuel Kant, *Allgemeine Naturgeschichte und Theorie des Himmels* ... [Universal Natural History and Theory of the Heavens ...], (Koenigsberg and Leipzig, (Germany): Johann Friederich Petersen, 1755).
Available in English translation by Ian Johnston at: Vancouver Island University, British Columbia, Canada

[33] William Herschel (1785) "On the Construction of the Heavens," *Philosophical Transactions of the Royal Society of London,* 75 : 213-266. Herschel's diagram of the galaxy appears immediately after the article's last page. See:
- Google Books
- The Royal Society of London

[34] Paul 1993, pp. 16–18

[35] Trimble, V. (1999). "Robert Trumpler and the (Non)transparency of Space". *Bulletin of the American Astronomical Society* 31 (31): 1479. Bibcode:1999AAS...195.7409T.

[36] Kepple & Sanner 1998, p. 18

[37] "Abd-al-Rahman Al Sufi (December 7, 903 – May 25, 986 A.D.)". Observatoire de Paris. Retrieved April 19, 2007.

[38] "The Large Magellanic Cloud, LMC". Observatoire de Paris. Retrieved April 19, 2007.

[39] See text quoted from Wright's *An original theory or new hypothesis of the Universe* in Dyson, F. (1979). *Disturbing the Universe.* Pan Books. p. 245. ISBN 0-330-26324-2.

[40] "Parsonstown | The genius of the Parsons family | William Rosse". *parsonstown.info.*

[41] Slipher, V. M. (1913). "The radial velocity of the Andromeda Nebula". *Lowell Observatory Bulletin* 1: 56–57. Bibcode:1913LowOB...2....56S.

[42] Slipher, V. M. (1915). "Spectrographic Observations of Nebulae". *Popular Astronomy* 23: 21–24. Bibcode:1915PA.....23...21S.

[43] Curtis, H. D. (1988). "Novae in Spiral Nebulae and the Island Universe Theory". *Publications of the Astronomical Society of the Pacific* **100**: 6. Bibcode:1988PASP..100....6C. doi:10.1086/132128.

[44] Weaver, H. F. "Robert Julius Trumpler". US National Academy of Sciences. Retrieved January 5, 2007.

[45] Öpik, E. (1922). "An estimate of the distance of the Andromeda Nebula". *Astrophysical Journal* **55**: 406. Bibcode:1922ApJ....55..406O. doi:10.1086/142680.

[46] Hubble, E. P. (1929). "A spiral nebula as a stellar system, Messier 31". *Astrophysical Journal* **69**: 103–158. Bibcode:1929ApJ....69..103H. doi:10.1086/143167.

[47] Sandage, A. (1989). "Edwin Hubble, 1889–1953". *Journal of the Royal Astronomical Society of Canada* **83** (6): 351–362. Bibcode:1989JRASC..83..351S. Retrieved January 8, 2007.

[48] Tenn, J. "Hendrik Christoffel van de Hulst". Sonoma State University. Retrieved January 5, 2007.

[49] López-Corredoira, M. et al. (2001). "Searching for the in-plane Galactic bar and ring in DENIS". *Astronomy and Astrophysics* **373** (1): 139–152. arXiv:astro-ph/0104307. Bibcode:2001A&A...373..139L. doi:10.1051/0004-6361:20010560.

[50] Rubin, V. C. (1983). "Dark matter in spiral galaxies". *Scientific American* **248** (6): 96–106. Bibcode:1983SciAm.248...96R. doi:10.1038/scientificamerican0683-96.

[51] Rubin, V. C. (2000). "One Hundred Years of Rotating Galaxies". *Publications of the Astronomical Society of the Pacific* **112** (772): 747–750. Bibcode:2000PASP..112..747R. doi:10.1086/316573.

[52] "Hubble Rules Out a Leading Explanation for Dark Matter". Hubble News Desk. October 17, 1994. Retrieved January 8, 2007.

[53] "How many galaxies are there?". NASA. November 27, 2002. Retrieved January 8, 2007.

[54] Kraan-Korteweg, R. C.; Juraszek, S. (2000). "Mapping the hidden Universe: The galaxy distribution in the Zone of Avoidance". *Publications of the Astronomical Society of Australia* **17** (1): 6–12. arXiv:astro-ph/9910572. Bibcode:1999astro.ph.10572K. doi:10.1071/AS00006.

[55] Barstow, M. A. (2005). "Elliptical Galaxies". Leicester University Physics Department. Retrieved June 8, 2006.

[56] "Galaxies". Cornell University. October 20, 2005. Retrieved August 10, 2006.

[57] "Galactic onion". *www.spacetelescope.org*. Retrieved 2015-05-11.

[58] Williams, M. J.; Bureau, M.; Cappellari, M. (2009). "Kinematic constraints on the stellar and dark matter content of spiral and S0 galaxies". *Monthly Notices of the Royal Astronomical Society* **400** (4): 1665. doi:10.1111/j.1365-2966.2009.15582.x.

[59] Smith, G. (March 6, 2000). "Galaxies — The Spiral Nebulae". University of California, San Diego Center for Astrophysics & Space Sciences. Retrieved November 30, 2006.

[60] Van den Bergh 1998, p. 17

[61] "Fat or flat: Getting galaxies into shape". *phys.org*. February 2014

[62] Bertin & Lin 1996, pp. 65–85

[63] Belkora 2003, p. 355

[64] Eskridge, P. B.; Frogel, J. A. (1999). "What is the True Fraction of Barred Spiral Galaxies?". *Astrophysics and Space Science*. 269/270: 427–430. Bibcode:1999Ap&SS.269..427E. doi:10.1023/A:1017025820201.

[65] Bournaud, F.; Combes, F. (2002). "Gas accretion on spiral galaxies: Bar formation and renewal". *Astronomy and Astrophysics* **392** (1): 83–102. arXiv:astro-ph/0206273. Bibcode:2002A&A...392...83B. doi:10.1051/0004-6361:20020920.

[66] Knapen, J. H.; Perez-Ramirez, D.; Laine, S. (2002). "Circumnuclear regions in barred spiral galaxies — II. Relations to host galaxies". *Monthly Notices of the Royal Astronomical Society* **337** (3): 808–828. arXiv:astro-ph/0207258. Bibcode:2002MNRAS.337..808K. doi:10.1046/j.1365-8711.2002.05840.x.

[67] Alard, C. (2001). "Another bar in the Bulge". *Astronomy and Astrophysics Letters* **379** (2): L44–L47. arXiv:astro-ph/0110491. Bibcode:2001A&A...379L..44A. doi:10.1051/0004-6361:20011487.

[68] Sanders, R. (January 9, 2006). "Milky Way galaxy is warped and vibrating like a drum". UCBerkeley News. Retrieved May 24, 2006.

[69] Bell, G. R.; Levine, S. E. (1997). "Mass of the Milky Way and Dwarf Spheroidal Stream Membership". *Bulletin of the American Astronomical Society* 29 (2): 1384. Bibcode:1997AAS...19110806B.

[70] Gerber, R. A.; Lamb, S. A.; Balsara, D. S. (1994). "Ring Galaxy Evolution as a Function of "Intruder" Mass". *Bulletin of the American Astronomical Society* 26: 911. Bibcode:1994AAS...184.3204G.

[71] "ISO unveils the hidden rings of Andromeda" (Press release). European Space Agency. October 14, 1998. Retrieved May 24, 2006.

[72] "Spitzer Reveals What Edwin Hubble Missed". Harvard-Smithsonian Center for Astrophysics. May 31, 2004. Retrieved December 6, 2006.

[73] Barstow, M. A. (2005). "Irregular Galaxies". University of Leicester. Retrieved December 5, 2006.

[74] Phillipps, S.; Drinkwater, M. J.; Gregg, M. D.; Jones, J. B. (2001). "Ultracompact Dwarf Galaxies in the Fornax Cluster". *Astrophysical Journal* 560 (1): 201–206. arXiv:astro-ph/0106377. Bibcode:2001ApJ...560..201P. doi:10.1086/322517.

[75] Groshong, K. (April 24, 2006). "Strange satellite galaxies revealed around Milky Way". New Scientist. Retrieved January 10, 2007.

[76] Schirber, M. (August 27, 2008). "No Slimming Down for Dwarf Galaxies". ScienceNOW. Retrieved August 27, 2008.

[77] "Galaxy Interactions". University of Maryland Department of Astronomy. Archived from the original on May 9, 2006. Retrieved December 19, 2006.

[78] "Interacting Galaxies". Swinburne University. Retrieved December 19, 2006.

[79] "Happy Sweet Sixteen, Hubble Telescope!". NASA. April 24, 2006. Retrieved August 10, 2006.

[80] "Starburst Galaxies". Harvard-Smithsonian Center for Astrophysics. August 29, 2006. Retrieved August 10, 2006.

[81] Kennicutt Jr., R. C. et al. (2005). *Demographics and Host Galaxies of Starbursts. Starbursts: From 30 Doradus to Lyman Break Galaxies* (Springer): 187. Bibcode:2005sdlb.proc..187K.

[82] Smith, G. (July 13, 2006). "Starbursts & Colliding Galaxies". University of California, San Diego Center for Astrophysics & Space Sciences. Retrieved August 10, 2006.

[83] Keel, B. (September 2006). "Starburst Galaxies". University of Alabama. Retrieved December 11, 2006.

[84] Keel, W. C. (2000). "Introducing Active Galactic Nuclei". University of Alabama. Retrieved December 6, 2006.

[85] Lochner, J.; Gibb, M. "A Monster in the Middle". NASA. Retrieved December 20, 2006.

[86] Heckman, T. M. (1980). "An optical and radio survey of the nuclei of bright galaxies — Activity in normal galactic nuclei". *Astronomy and Astrophysics* 87: 152–164. Bibcode:1980A&A....87..152H.

[87] Ho, L. C.; Filippenko, A. V.; Sargent, W. L. W. (1997). "A Search for "Dwarf" Seyfert Nuclei. V. Demographics of Nuclear Activity in Nearby Galaxies". *Astrophysical Journal* 487 (2): 568–578. arXiv:astro-ph/9704108. Bibcode:1997ApJ...487..568H. doi:10.1086/304638.

[88] "Construction Secrets of a Galactic Metropolis". *www.eso.org*. ESO Press Release. Retrieved October 15, 2014.

[89] "Search for Submillimeter Protogalaxies". Harvard-Smithsonian Center for Astrophysics. November 18, 1999. Retrieved January 10, 2007.

[90] Firmani, C.; Avila-Reese, V. (2003). "Physical processes behind the morphological Hubble sequence". *Revista Mexicana de Astronomía y Astrofísica* 17: 107–120. arXiv:astro-ph/0303543. Bibcode:2003RMxAC..17..107F.

[91] McMahon, R. (2006). "Journey to the birth of the Universe". *Nature* 443 (7108): 151–2. Bibcode:2006Natur.443..151M. doi:10.1038/443151a. PMID 16971933.

[92] Wall, Mike (December 12, 2012). "Ancient Galaxy May Be Most Distant Ever Seen". Space.com. Retrieved December 12, 2012.

[93] "Cosmic Detectives". The European Space Agency (ESA). April 2, 2013. Retrieved April 15, 2013.

[94] "HubbleSite - NewsCenter - Astronomers Set a New Galaxy Distance Record (05/05/2015) - Introduction". *hubblesite.org*. Retrieved 2015-05-07.

[95] "This Galaxy Far, Far Away Is the Farthest One Yet Found". Retrieved 2015-05-07.

[96] "Astronomers unveil the farthest galaxy". Retrieved 2015-05-07.

[97] Overbye, Dennis (2015-05-05). "Astronomers Measure Distance to Farthest Galaxy Yet". *The New York Times*. ISSN 0362-4331. Retrieved 2015-05-07.

[98] Oesch, P. A.; van Dokkum, P. G.; Illingworth, G. D.; Bouwens, R. J.; Momcheva, I.; Holden, B.; Roberts-Borsani, G. W.; Smit, R.; Franx, M. (2015-02-18). "A Spectroscopic Redshift Measurement for a Luminous Lyman Break Galaxy at z=7.730 using Keck/MOSFIRE". *arXiv:1502.05399 [astro-ph]*. Retrieved 2015-05-07.

[99] Eggen, O. J.; Lynden-Bell, D.; Sandage, A. R. (1962). "Evidence from the motions of old stars that the Galaxy collapsed". *Reports on Progress in Physics* 136: 748. Bibcode:1962ApJ...136..748E. doi:10.1086/147433.

[100] Searle, L.; Zinn, R. (1978). "Compositions of halo clusters and the formation of the galactic halo". *Astrophysical Journal* 225 (1): 357–379. Bibcode:1978ApJ...225..357S. doi:10.1086/156499.

[101] Heger, A.; Woosley, S. E. (2002). "The Nucleosynthetic Signature of Population III". *Astrophysical Journal* 567 (1): 532–543. arXiv:astro-ph/0107037. Bibcode:2002ApJ...567..532H. doi:10.1086/338487.

[102] Barkana, R.; Loeb, A. (1999). "In the beginning: the first sources of light and the reionization of the Universe". *Physics Reports* 349 (2): 125–238. arXiv:astro-ph/0010468. Bibcode:2001PhR...349..125B. doi:10.1016/S0370-1573(01)00019-9.

[103] Sobral, David; Matthee, Jorryt; Darvish, Behnam; Schaerer, Daniel; Mobasher, Bahram; Röttgering, Huub J. A.; Santos, Sérgio; Hemmati, Shoubaneh (4 June 2015). "Evidence For POPIII-Like Stellar Populations In The Most Luminous LYMAN-α Emitters At The Epoch Of Re-Ionisation: Spectroscopic Confirmation" (PDF). *The Astrophysical Journal*. Retrieved 17 June 2015.

[104] Overbye, Dennis (17 June 2015). "Astronomers Report Finding Earliest Stars That Enriched Cosmos". *New York Times*. Retrieved 17 June 2015.

[105] "Simulations Show How Growing Black Holes Regulate Galaxy Formation". Carnegie Mellon University. February 9, 2005. Retrieved January 7, 2007.

[106] Massey, R. (April 21, 2007). "Caught in the act; forming galaxies captured in the young Universe". Royal Astronomical Society. Retrieved April 20, 2007.

[107] Noguchi, M. (1999). "Early Evolution of Disk Galaxies: Formation of Bulges in Clumpy Young Galactic Disks". *Astrophysical Journal* 514 (1): 77–95. arXiv:astro-ph/9806355. Bibcode:1999ApJ...514...77N. doi:10.1086/306932.

[108] Baugh, C.; Frenk, C. (May 1999). "How are galaxies made?". PhysicsWeb. Retrieved January 16, 2007.

[109] Gonzalez, G. (1998). *The Stellar Metallicity — Planet Connection. Proceedings of a workshop on brown dwarfs and extrasolar planets*: 431. Bibcode:1998bdep.conf..431G.

[110] Moskowitz, Clara (September 25, 2012). "Hubble Telescope Reveals Farthest View Into Universe Ever". Space.com. Retrieved September 26, 2012.

[111] Conselice, C. J. (February 2007). "The Universe's Invisible Hand". *Scientific American* 296 (2): 35–41. doi:10.1038/scientificamerican0207-34.

[112] Ford, H. et al. (April 30, 2002). "Hubble's New Camera Delivers Breathtaking Views of the Universe". Hubble News Desk. Retrieved May 8, 2007.

[113] Struck, C. (1999). "Galaxy Collisions". *Physics Reports* 321: 1. arXiv:astro-ph/9908269. Bibcode:1999PhR...321... 1S. doi:10.1016/S0370-1573(99)00030-7.

[114] Wong, J. (April 14, 2000). "Astrophysicist maps out our own galaxy's end". University of Toronto. Archived from the original on January 8, 2007. Retrieved January 11, 2007.

[115] Panter, B.; Jimenez, R.; Heavens, A. F.; Charlot, S. (2007). "The star formation histories of galaxies in the Sloan Digital Sky Survey". *Monthly Notices of the Royal Astronomical Society* 378 (4): 1550–1564. arXiv:astro-ph/0608531. Bibcode:2007MNRAS.378.1550P. doi:10.1111/j.1365-2966.2007.11909.x.

[116] Kennicutt Jr., R. C.; Tamblyn, P.; Congdon, C. E. (1994). "Past and future star formation in disk galaxies". *Astrophysical Journal* 435 (1): 22–36. Bibcode:1994ApJ...435...22K. doi:10.1086/174790.

[117] Knapp, G. R. (1999). *Star Formation in Early Type Galaxies*. Astronomical Society of the Pacific. Bibcode:1998astro.ph..8266K. ISBN 1-886733-84-8. OCLC 41302839.

[118] Adams, Fred; Laughlin, Greg (July 13, 2006). "The Great Cosmic Battle". Astronomical Society of the Pacific. Retrieved January 16, 2007.

[119] "Cosmic 'Murder Mystery' Solved: Galaxies Are 'Strangled to Death'". Retrieved 2015-05-14.

[120] Pobojewski, S. (January 21, 1997). "Physics offers glimpse into the dark side of the Universe". University of Michigan. Retrieved January 13, 2007.

[121] McKee, M. (June 7, 2005). "Galactic loners produce more stars". New Scientist. Retrieved January 15, 2007.

[122] "Groups & Clusters of Galaxies". NASA/Chandra. Retrieved January 15, 2007.

[123] Ricker, P. "When Galaxy Clusters Collide". San Diego Supercomputer Center. Retrieved August 27, 2008.

[124] Dahlem, M. (November 24, 2006). "Optical and radio survey of Southern Compact Groups of galaxies". University of Birmingham Astrophysics and Space Research Group. Archived from the original on June 13, 2007. Retrieved January 15, 2007.

[125] Ponman, T. (February 25, 2005). "Galaxy Systems: Groups". University of Birmingham Astrophysics and Space Research Group. Retrieved January 15, 2007.

[126] Girardi, M.; Giuricin, G. (2000). "The Observational Mass Function of Loose Galaxy Groups". *The Astrophysical Journal* 540 (1): 45–56. arXiv:astro-ph/0004149. Bibcode:2000ApJ...540...45G. doi:10.1086/309314.

[127] "Hubble Pinpoints Furthest Protocluster of Galaxies Ever Seen". *ESA/Hubble Press Release*. Retrieved January 22, 2015.

[128] Dubinski, J. (1998). "The Origin of the Brightest Cluster Galaxies". *Astrophysical Journal* 502 (2): 141–149. arXiv:astro-ph/9709102. Bibcode:1998ApJ...502..141D. doi:10.1086/305901.

[129] Bahcall, N. A. (1988). "Large-scale structure in the Universe indicated by galaxy clusters". *Annual Review of Astronomy and Astrophysics* 26 (1): 631–686. Bibcode:1988ARA&A..26..631B. doi:10.1146/annurev.aa.26.090188.003215.

[130] Mandolesi, N. et al. (1986). "Large-scale homogeneity of the Universe measured by the microwave background". *Letters to Nature* 319 (6056): 751–753. Bibcode:1986Natur.319..751M. doi:10.1038/319751a0.

[131] van den Bergh, S. (2000). "Updated Information on the Local Group". *Publications of the Astronomical Society of the Pacific* 112 (770): 529–536. arXiv:astro-ph/0001040. Bibcode:2000PASP..112..529V. doi:10.1086/316548.

[132] Tully, R. B. (1982). "The Local Supercluster". *Astrophysical Journal* 257: 389–422. Bibcode:1982ApJ...257..389T. doi:10.1086/159999.

[133] "Near, Mid & Far Infrared". IPAC/NASA. Retrieved January 2, 2007.

[134] "The Effects of Earth's Upper Atmosphere on Radio Signals". NASA. Retrieved August 10, 2006.

[135] "Giant Radio Telescope Imaging Could Make Dark Matter Visible". ScienceDaily. December 14, 2006. Retrieved January 2, 2007.

[136] "NASA Telescope Sees Black Hole Munch on a Star". NASA. December 5, 2006. Retrieved January 2, 2007.

[137] Dunn, R. "An Introduction to X-ray Astronomy". Institute of Astronomy X-Ray Group. Retrieved January 2, 2007.

16.11.1   Other references

- "Unveiling the Secret of a Virgo Dwarf Galaxy". ESO. May 3, 2000. Retrieved January 3, 2007.

16.12 Bibliography

- Al-Biruni (2004). *The Book of Instruction in the Elements of the Art of Astrology*. R. Ramsay Wright (transl.). Kessinger Publishing. ISBN 0-7661-9307-1.

- Belkora, L. (2003). *Minding the Heavens: the Story of our Discovery of the Milky Way*. CRC Press. ISBN 0-7503-0730-7.

- Bertin, G.; Lin, C.-C. (1996). *Spiral Structure in Galaxies: a Density Wave Theory*. MIT Press. ISBN 0-262-02396-2.

- Binney, J.; Merrifield, M. (1998). *Galactic Astronomy*. Princeton University Press. ISBN 0-691-00402-1. OCLC 39108765.

- Dickinson, T. (2004). *The Universe and Beyond* (4th ed.). Firefly Books. ISBN 1-55297-901-6. OCLC 55596414.

- Heidarzadeh, T. (2008). *A History of Physical Theories of Comets, from Aristotle to Whipple*. Springer. ISBN 1-4020-8322-X.

- Ho, Houjun; van den Bosch, Frank; White, Simon (2010). *Galaxy Formation and Evolution* (1 ed.). Cambridge University Press. ISBN 978-0521857932.

- Kepple, G. R.; Sanner, G. W. (1998). *The Night Sky Observer's Guide, Volume 1*. Willmann-Bell. ISBN 0-943396-58-1.

- Merritt, D. (2013). *Dynamics and Evolution of Galactic Nuclei*. Princeton University Press. ISBN 9781400846122.

- Mohamed, M. (2000). *Great Muslim Mathematicians*. Penerbit UTM. ISBN 983-52-0157-9. OCLC 48759017.

- Paul, E. R. (1993). *The Milky Way Galaxy and Statistical Cosmology, 1890–1924*. Cambridge University Press. ISBN 0-521-35363-7.

- Sparke, L. S.; Gallagher III, J. S. (2000). *Galaxies in the Universe: An Introduction*. Cambridge University Press. ISBN 0-521-59740-4.

- Van den Bergh, S. (1998). *Galaxy Morphology and Classification*. Cambridge University Press. ISBN 0-521-62335-9.

- Waller, W. H.; Hodge, P. W. (2003). *Galaxies and the Cosmic Frontier*. Harvard University Press. ISBN 0-674-01079-5.

16.13 External links

-
- Galaxies on *In Our Time* at the BBC. (listen now)
- Galaxies, SEDS Messier pages
- An Atlas of The Universe
- Galaxies — Information and amateur observations
- The Oldest Galaxy Yet Found
- Galaxy classification project, harnessing the power of the internet and the human brain
- How many galaxies are in our Universe?
- The most beautiful galaxies on Astronoo
- 3-D Video (01:46) – Over a Million Galaxies of Billions of Stars each – BerkeleyLab/animated.

Chapter 17

Spacetime

For other uses of this term, see Spacetime (disambiguation).

In physics, **spacetime** (also **space–time**, **space time** or **space–time continuum**) is any mathematical model that combines space and time into a single interwoven continuum. The spacetime of our universe is usually interpreted from a Euclidean space perspective, which regards space as consisting of three dimensions, and time as consisting of one dimension, the "fourth dimension". By combining space and time into a single manifold called Minkowski space, physicists have significantly simplified a large number of physical theories, as well as described in a more uniform way the workings of the universe at both the supergalactic and subatomic levels.

In non-relativistic classical mechanics, the use of Euclidean space instead of spacetime is appropriate, because time is treated as universal with a constant rate of passage that is independent of the state of motion of an observer. In relativistic contexts, time cannot be separated from the three dimensions of space, because the observed rate at which time passes for an object depends on the object's velocity relative to the observer and also on the strength of gravitational fields, which can slow the passage of time for an object as seen by an observer outside the field.

In cosmology, the concept of spacetime combines space and time to a single abstract universe. Mathematically it is a manifold consisting of "events" which are described by some type of coordinate system. Typically **three spatial dimensions** (length, width, height), and one **temporal dimension** (time) are required. Dimensions are independent components of a coordinate grid needed to locate a point in a certain defined "space". For example, on the globe the latitude and longitude are two independent coordinates which together uniquely determine a location. In spacetime, a coordinate grid that spans the 3+1 dimensions locates events (rather than just points in space), i.e., time is added as another dimension to the coordinate grid. This way the coordinates specify *where* and *when* events occur. However, the unified nature of spacetime and the freedom of coordinate choice it allows imply that to express the temporal coordinate in one coordinate system requires both temporal and spatial coordinates in another coordinate system. Unlike in normal spatial coordinates, there are still restrictions for how measurements can be made spatially and temporally (see Spacetime intervals). These restrictions correspond roughly to a particular mathematical model which differs from Euclidean space in its manifest symmetry.

Until the beginning of the 20th century, time was believed to be independent of motion, progressing at a fixed rate in all reference frames; however, later experiments revealed that time slows at higher speeds of the reference frame relative to another reference frame. Such slowing, called time dilation, is explained in special relativity theory. Many experiments have confirmed time dilation, such as the relativistic decay of muons from cosmic ray showers and the slowing of atomic clocks aboard a Space Shuttle relative to synchronized Earth-bound inertial clocks. The duration of time can therefore vary according to events and reference frames.

When dimensions are understood as mere components of the grid system, rather than physical attributes of space, it is easier to understand the alternate dimensional views as being simply the result of coordinate transformations.

The term *spacetime* has taken on a generalized meaning beyond treating spacetime events with the normal 3+1 dimensions. It is really the combination of space and time. Other proposed spacetime theories include additional dimensions—

normally spatial but there exist some speculative theories that include additional temporal dimensions and even some that include dimensions that are neither temporal nor spatial (e.g., superspace). How many dimensions are needed to describe the universe is still an open question. Speculative theories such as string theory predict 10 or 26 dimensions (with M-theory predicting 11 dimensions: 10 spatial and 1 temporal), but the existence of more than four dimensions would only appear to make a difference at the subatomic level.[1]

17.1 Spacetime in literature

Incas regarded space and time as a single concept, referred to as *pacha* (Quechua: *pacha*, Aymara: *pacha*).[2][3] The peoples of the Andes maintain a similar understanding.[4]

Arthur Schopenhauer wrote in §18 of *On the Fourfold Root of the Principle of Sufficient Reason* (1813): "the representation of coexistence is impossible in Time alone; it depends, for its completion, upon the representation of Space; because, in mere Time, all things follow one another, and in mere Space all things are side by side; it is accordingly only by the combination of Time and Space that the representation of coexistence arises".

The idea of a unified spacetime is stated by Edgar Allan Poe in his essay on cosmology titled *Eureka* (1848) that "Space and duration are one". In 1895, in his novel *The Time Machine*, H. G. Wells wrote, "There is no difference between time and any of the three dimensions of space except that our consciousness moves along it", and that "any real body must have extension in four directions: it must have Length, Breadth, Thickness, and Duration".

Marcel Proust, in his novel *Swann's Way* (published 1913), describes the village church of his childhood's Combray as "a building which occupied, so to speak, four dimensions of space—the name of the fourth being Time".

17.1.1 Mathematical concept

In Encyclopedie under the term *dimension* Jean le Rond d'Alembert speculated that duration (time) might be considered a fourth dimension if the idea was not too novel.[5]

Another early venture was by Joseph Louis Lagrange in his *Theory of Analytic Functions* (1797, 1813). He said, "One may view mechanics as a geometry of four dimensions, and mechanical analysis as an extension of geometric analysis".[6]

The ancient idea of the cosmos gradually was described mathematically with differential equations, differential geometry, and abstract algebra. These mathematical articulations blossomed in the nineteenth century as electrical technology stimulated men like Michael Faraday and James Clerk Maxwell to describe the reciprocal relations of electric and magnetic fields. Daniel Siegel phrased Maxwell's role in relativity as follows:

> [...] the idea of the propagation of forces at the velocity of light through the electromagnetic field as described by Maxwell's equations—rather than instantaneously at a distance—formed the necessary basis for relativity theory.[7]

Maxwell used vortex models in his papers on On Physical Lines of Force, but ultimately gave up on any substance but the electromagnetic field. Pierre Duhem wrote:

> [Maxwell] was not able to create the theory that he envisaged except by giving up the use of any model, and by extending by means of analogy the abstract system of electrodynamics to displacement currents.[8]

In Siegel's estimation, "this very abstract view of the electromagnetic fields, involving no visualizable picture of what is going on out there in the field, is Maxwell's legacy."[9] Describing the behaviour of electric fields and magnetic fields led Maxwell to view the combination as an electromagnetic field. These fields have a value at every point of spacetime. It is the intermingling of electric and magnetic manifestations, described by Maxwell's equations, that give spacetime its structure. In particular, the rate of motion of an observer determines the electric and magnetic profiles of the electromagnetic field. The propagation of the field is determined by the electromagnetic wave equation, which requires spacetime for description.

Spacetime was described as an affine space with quadratic form in Minkowski space of 1908.[10] In his 1914 textbook *The Theory of Relativity*, Ludwik Silberstein used biquaternions to represent events in Minkowski space. He also exhibited the Lorentz transformations between observers of differing velocities as biquaternion mappings. Biquaternions were described in 1853 by W. R. Hamilton, so while the physical interpretation was new, the mathematics was well known in English literature, making relativity an instance of applied mathematics.

The first inkling of general relativity in spacetime was articulated by W. K. Clifford. Description of the effect of gravitation on space and time was found to be most easily visualized as a "warp" or stretching in the geometrical fabric of space and time, in a smooth and continuous way that changed smoothly from point-to-point along the spacetime fabric. In 1947 James Jeans provided a concise summary of the development of spacetime theory in his book *The Growth of Physical Science*.[11]

17.2 Basic concepts

Spacetimes are the arenas in which all physical events take place—an event is a point in spacetime specified by its time and place. For example, the motion of planets around the sun may be described in a particular type of spacetime, or the motion of light around a rotating star may be described in another type of spacetime. The basic elements of spacetime are events. In any given spacetime, an event is a unique position at a unique time. Because events are spacetime points, an example of an event in classical relativistic physics is (x, y, z, t), the location of an elementary (point-like) particle at a particular time. A spacetime itself can be viewed as the union of all events in the same way that a line is the union of all of its points, formally organized into a manifold, a space which can be described at small scales using coordinate systems.

A spacetime is independent of any observer.[12] However, in describing physical phenomena (which occur at certain moments of time in a given region of space), each observer chooses a convenient metrical coordinate system. Events are specified by four real numbers in any such coordinate system. The trajectories of elementary (point-like) particles through space and time are thus a continuum of events called the world line of the particle. Extended or composite objects (consisting of many elementary particles) are thus a union of many world lines twisted together by virtue of their interactions through spacetime into a "world-braid".

However, in physics, it is common to treat an extended object as a "particle" or "field" with its own unique (e.g., center of mass) position at any given time, so that the world line of a particle or light beam is the path that this particle or beam takes in the spacetime and represents the history of the particle or beam. The world line of the orbit of the Earth (in such a description) is depicted in two spatial dimensions x and y (the plane of the Earth's orbit) and a time dimension orthogonal to x and y. The orbit of the Earth is an ellipse in space alone, but its world line is a helix in spacetime.[13]

The unification of space and time is exemplified by the common practice of selecting a metric (the measure that specifies the interval between two events in spacetime) such that all four dimensions are measured in terms of units of distance: representing an event as $(x_0, x_1, x_2, x_3) = (ct, x, y, z)$ (in the Lorentz metric) or $(x_1, x_2, x_3, x_4) = (x, y, z, ict)$ (in the original Minkowski metric) where c is the speed of light.[14] The metrical descriptions of Minkowski Space and spacelike, lightlike, and timelike intervals given below follow this convention, as do the conventional formulations of the Lorentz transformation.

17.2.1 Spacetime intervals in flat space

In a Euclidean space, the separation between two points is measured by the distance between the two points. The distance is purely spatial, and is always positive. In spacetime, the displacement four-vector ΔR is given by the space displacement vector Δr and the time difference Δt between the events. The *spacetime interval*, also called *invariant interval*, between the two events, s^2,[15] is defined as:

$$s^2 = \Delta r^2 - c^2 \Delta t^2 \text{ (spacetime interval)},$$

where c is the speed of light. The choice of signs for s^2 above follows the space-like convention $(-+++)$.[16] Spacetime intervals may be classified into three distinct types, based on whether the temporal separation ($c^2 \Delta t^2$) or the spatial separation (Δr^2) of the two events is greater: time-like, light-like or space-like.

Certain types of world lines are called geodesics of the spacetime – straight lines in the case of Minkowski space and their closest equivalent in the curved spacetime of general relativity. In the case of purely time-like paths, geodesics are (locally) the paths of greatest separation (spacetime interval) as measured along the path between two events, whereas in Euclidean space and Riemannian manifolds, geodesics are paths of shortest distance between two points.[17][18] The concept of geodesics becomes central in general relativity, since geodesic motion may be thought of as "pure motion" (inertial motion) in spacetime, that is, free from any external influences.

Time-like interval

$$c^2 \Delta t^2 > \Delta r^2$$
$$s^2 < 0$$

For two events separated by a time-like interval, enough time passes between them that there could be a cause–effect relationship between the two events. For a particle traveling through space at less than the speed of light, any two events which occur to or by the particle must be separated by a time-like interval. Event pairs with time-like separation define a negative spacetime interval ($s^2 < 0$) and may be said to occur in each other's future or past. There exists a reference frame such that the two events are observed to occur in the same spatial location, but there is no reference frame in which the two events can occur at the same time.

The measure of a time-like spacetime interval is described by the proper time interval, $\Delta \tau$:

$$\Delta \tau = \sqrt{\Delta t^2 - \frac{\Delta r^2}{}}$$

The proper time interval would be measured by an observer with a clock traveling between the two events in an inertial reference frame, when the observer's path intersects each event as that event occurs. (The proper time interval defines a real number, since the interior of the square root is positive.)

Light-like interval

$$c^2 \Delta t^2 = \Delta r^2$$
$$s^2 = 0$$

In a light-like interval, the spatial distance between two events is exactly balanced by the time between the two events. The events define a spacetime interval of zero ($s^2 = 0$). Light-like intervals are also known as "null" intervals.

Events which occur to or are initiated by a photon along its path (i.e., while traveling at c , the speed of light) all have light-like separation. Given one event, all those events which follow at light-like intervals define the propagation of a light cone, and all the events which preceded from a light-like interval define a second (graphically inverted, which is to say "*pastward*") light cone.

Space-like interval

$$c^2 \Delta t^2 < \Delta r^2$$
$$s^2 > 0$$

When a space-like interval separates two events, not enough time passes between their occurrences for there to exist a causal relationship crossing the spatial distance between the two events at the speed of light or slower. Generally, the events are considered not to occur in each other's future or past. There exists a reference frame such that the two events are observed to occur at the same time, but there is no reference frame in which the two events can occur in the same spatial location.

For these space-like event pairs with a positive spacetime interval ($s^2 > 0$), the measurement of space-like separation is the proper distance, $\Delta\sigma$:

$$\Delta\sigma = \sqrt{s^2} = \sqrt{\Delta r^2 - c^2 \Delta t^2} \text{ (proper distance)}.$$

Like the proper time of time-like intervals, the proper distance of space-like spacetime intervals is a real number value.

17.2.2 Interval as area

The interval has been presented as the area of an oriented rectangle formed by two events and isotropic lines through them. Time-like or space-like separations correspond to oppositely oriented rectangles, one type considered to have rectangles of negative area. The case of two events separated by light corresponds to the rectangle degenerating to the segment between the events and zero area.[19] The transformations leaving interval-length invariant are the area-preserving squeeze mappings.

The parameters traditionally used rely on quadrature of the hyperbola, which is the natural logarithm. This transcendental function is essential in mathematical analysis as its inverse unites circular functions and hyperbolic functions: The exponential function, e^t, t a real number, used in the hyperbola (e^t, e^{-t}), generates hyperbolic sectors and the hyperbolic angle parameter. The functions cosh and sinh, used with rapidity as hyperbolic angle, provide the common representation of squeeze in the form $\begin{pmatrix} \cosh\phi & \sinh\phi \\ \sinh\phi & \cosh\phi \end{pmatrix}$, or as the split-complex unit $e^{j\phi} = \cosh\phi + j\sinh\phi$.

17.3 Mathematics of spacetimes

For physical reasons, a spacetime continuum is mathematically defined as a four-dimensional, smooth, connected Lorentzian manifold (M, g). This means the smooth Lorentz metric g has signature $(3, 1)$. The metric determines the geometry of spacetime, as well as determining the geodesics of particles and light beams. About each point (event) on this manifold, coordinate charts are used to represent observers in reference frames. Usually, Cartesian coordinates (x, y, z, t) are used. Moreover, for simplicity's sake, units of measurement are usually chosen such that the speed of light c is equal to 1.

A reference frame (observer) can be identified with one of these coordinate charts; any such observer can describe any event p. Another reference frame may be identified by a second coordinate chart about p. Two observers (one in each reference frame) may describe the same event p but obtain different descriptions.

Usually, many overlapping coordinate charts are needed to cover a manifold. Given two coordinate charts, one containing p (representing an observer) and another containing q (representing another observer), the intersection of the charts represents the region of spacetime in which both observers can measure physical quantities and hence compare results. The relation between the two sets of measurements is given by a non-singular coordinate transformation on this intersection. The idea of coordinate charts as local observers who can perform measurements in their vicinity also makes good physical sense, as this is how one actually collects physical data—locally.

For example, two observers, one of whom is on Earth, but the other one who is on a fast rocket to Jupiter, may observe a comet crashing into Jupiter (this is the event p). In general, they will disagree about the exact location and timing of this impact, i.e., they will have different 4-tuples (x, y, z, t) (as they are using different coordinate systems). Although their kinematic descriptions will differ, dynamical (physical) laws, such as momentum conservation and the first law of thermodynamics, will still hold. In fact, relativity theory requires more than this in the sense that it stipulates these (and all other physical) laws must take the same form in all coordinate systems. This introduces tensors into relativity, by which all physical quantities are represented.

Geodesics are said to be time-like, null, or space-like if the tangent vector to one point of the geodesic is of this nature. Paths of particles and light beams in spacetime are represented by time-like and null (light-like) geodesics, respectively.

17.3.1   Topology

Main article: Spacetime topology

The assumptions contained in the definition of a spacetime are usually justified by the following considerations.

The connectedness assumption serves two main purposes. First, different observers making measurements (represented by coordinate charts) should be able to compare their observations on the non-empty intersection of the charts. If the connectedness assumption were dropped, this would not be possible. Second, for a manifold, the properties of connectedness and path-connectedness are equivalent, and one requires the existence of paths (in particular, geodesics) in the spacetime to represent the motion of particles and radiation.

Every spacetime is paracompact. This property, allied with the smoothness of the spacetime, gives rise to a smooth linear connection, an important structure in general relativity. Some important theorems on constructing spacetimes from compact and non-compact manifolds include the following:

- A compact manifold can be turned into a spacetime if, and only if, its Euler characteristic is 0. (Proof idea: the existence of a Lorentzian metric is shown to be equivalent to the existence of a nonvanishing vector field.)

- Any non-compact 4-manifold can be turned into a spacetime.[20]

17.3.2   Spacetime symmetries

Main article: Spacetime symmetries

Often in relativity, spacetimes that have some form of symmetry are studied. As well as helping to classify spacetimes, these symmetries usually serve as a simplifying assumption in specialized work. Some of the most popular ones include:

- Axisymmetric spacetimes

- Spherically symmetric spacetimes

- Static spacetimes

- Stationary spacetimes

17.3.3   Causal structure

Main article: Causal structure
See also: Causality (physics) and Causality

The causal structure of a spacetime describes causal relationships between pairs of points in the spacetime based on the existence of certain types of curves joining the points.

17.4   Spacetime in special relativity

Main article: Minkowski space

The geometry of spacetime in special relativity is described by the Minkowski metric on R^4. This spacetime is called Minkowski space. The Minkowski metric is usually denoted by η and can be written as a four-by-four matrix:

$$\eta_{ab} = \text{diag}(1, -1, -1, -1)$$

where the Landau–Lifshitz space-like convention is being used. A basic assumption of relativity is that coordinate transformations must leave spacetime intervals invariant. Intervals are invariant under Lorentz transformations. This invariance property leads to the use of four-vectors (and other tensors) in describing physics.

Strictly speaking, one can also consider events in Newtonian physics as a single spacetime. This is Galilean–Newtonian relativity, and the coordinate systems are related by Galilean transformations. However, since these preserve spatial and temporal distances independently, such a spacetime can be decomposed into spatial coordinates plus temporal coordinates, which is not possible in the general case.

17.5   Spacetime in general relativity

In general relativity, it is assumed that spacetime is curved by the presence of matter (energy), this curvature being represented by the Riemann tensor. In special relativity, the Riemann tensor is identically zero, and so this concept of "non-curvedness" is sometimes expressed by the statement *Minkowski spacetime is flat.*

The earlier discussed notions of time-like, light-like and space-like intervals in special relativity can similarly be used to classify one-dimensional curves through curved spacetime. A time-like curve can be understood as one where the interval between any two infinitesimally close events on the curve is time-like, and likewise for light-like and space-like curves. Technically the three types of curves are usually defined in terms of whether the tangent vector at each point on the curve is time-like, light-like or space-like. The world line of a slower-than-light object will always be a time-like curve, the world line of a massless particle such as a photon will be a light-like curve, and a space-like curve could be the world line of a hypothetical tachyon. In the local neighborhood of any event, time-like curves that pass through the event will remain inside that event's past and future light cones, light-like curves that pass through the event will be on the surface of the light cones, and space-like curves that pass through the event will be outside the light cones. One can also define the notion of a three-dimensional "spacelike hypersurface", a continuous three-dimensional "slice" through the four-dimensional property with the property that every curve that is contained entirely within this hypersurface is a space-like curve.[21]

Many spacetime continua have physical interpretations which most physicists would consider bizarre or unsettling. For example, a compact spacetime has closed timelike curves, which violate our usual ideas of causality (that is, future events could affect past ones). For this reason, mathematical physicists usually consider only restricted subsets of all the possible spacetimes. One way to do this is to study "realistic" solutions of the equations of general relativity. Another way is to add some additional "physically reasonable" but still fairly general geometric restrictions and try to prove interesting things about the resulting spacetimes. The latter approach has led to some important results, most notably the Penrose–Hawking singularity theorems.

17.6   Quantized spacetime

Main article: Quantum spacetime

In general relativity, spacetime is assumed to be smooth and continuous—and not just in the mathematical sense. In the theory of quantum mechanics, there is an inherent discreteness present in physics. In attempting to reconcile these two theories, it is sometimes postulated that spacetime should be quantized at the very smallest scales. Current theory is focused on the nature of spacetime at the Planck scale. Causal sets, loop quantum gravity, string theory, causal dynamical triangulation, and black hole thermodynamics all predict a quantized spacetime with agreement on the order of magnitude. Loop quantum gravity makes precise predictions about the geometry of spacetime at the Planck scale.

17.7   See also

- Anthropic_principle § Applications of the principle §§ Spacetime
- Basic introduction to the mathematics of curved spacetime
- Four-vector
- Frame-dragging
- Global spacetime structure
- Hole argument
- List of mathematical topics in relativity
- Local spacetime structure
- Lorentz invariance
- Manifold
- Mathematics of general relativity
- Metric space
- Philosophy of space and time
- Relativity of simultaneity
- Strip photography
- World manifold

17.8   References

[1] Kopeikin, Sergei; Efroimsky, Michael; Kaplan, George (2011). *Relativistic Celestial Mechanics of the Solar System*. John Wiley & Sons. p. 157. ISBN 3527634576., Extract of page 157

[2] Atuq Eusebio Manga Qespi, Instituto de lingüística y Cultura Amerindia de la Universidad de Valencia. *Pacha: un concepto andino de espacio y tiempo.* Revísta española de Antropología Americana, 24, p. 155–189. Edit. Complutense, Madrid. 1994

[3] Paul Richard Steele, Catherine J. Allen, *Handbook of Inca mythology*, p. 86, (ISBN 1-57607-354-8)

[4] Shirley Ardener, University of Oxford, *Women and space: ground rules and social maps*, p. 36 (ISBN 0-85496-728-1)

[5] Jean d'Alembert (1754) Dimension from ARTFL Encyclopedie project

[6] R.C. Archibald (1914) *Time as a fourth dimension Bulletin of the American Mathematical Society* 20:409.

[7] Daniel M. Siegel (2014) "Maxwell's contributions to electricity and magnetism", chapter 10 in *James Clerk Maxwell: Perspectives on his Life and Work*, Raymond Flood, Mark McCartney, Andrew Whitaker, editors, Oxford University Press ISBN 978-0-19-966437-5

[8] Pierre Duhem (1954) *The Aim and Structure of Physical Theory*, page 98, Princeton University Press

[9] Siegel 2014 p 191

[10] Minkowski, Hermann (1909), "Raum und Zeit", *Physikalische Zeitschrift* 10: 75–88

   - Various English translations on Wikisource: Space and Time.

[11] James Jeans (1947) The Growth of Physical Science, "Space-time", pp. 205–301, link from Internet Archive

[12] Matolcsi, Tamás (1994). *Spacetime Without Reference Frames.* Budapest: Akadémiai Kiadó.

[13] Ellis, G. F. R.; Williams, Ruth M. (2000). *Flat and curved space–times* (2nd ed.). Oxford University Press. p. 9. ISBN 0-19-850657-0.

[14] Petkov, Vesselin (2010). *Minkowski Spacetime: A Hundred Years Later.* Springer. p. 70. ISBN 90-481-3474-9., Section 3.4, p. 70

[15] Note that the term *spacetime interval* is applied by several authors to the quantity s^2 and not to s. The reason that the quantity s^2 is used and not s is that s^2 can be positive, zero or negative, and is a more generally convenient and useful quantity than the Minkowski norm with a timelike/null/spacelike distinguisher: the pair $(\sqrt{|s^2|}, \operatorname{sgn}(s^2))$. Despite the notation, it should not be regarded as the square of a number, but as a symbol. The cost for this convenience is that this "interval" is quadratic in linear separation along a straight line.

[16] More generally the spacetime interval in flat space can be written as $s^2 = g_{\alpha\beta} \Delta x^\alpha \Delta x^\beta$ with metric tensor g independent of spacetime position.

[17] This characterization is not universal: both the arcs between two points of a great circle on a sphere are geodesics.

[18] Berry, Michael V. (1989). *Principles of Cosmology and Gravitation.* CRC Press. p. 58. ISBN 0-85274-037-9., Extract of page 58, caption of Fig. 25

[19] I. M. Yaglom (1979) *A Simple Non-Euclidean Geometry and its Physical Basis*, page 178, Springer, ISBN 0387-90332-1, MR 520230

[20] Geroch, Robert; Horowitz, Gary T. (1979). "Chapter 5. Global structure of spacetimes". In Hawking, S.W.; Israel, W. *General Relativity An Einstein Centenary Survey.* Cambridge University Press. p. 219. ISBN 0521299284.

[21] See "Quantum Spacetime and the Problem of Time in Quantum Gravity" by Leszek M. Sokolowski, where on this page he writes "Each of these hypersurfaces is spacelike, in the sense that every curve, which entirely lies on one of such hypersurfaces, is a spacelike curve." More commonly a space-like hypersurface is defined technically as a surface such that the normal vector at every point is time-like, but the definition above may be somewhat more intuitive.

17.9   External links

- http://universaltheory.org

- Barrow, John D.; Tipler, Frank J. (1988). *The Anthropic Cosmological Principle.* Oxford University Press. ISBN 978-0-19-282147-8. LCCN 87028148.

- Ehrenfest, Paul (1920) "How do the fundamental laws of physics make manifest that Space has 3 dimensions?" *Annalen der Physik 366*: 440.

- George F. Ellis and Ruth M. Williams (1992) *Flat and curved space–times.* Oxford Univ. Press. ISBN 0-19-851164-7

- Isenberg, J. A. (1981). "Wheeler–Einstein–Mach spacetimes". *Phys. Rev. D24* (2): 251–256. Bibcode:1981PhRvD..24..251I. doi:10.1103/PhysRevD.24.251.

- Kant, Immanuel (1929) "Thoughts on the true estimation of living forces" in J. Handyside, trans., *Kant's Inaugural Dissertation and Early Writings on Space.* Univ. of Chicago Press.

- Lorentz, H. A., Einstein, Albert, Minkowski, Hermann, and Weyl, Hermann (1952) *The Principle of Relativity: A Collection of Original Memoirs.* Dover.

- Lucas, John Randolph (1973) *A Treatise on Time and Space.* London: Methuen.

- Penrose, Roger (2004). *The Road to Reality.* Oxford: Oxford University Press. ISBN 0-679-45443-8. Chpts. 17–18.

- Poe, Edgar A. (1848). *Eureka; An Essay on the Material and Spiritual Universe*. Hesperus Press Limited. ISBN 1-84391-009-8.

- Robb, A. A. (1936). *Geometry of Time and Space*. University Press.

- Erwin Schrödinger (1950) *Space–time structure*. Cambridge Univ. Press.

- Schutz, J. W. (1997). *Independent axioms for Minkowski Space–time*. Addison-Wesley Longman. ISBN 0-582-31760-6.

- Tangherlini, F. R. (1963). "Schwarzschild Field in n Dimensions and the Dimensionality of Space Problem". *Nuovo Cimento* 14 (27): 636.

- Taylor, E. F.; Wheeler, John A. (1963). *Spacetime Physics*. W. H. Freeman. ISBN 0-7167-2327-1.

- Wells, H.G. (2004). *The Time Machine*. New York: Pocket Books. ISBN 0-671-57554-6. (pp. 5–6)

- Stanford Encyclopedia of Philosophy: "Space and Time: Inertial Frames" by Robert DiSalle.

Chapter 18

Steady State theory

This article is about the cosmological theory. For other uses, see Steady state (disambiguation).

In cosmology, the **Steady State theory** is a now-obsolete[1] expanding universe model alternative to the Big Bang theory of the universe and its origin. In steady state views, new matter is continuously created as the universe expands, thus adhering to the perfect cosmological principle (the principle that the observable universe is basically the same in any time as well as any place).

While the steady state model enjoyed some popularity in the mid-20th century, it is now rejected by the vast majority of cosmologists, astrophysicists and astronomers, as the observational evidence points to a hot Big Bang cosmology with a finite age of the universe, which the Steady State theory does not predict.[2]I

18.1 History

Cosmological expansion was originally discovered through observations by Edwin Hubble. Theoretical calculations also showed that a static universe was impossible under general relativity. The modern Big Bang theory is one in which the universe has a finite age and has evolved over time through cooling, expansion, and the formation of structures through gravitational collapse.

The steady state theory asserts that although the universe is expanding, it nevertheless does not change its appearance over time (the perfect cosmological principle); the universe has no beginning and no end. This requires that matter be continually created in order to keep the universe's density from decreasing. Influential papers on steady state cosmologies were published by Bondi, Gold, and Hoyle in 1948.[3][4]

18.2 Observational tests

18.2.1 Counts of radio sources

Problems with the steady-state theory began to emerge in the 1950s and 60s, when observations began to support the idea that the universe was in fact changing: bright radio sources (quasars and radio galaxies) were found only at large distances (therefore could have existed only in the distant past), not in closer galaxies. Whereas the Big Bang theory predicted as much, the Steady State theory predicted that such objects would be found throughout the universe, including close to our own galaxy. By 1961, statistical tests based on radio-source surveys[5] had ruled out the steady state model in the minds of most cosmologists, although some proponents of the steady state insisted that the radio data were suspect.

18.2.2   Cosmic microwave background

For most cosmologists, the definitive refutation of the steady-state theory came with the discovery of the cosmic microwave background radiation in 1965, which was predicted by the Big Bang theory. Stephen Hawking described this discovery as "the final nail in the coffin of the steady-state theory." The steady-state theory explained microwave background radiation as the result of light from ancient stars that has been scattered by galactic dust. However, the cosmic microwave background level is very even in all directions, making it difficult to explain how it could be generated by numerous point sources and the microwave background radiation shows no evidence of characteristics such as polarization that are normally associated with scattering. Furthermore, its spectrum is so close to that of an ideal black body that it could hardly be formed by the superposition of contributions from a multitude of dust clumps at different temperatures as well as at different redshifts. Steven Weinberg wrote in 1972,

> *The steady state model does not appear to agree with the observed dL versus z relation or with source counts ... In a sense, this disagreement is a credit to the model; alone among all cosmologies, the steady-state model makes such definite predictions that it can be disproved even with the limited observational evidence at our disposal. The steady-state model is so attractive that many of its adherents still retain hope that the evidence against it will eventually disappear as observations improve. However, if the cosmic microwave radiation . . . is really black-body radiation, it will be difficult to doubt that the universe has evolved from a hotter denser early stage.*[6]

Since this discovery, the Big Bang theory has been considered to provide the best explanation of the origin of the universe. In most astrophysical publications, the Big Bang is implicitly accepted and is used as the basis of more complete theories.

18.3   Quasi-steady state

Quasi-steady state cosmology (QSS) was proposed in 1993 by Fred Hoyle, Geoffrey Burbidge, and Jayant V. Narlikar as a new incarnation of the steady state ideas meant to explain additional features unaccounted for in the initial proposal. The theory suggests pockets of creation occurring over time within the universe, sometimes referred to as *minibangs, mini-creation events,* or *little bangs.* After the observation of an accelerating universe, further modifications of the model were made.[7]

Astrophysicist and cosmologist Ned Wright has pointed out flaws in the theory.[8] These first comments were soon rebutted by the proponents.[9] Wright and other mainstream cosmologists reviewing QSS have pointed out new flaws and discrepancies with observations left unexplained by proponents.[10]

18.4   Notes and citations

[1] "Steady State theory". *BBC.* Retrieved January 11, 2015. [T]he Steady State theorists' ideas are largely discredited today...

[2] Kragh, Helge (1999). *Cosmology and Controversy: The Historical Development of Two Theories of the Universe.* Princeton University Press. ISBN 0-691-02623-8.

[3] Bondi and Gold, "The Steady-State Theory of the Expanding Universe," MNRAS 108 (1948) 252. Bibcode: 1948MNRAS.108..252B

[4] Hoyle, "A New Model for the Expanding Universe," MNRAS 108 (1948) 372. Bibcode: 1948MNRAS.108..372H

[5] Ryle and Clarke, "An examination of the steady-state model in the light of some recent observations of radio sources," MNRAW 122 (1961) 349

[6] Weinberg, S. (1972). *Gravitation and Cosmology.* John Whitney & Sons. pp. 495–464. ISBN 978-0-471-92567-5.

[7] Narlikar, J. V.; Vishwakarma, R. G.; Burbidge, G. (2002). "Interpretations of the Accelerating Universe". arXiv:astro-ph/0205064 [astro-ph]. Hoyle, F.; Burbidge, G.; Narlikar, J. V. (1993). "A quasi-steady state cosmological model with creation of matter". *The Astrophysical Journal* **410**: 437–457. Bibcode:1993ApJ...410..437H. doi:10.1086/172761. Hoyle, F.; Burbidge, G.; Narlikar, J. V. (1994). "Astrophysical deductions from the quasi-steady state cosmology". *Monthly Notices of the*

Royal Astronomical Society **267**: 1007–1019. Bibcode:1994MNRAS.267.1007H. Hoyle, F.; Burbidge, G.; Narlikar, J. V. (1994). "Astrophysical deductions from the quasi-steady state : Erratum". *Monthly Notices of the Royal Astronomical Society* **269**: 1152. Bibcode:1994MNRAS.269.1152H. Hoyle, F.; Burbidge, G.; Narlikar, J. V. (1994). "Further astrophysical quantities expected in a quasi-steady state Universe". *Astronomy and Astrophysics* **289** (3): 729–739. Bibcode:1994A&A...289..729H. Hoyle, F.; Burbidge, G.; Narlikar, J. V. (1995). "The basic theory underlying the quasi-steady state cosmological model". *Proceedings of the Royal Society A* **448**: 191. Bibcode:1995RSPSA.448..191H. doi:10.1098/rspa.1995.0012.

[8] Wright, E. L. (1994). "Comments on the Quasi-Steady-State Cosmology". *arXiv*:astro-ph/9410070 [astro-ph].

[9] Hoyle, F.; Burbidge, G.; Narlikar, J. V. (1994). "Note on a Comment by Edward L. Wright". *arXiv*:astro-ph/9412045 [astro-ph].

[10] Wright, E. L. (20 December 2010). "Errors in the Steady State and Quasi-SS Models". UCLA, Physics & Astronomy Department.

18.5   Further reading

- Hoyle, F.; Burbidge, G.; Narlikar, J. V. (2000). *A Different Approach to Cosmology*. Cambridge University Press. ISBN 0-521-66223-0.

- Mitton, S. (2005). *Conflict in the Cosmos: Fred Hoyle's Life in Science*. Joseph Henry Press. ISBN 0-309-09313-9.

- Mitton, S. (2005). *Fred Hoyle: A Life in Science*. Aurum Press. ISBN 1-85410-961-8.

18.6   External links

- Historical note: the steady-state model – a brief presentation of the mathematical content of the Hoyle theory

Chapter 19

Empirical evidence

"Empirical" redirects here. For other uses, see Empirical (disambiguation).

Empirical evidence, data, or **knowledge,** also known as **sense experience,** is a collective term for the knowledge or source of knowledge acquired by means of the senses, particularly by observation and experimentation.[1] The term comes from the Greek word for experience, ἐμπειρία (*empeiría*). After Kant, it is common in philosophy to call the knowledge thus gained *a posteriori* knowledge. This is contrasted with *a priori* knowledge, the knowledge accessible from pure reason alone.

19.1 Meaning

Empirical evidence is information that justifies a belief in the truth or falsity of a claim. In the empiricist view, one can claim to have knowledge only when one has a true belief based on empirical evidence. This stands in contrast to the rationalist view under which reason or reflection alone is considered evidence for the truth or falsity of some propositions.[2] The senses are the primary source of empirical evidence. Although other sources of evidence, such as memory and the testimony of others, ultimately trace back to some sensory experience, they are considered secondary, or indirect.[2]

In another sense, empirical evidence may be synonymous with the outcome of an experiment. In this sense, an empirical result is a unified confirmation. In this context, the term *semi-empirical* is used for qualifying theoretical methods that use, in part, basic axioms or postulated scientific laws and experimental results. Such methods are opposed to theoretical *ab initio* methods, which are purely deductive and based on first principles.

In science, empirical evidence is required for a hypothesis to gain acceptance in the scientific community. Normally, this validation is achieved by the scientific method of hypothesis commitment, experimental design, peer review, adversarial review, reproduction of results, conference presentation and journal publication. This requires rigorous communication of hypothesis (usually expressed in mathematics), experimental constraints and controls (expressed necessarily in terms of standard experimental apparatus), and a common understanding of measurement.

Statements and arguments depending on empirical evidence are often referred to as *a posteriori* ("following experience") as distinguished from *a priori* (preceding it). *A priori* knowledge or justification is independent of experience (for example "All bachelors are unmarried"), whereas *a posteriori* knowledge or justification is dependent on experience or empirical evidence (for example "Some bachelors are very happy"). The notion of the distinction between *a priori* and *a posteriori* as tantamount to the distinction between empirical and non-empirical knowledge comes from Kant's *Critique of Pure Reason*.[3]

The standard positivist view of empirically acquired information has been that observation, experience, and experiment serve as neutral arbiters between competing theories. However, since the 1960s, a persistent critique most associated with Thomas Kuhn,[4] has argued that these methods are influenced by prior beliefs and experiences. Consequently it cannot be expected that two scientists when observing, experiencing, or experimenting on the same event will make the same theory-neutral observations. The role of observation as a theory-neutral arbiter may not be possible. Theory-dependence of observation means that, even if there were agreed methods of inference and interpretation, scientists may still disagree on the nature of empirical data.[5]

19.2   See also

- Anecdotal evidence

- Empirical distribution function

- Empirical formula

- Empirical measure

- Empirical research (more on the scientific usage)

- Phenomenology (science)

- Scientific evidence

- Scientific method

- Theory

19.3   Footnotes

[1] Pickett 2006, p. 585

[2] Feldman 2001, p. 293

[3] Craig 2005, p. 1

[4] Kuhn 1970

[5] Bird 2013

19.4   References

- Bird, Alexander (2013). Zalta, Edward N., ed. "Thomas Kuhn". *Stanford Encyclopedia of Philosophy*. Section 4.2 Perception, Observational Incommensurability, and World-Change. Retrieved 25 January 2012.

- Craig, Edward (2005). "a posteriori". *The Shorter Routledge Encyclopedia of Philosophy*. Routledge. ISBN 9780415324953.

- Feldman, Richard (2001) [1999]. "Evidence". In Audi, Robert. *The Cambridge Dictionary of Philosophy* (2nd ed.). Cambridge, UK: Cambridge University Press. pp. 293–294. ISBN 978-0521637220.

- Kuhn, Thomas S. (1970) [1962]. *The Structure of Scientific Revolutions* (2nd ed.). Chicago: University of Chicago Press. ISBN 978-0226458045.

- Pickett, Joseph P., ed. (2011). "Empirical". *The American Heritage Dictionary of the English Language* (5th ed.). Houghton Mifflin. ISBN 978-0-547-04101-8.

19.5 External links

- The dictionary definition of empirical at Wiktionary

- The dictionary definition of evidence at Wiktionary

- A Priori and A Posteriori entry in the *Internet Encyclopedia of Philosophy*

Chapter 20

Redshift

This article is about the astronomical phenomenon. For other uses, see Redshift (disambiguation).

In physics, **redshift** happens when light or other electromagnetic radiation from an object is increased in wavelength, or shifted to the red end of the spectrum. In general, whether or not the radiation is within the visible spectrum, "redder" means an increase in wavelength – equivalent to a lower frequency and a lower photon energy, in accordance with, respectively, the wave and quantum theories of light.

Some redshifts are an example of the Doppler effect, familiar in the change in the apparent pitches of sirens and frequency of the sound waves emitted by speeding vehicles. A redshift occurs whenever a light source moves away from an observer. Another kind of redshift is cosmological redshift, which is due to the expansion of the universe, and sufficiently distant light sources (generally more than a few million light years away) show redshift corresponding to the rate of increase in their distance from Earth. Finally, gravitational redshift is a relativistic effect observed in electromagnetic radiation moving out of gravitational fields. Conversely, a *decrease* in wavelength is called blueshift and is generally seen when a light-emitting object moves toward an observer or when electromagnetic radiation moves into a gravitational field. However, redshift is a more common term and sometimes blueshift is referred to as negative redshift.

Knowledge of redshifts and blueshifts has been applied to develop several terrestrial technologies such as Doppler radar and radar guns.[1] Redshifts are also seen in the spectroscopic observations of astronomical objects.[2] Its value is represented by the letter z.

A special relativistic redshift formula (and its classical approximation) can be used to calculate the redshift of a nearby object when spacetime is flat. However, in many contexts, such as black holes and Big Bang cosmology, redshifts must be calculated using general relativity.[3] Special relativistic, gravitational, and cosmological redshifts can be understood under the umbrella of frame transformation laws. There exist other physical processes that can lead to a shift in the frequency of electromagnetic radiation, including scattering and optical effects; however, the resulting changes are distinguishable from true redshift and are not generally referred to as such (see section on physical optics and radiative transfer).

20.1   History

The history of the subject began with the development in the 19th century of wave mechanics and the exploration of phenomena associated with the Doppler effect. The effect is named after Christian Doppler, who offered the first known physical explanation for the phenomenon in 1842.[4] The hypothesis was tested and confirmed for sound waves by the Dutch scientist Christophorus Buys Ballot in 1845.[5] Doppler correctly predicted that the phenomenon should apply to all waves, and in particular suggested that the varying colors of stars could be attributed to their motion with respect to the Earth.[6] Before this was verified, however, it was found that stellar colors were primarily due to a star's temperature, not motion. Only later was Doppler vindicated by verified redshift observations.

The first Doppler redshift was described by French physicist Hippolyte Fizeau in 1848, who pointed to the shift in spectral lines seen in stars as being due to the Doppler effect. The effect is sometimes called the "Doppler–Fizeau effect". In 1868, British astronomer William Huggins was the first to determine the velocity of a star moving away from the Earth by this method.[7] In 1871, optical redshift was confirmed when the phenomenon was observed in Fraunhofer lines using solar rotation, about 0.1 Å in the red.[8] In 1887, Vogel and Scheiner discovered the *annual Doppler effect*, the yearly change in the Doppler shift of stars located near the ecliptic due to the orbital velocity of the Earth.[9] In 1901, Aristarkh Belopolsky verified optical redshift in the laboratory using a system of rotating mirrors.[10]

The earliest occurrence of the term "red-shift" in print (in this hyphenated form) appears to be by American astronomer Walter S. Adams in 1908, in which he mentions "Two methods of investigating that nature of the nebular red-shift".[11] The word does not appear unhyphenated until about 1934 by Willem de Sitter, perhaps indicating that up to that point its German equivalent, *Rotverschiebung*, was more commonly used.[12]

Beginning with observations in 1912, Vesto Slipher discovered that most spiral galaxies, then mostly thought to be spiral nebulae, had considerable redshifts. Slipher first reports on his measurement in the inaugural volume of the *Lowell Observatory Bulletin*.[13] Three years later, he wrote a review in the journal *Popular Astronomy*.[14] In it he states, "[...] the early discovery that the great Andromeda spiral had the quite exceptional velocity of −300 km(/s) showed the means then available, capable of investigating not only the spectra of the spirals but their velocities as well."[15] Slipher reported the velocities for 15 spiral nebulae spread across the entire celestial sphere, all but three having observable "positive" (that is recessional) velocities. Subsequently, Edwin Hubble discovered an approximate relationship between the redshifts of such "nebulae" and the distances to them with the formulation of his eponymous Hubble's law.[16] These observations corroborated Alexander Friedmann's 1922 work, in which he derived the famous Friedmann equations.[17] They are today considered strong evidence for an expanding universe and the Big Bang theory.[18]

20.2   Measurement, characterization, and interpretation

The spectrum of light that comes from a single source (see idealized spectrum illustration top-right) can be measured. To determine the redshift, one searches for features in the spectrum such as absorption lines, emission lines, or other variations in light intensity. If found, these features can be compared with known features in the spectrum of various chemical compounds found in experiments where that compound is located on Earth. A very common atomic element in space is hydrogen. The spectrum of originally featureless light shone through hydrogen will show a signature spectrum specific to hydrogen that has features at regular intervals. If restricted to absorption lines it would look similar to the illustration (top right). If the same pattern of intervals is seen in an observed spectrum from a distant source but occurring at shifted wavelengths, it can be identified as hydrogen too. If the same spectral line is identified in both spectra—but at different wavelengths—then the redshift can be calculated using the table below. Determining the redshift of an object in this way requires a frequency- or wavelength-range. In order to calculate the redshift one has to know the wavelength of the emitted light in the rest frame of the source, in other words, the wavelength that would be measured by an observer located adjacent to and comoving with the source. Since in astronomical applications this measurement cannot be done directly, because that would require travelling to the distant star of interest, the method using spectral lines described here is used instead. Redshifts cannot be calculated by looking at unidentified features whose rest-frame frequency is unknown, or with a spectrum that is featureless or white noise (random fluctuations in a spectrum).[20]

Redshift (and blueshift) may be characterized by the relative difference between the observed and emitted wavelengths (or frequency) of an object. In astronomy, it is customary to refer to this change using a dimensionless quantity called z. If λ represents wavelength and f represents frequency (note, $\lambda f = c$ where c is the speed of light), then z is defined by the equations:[21]

After z is measured, the distinction between redshift and blueshift is simply a matter of whether z is positive or negative. See the formula section below for some basic interpretations that follow when either a redshift or blueshift is observed. For example, Doppler effect blueshifts ($z < 0$) are associated with objects approaching (moving closer to) the observer with the light shifting to greater energies. Conversely, Doppler effect redshifts ($z > 0$) are associated with objects receding (moving away) from the observer with the light shifting to lower energies. Likewise, gravitational blueshifts are associated with light emitted from a source residing within a weaker gravitational field as observed from within a stronger gravitational field, while gravitational redshifting implies the opposite conditions.

20.3 Redshift formulae

In general relativity one can derive several important special-case formulae for redshift in certain special spacetime geometries, as summarized in the following table. In all cases the magnitude of the shift (the value of z) is independent of the wavelength.[2]

20.3.1 Doppler effect

Main articles: Doppler effect and Relativistic Doppler effect

If a source of the light is moving away from an observer, then redshift ($z > 0$) occurs; if the source moves towards the observer, then blueshift ($z < 0$) occurs. This is true for all electromagnetic waves and is explained by the Doppler effect. Consequently, this type of redshift is called the *Doppler redshift*. If the source moves away from the observer with velocity v, which is much less than the speed of light ($v \ll c$), the redshift is given by

$$z \approx \frac{v}{c} \text{ (since } \gamma \approx 1 \text{)}$$

where c is the speed of light. In the classical Doppler effect, the frequency of the source is not modified, but the recessional motion causes the illusion of a lower frequency.

A more complete treatment of the Doppler redshift requires considering relativistic effects associated with motion of sources close to the speed of light. A complete derivation of the effect can be found in the article on the relativistic Doppler effect. In brief, objects moving close to the speed of light will experience deviations from the above formula due to the time dilation of special relativity which can be corrected for by introducing the Lorentz factor γ into the classical Doppler formula as follows (for motion solely in the line of sight):

$$1 + z = \left(1 + \frac{v}{c}\right)\gamma.$$

This phenomenon was first observed in a 1938 experiment performed by Herbert E. Ives and G.R. Stilwell, called the Ives–Stilwell experiment.[23]

Since the Lorentz factor is dependent only on the magnitude of the velocity, this causes the redshift associated with the relativistic correction to be independent of the orientation of the source movement. In contrast, the classical part of the formula is dependent on the projection of the movement of the source into the line-of-sight which yields different results for different orientations. If θ is the angle between the direction of relative motion and the direction of emission in the observer's frame[24] (zero angle is directly away from the observer), the full form for the relativistic Doppler effect becomes:

$$1 + z = \frac{1 + v\cos(\theta)/c}{\sqrt{1 - v^2/c^2}}$$

and for motion solely in the line of sight ($\theta = 0°$), this equation reduces to:

$$1 + z = \sqrt{\frac{1 + v/c}{1 - v/c}}$$

For the special case that the light is approaching at right angles ($\theta = 90°$) to the direction of relative motion in the observer's frame,[25] the relativistic redshift is known as the transverse redshift, and a redshift:

$$1 + z = \sqrt{\frac{1}{1 - v^2/c^2}}$$

is measured, even though the object is not moving away from the observer. Even when the source is moving towards the observer, if there is a transverse component to the motion then there is some speed at which the dilation just cancels the expected blueshift and at higher speed the approaching source will be redshifted.[26]

20.3.2 Expansion of space

Main article: Metric expansion of space

In the early part of the twentieth century, Slipher, Hubble and others made the first measurements of the redshifts and blueshifts of galaxies beyond the Milky Way. They initially interpreted these redshifts and blueshifts as due solely to the Doppler effect, but later Hubble discovered a rough correlation between the increasing redshifts and the increasing distance of galaxies. Theorists almost immediately realized that these observations could be explained by a different mechanism for producing redshifts. Hubble's law of the correlation between redshifts and distances is required by models of cosmology derived from general relativity that have a metric expansion of space.[18] As a result, photons propagating through the expanding space are stretched, creating the cosmological redshift.

There is a distinction between a redshift in cosmological context as compared to that witnessed when nearby objects exhibit a local Doppler-effect redshift. Rather than cosmological redshifts being a consequence of relative velocities, the photons instead increase in wavelength and redshift because of a feature of the spacetime through which they are traveling that causes space to expand.[27] Due to the expansion increasing as distances increase, the distance between two remote galaxies can increase at more than 3×10^8 m/s, but this does not imply that the galaxies move faster than the speed of light at their present location (which is forbidden by Lorentz covariance).

Mathematical derivation

The observational consequences of this effect can be derived using the equations from general relativity that describe a homogeneous and isotropic universe.

To derive the redshift effect, use the geodesic equation for a light wave, which is

$$ds^2 = 0 = -c^2 dt^2 + \frac{a^2 dr^2}{1 - kr^2}$$

where

- ds^2 is the spacetime interval
- dt^2 is the time interval
- dr^2 is the spatial interval
- c is the speed of light
- a is the time-dependent cosmic scale factor
- k is the curvature per unit area.

For an observer observing the crest of a light wave at a position $r = 0$ and time $t = t_0$, the crest of the light wave was emitted at a time $t = t_e$ in the past and a distant position $r = R$. Integrating over the path in both space and time that the light wave travels yields:

$$c\int_{t_{then}}^{t_{now}} \frac{dt}{a} = \int_R^0 \frac{dr}{\sqrt{1 - kr^2}}.$$

In general, the wavelength of light is not the same for the two positions and times considered due to the changing properties of the metric. When the wave was emitted, it had a wavelength λ_e. The next crest of the light wave was emitted at a time

$$t = t_{then} + \lambda_{then}/c.$$

The observer sees the next crest of the observed light wave with a wavelength λ_o to arrive at a time

$$t = t_{now} + \lambda_{now}/c.$$

Since the subsequent crest is again emitted from $r = R$ and is observed at $r = 0$, the following equation can be written:

$$c \int_{t_{then}+\lambda_{then}/c}^{t_{now}+\lambda_{now}/c} \frac{dt}{a} = \int_{R}^{0} \frac{dr}{\sqrt{1-kr^2}}.$$

The right-hand side of the two integral equations above are identical which means

$$c \int_{t_{then}+\lambda_{then}/c}^{t_{now}+\lambda_{now}/c} \frac{dt}{a} = c \int_{t_{then}}^{t_{now}} \frac{dt}{a}$$

Using the following manipulation:

$$0 = \frac{\int_{then}^{t_{now}} \frac{dt}{\int_{t_{then} \, t_{then}+\lambda_{then}/} dt} \int^{t_{now}+\lambda_{now}/c}}{a} \int \frac{dt}{dt} \int \frac{dt}{dt}$$

$$= \int_{now \, t_{now}+\lambda_{now}/c}^{t_{then}+\lambda_{then}/c} t \quad (\int^{t_{then}+\lambda_{then}/c} \quad \int^{t_{now}+\lambda_{now}/c}$$

$$= \frac{\int^{t_{then}+\lambda_{then}/c} \quad \int^{t_{now}+\lambda_{now}/c}}{a}$$

$$t_{then} \, t_{now}$$

we find that:

$$\int_{t_{now}}^{t_{now}+\lambda_{now}/c} \frac{dt}{a} = \int_{t_{then}}^{t_{then}+\lambda_{then}/c} \frac{dt}{a}.$$

For very small variations in time (over the period of one cycle of a light wave) the scale factor is essentially a constant ($a = a_o$ today and $a = a_e$ previously). This yields

$$\frac{t_{now} + \lambda_{now}/c}{a_{now}} \quad \frac{t_{now}}{} = \frac{t_{then} + \lambda_{then}/c}{a_{then}} \quad \frac{t_{then}}{}$$

which can be rewritten as

$$\frac{\lambda_{now}}{} \quad \lambda_{then} \quad \overset{=}{}$$

*a*how ,
*a*then ˚

Using the definition of redshift provided above, the equation

$$1 + z = \frac{a_{now}}{a_{then}}$$

is obtained. In an expanding universe such as the one we inhabit, the scale factor is monotonically increasing as time passes, thus, z is positive and distant galaxies appear redshifted.

Using a model of the expansion of the Universe, redshift can be related to the age of an observed object, the so-called *cosmic time–redshift relation*. Denote a density ratio as Ω_0:

$$\Omega_0 = \frac{\rho}{\rho_{crit}} ,$$

with ρ_{ri} the critical density demarcating a universe that eventually crunches from one that simply expands. This density is about three hydrogen atoms per thousand liters of space.[28] At large redshifts one finds:

$$t(z) = \frac{2}{3 H_0 \Omega_0^{1/2} (1 + z)^{3/2}} ,$$

where H_0 is the present-day Hubble constant, and z is the redshift.[29][30][31]

Distinguishing between cosmological and local eff ects

For cosmological redshifts of $z < 0.01$ additional Doppler redshifts and blueshifts due to the peculiar motions of the galaxies relative to one another cause a wide scatter from the standard Hubble Law.[32] The resulting situation can be illustrated by the Expanding Rubber Sheet Universe, a common cosmological analogy used to describe the expansion of space. If two objects are represented by ball bearings and spacetime by a stretching rubber sheet, the Doppler effect is caused by rolling the balls across the sheet to create peculiar motion. The cosmological redshift occurs when the ball bearings are stuck to the sheet and the sheet is stretched.[33][34][35]

The redshifts of galaxies include both a component related to recessional velocity from expansion of the Universe, and a component related to peculiar motion (Doppler shift).[36] The redshift due to expansion of the Universe depends upon the recessional velocity in a fashion determined by the cosmological model chosen to describe the expansion of the Universe, which is very different from how Doppler redshift depends upon local velocity.[37] Describing the cosmological expansion origin of redshift, cosmologist Edward Robert Harrison said, "Light leaves a galaxy, which is stationary in its local region of space, and is eventually received by observers who are stationary in their own local region of space. Between the galaxy and the observer, light travels through vast regions of expanding space. As a result, all wavelengths of the light are stretched by the expansion of space. It is as simple as that..."[38] Steven Weinberg clarified, "The increase of wavelength from emission to absorption of light does not depend on the rate of change of $a(t)$ [here $a(t)$ is the Robertson-Walker scale factor] at the times of emission or absorption, but on the increase of $a(t)$ in the whole period from emission to absorption."[39]

Popular literature often uses the expression "Doppler redshift" instead of "cosmological redshift" to describe the redshift of galaxies dominated by the expansion of spacetime, but the cosmological redshift is not found using the relativistic Doppler equation[40] which is instead characterized by special relativity; thus $v > c$ is impossible while, in contrast, $v > c$ is possible for cosmological redshifts because the space which separates the objects (for example, a quasar from the Earth) can expand faster than the speed of light.[41] More mathematically, the viewpoint that "distant galaxies are receding" and the viewpoint that "the space between galaxies is expanding" are related by changing coordinate systems. Expressing this precisely requires working with the mathematics of the Friedmann-Robertson-Walker metric.[42]

If the Universe were contracting instead of expanding, we would see distant galaxies blueshifted by an amount proportional to their distance instead of redshifted.[43]

20.3.3 Gravitational redshift

Main article: Gravitational redshift

In the theory of general relativity, there is time dilation within a gravitational well. This is known as the gravitational redshift or *Einstein Shift*.[44] The theoretical derivation of this effect follows from the Schwarzschild solution of the Einstein equations which yields the following formula for redshift associated with a photon traveling in the gravitational field of an uncharged, nonrotating, spherically symmetric mass:

$$1 + z = \sqrt{\frac{1}{1 - \frac{2GM}{rc^2}}},$$

where

- G is the gravitational constant,

- M is the mass of the object creating the gravitational field,

- r is the radial coordinate of the source (which is analogous to the classical distance from the center of the object, but is actually a Schwarzschild coordinate), and

- c is the speed of light.

This gravitational redshift result can be derived from the assumptions of special relativity and the equivalence principle; the full theory of general relativity is not required.[45]

The effect is very small but measurable on Earth using the Mössbauer effect and was first observed in the Pound–Rebka experiment.[46] However, it is significant near a black hole, and as an object approaches the event horizon the red shift becomes infinite. It is also the dominant cause of large angular-scale temperature fluctuations in the cosmic microwave background radiation (see Sachs-Wolfe effect).[47][48]

20.4 Observations in astronomy

The redshift observed in astronomy can be measured because the emission and absorption spectra for atoms are distinctive and well known, calibrated from spectroscopic experiments in laboratories on Earth. When the redshift of various absorption and emission lines from a single astronomical object is measured, z is found to be remarkably constant. Although distant objects may be slightly blurred and lines broadened, it is by no more than can be explained by thermal or mechanical motion of the source. For these reasons and others, the consensus among astronomers is that the redshifts they observe are due to some combination of the three established forms of Doppler-like redshifts. Alternative hypotheses and explanations for redshift such as tired light are not generally considered plausible.[49]

Spectroscopy, as a measurement, is considerably more difficult than simple photometry, which measures the brightness of astronomical objects through certain filters.[50] When photometric data is all that is available (for example, the Hubble Deep Field and the Hubble Ultra Deep Field), astronomers rely on a technique for measuring photometric redshifts.[51] Due to the broad wavelength ranges in photometric filters and the necessary assumptions about the nature of the spectrum at the light-source, errors for these sorts of measurements can range up to $\delta z = 0.5$, and are much less reliable than spectroscopic determinations.[52] However, photometry does at least allow a qualitative characterization of a redshift. For example, if a Sun-like spectrum had a redshift of $z = 1$, it would be brightest in the infrared rather than at the yellow-green color associated with the peak of its blackbody spectrum, and the light intensity will be reduced in the filter by a factor of four, $(1 + z)^2$. Both the photon count rate and the photon energy are redshifted. (See K correction for more details on the photometric consequences of redshift.)[53]

20.4.1   Local observations

In nearby objects (within our Milky Way galaxy) observed redshifts are almost always related to the line-of-sight velocities associated with the objects being observed. Observations of such redshifts and blueshifts have enabled astronomers to measure velocities and parametrize the masses of the orbiting stars in spectroscopic binaries, a method first employed in 1868 by British astronomer William Huggins.[7] Similarly, small redshifts and blueshifts detected in the spectroscopic measurements of individual stars are one way astronomers have been able to diagnose and measure the presence and characteristics of planetary systems around other stars and have even made very detailed differential measurements of redshifts during planetary transits to determine precise orbital parameters.[54] Finely detailed measurements of redshifts are used in helioseismology to determine the precise movements of the photosphere of the Sun.[55] Redshifts have also been used to make the first measurements of the rotation rates of planets,[56] velocities of interstellar clouds,[57] the rotation of galaxies,[2] and the dynamics of accretion onto neutron stars and black holes which exhibit both Doppler and gravitational redshifts.[58] Additionally, the temperatures of various emitting and absorbing objects can be obtained by measuring Doppler broadening – effectively redshifts and blueshifts over a single emission or absorption line.[59] By measuring the broadening and shifts of the 21-centimeter hydrogen line in different directions, astronomers have been able to measure the recessional velocities of interstellar gas, which in turn reveals the rotation curve of our Milky Way.[2] Similar measurements have been performed on other galaxies, such as Andromeda.[2] As a diagnostic tool, redshift measurements are one of the most important spectroscopic measurements made in astronomy.

20.4.2   Extragalactic observations

The most distant objects exhibit larger redshifts corresponding to the Hubble flow of the Universe. The largest observed redshift, corresponding to the greatest distance and furthest back in time, is that of the cosmic microwave background radiation; the numerical value of its redshift is about $z = 1089$ ($z = 0$ corresponds to present time), and it shows the state of the Universe about 13.8 billion years ago,[60] and 379,000 years after the initial moments of the Big Bang.[61]

The luminous point-like cores of quasars were the first "high-redshift" ($z > 0.1$) objects discovered before the improvement of telescopes allowed for the discovery of other high-redshift galaxies.

For galaxies more distant than the Local Group and the nearby Virgo Cluster, but within a thousand megaparsecs or so, the redshift is approximately proportional to the galaxy's distance. This correlation was first observed by Edwin Hubble and has come to be known as Hubble's law. Vesto Slipher was the first to discover galactic redshifts, in about the year 1912, while Hubble correlated Slipher's measurements with distances he measured by other means to formulate his Law. In the widely accepted cosmological model based on general relativity, redshift is mainly a result of the expansion of space: this means that the farther away a galaxy is from us, the more the space has expanded in the time since the light left that galaxy, so the more the light has been stretched, the more redshifted the light is, and so the faster it appears to be moving away from us. Hubble's law follows in part from the Copernican principle.[62] Because it is usually not known how luminous objects are, measuring the redshift is easier than more direct distance measurements, so redshift is sometimes in practice converted to a crude distance measurement using Hubble's law.

Gravitational interactions of galaxies with each other and clusters cause a significant scatter in the normal plot of the Hubble diagram. The peculiar velocities associated with galaxies superimpose a rough trace of the mass of virialized objects in the Universe. This effect leads to such phenomena as nearby galaxies (such as the Andromeda Galaxy) exhibiting blueshifts as we fall towards a common barycenter, and redshift maps of clusters showing a Fingers of God effect due to the scatter of peculiar velocities in a roughly spherical distribution.[62] This added component gives cosmologists a chance to measure the masses of objects independent of the *mass to light ratio* (the ratio of a galaxy's mass in solar masses to its brightness in solar luminosities), an important tool for measuring dark matter.[63]

The Hubble law's linear relationship between distance and redshift assumes that the rate of expansion of the Universe is constant. However, when the Universe was much younger, the expansion rate, and thus the Hubble "constant", was larger than it is today. For more distant galaxies, then, whose light has been travelling to us for much longer times, the approximation of constant expansion rate fails, and the Hubble law becomes a non-linear integral relationship and dependent on the history of the expansion rate since the emission of the light from the galaxy in question. Observations of the redshift-distance relationship can be used, then, to determine the expansion history of the Universe and thus the matter and energy content.

While it was long believed that the expansion rate has been continuously decreasing since the Big Bang, recent observations of the redshift-distance relationship using Type Ia supernovae have suggested that in comparatively recent times the expansion rate of the Universe has begun to accelerate.

20.4.3   Highest redshifts

See also: List of most distant objects by type

Currently, the objects with the highest known redshifts are galaxies and the objects producing gamma ray bursts. The most reliable redshifts are from spectroscopic data, and the highest confirmed spectroscopic redshift of a galaxy is that of UDFy-38135539 [64] at a redshift of $z = 8.6$, corresponding to just 600 million years after the Big Bang. The previous record was held by IOK-1,[65] at a redshift $z = 6.96$, corresponding to just 750 million years after the Big Bang. Slightly less reliable are Lyman-break redshifts, the highest of which is the lensed galaxy A1689-zD1 at a redshift $z = 7.6$[66] and the next highest being $z = 7.0$.[67] The most distant observed gamma ray burst with a spectroscopic redshift measurement was GRB 090423, which had a redshift of $z = 8.2$.[68] The most distant known quasar, ULAS J1120+0641, is at $z = 7.1$.[69][70] The highest known redshift radio galaxy (TN J0924-2201) is at a redshift $z = 5.2$[71] and the highest known redshift molecular material is the detection of emission from the CO molecule from the quasar SDSS J1148+5251 at $z = 6.42$[72]

Extremely red objects (EROs) are astronomical sources of radiation that radiate energy in the red and near infrared part of the electromagnetic spectrum. These may be starburst galaxies that have a high redshift accompanied by reddening from intervening dust, or they could be highly redshifted elliptical galaxies with an older (and therefore redder) stellar population.[73] Objects that are even redder than EROs are termed *hyper extremely red objects* (HEROs).[74]

The cosmic microwave background has a redshift of $z = 1089$, corresponding to an age of approximately 379,000 years after the Big Bang and a comoving distance of more than 46 billion light years.[75] The yet-to-be-observed first light from the oldest Population III stars, not long after atoms first formed and the CMB ceased to be absorbed almost completely, may have redshifts in the range of $20 < z < 100$.[76] Other high-redshift events predicted by physics but not presently observable are the cosmic neutrino background from about two seconds after the Big Bang (and a redshift in excess of $z > 10^{10}$)[77] and the cosmic gravitational wave background emitted directly from inflation at a redshift in excess of $z > 10^{25}$.[78]

In June 2015, astronomers reported evidence for Population III stars in the Cosmos Redshift 7 galaxy at $z = 6.60$. Such stars are likely to have existed in the very early universe (i.e., at high redshift), and may have started the production of chemical elements heavier than hydrogen that are needed for the later formation of planets and life as we know it.[79][80]

20.4.4   Redshift surveys

Main article: Redshift survey

With advent of automated telescopes and improvements in spectroscopes, a number of collaborations have been made to map the Universe in redshift space. By combining redshift with angular position data, a redshift survey maps the 3D distribution of matter within a field of the sky. These observations are used to measure properties of the large-scale structure of the Universe. The Great Wall, a vast supercluster of galaxies over 500 million light-years wide, provides a dramatic example of a large-scale structure that redshift surveys can detect.[81]

The first redshift survey was the CfA Redshift Survey, started in 1977 with the initial data collection completed in 1982.[82] More recently, the 2dF Galaxy Redshift Survey determined the large-scale structure of one section of the Universe, measuring redshifts for over 220,000 galaxies; data collection was completed in 2002, and the final data set was released 30 June 2003.[83] The Sloan Digital Sky Survey (SDSS), is ongoing as of 2013 and aims to measure the redshifts of around 3 million objects.[84] SDSS has recorded redshifts for galaxies as high as 0.8, and has been involved in the detection of quasars beyond $z = 6$. The DEEP2 Redshift Survey uses the Keck telescopes with the new "DEIMOS" spectrograph; a follow-up to the pilot program DEEP1, DEEP2 is designed to measure faint galaxies with redshifts 0.7 and above, and it is therefore planned to provide a high redshift complement to SDSS and 2dF.[85]

20.5   Eﬀects due to physical optics or radiative transfer

The interactions and phenomena summarized in the subjects of radiative transfer and physical optics can result in shifts in the wavelength and frequency of electromagnetic radiation. In such cases the shifts correspond to a physical energy transfer to matter or other photons rather than being due to a transformation between reference frames. These shifts can be due to such physical phenomena as coherence effects or the scattering of electromagnetic radiation whether from charged elementary particles, from particulates, or from fluctuations of the index of refraction in a dielectric medium as occurs in the radio phenomenon of radio whistlers.[2] While such phenomena are sometimes referred to as "redshifts" and "blueshifts", in astrophysics light-matter interactions that result in energy shifts in the radiation field are generally referred to as "reddening" rather than "redshifting" which, as a term, is normally reserved for the effects discussed above.[2]

In many circumstances scattering causes radiation to redden because entropy results in the predominance of many low-energy photons over few high-energy ones (while conserving total energy).[2] Except possibly under carefully controlled conditions, scattering does not produce the same relative change in wavelength across the whole spectrum; that is, any calculated z is generally a function of wavelength. Furthermore, scattering from random media generally occurs at many angles, and z is a function of the scattering angle. If multiple scattering occurs, or the scattering particles have relative motion, then there is generally distortion of spectral lines as well.[2]

In interstellar astronomy, visible spectra can appear redder due to scattering processes in a phenomenon referred to as interstellar reddening[2] – similarly Rayleigh scattering causes the atmospheric reddening of the Sun seen in the sunrise or sunset and causes the rest of the sky to have a blue color. This phenomenon is distinct from red*shifting* because the spectroscopic lines are not shifted to other wavelengths in reddened objects and there is an additional dimming and distortion associated with the phenomenon due to photons being scattered in and out of the line-of-sight.

For a list of scattering processes, see Scattering.

20.6   References

20.6.1   Notes

[1] See Feynman, Leighton and Sands (1989) or any introductory undergraduate (and many high school) physics textbooks. See Taylor (1992) for a relativistic discussion.

[2] See Binney and Merrifeld (1998), Carroll and Ostlie (1996), Kutner (2003) for applications in astronomy.

[3] See Misner, Thorne and Wheeler (1973) and Weinberg (1971) or any of the physical cosmology textbooks

[4] Doppler, Christian (1846). *Beiträge zur fixstemenkunde* 69. Prague: G. Haase Söhne. Bibcode:1846QB815.D69.....

[5] Maulik, Dev (2005). "Doppler Sonography: A Brief History". In Maulik, Dev; Zalud, Ivica. *Doppler Ultrasound in Obstetrics And Gynecology.* ISBN 978-3-540-23088-5.

[6] O'Connor, John J.; Robertson, Edmund F. (1998). "Christian Andreas Doppler". *MacTutor History of Mathematics archive.* University of St Andrews.

[7] Huggins, William (1868). "Further Observations on the Spectra of Some of the Stars and Nebulae, with an Attempt to Determine Therefrom Whether These Bodies are Moving towards or from the Earth, Also Observations on the Spectra of the Sun and of Comet II". *Philosophical Transactions of the Royal Society of London* 158: 529–564. Bibcode:1868RSPT..158..529H. doi:10.1098/rstl.1868.0022.

[8] Reber, G. (1995). "Intergalactic Plasma". *Astrophysics and Space Science* 227 (1–2): 93–96. Bibcode:1995Ap&SS.227...93R. doi:10.1007/BF00678069.

[9] Pannekoek, A (1961). *A History of Astronomy.* Dover. p. 451. ISBN 0-486-65994-1.

[10] Bélopolsky, A. (1901). "On an Apparatus for the Laboratory Demonstration of the Doppler-Fizeau Principle". *Astrophysical Journal* 13: 15. Bibcode:1901ApJ....13...15B. doi:10.1086/140786.

[11] Adams, Walter S. (1908). "Preliminary catalogue of lines affected in sun-spots". *Contributions from the Mount Wilson Observatory / Carnegie Institution of Washington* (Contributions from the Solar Observatory of the Carnegie Institution of Washington: Carnegie Institution of Washington) 22: 1–21. Bibcode:1908CMWCI..22....1A. Reprinted in Adams, Walter S. (1908). "Preliminary Catalogue of Lines Affected in Sun-Spots Region λ 4000 TO λ 4500". *Astrophysical Journal* 27: 45. Bibcode:1908ApJ....27...45A. doi:10.1086/141524.

[12] de Sitter, W. (1934). "On distance, magnitude, and related quantities in an expanding universe". *Bulletin of the Astronomical Institutes of the Netherlands* 7: 205. Bibcode:1934BAN.....7..205D. It thus becomes urgent to investigate the effect of the redshift and of the metric of the universe on the apparent magnitude and observed numbers of nebulae of given magnitude

[13] Slipher, Vesto (1912). "The radial velocity of the Andromeda Nebula". *Lowell Observatory Bulletin* 1: 2.56–2.57. Bibcode:1913LowOB...1b..56S. The magnitude of this velocity, which is the greatest hitherto observed, raises the question whether the velocity-like displacement might not be due to some other cause, but I believe we have at present no other interpretation for it

[14] Slipher, Vesto (1915). "Spectrographic Observations of Nebulae". *Popular Astronomy* 23: 21–24. Bibcode:1915PA.....23...21S.

[15] Slipher, Vesto (1915). "Spectrographic Observations of Nebulae". *Popular Astronomy* 23: 22. Bibcode:1915PA.....23...21S.

[16] Hubble, Edwin (1929). "A Relation between Distance and Radial Velocity among Extra-Galactic Nebulae". *Proceedings of the National Academy of Sciences of the United States of America* 15 (3): 168–173. Bibcode:1929PNAS...15..168H. doi:10.1073/pnas.15.3.168. PMC 522427. PMID 16577160.

[17] Friedman, A. A. (1922). "Über die Krümmung des Raumes". *Zeitschrift für Physik* 10 (1): 377–386. Bibcode:1922ZPhy...10..377F. doi:10.1007/BF01332580. English translation in Friedman, A. (1999). "On the Curvature of Space". *General Relativity and Gravitation* 31 (12): 1991–2000. Bibcode:1999GReGr..31.1991F. doi:10.1023/A:1026751225741.)

[18] This was recognized early on by physicists and astronomers working in cosmology in the 1930s. The earliest layman publication describing the details of this correspondence is Eddington, Arthur (1933). *The Expanding Universe: Astronomy's 'Great Debate', 1900–1931*. Cambridge University Press. (Reprint: ISBN 978-0-521-34976-5)

[19] "Hubble census finds galaxies at redshifts 9 to 12". *ESA/Hubble Press Release*. Retrieved 13 December 2012.

[20] See, for example, this 25 May 2004 press release from NASA's Swift space telescope that is researching gamma-ray bursts: "Measurements of the gamma-ray spectra obtained during the main outburst of the GRB have found little value as redshift indicators, due to the lack of well-defined features. However, optical observations of GRB afterglows have produced spectra with identifiable lines, leading to precise redshift measurements."

[21] See for a tutorial on how to define and interpret large redshift measurements.

[22] Where z = redshift; v|| = velocity parallel to line-of-sight (positive if moving away from receiver); c = speed of light; γ = Lorentz factor; a = scale factor; G = gravitational constant; M = object mass; r = radial Schwarzschild coordinate, g = t,t component of the metric tensor

[23] H. Ives and G. Stilwell, An Experimental study of the rate of a moving atomic clock, J. Opt. Soc. Am. 28, 215–226 (1938)

[24] Freund, Jurgen (2008). *Special Relativity for Beginners*. World Scientific. p. 120. ISBN 981-277-160-3.

[25] Ditchburn, R (1961). *Light*. Dover. p. 329. ISBN 0-12-218101-8.

[26] See "Photons, Relativity, Doppler shift" at the University of Queensland

[27] The distinction is made clear in Harrison, Edward Robert (2000). *Cosmology: The Science of the Universe* (2nd ed.). Cambridge University Press. pp. 306ff. ISBN 0-521-66148-X.

[28] Steven Weinberg (1993). *The First Three Minutes: A Modern View of the Origin of the Universe* (2nd ed.). Basic Books. p. 34. ISBN 0-465-02437-8.

[29] Lars Bergström; Ariel Goobar (2006). *Cosmology and Particle Astrophysics* (2nd ed.). Springer. p. 77, Eq.4.79. ISBN 3-540-32924-2.

[30] M.S. Longair (1998). *Galaxy Formation*. Springer. p. 161. ISBN 3-540-63785-0.

[31] Yu N Parijskij (2001). "The High Redshift Radio Universe". In Norma Sanchez. *Current Topics in Astrofundamental Physics*. Springer. p. 223. ISBN 0-7923-6856-8.

[32] Measurements of the peculiar velocities out to 5 Mpc using the Hubble Space Telescope were reported in 2003 by Karachentsev et al. *Local galaxy flows within 5 Mpc.* 02/2003 *Astronomy and Astrophysics,* **398,** 479–491.

[33] Theo Koupelis; Karl F. Kuhn (2007). *In Quest of the Universe* (5th ed.). Jones & Bartlett Publishers. p. 557. ISBN 0-7637-4387-9.

[34] "It is perfectly valid to interpret the equations of relativity in terms of an expanding space. The mistake is to push analogies too far and imbue space with physical properties that are not consistent with the equations of relativity." Geraint F. Lewis; Francis, Matthew J.; Barnes, Luke A.; Kwan, Juliana et al. (2008). "Cosmological Radar Ranging in an Expanding Universe". *Monthly Notices of the Royal Astronomical Society* **388** (3): 960–964. arXiv:0805.2197. Bibcode:2008MNRAS.388..960L. doi:10.1111/j.1365-2966.2008.13477.x.

[35] Michal Chodorowski (2007). "Is space really expanding? A counterexample". *Concepts Phys* **4**: 17–34. arXiv:astro-ph/0601171. Bibcode:2007ONCP....4...15C. doi:10.2478/v10005-007-0002-2.

[36] Bedran,M.L.(2002)http://www.df.uba.ar/users/sgil/physics_paper_doc/papers_phys/cosmo/doppler_redshift.pdf "A comparison between the Doppler and cosmological redshifts"; Am.J.Phys.**70**, 406–408 (2002)

[37] Edward Harrison (1992). "The redshift-distance and velocity-distance laws". *Astrophysical Journal, Part 1* **403**: 28–31. Bibcode:1993ApJ...403...28H. doi:10.1086/172179.. A pdf file can be found here .

[38] Harrison 2000, p. 315.

[39] Steven Weinberg (2008). *Cosmology.* Oxford University Press. p. 11. ISBN 978-0-19-852682-7.

[40] Odenwald & Fienberg 1993

[41] Speed faster than light is allowed because the expansion of the spacetime metric is described by general relativity in terms of sequences of only locally valid inertial frames as opposed to a global Minkowski metric. Expansion faster than light is an integrated effect over many local inertial frames and is allowed because no single inertial frame is involved. The speed-of-light limitation applies only locally. See Michal Chodorowski (2007). "Is space really expanding? A counterexample". *Concepts Phys* **4**: 17–34. arXiv:astro-ph/0601171. Bibcode:2007ONCP....4...15C. doi:10.2478/v10005-007-0002-2.

[42] M. Weiss, What Causes the Hubble Redshift?, entry in the Physics FAQ (1994), available via John Baez's website

[43] This is only true in a universe where there are no peculiar velocities. Otherwise, redshifts combine as

$$1 + z = (1 + z_{Doppler})(1 + z_{expansion})$$

which yields solutions where certain objects that "recede" are blueshifted and other objects that "approach" are redshifted. For more on this bizarre result see Davis, T. M., Lineweaver, C. H., and Webb, J. K. "Solutions to the tethered galaxy problem in an expanding universe and the observation of receding blueshifted objects", *American Journal of Physics* (2003), **71** 358–364.

[44] Chant, C. A. (1930). "Notes and Queries (Telescopes and Observatory Equipment – The Einstein Shift of Solar Lines)". *Journal of the Royal Astronomical Society of Canada* **24**: 390. Bibcode:1930JRASC..24..390C.

[45] Einstein, A (1907). "Über das Relativitätsprinzip und die aus demselben gezogenen Folgerungen". *Jahrbuch der Radioaktivität und Elektronik* **4**: 411–462.

[46] Pound, R.; Rebka, G. (1960). "Apparent Weight of Photons". *Physical Review Letters* **4** (7): 337. Bibcode:1960PhRvL...4..337P. doi:10.1103/PhysRevLett.4.337.. This paper was the first measurement.

[47] Sachs, R. K.; Wolfe, A. M. (1967). "Perturbations of a cosmological model and angular variations of the cosmic microwave background". *Astrophysical Journal* **147** (73): 73. Bibcode:1967ApJ...147...73S. doi:10.1086/148982.

[48] Brill, Dieter (19 January 2012). "Black Hole Horizons and How They Begin". *Astronomical Review.*

[49] When cosmological redshifts were first discovered, Fritz Zwicky proposed an effect known as tired light. While usually considered for historical interests, it is sometimes, along with intrinsic redshift suggestions, utilized by nonstandard cosmologies. In 1981, H. J. Reboul summarised many alternative redshift mechanisms that had been discussed in the literature since the 1930s. In 2001, Geoffrey Burbidge remarked in a review that the wider astronomical community has marginalized such discussions since the 1960s. Burbidge and Halton Arp, while investigating the mystery of the nature of quasars, tried to develop alternative redshift mechanisms, and very few of their fellow scientists acknowledged let alone accepted their work. Moreover, Goldhaber *et al.* 2001; "Timescale Stretch Parameterization of Type Ia Supernova B-Band Lightcurves", ApJ, 558:359–386, 2001 September 1 pointed out that alternative theories are unable to account for timescale stretch observed in type Ia supernovae

[50] For a review of the subject of photometry, consider Budding, E., *Introduction to Astronomical Photometry*, Cambridge University Press (September 24, 1993), ISBN 0-521-41867-4

[51] The technique was first described by Baum, W. A.: 1962, in G. C. McVittie (ed.), *Problems of extra-galactic research*, p. 390, IAU Symposium No. 15

[52] Bolzonella, M.; Miralles, J.-M.; Pelló, R., Photometric redshifts based on standard SED fitting procedures, *Astronomy and Astrophysics*, **363**, p.476–492 (2000).

[53] A pedagogical overview of the K-correction by David Hogg and other members of the SDSS collaboration can be found at astro-ph.

[54] The Exoplanet Tracker is the newest observing project to use this technique, able to track the redshift variations in multiple objects at once, as reported in Ge, Jian; Van Eyken, Julian; Mahadevan, Suvrath; Dewitt, Curtis et al. (2006). "The First Extrasolar Planet Discovered with a New-Generation High-Throughput Doppler Instrument". *The Astrophysical Journal* **648**: 683. arXiv:astro-ph/0605247. Bibcode:2006ApJ...648..683G. doi:10.1086/505699.

[55] Libbrecht, Keng (1988). "Solar and stellar seismology". *Space Science Reviews* **47** (3–4): 275–301. Bibcode:1988SSRv...47..275L. doi:10.1007/BF00243557.

[56] In 1871 Hermann Carl Vogel measured the rotation rate of Venus. Vesto Slipher was working on such measurements when he turned his attention to spiral nebulae.

[57] An early review by Oort, J. H. on the subject: Oort, J. H. (1970). "The formation of galaxies and the origin of the high-velocity hydrogen". *Astronomy and Astrophysics* **7**: 381. Bibcode:1970A&A.....7..381O.

[58] Asaoka, Ikuko (1989). "X-ray spectra at infinity from a relativistic accretion disk around a Kerr black hole". *Astronomical Society of Japan* **41** (4): 763–778. Bibcode:1989PASJ...41..763A. ISSN 0004-6264.

[59] Rybicki, G. B. and A. R. Lightman, *Radiative Processes in Astrophysics*, John Wiley & Sons, 1979, p. 288 ISBN 0-471-82759-2

[60] "Cosmic Detectives". The European Space Agency (ESA). 2013-04-02. Retrieved 2013-04-25.

[61] An accurate measurement of the cosmic microwave background was achieved by the COBE experiment. The final published temperature of 2.73 K was reported in this paper: Fixsen, D. J.; Cheng, E. S.; Cottingham, D. A.; Eplee, R. E., Jr.; Isaacman, R. B.; Mather, J. C.; Meyer, S. S.; Noerdlinger, P. D.; Shafer, R. A.; Weiss, R.; Wright, E. L.; Bennett, C. L.; Boggess, N. W.; Kelsall, T.; Moseley, S. H.; Silverberg, R. F.; Smoot, G. F.; Wilkinson, D. T.. (1994). "Cosmic microwave background dipole spectrum measured by the COBE FIRAS instrument", *Astrophysical Journal*, 420, 445. The most accurate measurement as of 2006 was achieved by the WMAP experiment.

[62] Peebles (1993).

[63] Binney, James; Scott Treimane (1994). *Galactic dynamics*. Princeton University Press. ISBN 0-691-08445-9.

[64] M.D.Lehnert; Nesvadba, NP; Cuby, JG; Swinbank, AM et al. (2010). "Spectroscopic Confirmation of a galaxy at redshift z = 8.6". *Nature* **467** (7318): 940–942. arXiv:1010.4312. Bibcode:2010Natur.467..940L. doi:10.1038/nature09462. PMID 20962840.

[65] Masanori Iye; Ota, Kazuaki; Kashikawa, Nobunari; Furusawa, Hisanori et al. (2006). "A galaxy at a redshift z = 6.96 url=http://www.nature.com/nature/journal/v443/n7108/abs/nature05104.html". *Nature* **443** (7108): 186–188. arXiv:astro-ph/0609393. Bibcode:2006Natur.443..186I. doi:10.1038/nature05104. PMID 16971942.

[66] Bradley, L.., et al., Discovery of a Very Bright Strongly Lensed Galaxy Candidate at z ~ 7.6, *The Astrophysical Journal* (2008), Volume 678, Issue 2, pp. 647-654. [ttp://adsabs.harvard.edu/abs/2008ApJ...678..647B

[67] Egami, E., et al., Spitzer and Hubble Space Telescope Constraints on the Physical Properties of the z~7 Galaxy Strongly Lensed by A2218, *The Astrophysical Journal* (2005), v. 618, Issue 1, pp. L5–L8 .

[68] Salvaterra, R.; Valle, M. Della; Campana, S.; Chincarini, G. et al. (2009). "GRB 090423 reveals an exploding star at the epoch of re-ionization". *Nature* **461** (7268): 1258–60. arXiv:0906.1578. Bibcode:2009Natur.461.1258S. doi:10.1038/nature08445. PMID 19865166.

[69] http://www.universetoday.com/87175/most-distant-quasar-opens-window-into-early-universe/

[70] Scientific American, "Brilliant, but Distant: Most Far-Flung Known Quasar Offers Glimpse into Early Universe", **John Matson,** *29 June 2011*

[71] Klamer, I. J.; Ekers, R. D.; Sadler, E. M.; Weiss, A. et al. (2005). "CO (1-0) and CO (5-4) Observations of the Most Distant Known Radio Galaxy at z = 5.2". *The Astrophysical Journal* **621**: L1. arXiv:astro-ph/0501447v1. Bibcode:2005ApJ...621L...1K. doi:10.1086/429147.

[72] Walter, Fabian; Bertoldi, Frank; Carilli, Chris; Cox, Pierre et al. (2003). "Molecular gas in the host galaxy of a quasar at redshift z = 6.42". *Nature* **424** (6947): 406–8. arXiv:astro-ph/0307410. Bibcode:2003Natur.424..406W. doi:10.1038/nature01821. PMID 12879063.

[73] Smail, Ian; Owen, F. N.; Morrison, G. E.; Keel, W. C. et al. (2002). "The Diversity of Extremely Red Objects". *The Astrophysical Journal* **581** (2): 844–864. arXiv:astro-ph/0208434. Bibcode:2002ApJ...581..844S. doi:10.1086/344440.

[74] Totani, Tomonori; Yoshii, Yuzuru; Iwamuro, Fumihide; Maihara, Toshinori et al. (2001). "Hyper Extremely Red Objects in the Subaru Deep Field: Evidence for Primordial Elliptical Galaxies in the Dusty Starburst Phase". *The Astrophysical Journal* **558** (2): L87–L91. arXiv:astro-ph/0108145. Bibcode:2001ApJ...558L..87T. doi:10.1086/323619.

[75] Lineweaver, Charles; Tamara M. Davis (2005). "Misconceptions about the Big Bang". Scientific American. Retrieved 2008-11-06.

[76] Naoz, S.; Noter, S.; Barkana, R. (2006). "The first stars in the Universe". *Monthly Notices of the Royal Astronomical Society: Letters* **373**: L98–L102. arXiv:astro-ph/0604050. Bibcode:2006MNRAS.373L..98N. doi:10.1111/j.1745-3933.2006.00251.x.

[77] Lesgourgues, J; Pastor, S (2006). "Massive neutrinos and cosmology". *Physics Reports* **429** (6): 307–379. arXiv:astro-ph/0603494. Bibcode:2006PhR...429..307L. doi:10.1016/j.physrep.2006.04.001.

[78] Grishchuk, Leonid P (2005). "Relic gravitational waves and cosmology". *Physics-Uspekhi* **48** (12): 1235–1247. arXiv:gr-qc/0504018. Bibcode:2005PhyU...48.1235G. doi:10.1070/PU2005v048n12ABEH005795.

[79] Sobral, David; Matthee, Jorryt; Darvish, Behnam; Schaerer, Daniel; Mobasher, Bahram; Röttgering, Huub J. A.; Santos, Sérgio; Hemmati, Shoubaneh (4 June 2015). "Evidence For POPIII-Like Stellar Populations In The Most Luminous LYMAN-α Emitters At The Epoch Of Re-Ionisation: Spectroscopic Confirmation" (PDF). *The Astrophysical Journal.* Retrieved 17 June 2015.

[80] Overbye, Dennis (17 June 2015). "Astronomers Report Finding Earliest Stars That Enriched Cosmos". *New York Times.* Retrieved 17 June 2015.

[81] M. J. Geller & J. P. Huchra, *Science* **246**, 897 (1989). online

[82] See the official CfA website for more details.

[83] Shaun Cole; Percival; Peacock; Norberg et al. (2005). "The 2dF galaxy redshift survey: Power-spectrum analysis of the final dataset and cosmological implications". *Mon. Not. Roy. Astron. Soc.* **362** (2): 505–34. arXiv:astro-ph/0501174. Bibcode:2005MNRAS.362..505C. doi:10.1111/j.1365-2966.2005.09318.x. 2dF Galaxy Redshift Survey homepage

[84] SDSS Homepage

[85] Marc Davis; DEEP2 collaboration (2002). "Science objectives and early results of the DEEP2 redshift survey". *Conference on Astronomical Telescopes and Instrumentation, Waikoloa, Hawaii, 22–28 Aug 2002.* arXiv:astro-ph/0209419.

20.6.2   Articles

- Odenwald, S. & Fienberg, RT. 1993; "Galaxy Redshifts Reconsidered" in *Sky & Telescope* Feb. 2003; pp31–35 (This article is useful further reading in distinguishing between the 3 types of redshift and their causes.)

- Lineweaver, Charles H. and Tamara M. Davis, "Misconceptions about the Big Bang", *Scientific American*, March 2005. (This article is useful for explaining the cosmological redshift mechanism as well as clearing up misconceptions regarding the physics of the expansion of space.)

20.6.3 Book references

- Nussbaumer, Harry; Lydia Bieri (2009). *Discovering the Expanding Universe*. Cambridge University Press. ISBN 978-0-521-51484-2.

- Binney, James; Michael Merrifeld (1998). *Galactic Astronomy*. Princeton University Press. ISBN 0-691-02565-7.

- Carroll, Bradley W. & Dale A. Ostlie (1996). *An Introduction to Modern Astrophysics*. Addison-Wesley Publishing Company, Inc. ISBN 0-201-54730-9.

- Feynman, Richard; Leighton, Robert; Sands, Matthew (1989). *Feynman Lectures on Physics. Vol. 1*. Addison-Wesley. ISBN 0-201-51003-0.

- Grøn, Øyvind; Hervik, Sigbjørn (2007). *Einstein's General Theory of Relativity*. New York: Springer. ISBN 978-0-387-69199-2.

- Kutner, Marc (2003). *Astronomy: A Physical Perspective*. Cambridge University Press. ISBN 0-521-52927-1.

- Misner, Charles; Thorne, Kip S.; Wheeler, John Archibald (1973). *Gravitation*. San Francisco: W. H. Freeman. ISBN 0-7167-0344-0.

- Peebles, P. J. E. (1993). *Principles of Physical Cosmology*. Princeton University Press. ISBN 0-691-01933-9.

- Taylor, Edwin F.; Wheeler, John Archibald (1992). *Spacetime Physics: Introduction to Special Relativity* (2nd ed.). W.H. Freeman. ISBN 0-7167-2327-1.

- Weinberg, Steven (1971). *Gravitation and Cosmology*. John Wiley. ISBN 0-471-92567-5.

- See also physical cosmology textbooks for applications of the cosmological and gravitational redshifts.

20.7 External links

- Ned Wright's Cosmology tutorial

- Cosmic reference guide entry on redshift

- Mike Luciuk's Astronomical Redshift tutorial

- Animated GIF of Cosmological Redshift by Wayne Hu

- Merrifeld, Michael; Hill, Richard (2009). "Z Redshift". *SIXTψ SYMBØLS*. Brady Haran for the University of Nottingham.

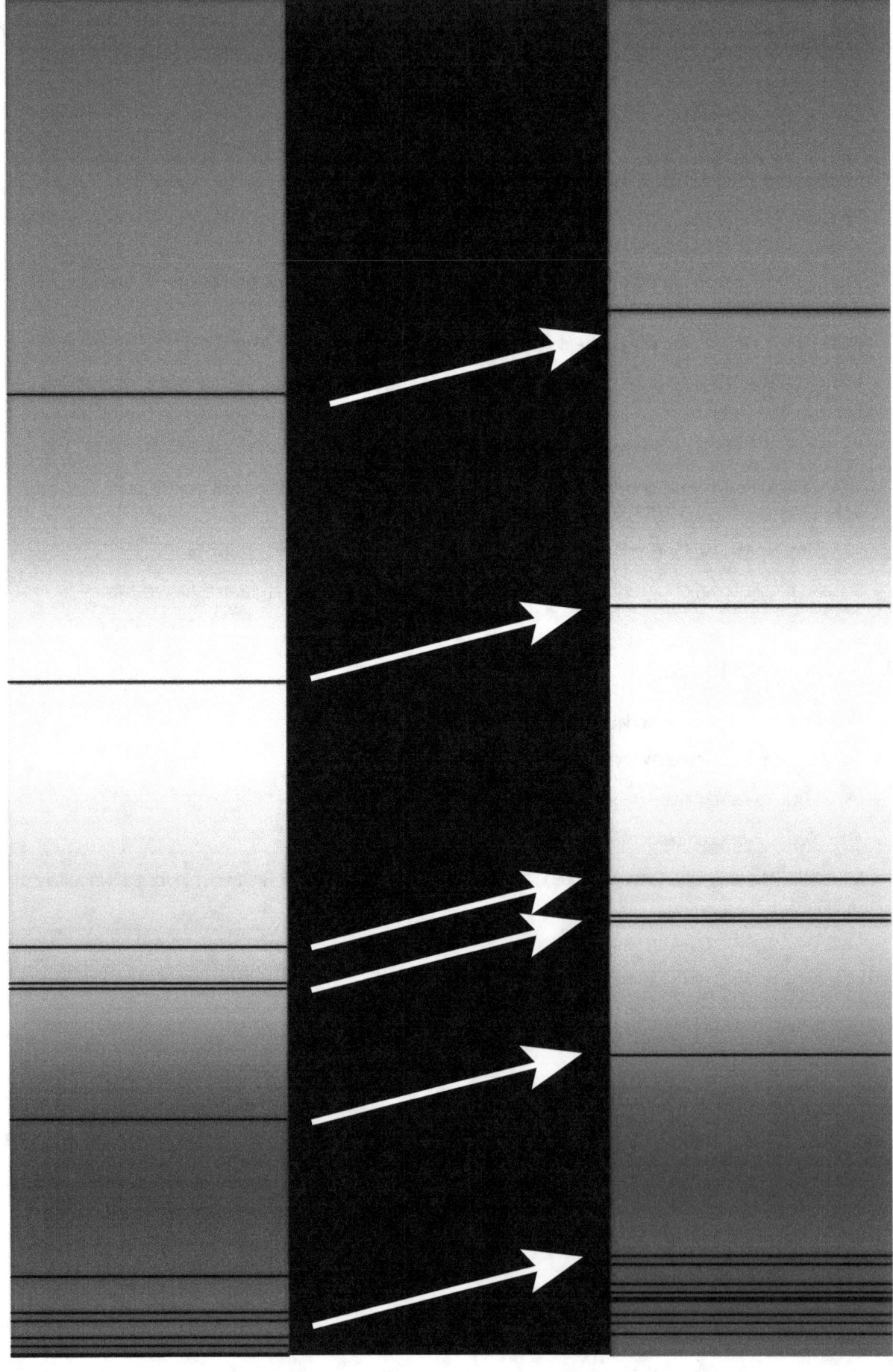

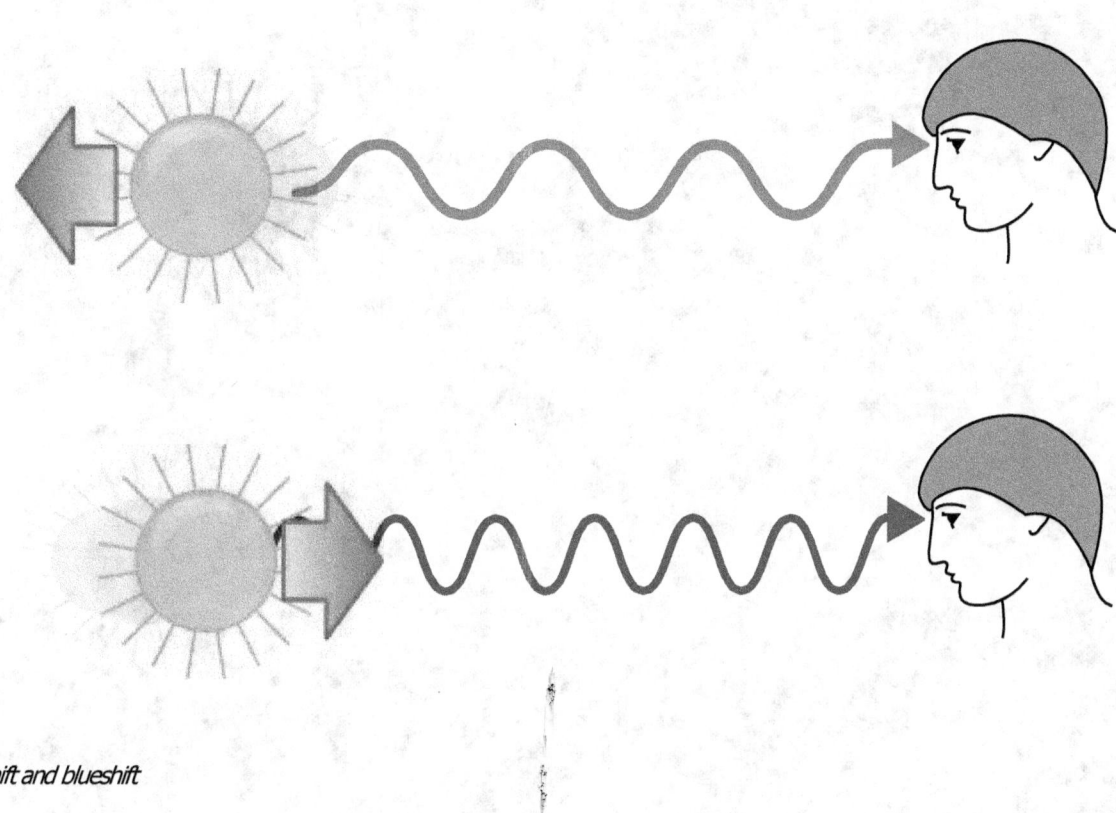

Redshift and blueshift

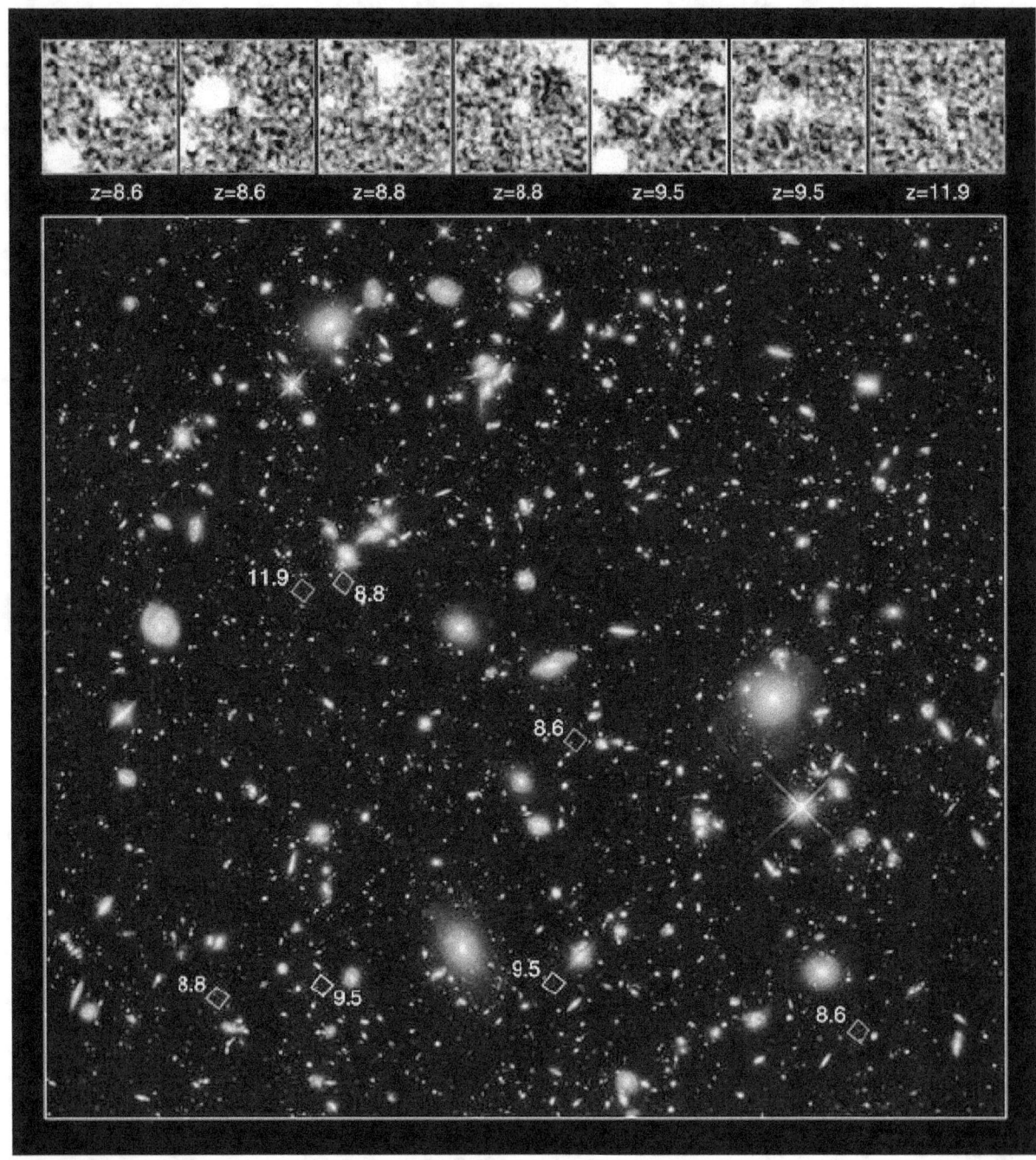

High-redshift galaxy candidates in the Hubble Ultra Deep Field 2012.[19]

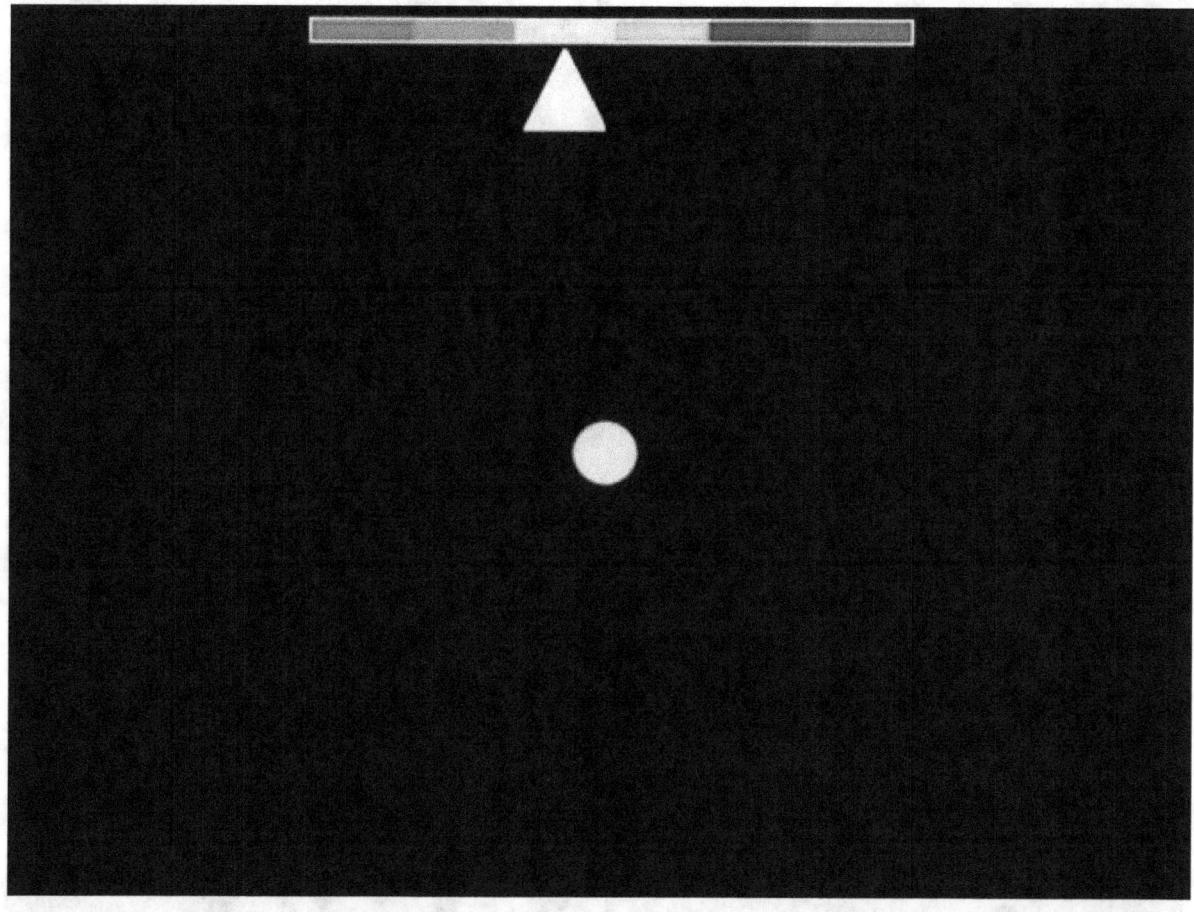

Doppler effect, yellow (~575 nm wavelength) ball appears greenish (blueshift to ~565 nm wavelength) approaching observer, turns orange (redshift to ~585 nm wavelength) as it passes, and returns to yellow when motion stops. To observe such a change in color, the object would have to be traveling at approximately 5200 km/s, or about 75 times faster than the speed record for the fastest manmade space probe.

A picture of the solar corona taken with the LASCO C1 coronagraph. The picture is a color-coded image of the doppler shift of the FeXIV 5308 Å line, caused by the coronal plasma velocity towards or away from the satellite.

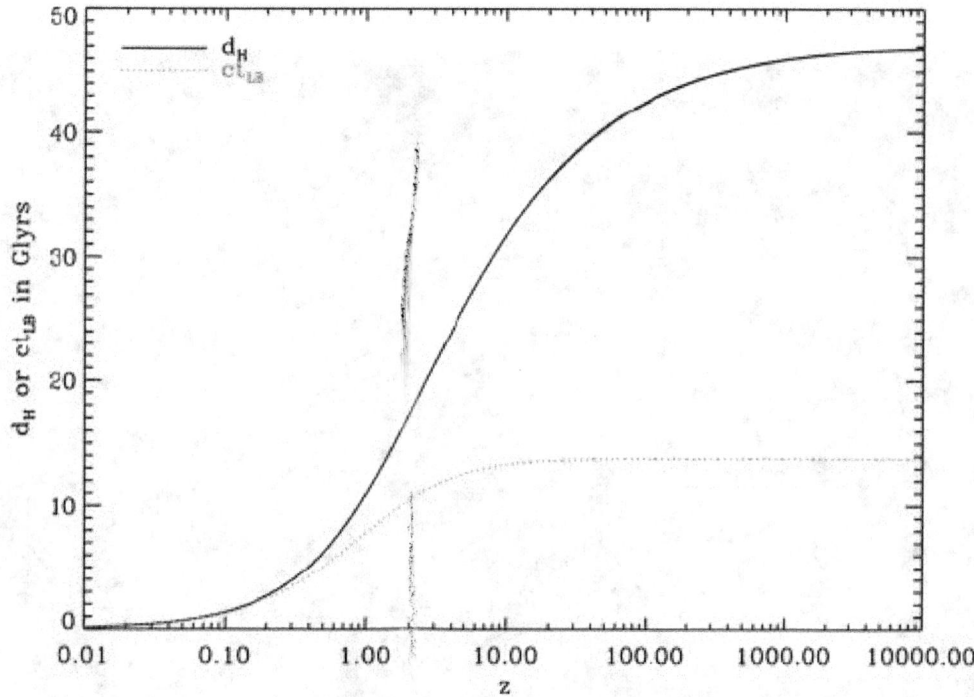

Plot of distance (in giga light-years) vs. redshift according to the Lambda-CDM model. dH (in solid black) is the comoving distance from Earth to the location with the Hubble redshift z while ct.LB (in dotted red) is the speed of light multiplied by the lookback time to Hubble redshift z. The comoving distance is the physical space-like distance between here and the distant location, asymptoting to the size of the observable universe at some 47 billion light years. The lookback time is the distance a photon traveled from the time it was emitted to now divided by the speed of light, with a maximum distance of 13.8 billion light years corresponding to the age of the universe.

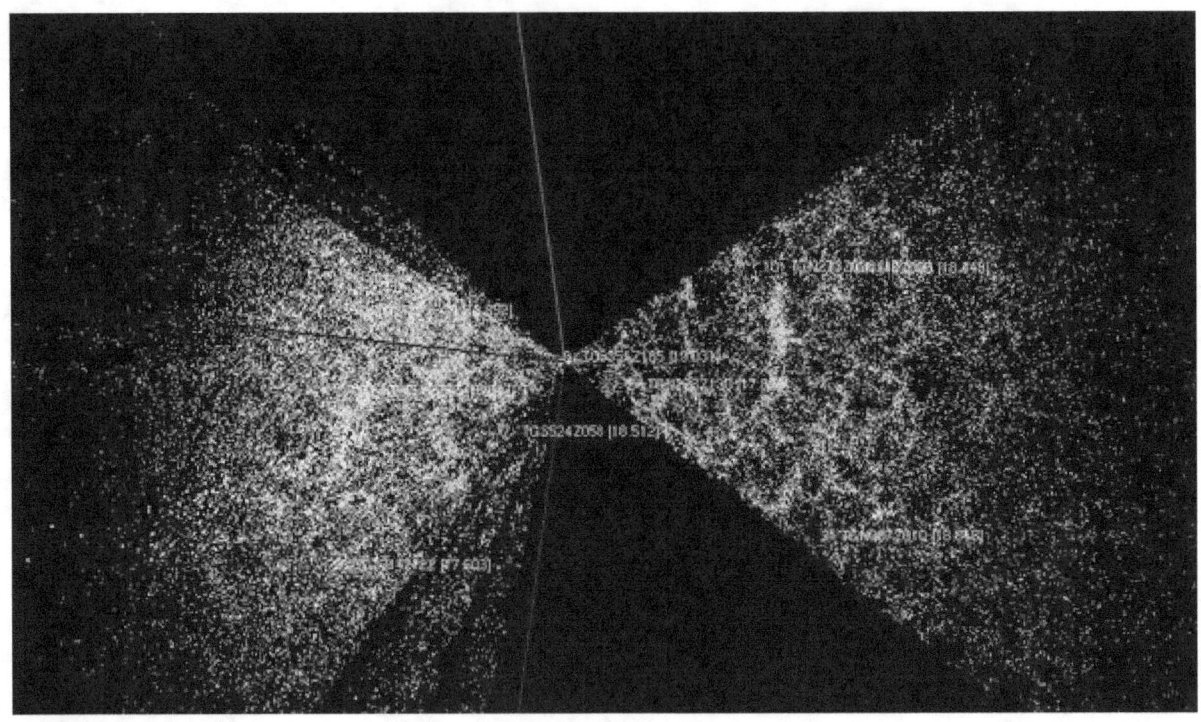

Rendering of the 2dFGRS data

Chapter 21

Accelerating universe

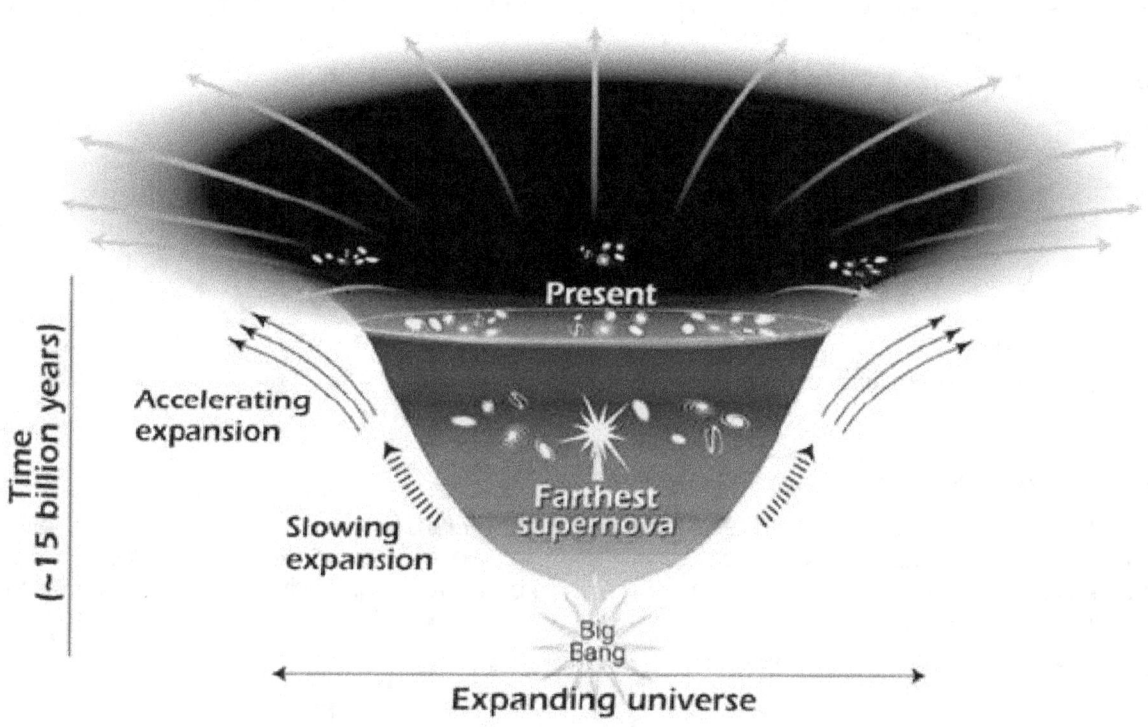

This diagram reveals changes in the rate of expansion since the universe's birth 15 billion years ago. The more shallow the curve, the faster the rate of expansion. The curve changes noticeably about 7.5 billion years ago, when objects in the universe began flying apart at a faster rate. Astronomers theorize that the faster expansion rate is due to a mysterious, dark force that is pushing galaxies apart.

The expansion of the Universe accelerating. Time flows from bottom to top

The **accelerating universe** is the observation that the universe appears to be expanding at an increasing rate. In formal terms, this means that the cosmic scale factor $a(t)$ has a positive second derivative,[1] so that the velocity at which a distant galaxy is receding from us should be continuously increasing with time.[2]

In 1998 two independent projects, the Supernova Cosmology Project and the High-Z Supernova Search Team simultaneously obtained results suggesting a totally unexpected acceleration in the expansion of the universe by using distant type Ia supernovae as standard candles.[3][4][5] Three members of these two groups have subsequently been awarded Nobel Prizes for their discovery.[6] Confirmatory evidence has been found in baryon acoustic oscillations and other new results about the clustering of galaxies.

21.1   Background

Since Hubble's discovery of the expansion of the universe in 1929,[7] the Big Bang model has become the accepted explanation for the origin of our universe. The Friedmann equation defines how the energy in the universe drives its expansion.

$$H^2 = \left(\frac{\dot{a}}{a}\right)^2 = \frac{8\pi G}{} \quad \frac{Kc^2}{R^2 a^2}$$

where K represents the curvature of the universe, $a(t)$ is the scale factor, ρ is the total energy density of the universe, and H is the Hubble parameter.[8]

We define a critical density

$$\rho_c = \frac{3H^2}{8\pi G}$$

and the density parameter

$$\Omega = \frac{\rho}{\rho_c}$$

We can then rewrite the Hubble parameter as

$$H(a) = H_0 \sqrt{\Omega_k a^{-2} + \Omega_m a^{-3} + \Omega_r a^{-4} + \Omega_{DE} a^{-3(1+w)}}$$

where the four currently hypothesized contributors to the energy density of the universe are curvature, matter, radiation and dark energy.[9] Each of the components decreases with the expansion of the universe (increasing scale factor), except perhaps the dark energy term. It is the values of these cosmological parameters which physicists use to determine the acceleration of the universe.

The acceleration equation describes the evolution of the scale factor with time

$$\ddot{a} \quad \frac{4\pi G}{3} \left(\rho + \frac{3P}{c^2}\right)$$

where the pressure P is defined by the cosmological model chosen. (see explanatory models below)

Physicists at one time were so assured of the deceleration of the universe's expansion that they introduced a so-called deceleration parameter q_0.[10] Current observations point towards this deceleration parameter being negative.

21.2   Evidence for acceleration

To learn about the rate of expansion of the universe we look at the magnitude-redshift relationship of astronomical objects using standard candles, or their distance-redshift relationship using standard rulers. We can also look at the growth of large-scale structure, and find that the observed values of the cosmological parameters are best described by models which include an accelerating expansion.

21.2.1   Supernova observation

Artist's impression of a Type Ia supernovae, as revealed by spectro-polarimetry observations

The first evidence for acceleration came from the observation of Type Ia supernovae, which are exploding white dwarfs who have exceeded their stability limit. Because they all have similar masses, their intrinsic luminosity is standardizable. Repeated imaging of selected areas of sky is used to discover the supernovae, then follow-up observations give their peak brightness, which is converted into a quantity known as luminosity distance (see distance measures in cosmology for details).[11] Spectral lines of their light can be used to determine their redshift.

For supernovae at redshift less than around 0.1, or light travel time less than 10 percent of the age of the universe, this gives a nearly linear distance-redshift relation due to Hubble's law. At larger distances, since the expansion rate of the universe has changed over time, the distance-redshift relation deviates from linearity, and this deviation depends on how the expansion rate has changed over time. The full calculation requires integration of the Friedmann equation, but a simple derivation can be given as follows: the redshift z directly gives the cosmic scale factor at the time the supernova exploded.

$$a(t) = \frac{1}{1+z}$$

So a supernova with a measured redshift z = 0.5 implies the universe was $1/(1+0.5) = 2/3$ of its present size when the supernova exploded. In an accelerating universe, the universe was expanding more slowly in the past than it is today, which means it took a longer time to expand from 2/3 to 1.0 times its present size compared to a non-accelerating universe. This results in a larger light-travel time, larger distance and fainter supernovae, which corresponds to the actual observations. Riess found that "the distances of the high-redshift SNe Ia were, on average, 10% to 15% farther than expected in a low mass density Ω_M = 0.2 universe without a cosmological constant".[12] This means that the measured high-redshift distances were too large, compared to nearby ones, for a decelerating universe.[13]

21.2.2   Baryon acoustic oscillations

Main article: Baryon acoustic oscillations

In the early universe before recombination and decoupling took place, photons and matter existed in a primordial plasma. Points of higher density in the photon-baryon plasma would contract, being compressed by gravity until the pressure became too large and they expanded again.[10] This contraction and expansion created vibrations in the plasma analogous to sound waves. Since dark matter only interacts gravitationally it stayed at the centre of the sound wave, the origin of the original overdensity. When decoupling occurred, approximately 380,000 years after the Big Bang,[14] photons separated from matter and were able to stream freely through the universe, creating the cosmic microwave background as we know it. This left shells of baryonic matter at a fixed radius from the overdensities of dark matter, a distance known as the **sound horizon**. As time passed and the universe expanded, it was at these anisotropies of matter density where galaxies started to form. So by looking at the distances at which galaxies at different redshifts tend to cluster, it is possible to determine a standard angular diameter distance and use that to compare to the distances predicted by different cosmological models.

Peaks have been found in the correlation function (the probability that two galaxies will be a certain distance apart) at $100h^{-1}$ Mpc,[9] indicating that this is the size of the sound horizon today, and by comparing this to the sound horizon at the time of decoupling (using the CMB), we can confirm that the expansion of the universe is accelerating.[15]

21.2.3   Clusters of galaxies

Measuring the mass functions of galaxy clusters, which describe the number density of the clusters above a threshold mass, also provides evidence for dark energy.[16] By comparing these mass functions at high and low redshifts to those predicted by different cosmological models, values for w and Ω_m are obtained which confirm a low matter density and a non zero amount of dark energy.[13]

21.2.4   Age of the universe

See also: Age of the universe

Given a cosmological model with certain values of the cosmological density parameters, it is possible to integrate the Friedmann equations and derive the age of the universe.

$$t_0 = \int_0^1 \frac{da}{a}$$

By comparing this to actual measured values of the cosmological parameters, we can confirm the validity of a model which is accelerating now, and had a slower expansion in the past.[13]

21.3   Explanatory models

21.3.1   Dark energy

Main article: Dark energy

The most important property of dark energy is that it has negative pressure which is distributed relatively homogeneously in space.

$$P = wc^2\rho$$

where c is the speed of light, ρ is the energy density. Different theories of dark energy suggest different values of w, with w < −1/3 for cosmic acceleration (this leads to a positive value of *a* in the acceleration equation above).

The simplest explanation for dark energy is that it is a cosmological constant or vacuum energy; in this case w = −1. This leads to the Lambda-CDM model, which has generally been known as the Standard Model of Cosmology from 2003 through the present, since it is the simplest model in good agreement with a variety of recent observations. Riess found that their results from supernovae observations favoured expanding models with positive cosmological constant ($\Omega_\lambda > 0$) and a current acceleration of the expansion ($q_0 < 0$).[12]

21.3.2   Phantom energy

Main article: Phantom energy

Current observations allow the possibility of a cosmological model containing a dark energy component with equation of state:

$$w < -1$$

This phantom energy density would become infinite in finite time, causing such a huge gravitational repulsion that the universe would lose all structure and end in a Big Rip.[17] For example, for w = −3/2 and H_0 = 70 km·s^{-1}·Mpc^{-1}, the time remaining before the universe ends in this "Big Rip" is 22 billion years.[18]

21.3.3   Alternative theories

Other explanations for the accelerating universe include quintessence, a proposed form of dark energy with a non-constant state equation, whose density decreases with time. Dark fluid is an alternative explanation for accelerating expansion which attempts to unite dark matter and dark energy into a single framework.[19] Alternatively, some authors have argued that the universe expansion acceleration could be due to a repulsive gravitational interaction of antimatter.[20][21][22]

Another type of model, the backreaction conjecture,[23][24] was proposed by cosmologist Syksy Räsänen: the rate of expansion is not homogenous, but we are coincidentally in a region where expansion is faster than the background. In this model, inhomogeneities in the early universe cause the formation of walls and bubbles, where the inside of a bubble has less matter than on average. According to general relativity, space is less curved than on the walls, and thus appears to have more volume and a higher expansion rate. If we live inside such a bubble, then it would appear that the universe is expanding at an accelerating rate.[25] The benefit is that it does not require any new physics such as dark energy. Räsänen does not consider the model likely, but without any falsification, it must remain a possibility.

21.4   Theories for the consequences to the universe

See also: Future of an expanding universe

As the universe expands, the density of radiation and ordinary and dark matter declines more quickly than the density of dark energy (see equation of state) and, eventually, dark energy dominates. Specifically, when the scale of the universe doubles, the density of matter is reduced by a factor of 8, but the density of dark energy is nearly unchanged (it is exactly constant if the dark energy is a cosmological constant).[10]

In models where dark energy is a cosmological constant, the universe will expand exponentially with time from now on, coming closer and closer to a de Sitter spacetime. This will eventually lead to all evidence for the Big Bang disappearing, as the cosmic microwave background is redshifted to lower intensities and longer wavelengths. Eventually its frequency will be small enough that it will be absorbed by the interstellar medium, and so be screened from any observer within the galaxy. This will occur when the universe is less than 50 times its current age, leading to the end of cosmology as we know it as the distant universe turns dark.[26]

Alternatives for the ultimate fate of the universe include the Big Rip mentioned above, a Big Bounce, Big Freeze, or Big Crunch.

21.5   See also

- Metric expansion of space
- High-z Supernova Search Team
- Supernova Cosmology Project
- List of multiple discoveries
- Cosmological constant
- Scale factor (cosmology)
- Friedmann–Lemaître–Robertson–Walker metric

21.6   References

[1] Jones, Mark H.; Robert J. Lambourne (2004). *An Introduction to Galaxies and Cosmology.* Cambridge University Press. p. 244. ISBN 978-0-521-83738-5.

[2] Is the universe expanding faster than the speed of light? (see final paragraph)

[3] "Nobel physics prize honours accelerating universe find". BBC News. October 4, 2011.

[4] "The Nobel Prize in Physics 2011". Nobelprize.org. Retrieved 2011-10-06.

[5] Peebles, P. J. E. and Ratra, Bharat (2003). "The cosmological constant and dark energy". *Reviews of Modern Physics* 75 (2): 559–606. arXiv:astro-ph/0207347. Bibcode:2003RvMP...75..559P. doi:10.1103/RevModPhys.75.559.

[6] *Cosmology,* Steven Weinberg, Oxford University Press, 2008

[7] Hubble, Edwin (1929). "A relation between distance and radial velocity among extra-galactic nebulae". *PNAS* 15 (3): 168–173. Bibcode:1929PNAS...15..168H. doi:10.1073/pnas.15.3.168. PMC 522427. PMID 16577160.

[8] Nemiroff, Robert J.; Patla, Bijunath. "Adventures in Friedmann cosmology: A detailed expansion of the cosmological Friedmann equations". *American Journal of Physics* 76 (3): 265. arXiv:astro-ph/0703739. Bibcode:2008AmJPh..76..265N. doi:10.1119/1.2830536.

[9] Lapuente, P.. "Baryon Acoustic Oscillations." Dark energy: observational and theoretical approaches. Cambridge, UK: Cambridge University Press, 2010.

[10] Ryden, Barbara. "Introduction to Cosmology." Physics Today: 77. Print.

[11] Albrecht, A., Bernstein, G., Cahn, R., et al. Report of the Dark Energy TaskForce. ArXiv Astrophysics e-prints, September 2006.

[12] Riess, Adam G.; Filippenko, Alexei V.; Challis, Peter; Clocchiatti, Alejandro; Diercks, Alan; Garnavich, Peter M.; Gilliland, Ron L.; Hogan, Craig J.; Jha, Saurabh; Kirshner, Robert P.; Leibundgut, B.; Phillips, M. M.; Reiss, David; Schmidt, Brian P.; Schommer, Robert A.; Smith, R. Chris; Spyromilio, J.; Stubbs, Christopher; Suntzeff, Nicholas B.; Tonry, John. "Observational Evidence from Supernovae for an Accelerating Universe and a Cosmological Constant". *The Astronomical Journal* 116 (3): 1009–1038. arXiv:astro-ph/9805201. Bibcode:1998AJ....116.1009R. doi:10.1086/300499.

[13] Pain, Reynald. "Observational evidence of the accelerated expansion of the Universe." Comptes Rendus Physique: 521-538.

[14] Hinshaw, G. (2014). "Five-Year Wilkinson Microwave Anisotropy Probe (WMAP) Observations: Data Processing, Sky Maps, and Basic Results". *Astrophysical Journal Supplement* **180**: 225–245. arXiv:0803.0732. Bibcode:2009ApJS..180..225H. doi:10.1088/0067-0049/180/2/225.

[15] Eisenstein, Daniel J.; Zehavi, Idit; Hogg, David W.; Scoccimarro, Roman; Blanton, Michael R.; Nichol, Robert C.; Scranton, Ryan; Seo, Hee-Jong; Tegmark, Max; Zheng, Zheng; Anderson, Scott F.; Annis, Jim; Bahcall, Neta; Brinkmann, Jon; Burles, Scott; Castander, Francisco J.; Connolly, Andrew; Csabai, Istvan; Doi, Mamoru; Fukugita, Masataka; Frieman, Joshua A.; Glazebrook, Karl; Gunn, James E.; Hendry, John S.; Hennessy, Gregory; Ivezić, Zeljko; Kent, Stephen; Knapp, Gillian R.; Lin, Huan; Loh, Yeong-Shang; Lupton, Robert H.; Margon, Bruce; McKay, Timothy A.; Meiksin, Avery; Munn, Jeffery A.; Pope, Adrian; Richmond, Michael W.; Schlegel, David; Schneider, Donald P.; Shimasaku, Kazuhiro; Stoughton, Christopher; Strauss, Michael A.; SubbaRao, Mark; Szalay, Alexander S.; Szapudi, Istvan; Tucker, Douglas L.; Yanny, Brian; York, Donald G. (10 November 2005). "Detection of the Baryon Acoustic Peak in the Large-Scale Correlation Function of SDSS Luminous Red Galaxies". *The Astrophysical Journal* **633** (2): 560–574. arXiv:astro-ph/0501171. Bibcode:2005ApJ...633..560E. doi:10.1086/466512.

[16] Dekel, Avishai. Formation of structure in the Universe. New York: Cambridge University Press, 1999.

[17] Caldwell, Robert; Kamionkowski, Marc; Weinberg, Nevin (August 2003). "Phantom Energy: Dark Energy with w-1 Causes a Cosmic Doomsday". *Physical Review Letters* **91** (7): 071301. arXiv:astro-ph/0302506. Bibcode:2003PhRvL..91g1301C. doi:10.1103/PhysRevLett.91.071301. PMID 12935004.

[18] Caldwell, R.R. "A phantom menace? Cosmological consequences of a dark energy component with super-negative equation of state". *Physics Letters B* **545** (1-2): 23–29. arXiv:astro-ph/9908168. Bibcode:2002PhLB..545...23C. doi:10.1016/S0370-2693(02)02589-3.

[19] Anaelle Halle, HongSheng Zhao, Baojiu Li (2008) "Perturbations in a non-uniform dark energy fluid: equations reveal effects of modified gravity and dark matter "

[20] A. Benoit-Lévy and G. Chardin, Introducing the Dirac-Milne universe, Astronomy and Astrophysics 537, A78 (2012)

[21] D.S. Hajdukovic, Quantum vacuum and virtual gravitational dipoles: the solution to the dark energy problem?, Astrophysics and Space Science 339(1), 1-–5 (2012)

[22] M. Villata, On the nature of dark energy: the lattice Universe, 2013, Astrophysics and Space Science 345, 1. Also available here

[23] "Backreaction: directions of progress". *Classical and Quantum Gravity* **28**: 164008. arXiv:1102.0408. Bibcode:2011CQGra..28p4008R. doi:10.1088/0264-9381/28/16/164008.

[24] "Backreaction in Late-Time Cosmology". *Annual Review of Nuclear and Particle Science* **62**: 57–79. arXiv:1112.5335. Bibcode:2012ARNPS..62...57B. doi:10.1146/annurev.nucl.012809.104435.

[25] http://www.space.com/23025-doctor-who-tardis-regions-universe.html

[26] Krauss, Lawrence M.; Scherrer, Robert J. (28 June 2007). "The return of a static universe and the end of cosmology". *General Relativity and Gravitation* **39** (10): 1545–1550. arXiv:0704.0221. Bibcode:2007GReGr..39.1545K. doi:10.1007/s10714-007-0472-9.

Chapter 22

Physical law

This article is about the philosophy of scientific laws. For the scientific and mathematical aspects, see Laws of science.

A **physical law** or **scientific law** "is a theoretical principle deduced from particular facts, applicable to a defined group or class of phenomena, and expressible by the statement that a particular phenomenon always occurs if certain conditions be present."[1] Physical laws are typically conclusions based on repeated scientific experiments and observations over many years and which have become accepted universally within the scientific community. The production of a summary description of our environment in the form of such laws is a fundamental aim of science. These terms are not used the same way by all authors.

The distinction between natural law in the political-legal sense and law of nature or physical law in the scientific sense is a modern one, both concepts being equally derived from *physis*, the Greek word (translated into Latin as *natura*) for *nature*.[2]

22.1   Description

Several general properties of physical laws have been identified (see Davies (1992) and Feynman (1965) as noted, although each of the characterizations are not necessarily original to them). Physical laws are:

- True, at least within their regime of validity. By definition, there have never been repeatable contradicting observations.

- Universal. They appear to apply everywhere in the universe. (Davies, 1992:82)

- Simple. They are typically expressed in terms of a single mathematical equation. (Davies)

- Absolute. Nothing in the universe appears to affect them. (Davies, 1992:82)

- Stable. Unchanged since first discovered (although they may have been shown to be approximations of more accurate laws—see "Laws as approximations" below),

- Omnipotent. Everything in the universe apparently must comply with them (according to observations). (Davies, 1992:83)

- Generally conservative of quantity. (Feynman, 1965:59)

- Often expressions of existing homogeneities (symmetries) of space and time. (Feynman)

- Typically theoretically reversible in time (if non-quantum), although time itself is irreversible. (Feynman)

Physical laws are distinguished from scientific theories by their simplicity. Scientific theories are generally more complex than laws; they have many component parts, and are more likely to be changed as the body of available experimental data and analysis develops. This is because a physical law is a summary observation of strictly empirical matters, whereas a theory is a model that accounts for the observation, explains it, relates it to other observations, and makes testable predictions based upon it. Simply stated, while a law notes *that* something happens, a theory explains *why* and *how* something happens.

22.2   Examples

Main article: List of laws in science
See also: scientific laws named after people

Some of the more famous laws of nature are found in Isaac Newton's theories of (now) classical mechanics, presented in his *Philosophiae Naturalis Principia Mathematica*, and in Albert Einstein's theory of relativity. Other examples of laws of nature include Boyle's law of gases, conservation laws, the four laws of thermodynamics, etc.

22.3   Laws as definitions

Some "scientific laws" appear to be mathematical definitions (e.g., Newton's Second law $F = {}^{dp}/$, or the uncertainty principle, or the principle of least action, or causality). While these "scientific laws" explain what our senses perceive, they are still empirical and, thus, they are not "mathematical" facts. (Reference to a "law" often suggests a "fact", although "facts" do not exist scientifically *a priori*.)

22.4   Laws being consequences of mathematical symmetries

Other laws reflect mathematical symmetries found in Nature (say, Pauli exclusion principle reflects identity of electrons, conservation laws reflect homogeneity of space, time, Lorentz transformations reflect rotational symmetry of space–time). Laws are constantly being checked experimentally to higher and higher degrees of precision. This is one of the main goals of science. The fact that laws have never been seen to be violated does not preclude testing them at increased accuracy or new kinds of conditions to confirm whether they continue to hold, or whether they break, and what can be discovered in the process. It is always possible for laws to be invalidated or proven to have limitations, by repeatable experimental evidence; should any be seen. However, fundamental changes to the laws are extremely unlikely, since this would imply a change to experimental facts they were derived from in the first place.

Well-established laws have indeed been invalidated in some special cases, but the new formulations created to explain the discrepancies can be said to generalize upon, rather than overthrow, the originals. That is, the invalidated laws have been found to be only close approximations (see below), to which other terms or factors must be added to cover previously unaccounted-for conditions, e.g., very large or very small scales of time or space, enormous speeds or masses, etc. Thus, rather than unchanging knowledge, physical laws are better viewed as a series of improving and more precise generalizations.

22.5   Laws as approximations

Some laws are only approximations of other more general laws, and are good approximations with a restricted domain of applicability. For example, Newtonian dynamics (which is based on Galilean transformations) is the low speed limit of special relativity (since the Galilean transformation is the low-speed approximation to the Lorentz transformation). Similarly, the Newtonian gravitation law is a low-mass approximation of general relativity, and Coulomb's law is an approximation to Quantum Electrodynamics at large distances (compared to the range of weak interactions). In such cases it is common to use the simpler, approximate versions of the laws, instead of the more accurate general laws.

22.6   Physical laws derived from symmetry principles

Many fundamental physical laws are mathematical consequences of various symmetries of space, time, or other aspects of nature. Specifically, Noether's theorem connects some conservation laws to certain symmetries. For example, conservation of energy is a consequence of the shift symmetry of time (no moment of time is different from any other), while conservation of momentum is a consequence of the symmetry (homogeneity) of space (no place in space is special, or different than any other). The indistinguishability of all particles of each fundamental type (say, electrons, or photons) results in the Dirac and Bose quantum statistics which in turn result in the Pauli exclusion principle for fermions and in Bose–Einstein condensation for bosons. The rotational symmetry between time and space coordinate axes (when one is taken as imaginary, another as real) results in Lorentz transformations which in turn result in special relativity theory. Symmetry between inertial and gravitational mass results in general relativity.

The inverse square law of interactions mediated by massless bosons is the mathematical consequence of the 3-dimensionality of space.

One strategy in the search for the most fundamental laws of nature is to search for the most general mathematical symmetry group that can be applied to the fundamental interactions.

22.7   History

Compared to pre-modern accounts of causality, laws of nature fill the role played by divine causality on the one hand, and accounts such as Plato's theory of forms on the other.

The observation that there are underlying regularities in nature dates to prehistoric times, since the recognition of cause-and-effect relationships is an implicit recognition that there are laws of nature. The recognition of such regularities as independent scientific laws *per se*, though, was limited by their entanglement in animism, and by the attribution of many effects that do not have readily obvious causes—such as meteorological, astronomical and biological phenomena—to the actions of various gods, spirits, supernatural beings, etc. Observation and speculation about nature were intimately bound up with metaphysics and morality.

In Europe, systematic theorizing about nature (*physis*) began with the early Greek philosophers and scientists and continued into the Hellenistic and Roman imperial periods, during which times the intellectual influence of Roman law increasingly became paramount.

> The formula "law of nature" first appears as "a live metaphor" favored by Latin poets Lucretius, Virgil, Ovid, Manilius, in time gaining a firm theoretical presence in the prose treatises of Seneca and Pliny. Why this Roman origin? According to [historian and classicist Daryn] Lehoux's persuasive narrative,[3] the idea was made possible by the pivotal role of codified law and forensic argument in Roman life and culture.
>
> For the Romans . . . the place par excellence where ethics, law, nature, religion and politics overlap is the law court. When we read Seneca's *Natural Questions*, and watch again and again just how he applies standards of evidence, witness evaluation, argument and proof, we can recognize that we are reading one of the great Roman rhetoricians of the age, thoroughly immersed in forensic method. And not Seneca alone. Legal models of scientific judgment turn up all over the place, and for example prove equally integral to Ptolemy's approach to verification, where the mind is assigned the role of magistrate, the senses that of disclosure of evidence, and dialectical reason that of the law itself.[4]

The precise formulation of what are now recognized as modern and valid statements of the laws of nature dates from the 17th century in Europe, with the beginning of accurate experimentation and development of advanced form of mathematics.. The modern scientific method which took shape at this time (with Francis Bacon and Galileo) aimed at total separation of science from theology, with minimal speculation about metaphysics and ethics. Natural law in the political sense, conceived as universal (i.e., divorced from sectarian religion and accidents of place), was also elaborated in this period (by Grotius, Spinoza, and Hobbes, to name a few).

22.8 Other fields

Some mathematical theorems and axioms are referred to as laws because they provide logical foundation to empirical laws.

Examples of other observed phenomena sometimes described as laws include the Titius-Bode law of planetary positions, Zipf's law of linguistics, Moore's law of technological growth. Many of these laws fall within the scope of uncomfortable science. Other laws are pragmatic and observational, such as the law of unintended consequences. By analogy, principles in other fields of study are sometimes loosely referred to as "laws". These include Occam's razor as a principle of philosophy and the Pareto principle of economics.

22.9 See also

- Philosophy of science

- Scientific method

- Inductive reasoning

- Physical constant

- Laws of science

22.10 Notes

[1] "Law of Nature". *Oxford English Dictionary* (3rd ed.). Oxford University Press. September 2005.

[2] Some modern philosophers, e.g. Norman Swartz, use "physical law" to mean the laws of nature as they truly are and not as they are inferred by scientists. See Norman Swartz, *The Concept of Physical Law* (New York: Cambridge University Press), 1985. Second edition available online .

[3] in Daryn Lehoux, *What Did the Romans Know? An Inquiry into Science and Worldmaking* (Chicago: University of Chicago Press, 2012), reviewed by David Sedley, "When Nature Got its Laws", *Times Literary Supplement* (October 12, 2012).

[4] Sedley, "When Nature Got Its Laws", *Times Literary Supplement* (October 12, 2012).

22.11 References

- Francis Bacon (1620). *Novum Organum*.

- John Barrow (1991). *Theories of Everything: The Quest for Ultimate Explanations*. (ISBN 0-449-90738-4)

- Davies, Paul (1992) *The Mind of God*. (ISBN 0-671-79718-2)

- Feynman, Richard (1965) *The Character of Physical Law*. (ISBN 0-679-60127-9)

- Daryn Lehoux (2012). *What Did the Romans Know? An Inquiry into Science and Worldmaking*. University of Chicago Press. (ISBN 9780226471143)

22.12   External links

- Stanford Encyclopedia of Philosophy: "Laws of Nature" by John W. Carroll.

- Baaquie, Belal E. "Laws of Physics : "A Primer". Core Curriculum, National University of Singapore.

- Francis, Erik Max. "The laws list".. Physics. Alcyone Systems

- Pazameta, Zoran. "The laws of nature". Committee for the scientific investigation of Claims of the Paranormal.

- The Internet Encyclopedia of Philosophy. "Laws of Nature" – By Norman Swartz

Chapter 23

Dark energy

Not to be confused with Dark flow, Dark fluid, or Dark matter.

In physical cosmology and astronomy, **dark energy** is an unknown form of energy which is hypothesized to permeate all of space, tending to accelerate the expansion of the universe.[1] Dark energy is the most accepted hypothesis to explain the observations since the 1990s indicating that the universe is expanding at an accelerating rate. According to the Planck mission team, and based on the standard model of cosmology, on a mass–energy equivalence basis, the observable universe contains 26.8% dark matter, 68.3% dark energy (for a total of 95.1%) and 4.9% ordinary matter.[2][3][4][5] Again on a mass–energy equivalence basis, the density of dark energy (6.91×10^{-27} kg/m^3) is very low, much less than the density of ordinary matter or dark matter within galaxies. However, it comes to dominate the mass–energy of the universe because it is uniform across space.[6][7]

Two proposed forms for dark energy are the cosmological constant, a *constant* energy density filling space homogeneously,[8] and scalar fields such as quintessence or moduli, *dynamic* quantities whose energy density can vary in time and space. Contributions from scalar fields that are constant in space are usually also included in the cosmological constant. The cosmological constant can be formulated to be equivalent to vacuum energy. Scalar fields that do change in space can be difficult to distinguish from a cosmological constant because the change may be extremely slow.

High-precision measurements of the expansion of the universe are required to understand how the expansion rate changes over time and space. In general relativity, the evolution of the expansion rate is parameterized by the cosmological equation of state (the relationship between temperature, pressure, and combined matter, energy, and vacuum energy density for any region of space). Measuring the equation of state for dark energy is one of the biggest efforts in observational cosmology today.

Adding the cosmological constant to cosmology's standard FLRW metric leads to the Lambda-CDM model, which has been referred to as the "*standard model of cosmology*" because of its precise agreement with observations. Dark energy has been used as a crucial ingredient in a recent attempt to formulate a cyclic model for the universe.[9]

23.1 Nature of dark energy

Many things about the nature of dark energy remain matters of speculation. The evidence for dark energy is indirect but comes from three independent sources:

- Distance measurements and their relation to redshift, which suggest the universe has expanded more in the last half of its life.[10]

- The theoretical need for a type of additional energy that is not matter or dark matter to form the observationally flat universe (absence of any detectable global curvature).

- It can be inferred from measures of large scale wave-patterns of mass density in the universe.

Dark energy is thought to be very homogeneous, not very dense and is not known to interact through any of the fundamental forces other than gravity. Since it is quite rarefied—roughly 10^{-30} g/cm^3—it is unlikely to be detectable in laboratory experiments. Dark energy can have such a profound effect on the universe, making up 68% of universal density, only because it uniformly fills otherwise empty space. The two leading models are a cosmological constant and quintessence. Both models include the common characteristic that dark energy must have negative pressure.

23.1.1   Eff ect of dark energy: a small constant negative pressure of vacuum

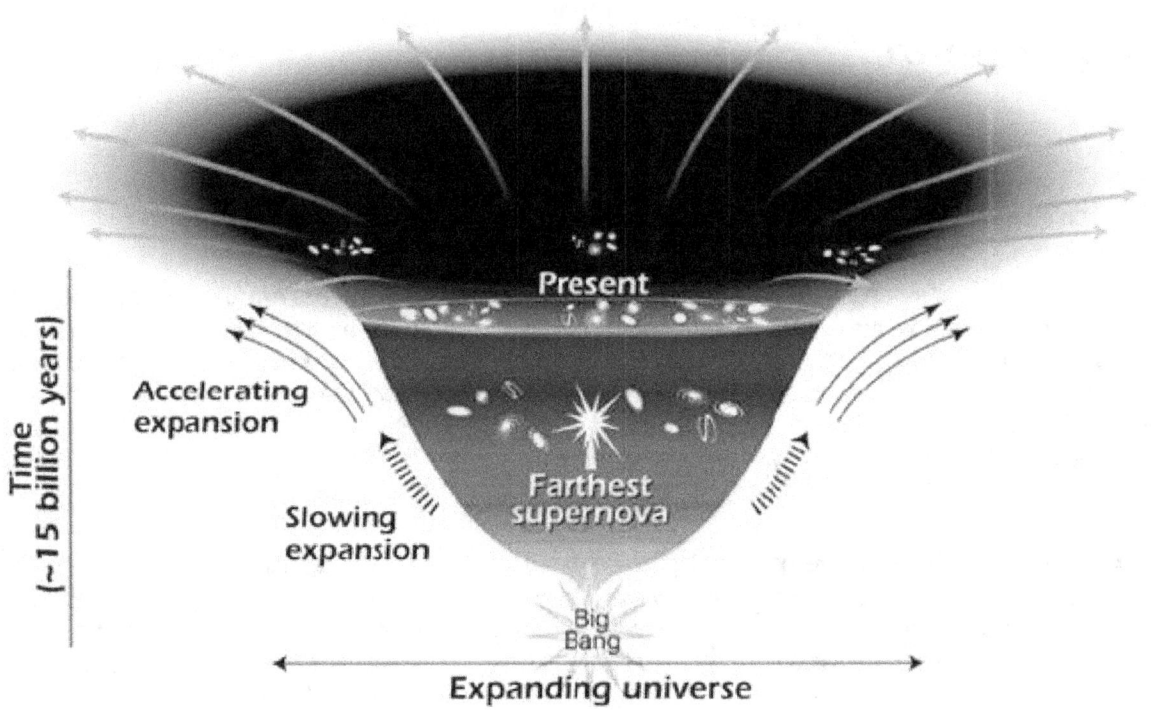

This diagram reveals changes in the rate of expansion since the universe's birth 15 billion years ago. The more shallow the curve, the faster the rate of expansion. The curve changes noticeably about 7.5 billion years ago, when objects in the universe began flying apart at a faster rate. Astronomers theorize that the faster expansion rate is due to a mysterious, dark force that is pushing galaxies apart.

Diagram representing the accelerated expansion of the universe due to dark energy.

Independently of its actual nature, dark energy would need to have a strong negative pressure (acting repulsively) in order to explain the observed acceleration of the expansion of the universe.

According to general relativity, the pressure within a substance contributes to its gravitational attraction for other things just as its mass density does. This happens because the physical quantity that causes matter to generate gravitational effects is the stress–energy tensor, which contains both the energy (or matter) density of a substance and its pressure and viscosity.

In the Friedmann–Lemaître–Robertson–Walker metric, it can be shown that a strong constant negative pressure in all the universe causes an acceleration in universe expansion if the universe is already expanding, or a deceleration in universe contraction if the universe is already contracting. More exactly, the second derivative of the universe scale factor, a, is positive if the equation of state of the universe is such that $w < -1/3$ (see Friedmann equations).

This accelerating expansion effect is sometimes labeled "gravitational repulsion", which is a colorful but possibly confusing expression. In fact a negative pressure does not influence the gravitational interaction between masses—which remains attractive—but rather alters the overall evolution of the universe at the cosmological scale, typically resulting in the accelerating expansion of the universe despite the attraction among the masses present in the universe.

The acceleration is simply a function of dark energy density. Dark energy is persistent: its density remains constant (experimentally, within a factor of 1:10), i.e. it does not get diluted when space expands.

23.2 Evidence of existence

23.2.1 Supernovae

A Type Ia supernova (bright spot on the bottom-left) near a galaxy

In 1998, published observations of Type Ia supernovae ("one-A") by the High-Z Supernova Search Team[11] followed in

1999 by the Supernova Cosmology Project[12] suggested that the expansion of the universe is accelerating.[13] The 2011 Nobel Prize in Physics was awarded to Saul Perlmutter, Brian P. Schmidt and Adam G. Riess for their leadership in the discovery.[14][15]

Since then, these observations have been corroborated by several independent sources. Measurements of the cosmic microwave background, gravitational lensing, and the large-scale structure of the cosmos as well as improved measurements of supernovae have been consistent with the Lambda-CDM model.[16] Some people argue that the only indication for the existence of dark energy is observations of distance measurements and associated redshifts. Cosmic microwave background anisotropies and baryon acoustic oscillations are only observations that distances to a given redshift are larger than expected from a "dusty" Friedmann–Lemaître universe and the local measured Hubble constant.[17]

Supernovae are useful for cosmology because they are excellent standard candles across cosmological distances. They allow the expansion history of the universe to be measured by looking at the relationship between the distance to an object and its redshift, which gives how fast it is receding from us. The relationship is roughly linear, according to Hubble's law. It is relatively easy to measure redshift, but finding the distance to an object is more difficult. Usually, astronomers use standard candles: objects for which the intrinsic brightness, the absolute magnitude, is known. This allows the object's distance to be measured from its actual observed brightness, or apparent magnitude. Type Ia supernovae are the best-known standard candles across cosmological distances because of their extreme and consistent luminosity.

Recent observations of supernovae are consistent with a universe made up 71.3% of dark energy and 27.4% of a combination of dark matter and baryonic matter.[18]

23.2.2   Cosmic microwave background

The existence of dark energy, in whatever form, is needed to reconcile the measured geometry of space with the total amount of matter in the universe. Measurements of cosmic microwave background (CMB) anisotropies indicate that the universe is close to flat. For the shape of the universe to be flat, the mass/energy density of the universe must be equal to the critical density. The total amount of matter in the universe (including baryons and dark matter), as measured from the CMB spectrum, accounts for only about 30% of the critical density. This implies the existence of an additional form of energy to account for the remaining 70%.[16] The Wilkinson Microwave Anisotropy Probe (WMAP) spacecraft seven-year analysis estimated a universe made up of 72.8% dark energy, 22.7% dark matter and 4.5% ordinary matter.[4] Work done in 2013 based on the Planck spacecraft observations of the CMB gave a more accurate estimate of 68.3% of dark energy, 26.8% of dark matter and 4.9% of ordinary matter.[20]

23.2.3   Large-scale structure

The theory of large-scale structure, which governs the formation of structures in the universe (stars, quasars, galaxies and galaxy groups and clusters), also suggests that the density of matter in the universe is only 30% of the critical density.

A 2011 survey, the WiggleZ galaxy survey of more than 200,000 galaxies, provided further evidence towards the existence of dark energy, although the exact physics behind it remains unknown.[21][22] The WiggleZ survey from Australian Astronomical Observatory scanned the galaxies to determine their redshift. Then, by exploiting the fact that baryon acoustic oscillations have left voids regularly of ~150 Mpc diameter, surrounded by the galaxies, the voids were used as standard rulers to determine distances to galaxies as far as 2,000 Mpc (redshift 0.6), which allowed astronomers to determine more accurately the speeds of the galaxies from their redshift and distance. The data confirmed cosmic acceleration up to half of the age of the universe (7 billion years) and constrain its inhomogeneity to 1 part in 10.[22] This provides a confirmation to cosmic acceleration independent of supernovae.

23.2.4   Late-time integrated Sachs-Wolfe effect

Accelerated cosmic expansion causes gravitational potential wells and hills to flatten as photons pass through them, producing cold spots and hot spots on the CMB aligned with vast supervoids and superclusters. This so-called late-time Integrated Sachs–Wolfe effect (ISW) is a direct signal of dark energy in a flat universe.[23] It was reported at high significance in 2008 by Ho *et al.*[24] and Giannantonio *et al.*[25]

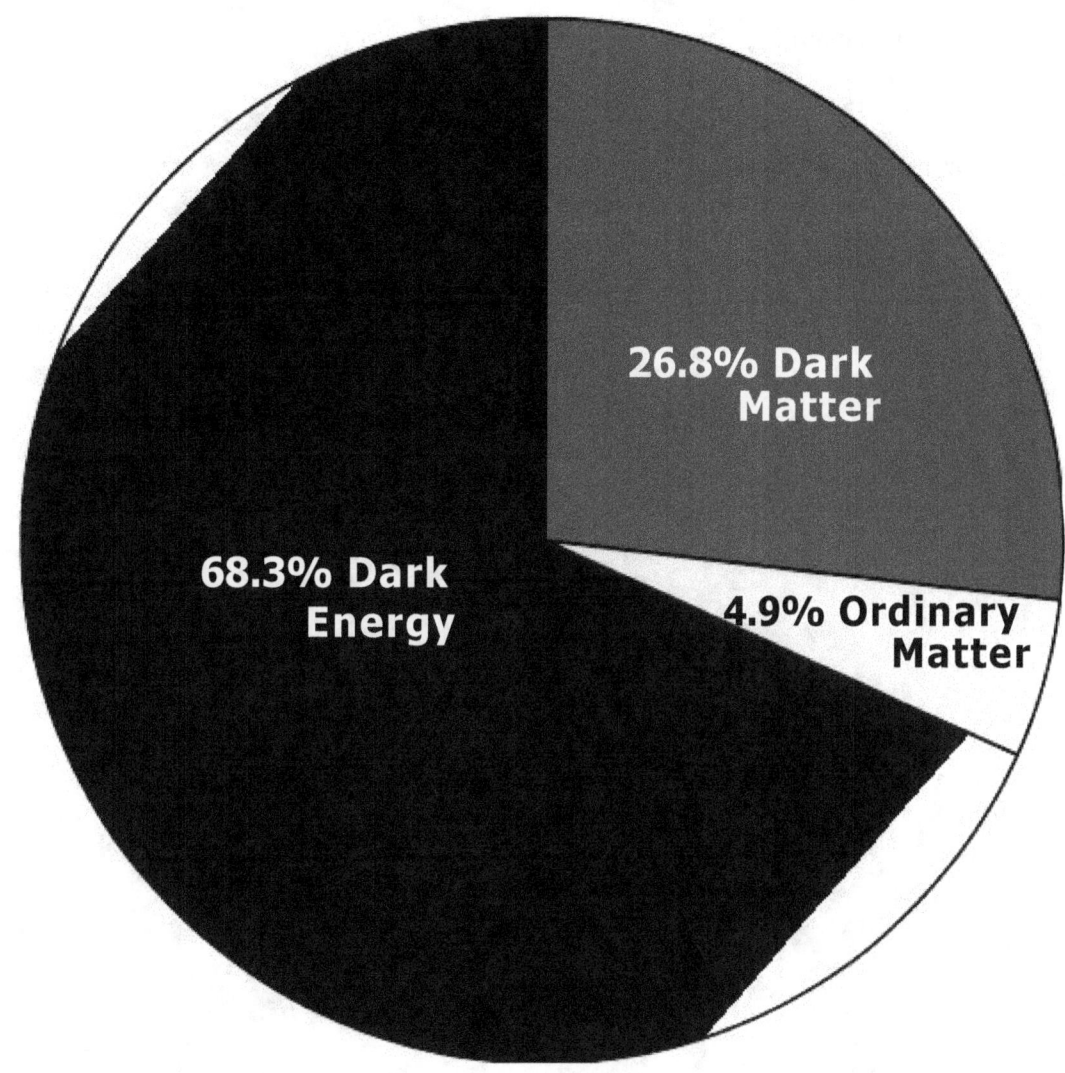

Estimated distribution of matter and energy in the universe[19]

23.2.5   Observational Hubble constant data

A new approach to test evidence of dark energy through observational Hubble constant (H(z)) data (OHD) has gained significant attention in recent years.[26][27][28][29] The Hubble constant is measured as a function of cosmological redshift. OHD directly tracks the expansion history of the universe by taking passively evolving early-type galaxies as "cosmic chronometers".[30] From this point, this approach provides standard clocks in the universe. The core of this idea is the measurement of the differential age evolution as a function of redshift of these cosmic chronometers. Thus, it provides a direct estimate of the Hubble parameter $H(z) = -1/(1+z)dz/dt \approx -1/(1+z)\Delta z/\Delta t$. The merit of this approach is clear: the reliance on a differential quantity, $\Delta z/\Delta t$, can minimize many common issues and systematic effects; and as a direct measurement of the Hubble parameter instead of its integral, like supernovae and baryon acoustic oscillations (BAO), it brings more information and is appealing in computation. For these reasons, it has been widely used to examine the accelerated cosmic expansion and study properties of dark energy.

23.3   Theories of explanation

23.3.1   Cosmological constant

Main article: Cosmological constant

For more details on this topic, see Equation of state (cosmology).

 The simplest explanation for dark energy is that it is simply the "cost of having space": that is, a volume of space has

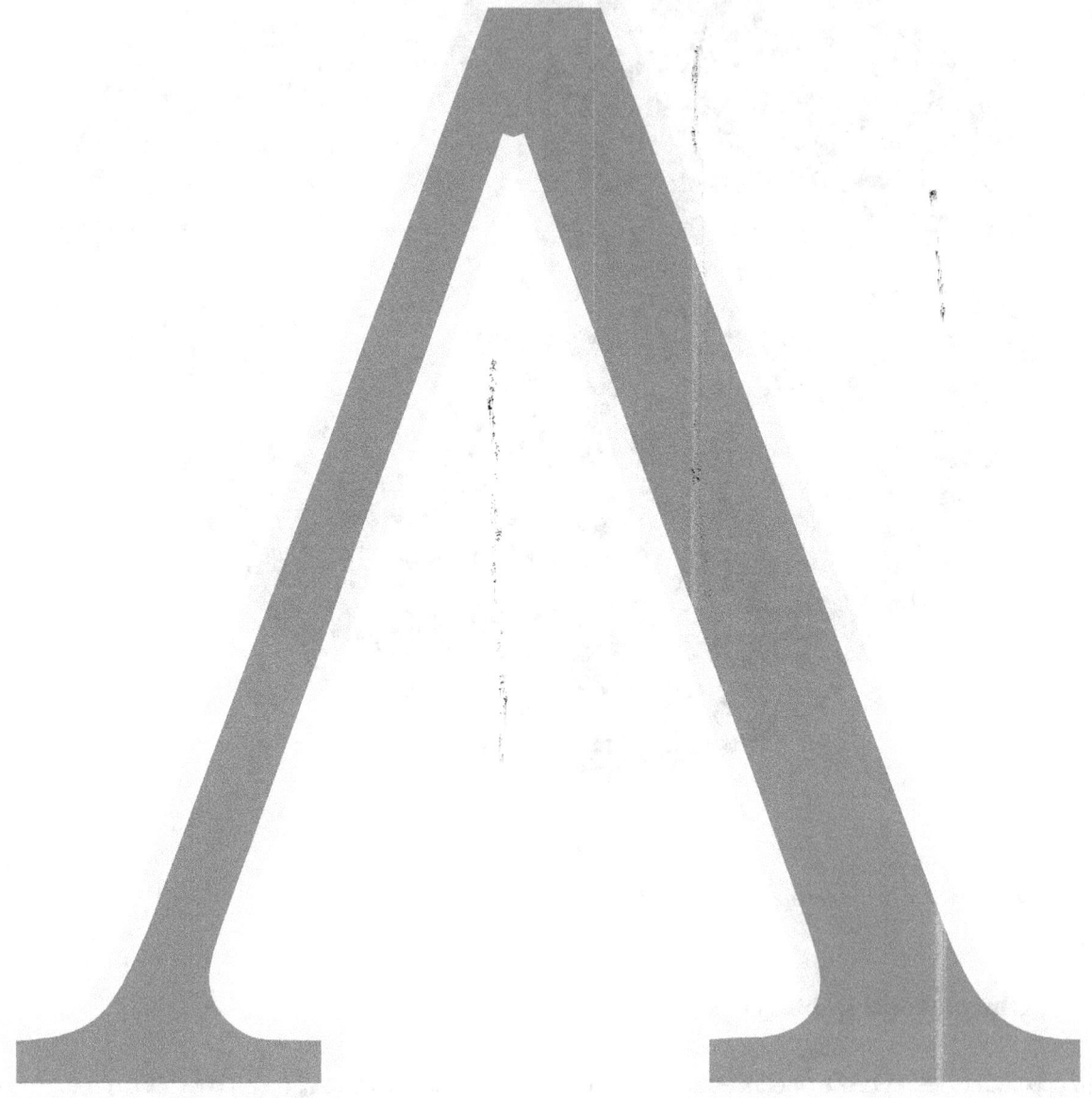

Lambda, the letter that represents the cosmological constant

some intrinsic, fundamental energy. This is the cosmological constant, sometimes called Lambda (hence Lambda-CDM model) after the Greek letter Λ, the symbol used to represent this quantity mathematically. Since energy and mass are related by $E = mc^2$, Einstein's theory of general relativity predicts that this energy will have a gravitational effect. It is sometimes called a vacuum energy because it is the energy density of empty vacuum. In fact, most theories of particle physics predict vacuum fluctuations that would give the vacuum this sort of energy. This is related to the Casimir effect, in which there is a small suction into regions where virtual particles are geometrically inhibited from forming (e.g. between plates with tiny separation). The cosmological constant is estimated by cosmologists to be on the order of 10^{-29} g/cm^3, or about 10^{-120} in reduced Planck units. Particle physics predicts a natural value of 1 in reduced Planck units, leading to a large discrepancy.

The cosmological constant has negative pressure equal to its energy density and so causes the expansion of the universe to accelerate. The reason why a cosmological constant has negative pressure can be seen from classical thermodynamics; Energy must be lost from inside a container to do work on the container. A change in volume dV requires work done equal to a change of energy $-P\,dV$, where P is the pressure. But the amount of energy in a container full of vacuum actually increases when the volume increases (dV is positive), because the energy is equal to ρV, where ρ (rho) is the energy density of the cosmological constant. Therefore, P is negative and, in fact, $P = -\rho$.

A major outstanding problem is that most quantum field theories predict a huge cosmological constant from the energy of the quantum vacuum, more than 100 orders of magnitude too large.[8] This would need to be cancelled almost, but not exactly, by an equally large term of the opposite sign. Some supersymmetric theories require a cosmological constant that is exactly zero,[31] which does not help because supersymmetry must be broken. The present scientific consensus amounts to extrapolating the empirical evidence where it is relevant to predictions, and fine-tuning theories until a more elegant solution is found. Technically, this amounts to checking theories against macroscopic observations. Unfortunately, as the known error-margin in the constant predicts the fate of the universe more than its present state, many such "deeper" questions remain unknown.

In spite of its problems, the cosmological constant is in many respects the most economical solution to the problem of cosmic acceleration. One number successfully explains a multitude of observations. Thus, the current standard model of cosmology, the Lambda-CDM model, includes the cosmological constant as an essential feature.

23.3.2 Quintessence

Main article: Quintessence (physics)

In quintessence models of dark energy, the observed acceleration of the scale factor is caused by the potential energy of a dynamical field, referred to as quintessence field. Quintessence differs from the cosmological constant in that it can vary in space and time. In order for it not to clump and form structure like matter, the field must be very light so that it has a large Compton wavelength.

No evidence of quintessence is yet available, but it has not been ruled out either. It generally predicts a slightly slower acceleration of the expansion of the universe than the cosmological constant. Some scientists think that the best evidence for quintessence would come from violations of Einstein's equivalence principle and variation of the fundamental constants in space or time.[32] Scalar fields are predicted by the *Standard Model of particle physics* and string theory, but an analogous problem to the cosmological constant problem (or the problem of constructing models of cosmic inflation) occurs: renormalization theory predicts that scalar fields should acquire large masses.

The **cosmic coincidence problem** asks why the cosmic acceleration began when it did. If cosmic acceleration began earlier in the universe, structures such as galaxies would never have had time to form and life, at least as we know it, would never have had a chance to exist. Proponents of the anthropic principle view this as support for their arguments. However, many models of quintessence have a so-called **tracker** behavior, which solves this problem. In these models, the quintessence field has a density which closely tracks (but is less than) the radiation density until matter-radiation equality, which triggers quintessence to start behaving as dark energy, eventually dominating the universe. This naturally sets the low energy scale of the dark energy.[33]

In 2004, when scientists fit the evolution of dark energy with the cosmological data, they found that the equation of state had possibly crossed the cosmological constant boundary ($w=-1$) from above to below. A No-Go theorem has been proved that gives this scenario at least two degrees of freedom as required for dark energy models. This scenario is so-called Quintom scenario.

Some special cases of quintessence are phantom energy, in which the energy density of quintessence actually increases with time, and k-essence (short for kinetic quintessence) which has a non-standard form of kinetic energy. They can have unusual properties: phantom energy, for example, can cause a Big Rip.

23.4   Alternative ideas

Some alternatives to dark energy aim to explain the observational data by a more refined use of established theories, focusing, for example, on the gravitational effects of density inhomogeneities, or on consequences of electroweak symmetry breaking in the early universe. If we are located in an emptier-than-average region of space, the observed cosmic expansion rate could be mistaken for a variation in time, or acceleration.[34][35][36][37] A different approach uses a cosmological extension of the equivalence principle to show how space might appear to be expanding more rapidly in the voids surrounding our local cluster. While weak, such effects considered cumulatively over billions of years could become significant, creating the illusion of cosmic acceleration, and making it appear as if we live in a Hubble bubble.[38][39][40]

Another class of theories attempts to come up with an all-encompassing theory of both dark matter and dark energy as a single phenomenon that modifies the laws of gravity at various scales. An example of this type of theory is the theory of dark fluid. Another class of theories that unifies dark matter and dark energy are suggested to be covariant theories of modified gravities. These theories alter the dynamics of the space-time such that the modified dynamic stems what have been assigned to the presence of dark energy and dark matter.[41]

A 2011 paper in the journal *Physical Review D* by Christos Tsagas, a cosmologist at Aristotle University of Thessaloniki in Greece, argued that it is likely that the accelerated expansion of the universe is an illusion caused by the relative motion of us to the rest of the universe. The paper cites data showing that the 2.5 billion ly wide region of space we are inside of is moving very quickly relative to everything around it. If the theory is confirmed, then dark energy would not exist (but the "dark flow" still might).[42][43]

Some theorists think that dark energy and cosmic acceleration are a failure of general relativity on very large scales, larger than superclusters. However most attempts at modifying general relativity have turned out to be either equivalent to theories of quintessence, or inconsistent with observations. Other ideas for dark energy have come from string theory, brane cosmology and the holographic principle, but have not yet proved as compelling as quintessence and the cosmological constant.

On string theory, an article in the journal *Nature* described:[44]

> *String theories, popular with many particle physicists, make it possible, even desirable, to think that the observable universe is just one of 10^{500} universes in a grander multiverse, says Leonard Susskind, a cosmologist at Stanford University in California. The vacuum energy will have different values in different universes, and in many or most it might indeed be vast. But it must be small in ours because it is only in such a universe that observers such as ourselves can evolve.*

Paul Steinhardt in the same article criticizes string theory's explanation of dark energy stating "...Anthropics and randomness don't explain anything... I am disappointed with what most theorists are willing to accept".[44]

Another set of proposals is based on the possibility of a double metric tensor for space-time.[45][46] It has been argued that time reversed solutions in general relativity require such double metric for consistency, and that both dark matter and dark energy can be understood in terms of time reversed solutions of general relativity.[47]

It has been shown that if inertia is assumed to be due to the effect of horizons on Unruh radiation then this predicts galaxy rotation and a cosmic acceleration similar to that observed.[48]

23.4.1   Variable Dark Energy models

In general, the dark energy can be variable. Modern observational data have determined the density of dark energy in the present. Using baryon acoustic oscillations, we can investigate the effect of dark energy in the history of the Universe and we can constrain parameters of the equation of state of dark energy. One of the proposed solutions to get closer to answering the question of dark energy, is to assume that it is variable. To that end, several models have been proposed. One of their most popular models is Chevallier–Polarski–Linder model (CPL).[50][51] Some other common models are, (Barboza & Alcaniz. 2008),[52] (Jassal et al. 2005),[53] (Wetterich. 2004).[54]

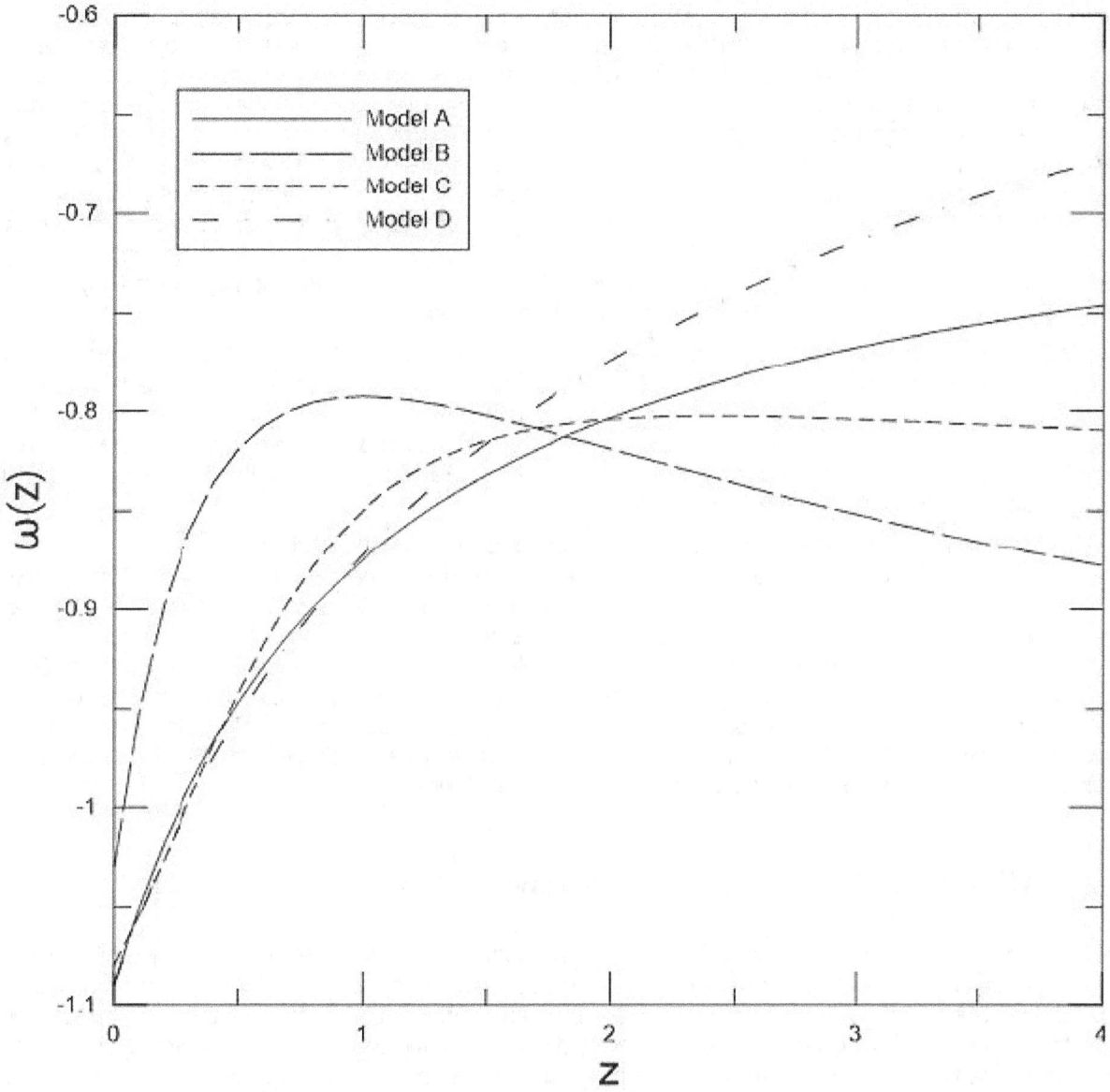

The equation of state of Dark Energy for 4 common models by Redshift.[49]
A: CPL Model,
B: Jassal Model,
C: Barboza & Alcaniz Model,
D: Wetterich Model

23.5 Implications for the fate of the universe

Cosmologists estimate that the acceleration began roughly 5 billion years ago. Before that, it is thought that the expansion was decelerating, due to the attractive influence of dark matter and baryons. The density of dark matter in an expanding universe decreases more quickly than dark energy, and eventually the dark energy dominates. Specifically, when the volume of the universe doubles, the density of dark matter is halved, but the density of dark energy is nearly unchanged (it is exactly constant in the case of a cosmological constant).

If the acceleration continues indefinitely, the ultimate result will be that galaxies outside the local supercluster will have a line-of-sight velocity that continually increases with time, eventually far exceeding the speed of light.[55] This is not a violation of special relativity because the notion of "velocity" used here is different from that of velocity in a local inertial frame of reference, which is still constrained to be less than the speed of light for any massive object (see Uses of the proper distance for a discussion of the subtleties of defining any notion of relative velocity in cosmology). Because the Hubble parameter is decreasing with time, there can actually be cases where a galaxy that is receding from us faster than light does manage to emit a signal which reaches us eventually.[56][57] However, because of the accelerating expansion, it is projected that most galaxies will eventually cross a type of cosmological event horizon where any light they emit past that point will never be able to reach us at any time in the infinite future[58] because the light never reaches a point where its "peculiar velocity" toward us exceeds the expansion velocity away from us (these two notions of velocity are also discussed in Uses of the proper distance). Assuming the dark energy is constant (a cosmological constant), the current distance to this cosmological event horizon is about 16 billion light years, meaning that a signal from an event happening *at present* would eventually be able to reach us in the future if the event were less than 16 billion light years away, but the signal would never reach us if the event were more than 16 billion light years away.[57]

As galaxies approach the point of crossing this cosmological event horizon, the light from them will become more and more redshifted, to the point where the wavelength becomes too large to detect in practice and the galaxies appear to vanish completely[59][60] (*see* Future of an expanding universe). The Earth, the Milky Way, and the Virgo Supercluster, however, would remain virtually undisturbed while the rest of the universe recedes and disappears from view. In this scenario, the local supercluster would ultimately suffer heat death, just as was thought for the flat, matter-dominated universe before measurements of cosmic acceleration.

There are some very speculative ideas about the future of the universe. One suggests that phantom energy causes *divergent* expansion, which would imply that the effective force of dark energy continues growing until it dominates all other forces in the universe. Under this scenario, dark energy would ultimately tear apart all gravitationally bound structures, including galaxies and solar systems, and eventually overcome the electrical and nuclear forces to tear apart atoms themselves, ending the universe in a "Big Rip". On the other hand, dark energy might dissipate with time or even become attractive. Such uncertainties leave open the possibility that gravity might yet rule the day and lead to a universe that contracts in on itself in a "Big Crunch".[61] Some scenarios, such as the cyclic model, suggest this could be the case. It is also possible the universe may never have an end and continue in its present state forever (see The Second Law as a law of disorder). While these ideas are not supported by observations, they are not ruled out.

23.6   History of discovery and previous speculation

The cosmological constant was first proposed by Einstein as a mechanism to obtain a solution of the gravitational field equation that would lead to a static universe, effectively using dark energy to balance gravity.[62] Not only was the mechanism an inelegant example of fine-tuning but it was also later realized that Einstein's static universe would actually be unstable because local inhomogeneities would ultimately lead to either the runaway expansion or contraction of the universe. The equilibrium is unstable: If the universe expands slightly, then the expansion releases vacuum energy, which causes yet more expansion. Likewise, a universe which contracts slightly will continue contracting. These sorts of disturbances are inevitable, due to the uneven distribution of matter throughout the universe. More importantly, observations made by Edwin Hubble in 1929 showed that the universe appears to be expanding and not static at all. Einstein reportedly referred to his failure to predict the idea of a dynamic universe, in contrast to a static universe, as his greatest blunder.[63]

Alan Guth and Alexei Starobinsky proposed in 1980 that a negative pressure field, similar in concept to dark energy, could drive cosmic inflation in the very early universe. Inflation postulates that some repulsive force, qualitatively similar to dark energy, resulted in an enormous and exponential expansion of the universe slightly after the Big Bang. Such expansion is an essential feature of most current models of the Big Bang. However, inflation must have occurred at a much higher energy density than the dark energy we observe today and is thought to have completely ended when the universe was just a fraction of a second old. It is unclear what relation, if any, exists between dark energy and inflation. Even after inflationary models became accepted, the cosmological constant was thought to be irrelevant to the current universe.

Nearly all inflation models predict that the total (matter+energy) density of the universe should be very close to the critical density. During the 1980s, most cosmological research focused on models with critical density in matter only, usually 95% cold dark matter and 5% ordinary matter (baryons). These models were found to be successful at forming realistic galaxies and clusters, but some problems appeared in the late 1980s: notably, the model required a value for the Hubble constant lower than preferred by observations, and the model under-predicted observations of large-scale galaxy clustering. These difficulties became stronger after the discovery of anisotropy in the cosmic microwave background by the COBE spacecraft in 1992, and several modified CDM models came under active study through the mid-1990s: these included the Lambda-CDM model and a mixed cold/hot dark matter model. The first direct evidence for dark energy came from supernova observations in 1998 of accelerated expansion in Riess *et al.*[11] and in Perlmutter *et al.*,[12] and the Lambda-CDM model then became the leading model. Soon after, dark energy was supported by independent observations: in 2000, the BOOMERanG and Maxima cosmic microwave background experiments observed the first acoustic peak in the CMB, showing that the total (matter+energy) density is close to 100% of critical density. Then in 2001, the 2dF Galaxy Redshift Survey gave strong evidence that the matter density is around 30% of critical. The large difference between these two supports a smooth component of dark energy making up the difference. Much more precise measurements from WMAP in 2003–2010 have continued to support the standard model and give more accurate measurements of the key parameters.

The term "dark energy", echoing Fritz Zwicky's "dark matter" from the 1930s, was coined by Michael Turner in 1998.[64]

As of 2013, the Lambda-CDM model is consistent with a series of increasingly rigorous cosmological observations, including the Planck spacecraft and the Supernova Legacy Survey. First results from the SNLS reveal that the average behavior (i.e., equation of state) of dark energy behaves like Einstein's cosmological constant to a precision of 10%.[65] Recent results from the Hubble Space Telescope Higher-Z Team indicate that dark energy has been present for at least 9 billion years and during the period preceding cosmic acceleration.

23.7   See also

- Conformal gravity
- De Sitter relativity
- Illustris project
- The Dark Energy Survey
- Vacuum state

23.8   References

[1] Peebles, P. J. E. and Ratra, Bharat (2003). "The cosmological constant and dark energy". *Reviews of Modern Physics* 75 (2): 559–606. arXiv:astro-ph/0207347. Bibcode:2003RvMP...75..559P. doi:10.1103/RevModPhys.75.559.

[2] Ade, P. A. R.; Aghanim, N.; Armitage-Caplan, C.; et al. (Planck Collaboration), C.; Arnaud, M.; Ashdown, M.; Atrio-Barandela, F.; Aumont, J.; Aussel, H.; Baccigalupi, C.; Banday, A. J.; Barreiro, R. B.; Barrena, R.; Bartelmann, M.; Bartlett, J. G.; Bartolo, N.; Basak, S.; Battaner, E.; Battye, R.; Benabed, K.; Benoît, A.; Benoit-Lévy, A.; Bernard, J.-P.; Bersanelli, M.; Bertincourt, B.; Bethermin, M.; Bielewicz, P.; Bikmaev, I.; Blanchard, A. et al. (22 March 2013). "Planck 2013 results. I. Overview of products and scientific results – Table 9". *Astronomy and Astrophysics* 571: A1. arXiv:1303.5062. Bibcode:2014A&A...571A...1P. doi:10.1051/0004-6361/201321529.

[3] Ade, P. A. R.; Aghanim, N.; Armitage-Caplan, C.; et al. (Planck Collaboration), C.; Arnaud, M.; Ashdown, M.; Atrio-Barandela, F.; Aumont, J.; Aussel, H.; Baccigalupi, C.; Banday, A. J.; Barreiro, R. B.; Barrena, R.; Bartelmann, M.; Bartlett, J. G.; Bartolo, N.; Basak, S.; Battaner, E.; Battye, R.; Benabed, K.; Benoît, A.; Benoit-Lévy, A.; Bernard, J.-P.; Bersanelli, M.; Bertincourt, B.; Bethermin, M.; Bielewicz, P.; Bikmaev, I.; Blanchard, A. et al. (31 March 2013). "Planck 2013 Results Papers". *Astronomy and Astrophysics* 571: A1. arXiv:1303.5062. Bibcode:2014A&A...571A...1P. doi:10.1051/0004-6361/201321529.

[4] "First Planck results: the Universe is still weird and interesting".

[5] Sean Carroll, Ph.D., Cal Tech, 2007, The Teaching Company, *Dark Matter, Dark Energy: The Dark Side of the Universe*, Guidebook Part 2 page 46. Retrieved Oct. 7, 2013, "...dark energy: A smooth, persistent component of invisible energy, thought to make up about 70 percent of the current energy density of the universe. Dark energy is known to be smooth because it doesn't accumulate preferentially in galaxies and clusters..."

[6] "Dark Energy". *Hyperphysics*. Retrieved January 4, 2014.

[7] Ferris, Timothy. "Dark Matter(Dark Energy)". Retrieved 2015-06-10.

[8] Carroll, Sean (2001). "The cosmological constant". *Living Reviews in Relativity* 4. Retrieved 2006-09-28.

[9] Baum, L. and Frampton, P.H. (2007). "Turnaround in Cyclic Cosmology". *Physical Review Letters* 98 (7): 071301. arXiv:hep-th/0610213. Bibcode:2007PhRvL..98g1301B. doi:10.1103/PhysRevLett.98.071301. PMID 17359014.

[10] Durrer, R. (2011). "What do we really know about Dark Energy?". *Philosophical Transactions of the Royal Society A: Mathematical, Physical and Engineering Sciences* 369 (1957): 5102. arXiv:1103.5331. Bibcode:2011RSPTA.369.5102D. doi:10.1098/rsta.2011.0285.

[11] Riess, Adam G.; Filippenko; Challis; Clocchiatti; Diercks; Garnavich; Gilliland; Hogan; Jha; Kirshner; Leibundgut; Phillips; Reiss; Schmidt; Schommer; Smith; Spyromilio; Stubbs; Suntzeff; Tonry (1998). "Observational evidence from supernovae for an accelerating universe and a cosmological constant". *Astronomical J.* 116 (3): 1009-38. arXiv:astro-ph/9805201. Bibcode:1998AJ....116.1009R. doi:10.1086/300499.

[12] Perlmutter, S.; Aldering; Goldhaber; Knop; Nugent; Castro; Deustua; Fabbro; Goobar; Groom; Hook; Kim; Kim; Lee; Nunes; Pain; Pennypacker; Quimby; Lidman; Ellis; Irwin; McMahon; Ruiz-Lapuente; Walton; Schaefer; Boyle; Filippenko; Matheson; Fruchter et al. (1999). "Measurements of Omega and Lambda from 42 high redshift supernovae". *Astrophysical Journal* 517 (2): 565-86. arXiv:astro-ph/9812133. Bibcode:1999ApJ...517..565P. doi:10.1086/307221.

[13] The first paper, using observed data, which claimed a positive Lambda term was Paál, G. et al. (1992). "Inflation and compactification from galaxy redshifts?". *Astrophysics and Space Science* 191: 107-24. Bibcode:1992Ap&SS.191..107P. doi:10.1007/BF00644200.

[14] "The Nobel Prize in Physics 2011". Nobel Foundation. Retrieved 2011-10-04.

[15] The Nobel Prize in Physics 2011. Perlmutter got half the prize, and the other half was shared between Schmidt and Riess.

[16] Spergel, D. N. (WMAP collaboration) et al. (March 2006). "Wilkinson Microwave Anisotropy Probe (WMAP) three year results: implications for cosmology".

[17] Durrer, R. (2011). "What do we really know about dark energy?". *Philosophical Transactions of the Royal Society A* 369 (1957): 5102-5114. arXiv:1103.5331. Bibcode:2011RSPTA.369.5102D. doi:10.1098/rsta.2011.0285.

[18] Kowalski, Marek; Rubin, David; Aldering, G.; Agostinho, R. J.; Amadon, A.; Amanullah, R.; Balland, C.; Barbary, K.; Blanc, G.; Challis, P. J.; Conley, A.; Connolly, N. V.; Covarrubias, R.; Dawson, K. S.; Deustua, S. E.; Ellis, R.; Fabbro, S.; Fadeyev, V.; Fan, X.; Farris, B.; Folatelli, G.; Frye, B. L.; Garavini, G.; Gates, E. L.; Germany, L.; Goldhaber, G.; Goldman, B.; Goobar, A.; Groom, D. E. et al. (October 27, 2008). "Improved Cosmological Constraints from New, Old and Combined Supernova Datasets". *The Astrophysical Journal* (Chicago: University of Chicago Press) 686 (2): 749-778. arXiv:0804.4142. Bibcode:2008ApJ...686..749K. doi:10.1086/589937.. They find a best fit value of the dark energy density, $\Omega\Lambda$ of 0.713+0.027-0.029(stat)+0.036-0.039(sys), of the total matter density, ΩM, of 0.274+0.016-0.016(stat)+0.013-0.012(sys) with an equation of state parameter w of -0.969+0.059-0.063(stat)+0.063-0.066(sys).

[19] "Planck reveals an almost perfect universe". *Planck.* ESA. 2013-03-21. Retrieved 2013-03-21.

[20] "Big Bang's afterglow shows universe is 80 million years older than scientists first thought". *The Washington Post.* Retrieved 22 March 2013.

[21] "New method 'confirms dark energy'". BBC News. 2011-05-19.

[22] Dark energy is real, Swinburne University of Technology, 19 May 2011

[23] Crittenden; Neil Turok (1995). "Looking for Λ with the Rees-Sciama Effect". *Physical Review Letters* 76 (4): 575-578. arXiv:astro-ph/9510072. Bibcode:1996PhRvL..76..575C. doi:10.1103/PhysRevLett.76.575. PMID 10061494.

[24] Shirley Ho; Hirata; Nikhil Padmanabhan; Uros Seljak; Neta Bahcall (2008). "Correlation of CMB with large-scale structure: I. ISW Tomography and Cosmological Implications". *Physical Review D* 78 (4): 043519. arXiv:0801.0642. Bibcode:2008PhRvD..78d3519H. doi:10.1103/PhysRevD.78.043519.

[25] Tommaso Giannantonio; Ryan Scranton; Crittenden; Nichol; Boughn; Myers; Richards (2008). "Combined analysis of the integrated Sachs-Wolfe effect and cosmological implications". *Physical Review D* 77 (12): 123520. arXiv:0801.4380. Bibcode:2008PhRvD..77l3520G. doi:10.1103/PhysRevD.77.123520.

[26] Zelong Yi; Tongjie Zhang (2007). "Constraints on holographic dark energy models using the differential ages of passively evolving galaxies". *Modern Physics Letters A* 22 (1): 41. arXiv:astro-ph/0605596. Bibcode:2007MPLA...22...41Y. doi:10.1142/S0217732307020889.

[27] Haoyi Wan; Zelong Yi; Tongjie Zhang; Jie Zhou (2007). "Constraints on the DGP Universe Using Observational Hubble parameter". *Physics Letters B* 651 (5): 352. arXiv:0706.2723. Bibcode:2007PhLB..651..352W. doi:10.1016/j.physletb.2007.06.053.

[28] Cong Ma; Tongjie Zhang (2010). "Power of Observational Hubble Parameter Data: a Figure of Merit Exploration". *Astrophysical Journal* 730 (2): 74. arXiv:1007.3787. Bibcode:2011ApJ...730...74M. doi:10.1088/0004-637X/730/2/74.

[29] Tongjie Zhang; Cong Ma; Tian Lan (2010). "Constraints on the Dark Side of the Universe and Observational Hubble Parameter Data". *Advances in Astronomy* 2010 (1): 1. arXiv:1010.1307. Bibcode:2010AdAst2010E..81Z. doi:10.1155/2010/184284.

[30] Joan Simon; Licia Verde; Raul Jimenez (2005). "Constraints on the redshift dependence of the dark energy potential". *Physical Review D* 71 (12): 123001. arXiv:astro-ph/0412269. Bibcode:2005PhRvD..71l3001S. doi:10.1103/PhysRevD.71.123001.

[31] Wess, Julius; Bagger, Jonathan. *Supersymmetry and Supergravity*. ISBN 978-0691025308.

[32] Carroll, Sean M. (1998). "Quintessence and the Rest of the World: Suppressing Long-Range Interactions". *Physical Review Letters* 81 (15): 3067–3070. doi:10.1103/PhysRevLett.81.3067. ISSN 0031-9007.

[33] Steinhardt, Paul J.; Wang, Li-Min; Zlatev, Ivaylo. "Cosmological tracking solutions". *Phys. Rev.* D59: 123504. doi:10.1103/PhysRevD.59.123504.

[34] Wiltshire, David L. (2007). "Exact Solution to the Averaging Problem in Cosmology". *Physical Review Letters* 99 (25): 251101. arXiv:0709.0732. Bibcode:2007PhRvL..99y1101W. doi:10.1103/PhysRevLett.99.251101. PMID 18233512.

[35] Ishak, Mustapha; Richardson, James; Garred, David; Whittington, Delilah; Nwankwo, Anthony; Sussman, Roberto (2007). "Dark Energy or Apparent Acceleration Due to a Relativistic Cosmological Model More Complex than FLRW?". *Physical Review D* 78 (12): 123531. arXiv:0708.2943. Bibcode:2008PhRvD..78l3531I. doi:10.1103/PhysRevD.78.123531.

[36] Mattsson, Teppo (2007). "Dark energy as a mirage". *Gen. Rel. Grav.* 42 (3): 567–599. arXiv:0711.4264. Bibcode:2010GReGr..42..567M. doi:10.1007/s10714-009-0873-z.

[37] Clifton, Timothy; Ferreira, Pedro (April 2009). "Does Dark Energy Really Exist?". *Scientific American* 300 (4): 48–55. doi:10.1038/scientificamerican0409-48. PMID 19363920. Retrieved April 30, 2009.

[38] Wiltshire, D. (2008). "Cosmological equivalence principle and the weak-field limit". *Physical Review D* 78 (8): 084032. arXiv:0809.1183. Bibcode:2008PhRvD..78h4032W. doi:10.1103/PhysRevD.78.084032.

[39] Gray, Stuart. "Dark questions remain over dark energy". ABC Science Australia. Retrieved 27 January 2013.

[40] Merali, Zeeya (March 2012). "Is Einstein's Greatest Work All Wrong—Because He Didn't Go Far Enough?". *Discover magazine*. Retrieved 27 January 2013.

[41] Exirifard, Q. (2010). "Phenomenological covariant approach to gravity". *General Relativity and Gravitation* 43: 93–106. arXiv:0808.1962. Bibcode:2011GReGr..43...93E. doi:10.1007/s10714-010-1073-6.

[42] Wolchover, Natalie (27 September 2011) 'Accelerating universe' could be just an illusion, MSNBC

[43] Tsagas, Christos G. (2011). "Peculiar motions, accelerated expansion, and the cosmological axis". *Physical Review D* 84 (6): 063503. arXiv:1107.4045. Bibcode:2011PhRvD..84f3503T. doi:10.1103/PhysRevD.84.063503.

[44] Hogan, Jenny (2007). "Unseen Universe: Welcome to the dark side". *Nature* 448 (7151): 240–245. Bibcode:2007Natur.448..240H. doi:10.1038/448240a. PMID 17637630.

[45] Hossenfelder, S. (2008). "A Bi-Metric Theory with Exchange Symmetry". *Physical Review D* 78 (4): 044015. arXiv:0807.2838. Bibcode:2008PhRvD..78d4015H. doi:10.1103/PhysRevD.78.044015.

[46] Henry-Couannier, F. (2005). "Discrete Symmetries and General Relativity, the Dark Side of Gravity". *International Journal of Modern Physics A* 20 (11): 2341. arXiv:gr-qc/0410055. Bibcode:2005IJMPA..20.2341H. doi:10.1142/S0217751X05024602.

[47] Ripalda, Jose M. (1999). "Time reversal and negative energies in general relativity". arXiv:gr-qc/9906012.

[48] McCulloch, M.E. (2010). "Minimum accelerations from quantised inertia". *EPL* 90 (2): 29001. arXiv:1004.3303. Bibcode:2010EL.....9029001M. doi:10.1209/0295-5075/90/29001.

[49] by Ehsan Sadri M.A Ap

[50] Chevallier, M; Polarski, D (2001). "Accelerating Universes with Scaling Dark Matter". *International Journal of Modern Physics D* 10: 213–224. arXiv:gr-qc/0009008.

[51] Linder, Eric V. (3 March 2003). "Exploring the Expansion History of the Universe". *Physical Review Letters* 90 (9). arXiv:astro-ph/0208512v1. Bibcode:2003PhRvL..90i1301L. doi:10.1103/PhysRevLett.90.091301.

[52] Alcaniz, E.M.; Alcaniz, J.S. (2008). "A parametric model for dark energy". *Physics Letters B* 666: 415–419. arXiv:0805.1713. Bibcode:2008PhLB..666..415B. doi:10.1016/j.physletb.2008.08.012.

[53] Jassal, H.K; Bagla, J.S (2010). "Understanding the origin of CMB constraints on Dark Energy". *Monthly Notices of the Royal Astronomical Society* 405: 2639–2650. arXiv:astro-ph/0601389. Bibcode:2010MNRAS.405.2639J. doi:10.1111/j.1365-2966.2010.16647.x.

[54] Wetterich, C. (2004). "Phenomenological parameterization of quintessence". arXiv:astro-ph/0403289v1.

[55] Krauss, Lawrence M. and Scherrer, Robert J. (March 2008). "The End of Cosmology?". *Scientific American* 82. Retrieved 2011-01-06.

[56] Is the universe expanding faster than the speed of light? (see the last two paragraphs)

[57] Lineweaver, Charles; Tamara M. Davis (2005). "Misconceptions about the Big Bang" (PDF). *Scientific American*. Retrieved 2008-11-06.

[58] Loeb, Abraham (2002). "The Long-Term Future of Extragalactic Astronomy". *Physical Review D* 65 (4): 047301. arXiv:astro-ph/0107568. Bibcode:2002PhRvD..65d7301L. doi:10.1103/PhysRevD.65.047301.

[59] Krauss, Lawrence M.; Robert J. Scherrer (2007). "The Return of a Static Universe and the End of Cosmology". *General Relativity and Gravitation* 39 (10): 1545–1550. arXiv:0704.0221. Bibcode:2007GReGr..39.1545K. doi:10.1007/s10714-007-0472-9.

[60] Using Tiny Particles To Answer Giant Questions. Science Friday, 3 Apr 2009. According to the transcript, Brian Greene makes the comment "And actually, in the far future, everything we now see, except for our local galaxy and a region of galaxies will have disappeared. The entire universe will disappear before our very eyes, and it's one of my arguments for actually funding cosmology. We've got to do it while we have a chance."

[61] *How the Universe Works 3.* End of the Universe. Discovery Channel. 2014.

[62] Harvey, Alex (2012). "How Einstein Discovered Dark Energy". arXiv:1211.6338.

[63] Gamow, George (1970) *My World Line: An Informal Autobiography.* p. 44: "Much later, when I was discussing cosmological problems with Einstein, he remarked that the introduction of the cosmological term was the biggest blunder he ever made in his life." – Here the "cosmological term" refers to the cosmological constant in the equations of general relativity, whose value Einstein initially picked to ensure that his model of the universe would neither expand nor contract; if he hadn't done this he might have theoretically predicted the universal expansion that was first observed by Edwin Hubble.

[64] The first appearance of the term "dark energy" is in the article with another cosmologist and Turner's student at the time, Dragan Huterer, "Prospects for Probing the Dark Energy via Supernova Distance Measurements", which was posted to the ArXiv.org e-print archive in August 1998 and published in Huterer, D.; Turner, M. (1999). "Prospects for probing the dark energy via supernova distance measurements". *Physical Review D* 60 (8). doi:10.1103/PhysRevD.60.081301., although the manner in which the term is treated there suggests it was already in general use. Cosmologist Saul Perlmutter has credited Turner with coining the term in an article they wrote together with Martin White, where it is introduced in quotation marks as if it were a neologism. Perlmutter, S.; Turner, M.; White, M. (1999). "Constraining Dark Energy with Type Ia Supernovae and Large-Scale Structure". *Physical Review Letters* 83 (4): 670. doi:10.1103/PhysRevLett.83.670.

[65] Astier, Pierre (Supernova Legacy Survey); Guy; Regnault; Pain; Aubourg; Balam; Basa; Carlberg; Fabbro; Fouchez; Hook; Howell; Lafoux; Neill; Palanque-Delabrouille; Perrett; Pritchet; Rich; Sullivan; Taillet; Aldering; Antilogus; Arsenijevic; Balland; Baumont; Bronder; Courtois; Ellis; Filiol et al. (2006). "The Supernova legacy survey: Measurement of ΩM, $\Omega\Lambda$ and W from the first year data set". *Astronomy and Astrophysics* 447: 31–48. arXiv:astro-ph/0510447. Bibcode:2006A&A...447...31A. doi:10.1051/0004-6361:20054185.

23.9 External links

-

- Dark Energy on *In Our Time* at the BBC. (listen now)

- Dark energy: how the paradigm shifted Physicsworld.com

- Dennis Overbye (November 2006). "9 Billion-Year-Old 'Dark Energy' Reported". *The New York Times*.

- "Mysterious force's long presence" BBC News online (2006) More evidence for dark energy being the cosmological constant

- "Astronomy Picture of the Day" one of the images of the Cosmic Microwave Background which confirmed the presence of dark energy and dark matter

- SuperNova Legacy Survey home page The Canada-France-Hawaii Telescope Legacy Survey Supernova Program aims primarily at measuring the equation of state of Dark Energy. It is designed to precisely measure several hundred high-redshift supernovae.

- "Report of the Dark Energy Task Force"

- "HubbleSite.org – Dark Energy Website" Multimedia presentation explores the science of dark energy and Hubble's role in its discovery.

- "Surveying the dark side"

- "Dark energy and 3-manifold topology" Acta Physica Polonica 38 (2007), p. 3633–3639

- The Dark Energy Survey

- The Joint Dark Energy Mission

- Harvard: Dark Energy Found Stifling Growth in Universe, primary source

- April 2010 Smithsonian Magazine Article

- HETDEX Dark energy experiment

- Dark Energy FAQ

- "The Hunt for Dark Energy" George FR Ellis, Peter Cameron and David Tong discuss the presence of dark energy in the Universe

April 2010 Smithsonian Magazine Article]

- Euclid ESA Satellite, a mission to map the geometry of the dark universe

Chapter 24

Gravitational wave

Not to be confused with gravity wave.

In physics, **gravitational waves** are ripples in the curvature of spacetime which propagate as waves, travelling outward from the source. Predicted in 1916[1][2] by Albert Einstein to exist on the basis of his theory of general relativity,[3][4] gravitational waves theoretically transport energy as **gravitational radiation**. Sources of detectable gravitational waves could possibly include binary star systems composed of white dwarfs, neutron stars, or black holes. The existence of gravitational waves is a possible consequence of the Lorentz invariance of general relativity since it brings the concept of a limiting speed of propagation of the physical interactions with it. Gravitational waves cannot exist in the Newtonian theory of gravitation, in which physical interactions propagate at infinite speed.

Although gravitational radiation has not been *directly* detected, there is *indirect* evidence for its existence.[5] For example, the 1993 Nobel Prize in Physics was awarded for measurements of the Hulse–Taylor binary system which suggest that gravitational waves are more than mathematical anomalies. Various gravitational wave detectors exist and on 17 March 2014, astronomers at the Harvard–Smithsonian Center for Astrophysics claimed that they had detected and produced "the first direct image of gravitational waves across the primordial sky" within the cosmic microwave background, providing strong evidence for inflation and the Big Bang.[6][7][8][9][5] Peer review will be needed before there can be any scientific consensus about these new findings.[10][11] On 19 June 2014, lowered confidence in confirming the cosmic inflation findings was reported;[12][13][14] on 19 September 2014, a further reduction in confidence was reported[15][16] and, on 30 January 2015, even less confidence yet was reported.[17][18]

24.1 Introduction

In Einstein's theory of general relativity, gravity is treated as a phenomenon resulting from the curvature of spacetime. This curvature is caused by the presence of mass. Generally, the more mass that is contained within a given volume of space, the greater the curvature of spacetime will be at the boundary of this volume.[5] As objects with mass move around in spacetime, the curvature changes to reflect the changed locations of those objects. In certain circumstances, accelerating objects generate changes in this curvature, which propagate outwards at the speed of light in a wave-like manner. These propagating phenomena are known as gravitational waves.

As a gravitational wave passes a distant observer, that observer will find spacetime distorted by the effects of strain. Distances between free objects increase and decrease rhythmically as the wave passes, at a frequency corresponding to that of the wave. This occurs despite such free objects never being subjected to an unbalanced force. The magnitude of this effect decreases inversely with distance from the source. Inspiralling binary neutron stars are predicted to be a powerful source of gravitational waves as they coalesce, due to the very large acceleration of their masses as they orbit close to one another. However, due to the astronomical distances to these sources the effects when measured on Earth are predicted to be very small, having strains of less than 1 part in 10^{20}. Scientists are attempting to demonstrate the existence of these waves with ever more sensitive detectors. The current most sensitive measurement is about one part in 5×10^{22} (as

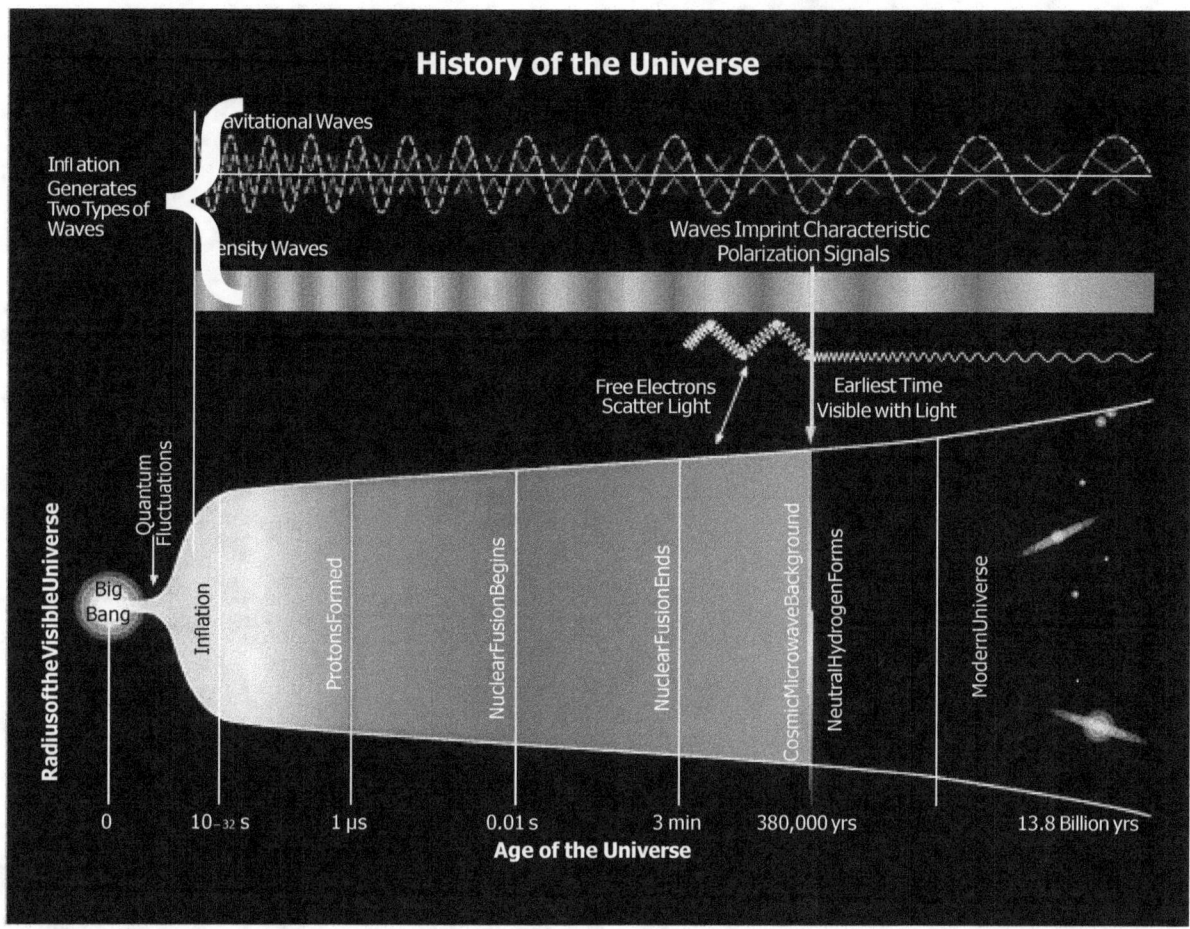

History of the Universe - gravitational waves are hypothesized to arise from cosmic inflation, a faster-than-light expansion just after the Big Bang (17 March 2014).[6][8][9]

of 2012) provided by the LIGO and VIRGO observatories.[19] The lack of detection in these observatories provides an upper limit on the frequency of such powerful sources.[20][21] A space based observatory, the Laser Interferometer Space Antenna, is currently under development by ESA.

Gravitational waves should penetrate regions of space that electromagnetic waves cannot. It is hypothesized that they will be able to provide observers on Earth with information about black holes and other exotic objects in the distant Universe. Such systems cannot be observed with more traditional means such as optical telescopes and radio telescopes. In particular, gravitational waves could be of interest to cosmologists as they offer a possible way of observing the very early universe. This is not possible with conventional astronomy, since before recombination the universe was opaque to electromagnetic radiation.[22] Precise measurements of gravitational waves will also allow scientists to test the general theory of relativity more thoroughly.

In principle, gravitational waves could exist at any frequency. However, very low frequency waves would be impossible to detect and there is no credible source for detectable waves of very high frequency. Stephen W. Hawking and Werner Israel list different frequency bands for gravitational waves that could be plausibly detected, ranging from 10^{-7} Hz up to 10^{11} Hz.[23]

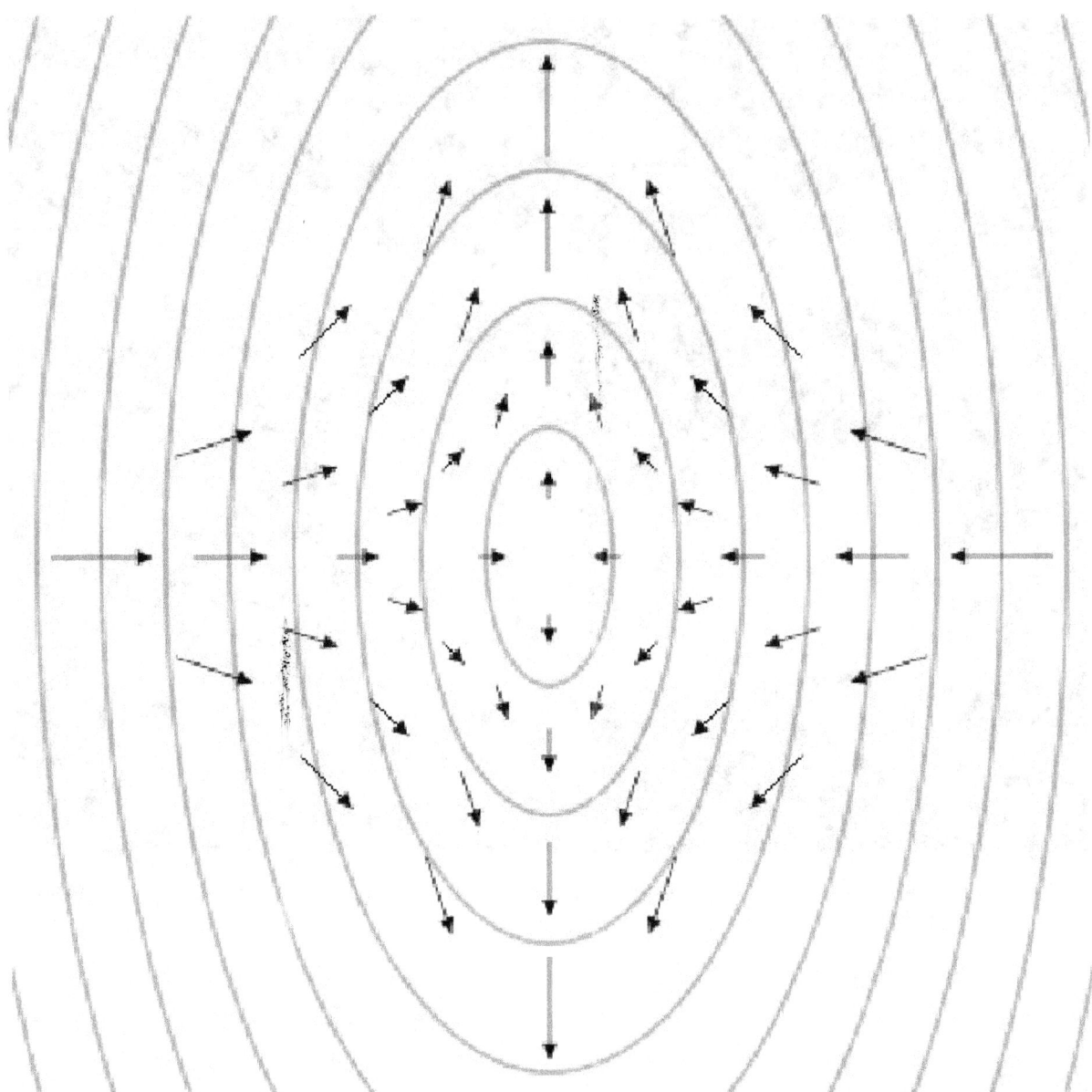

Linearly polarised gravitational wave

24.2   Eﬀ ects of a passing gravitational wave

The effects of a passing gravitational wave can be visualized by imagining a perfectly flat region of spacetime with a group of motionless test particles lying in a plane (the surface of your screen). As a gravitational wave passes through the particles along a line perpendicular to the plane of the particles (i.e. following your line of vision into the screen), the particles will follow the distortion in spacetime, oscillating in a "cruciform" manner, as shown in the animations. The area enclosed by the test particles does not change and there is no motion along the direction of propagation.

The oscillations depicted here in the animation are exaggerated for the purpose of discussion—in reality a gravitational wave has a very small amplitude (as formulated in linearized gravity). However they enable us to visualize the kind of oscillations associated with gravitational waves as produced for example by a pair of masses in a circular orbit. In this case the amplitude of the gravitational wave is constant, but its plane of polarization changes or rotates at twice the orbital rate and so the time-varying gravitational wave size (or 'periodic spacetime strain') exhibits a variation as shown in the

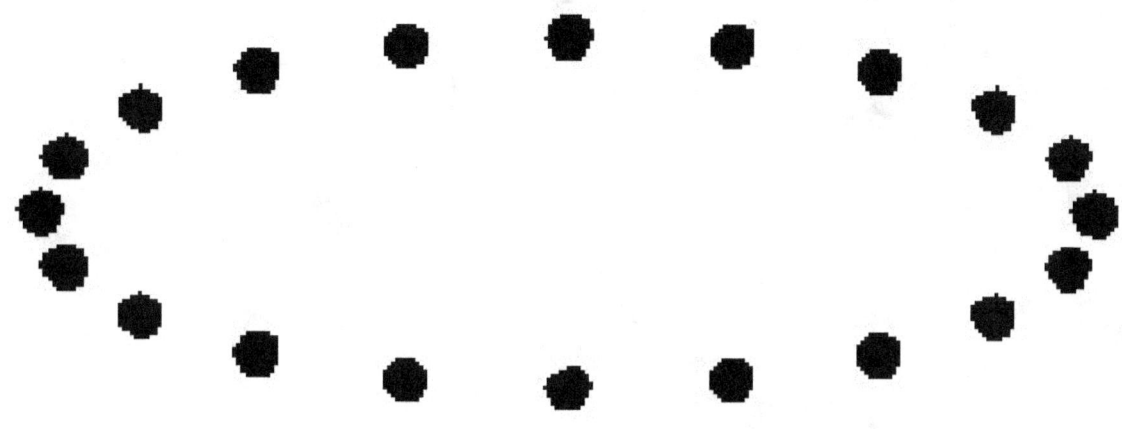

The effect of a plus-polarized gravitational wave on a ring of particles.

animation.[24] If the orbit is elliptical then the gravitational wave's amplitude also varies with time according to Einstein's quadrupole formula.[25]

Like other waves, there are a few useful characteristics describing a gravitational wave:

- **Amplitude:** Usually denoted h, this is the size of the wave — the fraction of stretching or squeezing in the animation. The amplitude shown here is roughly $h = 0.5$ (or 50%). Gravitational waves passing through the Earth are many billions times weaker than this — $h \approx 10^{-20}$. Note that this is not the quantity that would be analogous to what is usually called the amplitude of an electromagnetic wave, which would be $\frac{dh}{dt}$.

- **Frequency:** Usually denoted f, this is the frequency with which the wave oscillates (1 divided by the amount of time between two successive maximum stretches or squeezes)

- **Wavelength:** Usually denoted λ, this is the distance along the wave between points of maximum stretch or squeeze.

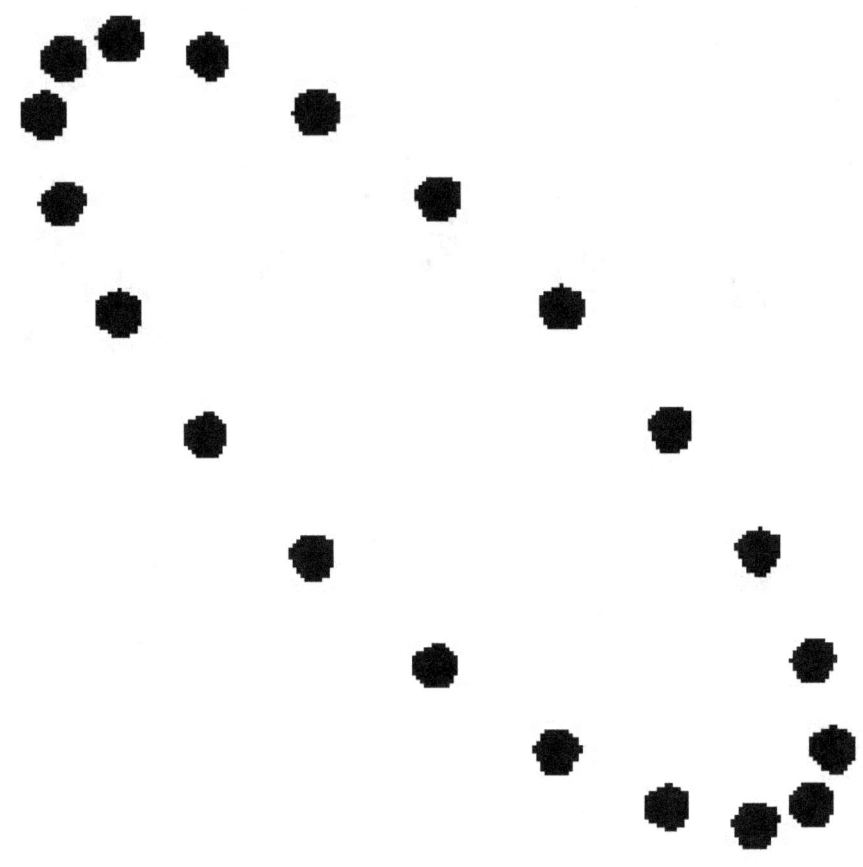

The effect of a cross-polarized gravitational wave on a ring of particles.

- **Speed**: This is the speed at which a point on the wave (for example, a point of maximum stretch or squeeze) travels. For gravitational waves with small amplitudes, this is equal to the speed of light, c.

The speed, wavelength, and frequency of a gravitational wave are related by the equation $c = \lambda f$, just like the equation for a light wave. For example, the animations shown here oscillate roughly once every two seconds. This would correspond to a frequency of 0.5 Hz, and a wavelength of about 600,000 km, or 47 times the diameter of the Earth.

In the example just discussed, we actually assume something special about the wave. We have assumed that the wave is linearly polarized, with a "plus" polarization, written h_+. Polarization of a gravitational wave is just like polarization of a light wave except that the polarizations of a gravitational wave are at 45 degrees, as opposed to 90 degrees. In particular, if we had a "cross"-polarized gravitational wave, $h_\times$, the effect on the test particles would be basically the same, but rotated by 45 degrees, as shown in the second animation. Just as with light polarization, the polarizations of gravitational waves may also be expressed in terms of circularly polarized waves. Gravitational waves are polarized because of the nature of their sources. The polarization of a wave depends on the angle from the source, as we will see in the next section.

24.3   Sources of gravitational waves

In general terms, gravitational waves are radiated by objects whose motion involves acceleration, provided that the motion is not perfectly spherically symmetric (like an expanding or contracting sphere) or cylindrically symmetric (like a spinning disk or sphere). A simple example of this principle is provided by the spinning dumbbell. If the dumbbell spins like wheels on an axle, it will not radiate gravitational waves; if it tumbles end over end like two planets orbiting each other, it will radiate gravitational waves. The heavier the dumbbell, and the faster it tumbles, the greater is the gravitational radiation it will give off. If we imagine an extreme case in which the two weights of the dumbbell are massive stars like neutron stars or black holes, orbiting each other quickly, then significant amounts of gravitational radiation would be given off.

Some more detailed examples:

- Two objects orbiting each other in a quasi-Keplerian planar orbit (basically, as a planet would orbit the Sun) *will* radiate.

- A spinning non-axisymmetric planetoid — say with a large bump or dimple on the equator — *will* radiate.

- A supernova *will* radiate except in the unlikely event that the explosion is perfectly symmetric.

- An isolated non-spinning solid object moving at a constant velocity *will not* radiate. This can be regarded as a consequence of the principle of conservation of linear momentum.

- A spinning disk *will not* radiate. This can be regarded as a consequence of the principle of conservation of angular momentum. However, it *will* show gravitomagnetic effects.

- A spherically pulsating spherical star (non-zero monopole moment or mass, but zero quadrupole moment) *will not* radiate, in agreement with Birkhoff 's theorem.

More technically, the third time derivative of the quadrupole moment (or the *l*-th time derivative of the *l*-th multipole moment) of an isolated system's stress–energy tensor must be nonzero in order for it to emit gravitational radiation. This is analogous to the changing dipole moment of charge or current necessary for electromagnetic radiation.

24.3.1   Power radiated by orbiting bodies

Gravitational waves carry energy away from their sources and, in the case of orbiting bodies, this is associated with an inspiral or decrease in orbit. Imagine for example a simple system of two masses — such as the Earth-Sun system — moving slowly compared to the speed of light in circular orbits. Assume that these two masses orbit each other in a circular orbit in the $x - y$ plane. To a good approximation, the masses follow simple Keplerian orbits. However, such an orbit represents a changing quadrupole moment. That is, the system will give off gravitational waves.

Suppose that the two masses are m_1 and m_2, and they are separated by a distance r. The power given off (radiated) by this system is:

$$P = \frac{dE}{t} = -\frac{32}{5} \; \frac{G^4}{c^5} \frac{(m_1 m_2)^2 (m_1 + m_2)}{r^5} \quad [26]$$

where G is the gravitational constant, c is the speed of light in vacuum and where the negative sign means that power is being given off by the system, rather than received. For a system like the Sun and Earth, r is about 1.5×10^{11} m and m_1 and m_2 are about 2×10^{30} and 6×10^{24} kg respectively. In this case, the power is about 200 watts. This is truly tiny compared to the total electromagnetic radiation given off by the Sun (roughly 3.86×10^{26} watts).

In theory, the loss of energy through gravitational radiation could eventually drop the Earth into the Sun. However, the total energy of the Earth orbiting the Sun (kinetic energy + gravitational potential energy) is about 1.14×10^{36} joules of which only 200 joules per second is lost through gravitational radiation, leading to a decay in the orbit by about 1×10^{-15} meters per day or roughly the diameter of a proton. At this rate, it would take the Earth approximately 1×10^{13} times more than the current age of the Universe to spiral onto the Sun. This estimate overlooks the decrease in r over time, but the majority of the time the bodies are far apart and only radiating slowly, so the difference is unimportant in this example.

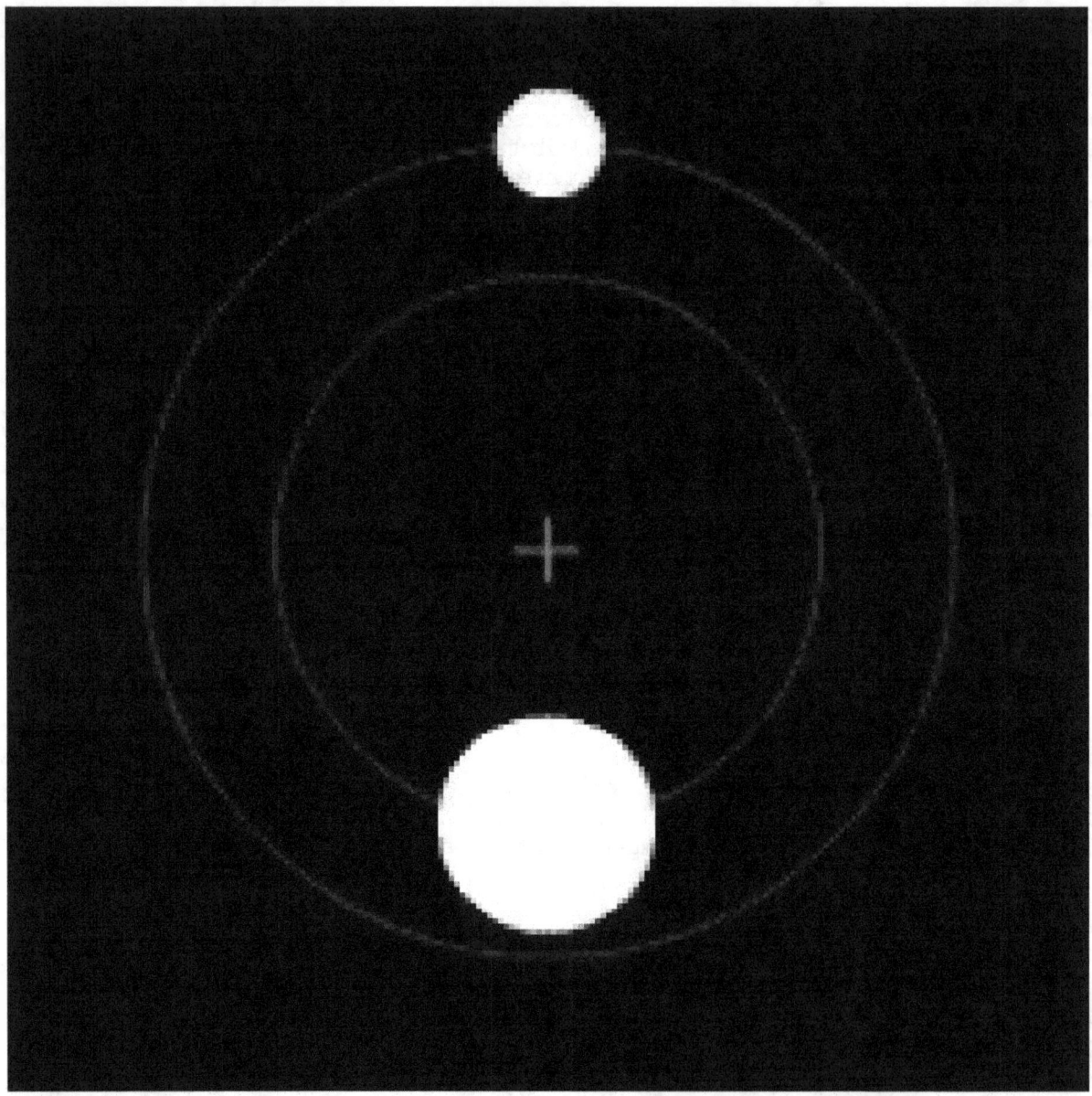

Two stars of dissimilar mass are in circular orbits. Each revolves about their common center of mass (denoted by the small red cross) in a circle with the larger mass having the smaller orbit.

A more dramatic example of radiated gravitational energy is represented by two solar mass ($M\odot$) neutron stars orbiting at a distance from each other of 1.89×10^8 m (only 0.63 light-seconds apart). [The Sun is 8 light minutes from the Earth.] Plugging their masses into the above equation shows that the gravitational radiation from them would be 1.38×10^{28} watts, which is about 100 times more than the Sun's electromagnetic radiation.

24.3.2   Orbital decay from gravitational radiation

See also: Two-body problem in general relativity

Gravitational radiation robs the orbiting bodies of energy. It first circularizes their orbits and then gradually shrinks their radius. As the energy of the orbit is reduced, the distance between the bodies decreases, and they rotate more rapidly.

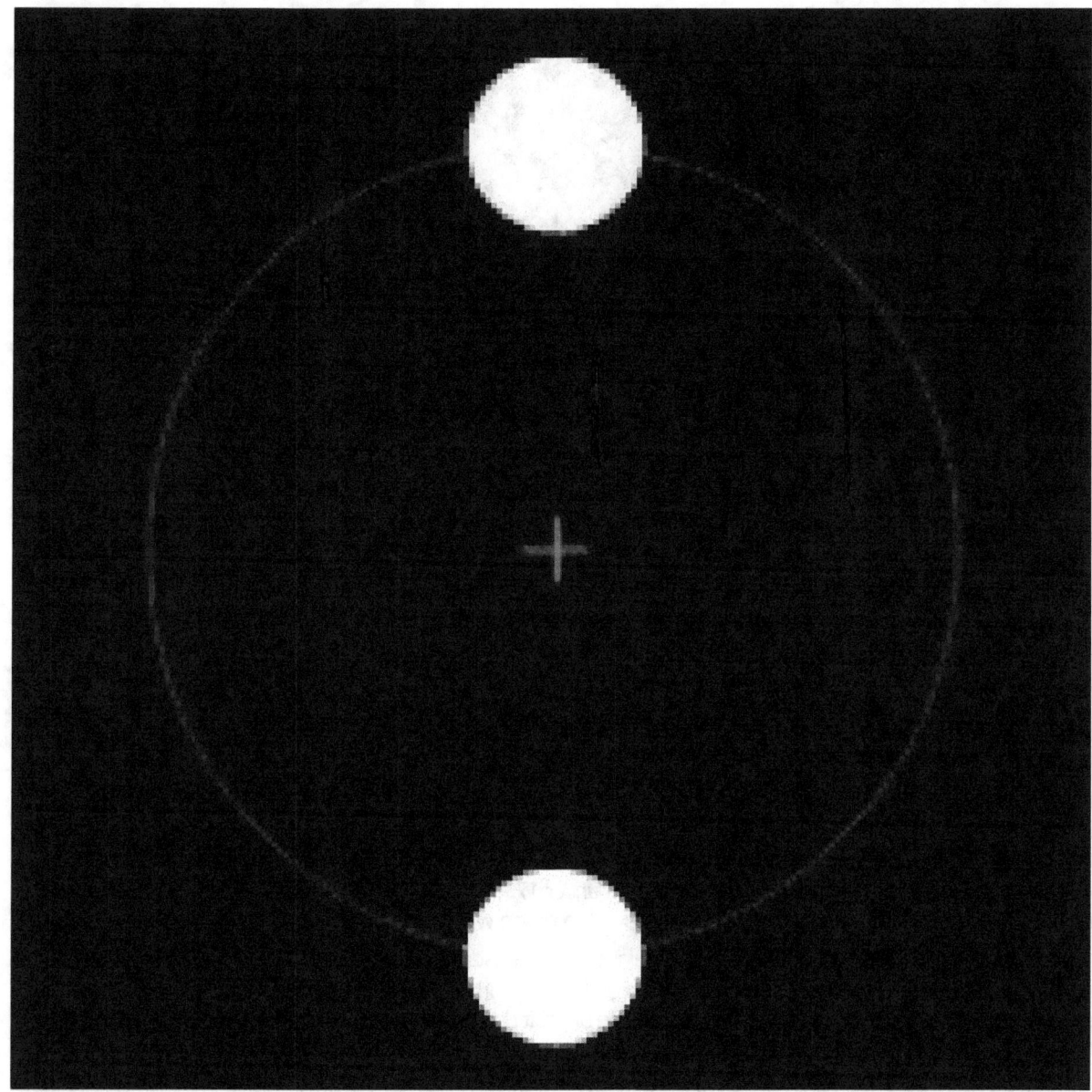

Two stars of similar mass are in circular orbits about their center of mass

The overall angular momentum is reduced however. This reduction corresponds to the angular momentum carried off by gravitational radiation. The rate of decrease of distance between the bodies versus time is given by:[26]

$$\frac{dr}{-} \frac{}{\frac{5}{64}\frac{c^5}{G^3}} \overline{(m_1 m_2)(m_1 + m_2)}$$

where the variables are the same as in the previous equation.

The orbit decays at a rate proportional to the inverse third power of the radius. When the radius has shrunk to half its initial value, it is shrinking eight times faster than before. By Kepler's Third Law, the new rotation rate at this point will be faster by $\sqrt{8} = 2.828$, or nearly three times the previous orbital frequency. As the radius decreases, the power lost to gravitational radiation increases even more. As can be seen from the previous equation, power radiated varies as the inverse fifth power of the radius, or 32 times more in this case.

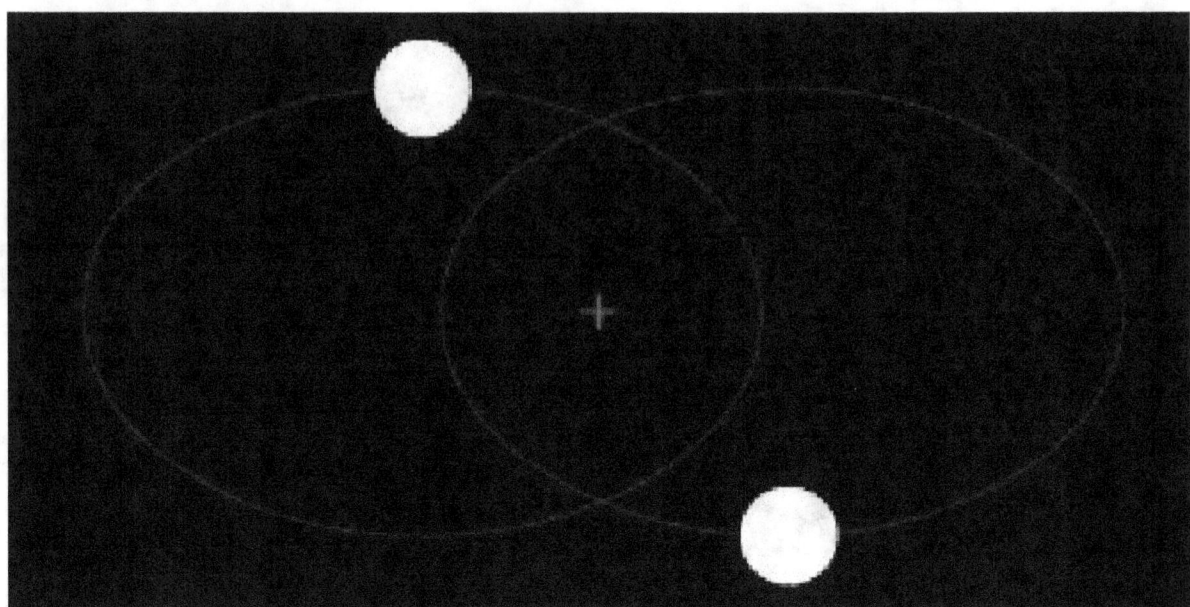

Two stars of similar mass are in highly elliptical orbits about their center of mass

If we use the previous values for the Sun and the Earth, we find that the Earth's orbit shrinks by 1.1×10^{-20} meter per second. This is 3.5×10^{-13} m per year, which is about 1/300 the diameter of a hydrogen atom. The effect of gravitational radiation on the size of the Earth's orbit is negligible over the age of the universe. This is not true for closer orbits.

A more practical example is the orbit of a Sun-like star around a heavy black hole. Our Milky Way is believed to have a 4 million $M\odot$ black hole at its center in Sagittarius A. Such supermassive black holes are being found in the center of almost all galaxies. For this example take a 2 million $M\odot$ black hole with a solar-mass star orbiting it at a radius of 1.89×10^{10} m (63 light-seconds). The mass of the black hole will be 4×10^{36} kg and its gravitational radius will be 6×10^9 m. The orbital period will be 1,000 seconds, or a little under 17 minutes. The solar-mass star will draw closer to the black hole by 7.4 meters per second or 7.4 km per orbit. A collision will not be long in coming.

Assume that a pair of 1 $M\odot$ neutron stars are in circular orbits at a distance of 1.89×10^8 m (189,000 km). This is a little less than 1/7 the diameter of the Sun or 0.63 light-seconds. Their orbital period would be 1,000 seconds. Substituting the new mass and radius in the above formula gives a rate of orbit decrease of 3.7×10^{-6} m/s or 3.7 mm per orbit. This is 116 meters per year and is not negligible over cosmic time scales.

Suppose instead that these two neutron stars were orbiting at a distance of 1.89×10^6 m (1890 km). Their period would be 1 second and their orbital velocity would be about 1/50 of the speed of light. Their orbit would now shrink by 3.7 meters per orbit. A collision is imminent. A runaway loss of energy from the orbit results in an ever more rapid decrease in the distance between the stars. They will eventually merge to form a black hole and cease to radiate gravitational waves. This is referred to as the inspiral.

The above equation can not be applied directly for calculating the lifetime of the orbit, because the rate of change in radius depends on the radius itself, and is thus non-constant with time. The lifetime can be computed by integration of this equation (see next section).

24.3.3   Orbital lifetime limits from gravitational radiation

Orbital lifetime is one of the most important properties of gravitational radiation sources. It determines the average number of binary stars in the universe that are close enough to be detected. Short lifetime binaries are strong sources of gravitational radiation but are few in number. Long lifetime binaries are more plentiful but they are weak sources of gravitational waves. LIGO is most sensitive in the frequency band where two neutron stars are about to merge. This time frame is only a few seconds. It takes luck for the detector to see this blink in time out of a million year orbital lifetime.

It is predicted that such a merger will only be seen once per decade or so.

The lifetime of an orbit is given by:[26]

$$t = \frac{5}{256}\frac{c^5}{G^3}\frac{r^4}{(m_1 m_2)(m_1 + m_2)}$$

where r is the initial distance between the orbiting bodies. This equation can be derived by integrating the previous equation for the rate of radius decrease. It predicts the time for the radius of the orbit to shrink to zero. As the orbital speed becomes a significant fraction of the speed of light, this equation becomes inaccurate. It is useful for inspirals until the last few milliseconds before the merger of the objects.

Substituting the values for the mass of the Sun and Earth as well as the orbital radius gives a very large lifetime of 3.44×10^{30} seconds or 1.09×10^{23} years (that is approximately 10^{13} times larger than the age of the universe). The actual figure would be slightly less than that. The Earth will break apart from tidal forces if it orbits closer than a few radii from the Sun. This would form a ring around the Sun and instantly stop the emission of gravitational waves.

If we use a 2 million $M\odot$ black hole with a solar mass star orbiting it at 1.89×10^{10} meters, we get a lifetime of 6.50×10^8 seconds or 20.7 years.

Assume that a pair of solar mass neutron stars with a diameter of 10 kilometers are in circular orbits at a distance of 1.89×10^8 m (189,000 km). Their lifetime is 1.30×10^{13} seconds or about 414,000 years. Their orbital period will be 1,000 seconds and it could be observed by LISA if they were not too far away. A far greater number of white dwarf binaries exist with orbital periods in this range. White dwarf binaries have masses on the order of our Sun and diameters on the order of our Earth. They cannot get much closer together than 10,000 km before they will merge and cease to radiate gravitational waves. This results in the creation of either a neutron star or a black hole. Until then, their gravitational radiation will be comparable to that of a neutron star binary. LISA is the only gravitational wave experiment that is likely to succeed in detecting such types of binaries.

If the orbit of a neutron star binary has decayed to 1.89×10^6 m (1890 km), its remaining lifetime is 130,000 seconds or about 36 hours. The orbital frequency will vary from 1 revolution per second at the start and 918 revolutions per second when the orbit has shrunk to 20 km at merger. The gravitational radiation emitted will be at twice the orbital frequency. Just before merger, the inspiral can be observed by LIGO if the binary is close enough. LIGO has only a few minutes to observe this merger out of a total orbital lifetime that may have been billions of years. The chance of success with LIGO as initially constructed is quite low despite the large number of such mergers occurring in the universe, because the sensitivity of the instrument does not 'reach' out to enough systems to see events frequently. No mergers have been seen in the few years that initial LIGO has been in operation, and it is thought that a merger should be seen about once per several tens of years of observing time with initial LIGO.[27] The upgraded Advanced LIGO detector, with a ten times greater sensitivity, 'reaches' out 10 times further—encompassing a volume 1000 times greater, and seeing 1000 times as many candidate sources. Thus, the expectation is that detections will be made at the rate of tens per year.

24.3.4   Wave amplitudes from the Earth–Sun system

We can also think in terms of the amplitude of the wave from a system in circular orbits. Let θ be the angle between the perpendicular to the plane of the orbit and the line of sight of the observer. Suppose that an observer is outside the system at a distance R from its center of mass. If R is much greater than a wavelength, the two polarizations of the wave will be

$$h_+ = -\frac{1}{R}\frac{G^2}{c^4}\frac{2m_1 m_2}{r}(1 + \cos^2 \theta)\cos[2\omega(t - R)],$$

$$h_\times = -\frac{1}{R}\frac{G^2}{c^4}\frac{4m_1 m_2}{r}(\cos \theta)\sin[2\omega(t - R)].$$

Here, we use the constant angular velocity of a circular orbit in Newtonian physics:

$$\omega = \sqrt{G(m_1 + m_2)/r^3}.$$

For example, if the observer is in the x - y plane then $\theta = \pi/2$, and $\cos(\theta) = 0$, so the $h_\times$ polarization is always zero. We also see that the frequency of the wave given off is twice the rotation frequency. If we put in numbers for the Earth-Sun system, we find:

$$h_+ = -\frac{1}{R}\frac{G^2}{c^4}\frac{4m_1 m_2}{r} = -\frac{1}{1.7 \cdot 10^{-10} \text{ m}}$$

In this case, the minimum distance to find waves is $R \approx 1$ light-year, so typical amplitudes will be $h \approx 10^{-26}$. That is, a ring of particles would stretch or squeeze by just one part in 10^{26}. This is well under the detectability limit of all conceivable detectors.

24.3.5 Radiation from other sources

Although the waves from the Earth-Sun system are minuscule, astronomers can point to other sources for which the radiation should be substantial. One important example is the Hulse-Taylor binary — a pair of stars, one of which is a pulsar.[28] The characteristics of their orbit can be deduced from the Doppler shifting of radio signals given off by the pulsar. Each of the stars are about 1.4 $M\odot$ and the size of their orbit is about 1/75 of the Earth-Sun orbit. This means the distance between the two stars is just a few times larger than the diameter of our own Sun. The combination of greater masses and smaller separation means that the energy given off by the Hulse-Taylor binary will be far greater than the energy given off by the Earth-Sun system — roughly 10^{22} times as much.

The information about the orbit can be used to predict just how much energy (and angular momentum) should be given off in the form of gravitational waves. As the energy is carried off, the stars should draw closer to each other. This effect is called an inspiral, and it can be observed in the pulsar's signals. The measurements on the Hulse-Taylor system have been carried out over more than 30 years. It has been shown that the gravitational radiation predicted by general relativity allows these observations to be matched within 0.2 percent. In 1993, Russell Hulse and Joe Taylor were awarded the Nobel Prize in Physics for this work, which was the first indirect evidence for gravitational waves. The orbital lifetime of this binary system before merger is a few hundred million years.[29]

Inspirals are very important sources of gravitational waves. Any time two compact objects (white dwarfs, neutron stars, or black holes) are in close orbits, they send out intense gravitational waves. As they spiral closer to each other, these waves become more intense. At some point they should become so intense that direct detection by their effect on objects on Earth or in space is possible. This direct detection is the goal of several large scale experiments.[30]

The only difficulty is that most systems like the Hulse-Taylor binary are so far away. The amplitude of waves given off by the Hulse-Taylor binary as seen on Earth would be roughly $h \approx 10^{-26}$. There are some sources, however, that astrophysicists expect to find with much larger amplitudes of $h \approx 10^{-20}$. At least eight other binary pulsars have been discovered.[31]

24.4 Astrophysics and gravitational waves

During the past century, astronomy has been revolutionized by the use of new methods for observing the universe. Astronomical observations were originally made using visible light. Galileo Galilei pioneered the use of telescopes to enhance these observations. However, visible light is only a small portion of the electromagnetic spectrum, and not all objects in the distant universe shine strongly in this particular band. More useful information may be found, for example, in radio wavelengths. Using radio telescopes, astronomers have found pulsars, quasars, and other extreme objects that push the limits of our understanding of physics. Observations in the microwave band have opened our eyes to the faint imprints of the Big Bang, a discovery Stephen Hawking called the "greatest discovery of the century, if not all time". Similar advances in observations using gamma rays, x-rays, ultraviolet light, and infrared light have also brought new insights to astronomy. As each of these regions of the spectrum has opened, new discoveries have been made that could not have been made otherwise. Astronomers hope that the same holds true of gravitational waves.

Gravitational waves have two important and unique properties. First, there is no need for any type of matter to be present nearby in order for the waves to be generated by a binary system of uncharged black holes, which would emit no

Two-dimensional representation of gravitational waves generated by two neutron stars orbiting each other.

electromagnetic radiation. Second, gravitational waves can pass through any intervening matter without being scattered significantly. Whereas light from distant stars may be blocked out by interstellar dust, for example, gravitational waves will pass through essentially unimpeded. These two features allow gravitational waves to carry information about astronomical phenomena never before observed by humans.

The sources of gravitational waves described above are in the low-frequency end of the gravitational-wave spectrum (10^{-7} to 10^5 Hz). An astrophysical source at the high-frequency end of the gravitational-wave spectrum (above 10^5 Hz and probably 10^{10} Hz) generates relic gravitational waves that are theorized to be faint imprints of the Big Bang like the cosmic microwave background (see gravitational wave background).[32] At these high frequencies it is potentially possible that the sources may be "man made"[23] that is, gravitational waves generated and detected in the laboratory.[33][34]

24.4.1   Energy, momentum, and angular momentum carried by gravitational waves

Waves familiar from other areas of physics such as water waves, sound waves, and electromagnetic waves are able to carry energy, momentum, and angular momentum. By carrying these away from a source, waves are able to rob that source of its energy as well as its linear and angular momentum. Gravitational waves perform the same function. Thus, for example, a binary system loses angular momentum as the two orbiting objects spiral towards each other—the angular momentum is radiated away by gravitational waves.

The waves can also carry off linear momentum, a possibility that has some interesting implications for astrophysics.[35] After two supermassive black holes coalesce, emission of linear momentum can produce a "kick" with amplitude as large as 4000 km/s. This is fast enough to eject the coalesced black hole completely from its host galaxy. Even if the kick is too small to eject the black hole completely, it can remove it temporarily from the nucleus of the galaxy, after which it will oscillate about the center, eventually coming to rest.[36] A kicked black hole can also carry a star cluster with it, forming a hyper-compact stellar system.[37] Or it may carry gas, allowing the recoiling black hole to appear temporarily as a "naked quasar". The quasar SDSS J092712.65+294344.0 is believed to contain a recoiling supermassive black hole.[38]

24.5   Detecting gravitational waves

24.5.1   Difficulties in detection

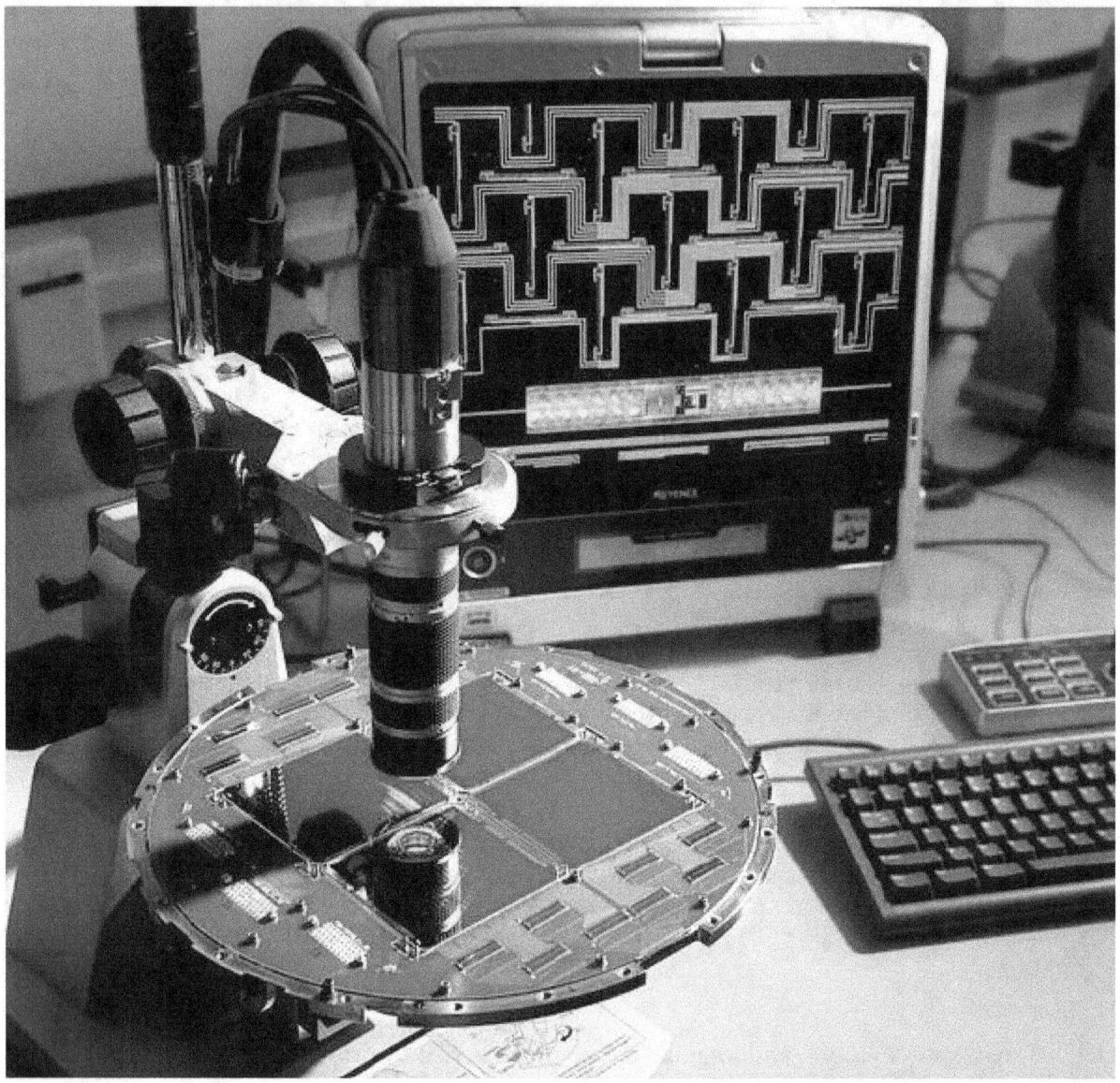

Evidence of gravitational waves in the infant universe may have been uncovered by the BICEP2 radio telescope. The microscopic examination of the focal plane of the BICEP2 detector is shown here.[6][7][8][9][11]

Gravitational waves are not easily detectable. This knowledge gap is primarily due to the massive presence of noise in the low frequencies where antennas currently operate. Gravitational waves are expected to have frequencies 10^{-16} Hz $< f < 10^4$ Hz .[39]

24.5.2   Ground-based interferometers

Main article: Gravitational wave detector

Though the Hulse-Taylor observations were very important, they give only *indirect* evidence for gravitational waves. A more conclusive observation would be a *direct* measurement of the effect of a passing gravitational wave, which could also provide more information about the system that generated it. Any such direct detection is complicated by the extraordinarily small effect the waves would produce on a detector. The amplitude of a spherical wave will fall off as the inverse of the distance from the source (the $1/R$ term in the formulas for h above). Thus, even waves from extreme systems like merging binary black holes die out to very small amplitude by the time they reach the Earth. Astrophysicists expect that some gravitational waves passing the Earth may be as large as $h \approx 10^{-20}$, but generally no bigger.[40]

A simple device theorised to detect the expected wave motion is called a Weber bar — a large, solid bar of metal isolated from outside vibrations. This type of instrument was the first type of gravitational wave detector. Strains in space due to an incident gravitational wave excite the bar's resonant frequency and could thus be amplified to detectable levels. Conceivably, a nearby supernova might be strong enough to be seen without resonant amplification. With this instrument, Joseph Weber claimed to have detected daily signals of gravitational waves. His results, however, were contested in 1974 by physicists Richard Garwin and David Douglass. Modern forms of the Weber bar are still operated, cryogenically cooled, with superconducting quantum interference devices to detect vibration. Weber bars are not sensitive enough to detect anything but extremely powerful gravitational waves.[41]

MiniGRAIL is a spherical gravitational wave antenna using this principle. It is based at Leiden University, consisting of an exactingly machined 1150 kg sphere cryogenically cooled to 20 mK.[42] The spherical configuration allows for equal sensitivity in all directions, and is somewhat experimentally simpler than larger linear devices requiring high vacuum. Events are detected by measuring deformation of the detector sphere. MiniGRAIL is highly sensitive in the 2–4 kHz range, suitable for detecting gravitational waves from rotating neutron star instabilities or small black hole mergers.[43]

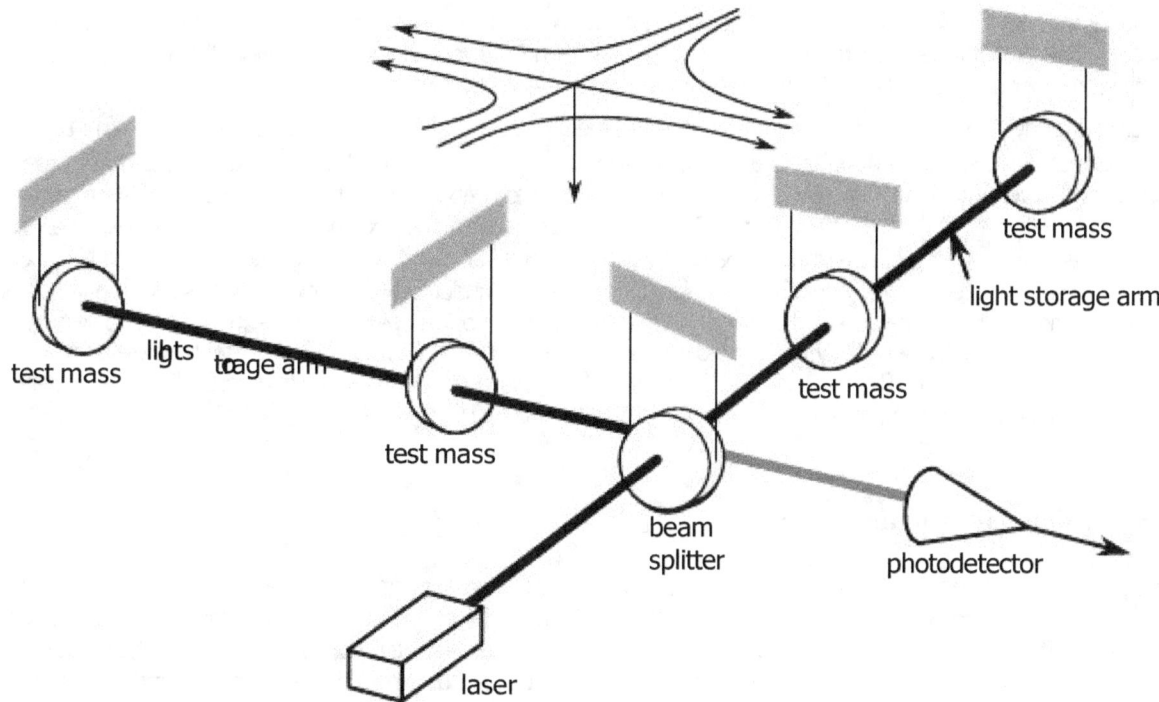

test mass

light storage arm

test mass

light storage arm

test mass

test mass

beam splitter

photodetector

laser

A schematic diagram of a laser interferometer

A more sensitive class of detector uses laser interferometry to measure gravitational-wave induced motion between separated 'free' masses.[44] This allows the masses to be separated by large distances (increasing the signal size); a further advantage is that it is sensitive to a wide range of frequencies (not just those near a resonance as is the case for Weber bars). Ground-based interferometers are now operational. Currently, the most sensitive is LIGO — the Laser Interferometer Gravitational Wave Observatory. LIGO has three detectors: one in Livingston, Louisiana; the other two (in the same vacuum tubes) at the Hanford site in Richland, Washington. Each consists of two light storage arms that are 2 to 4 kilometers in length. These are at 90 degree angles to each other, with the light passing through 1m diameter vacuum tubes running the entire 4 kilometers. A passing gravitational wave will slightly stretch one arm as it shortens the other. This is precisely the motion to which an interferometer is most sensitive.

Even with such long arms, the strongest gravitational waves will only change the distance between the ends of the arms by at most roughly 10^{-18} meters. LIGO should be able to detect gravitational waves as small as $h \sim 5 \times 10^{-20}$. Upgrades to LIGO and other detectors such as Virgo, GEO 600, and TAMA 300 should increase the sensitivity still further; the next generation of instruments (Advanced LIGO and Advanced Virgo) will be more than ten times more sensitive. Another highly sensitive interferometer (LCGT) is currently in the design phase. A key point is that a tenfold increase in sensitivity (radius of 'reach') increases the volume of space accessible to the instrument by one thousand times. This increases the rate at which detectable signals should be seen from one per tens of years of observation, to tens per year.[27]

Interferometric detectors are limited at high frequencies by shot noise, which occurs because the lasers produce photons randomly; one analogy is to rainfall—the rate of rainfall, like the laser intensity, is measurable, but the raindrops, like photons, fall at random times, causing fluctuations around the average value. This leads to noise at the output of the detector, much like radio static. In addition, for sufficiently high laser power, the random momentum transferred to the test masses by the laser photons shakes the mirrors, masking signals at low frequencies. Thermal noise (e.g., Brownian motion) is another limit to sensitivity. In addition to these 'stationary' (constant) noise sources, all ground-based detectors are also limited at low frequencies by seismic noise and other forms of environmental vibration, and other 'non-stationary' noise sources; creaks in mechanical structures, lightning or other large electrical disturbances, etc. may also create noise masking an event or may even imitate an event. All these must be taken into account and excluded by analysis before a detection may be considered a true gravitational wave event.

Space-based interferometers, such as LISA and DECIGO, are also being developed. LISA's design calls for three test masses forming an equilateral triangle, with lasers from each spacecraft to each other spacecraft forming two independent interferometers. LISA is planned to occupy a solar orbit trailing the Earth, with each arm of the triangle being five million kilometers. This puts the detector in an excellent vacuum far from Earth-based sources of noise, though it will still be susceptible to shot noise, as well as artifacts caused by cosmic rays and solar wind.

There are currently two detectors focusing on detection at the higher end of the gravitational wave spectrum (10^{-7} to 10^5 Hz): one at University of Birmingham, England, and the other at INFN Genoa, Italy. A third is under development at Chongqing University, China. The Birmingham detector measures changes in the polarization state of a microwave beam circulating in a closed loop about one meter across. $\sqrt{}$Two have been fabricated and they are currently expected to be sensitive to periodic spacetime strains of $h \sim 2 \times 10^{-13}/\sqrt{\mathrm{Hz}}$, given as an amplitude spectral density. The INFN Genoa detector is a resonant antenna consisting of two coupled spherical superconducting harmonic oscillators a few centimeters in diameter. The oscillators are designed to have (when uncoupled) almost equal resonant frequencies. The system is currently expected to have a sensitivity to periodic spacetime strains of $h \sim 2 \times 10^{-17}/\sqrt{\mathrm{Hz}}$, with an expectation to reach a sensitivity of $h \sim 2 \times 10^{-20}/\sqrt{\mathrm{Hz}}$. The Chongqing University detector is planned to detect relic high-frequency gravitational waves with the predicted typical parameters $? \sim 10^{10}$ Hz (10 GHz) and $h \sim 10^{-30} - 10^{-31}$.

24.5.3   Using pulsar timing arrays

Pulsars are rapidly rotating stars. A pulsar emits beams of radio waves that, like lighthouse beams, sweep through the sky as the pulsar rotates. The signal from a pulsar can be detected by radio telescopes as a series of regularly spaced pulses, essentially like the ticks of a clock. Gravitational waves affect the time it takes the pulses to travel from the pulsar to a telescope on Earth. A pulsar timing array uses millisecond pulsars to seek out perturbations due to gravitational waves in measurements of pulse arrival times at a telescope, in other words, to look for deviations in the clock ticks. In particular, pulsar timing arrays can search for a distinct pattern of correlation and anti-correlation between the signals over an array of different pulsars (resulting in the name "pulsar timing array"). Although pulsar pulses travel through space for hundreds or thousands of years to reach us, pulsar timing arrays are sensitive to perturbations in their travel time of much less than a millionth of a second.

Globally there are three active pulsar timing array projects. The North American Nanohertz Gravitational Wave Observatory uses data collected by the Arecibo Radio Telescope and Green Bank Telescope. The Parkes Pulsar Timing Array at the Parkes radio-telescope has been collecting data since March 2005. The European Pulsar Timing Array uses data from the four largest telescopes in Europe: the Lovell Telescope, the Westerbork Synthesis Radio Telescope, the Effelsberg Telescope and the Nancay Radio Telescope. (Upon completion the Sardinia Radio Telescope will be added to the EPTA also.) These three projects have begun collaborating under the title of the International Pulsar Timing Array project.

24.5.4 Einstein@Home

Main article: Einstein@Home

In some sense, the easiest signals to detect should be constant sources. Supernovae and neutron star or black hole mergers should have larger amplitudes and be more interesting, but the waves generated will be more complicated. The waves given off by a spinning, aspherical neutron star would be "monochromatic"—like a pure tone in acoustics. It would not change very much in amplitude or frequency.

The Einstein@Home project is a distributed computing project similar to SETI@home intended to detect this type of simple gravitational wave. By taking data from LIGO and GEO, and sending it out in little pieces to thousands of volunteers for parallel analysis on their home computers, Einstein@Home can sift through the data far more quickly than would be possible otherwise.[45]

24.5.5 Primordial gravitational waves

Main article: Primordial gravitational wave

Primordial gravitational waves are gravitational waves observed in the cosmic microwave background. They were allegedly detected by the BICEP2 instrument, an announcement which was made on 17 March 2014.

24.6 Mathematics

Einstein's equations form the fundamental law of general relativity. The curvature of spacetime can be expressed mathematically using the metric tensor — denoted $g_{\mu\nu}$. The metric holds information regarding how distances are measured in the space under consideration. Because the propagation of gravitational waves through space and time change distances, we will need to use this to find the solution to the wave equation.

Spacetime curvature is also expressed with respect to a covariant derivative, ∇ , in the form of the Einstein tensor, $G_{\mu\nu}$. This curvature is related to the stress–energy tensor, $T_{\mu\nu}$, by the key equation

$$G_{\mu\nu} = \frac{8\pi G_N}{c^4} T_{\mu\nu},$$

where G_N is Newton's gravitational constant, and c is the speed of light. We assume geometrized units, so $G_N = 1 = c$.

With some simple assumptions, Einstein's equations can be rewritten to show explicitly that they are wave equations. To begin with, we adopt some coordinate system, like (t, r, θ, ϕ) . We define the "flat-space metric" $\eta_{\mu\nu}$ to be the quantity that — in this coordinate system — has the components we would expect for the flat space metric. For example, in these spherical coordinates, we have

$$\eta_{\mu\nu} = \begin{bmatrix} -1 & 0 & 0 & 0 \\ 0 & 1 & 0 & 0 \\ 0 & 0 & r^2 & 0 \\ 0 & 0 & 0 & r^2 \sin^2 \theta \end{bmatrix}.$$

This flat-space metric has no physical significance; it is a purely mathematical device necessary for the analysis. Tensor indices are raised and lowered using this "flat-space metric".

Now, we can also think of the physical metric $g_{\mu\nu}$ as a matrix, and find its determinant, $\det g$. Finally, we define a quantity

$$h^{\alpha\beta} \equiv \eta^{\alpha\beta} - \sqrt{|\det g|}g^{\alpha\beta}$$

This is the crucial field, which will represent the radiation. It is possible (at least in an asymptotically flat spacetime) to choose the coordinates in such a way that this quantity satisfies the "de Donder" gauge conditions (conditions on the coordinates):

$$\nabla_\beta h^{\alpha\beta} = 0,$$

where ∇ represents the flat-space derivative operator. These equations say that the divergence of the field is zero. The linear Einstein equations can now be written[46] as

$$\Box h^{\alpha\beta} = -16\pi\tau^{\alpha\beta}$$

where $\Box = -\partial_t^2 + \Delta$ represents the flat-space d'Alembertian operator, and $\tau^{\alpha\beta}$ represents the stress–energy tensor plus quadratic terms involving $h^{\alpha\beta}$. This is just a wave equation for the field with a source, despite the fact that the source involves terms quadratic in the field itself. That is, it can be shown that solutions to this equation are waves traveling with velocity 1 in these coordinates.

24.6.1   Linear approximation

The equations above are valid everywhere — near a black hole, for instance. However, because of the complicated source term, the solution is generally too difficult to find analytically. We can often assume that space is nearly flat, so the metric is nearly equal to the $\eta^{\alpha\beta}$ tensor. In this case, we can neglect terms quadratic in $h^{\alpha\beta}$, which means that the $\tau^{\alpha\beta}$ field reduces to the usual stress–energy tensor $T^{\alpha\beta}$. That is, Einstein's equations become

$$\Box h^{\alpha\beta} = -16\pi T^{\alpha\beta}$$

If we are interested in the field far from a source, however, we can treat the source as a point source; everywhere else, the stress–energy tensor would be zero, so

$$\Box h^{\alpha\beta} = 0$$

Now, this is the usual homogeneous wave equation — one for each component of $h^{\alpha\beta}$. Solutions to this equation are well known. For a wave moving away from a point source, the radiated part (meaning the part that dies off as $1/r$ far from the source) can always be written in the form $A(t-r,\theta,\phi)/r$, where A is just some function. It can be shown[47] that — to a linear approximation — it is always possible to make the field traceless. Now, if we further assume that the source is positioned at $r=0$, the general solution to the wave equation in spherical coordinates is

$$h^{\alpha\beta} = \frac{1}{r}\begin{bmatrix} 0 & 0 & 0 & 0 \\ 0 & 0 & 0 & 0 \\ 0 & 0 & A_+(t-r,\theta,\phi) & A_\times(t-r,\theta,\phi) \\ 0 & 0 & A_\times(t-r,\theta,\phi) & -A_+(t-r,\theta,\phi) \end{bmatrix}$$

$$\equiv \begin{bmatrix} 0 & 0 & 0 & 0 \\ 0 & 0 & 0 & 0 \\ 0 & 0 & h_+(t-r,r,\theta,\phi) & h_\times(t-r,r,\theta,\phi) \\ 0 & 0 & h_\times(t-r,r,\theta,\phi) & -h_+(t-r,r,\theta,\phi) \end{bmatrix}$$

where we now see the origin of the two polarizations.

24.6.2   Relation to the source

If we know the details of a source — for instance, the parameters of the orbit of a binary — we can relate the source's motion to the gravitational radiation observed far away. With the relation

$$\Box h^{\alpha\beta} = -16\pi T^{\alpha\beta}$$

we can write the solution in terms of the tensorial Green's function for the d'Alembertian operator:[46]

$$h^{\alpha\beta}(t, x) = -16\pi \int G^{\alpha\beta}_{\gamma\delta}(t, x; t', x') \, T^{\gamma\delta}(t', x') \, dt' \, d^3x'$$

Though it is possible to expand the Green's function in tensor spherical harmonics, it is easier to simply use the form

$$\frac{1}{4\pi}^{\gamma}_{\alpha}{}^{\delta}_{\beta} \overline{\delta t + \pi |x - x'| / L t}$$

where the positive and negative signs correspond to ingoing and outgoing solutions, respectively. Generally, we are interested in the outgoing solutions, so

$$h^{\alpha\beta}(t, x) = -4 \int \frac{T^{\alpha\beta}(t - |x - x'| / x)}{|x - x'|} \, d^3x'$$

If the source is confined to a small region very far away, to an excellent approximation we have:

$$\bar{h}^{\alpha\beta}(t, x) \approx -\frac{4}{r} \int T^{\alpha\beta}(t - r, x) \, d^3x'$$

where $r = |x|$.

Now, because we will eventually only be interested in the spatial components of this equation (time components can be set to zero with a coordinate transformation), and we are integrating this quantity — presumably over a region of which there is no boundary — we can put this in a different form. Ignoring divergences with the help of Stokes' theorem and an empty boundary, we can see that

$$\int T^{ij}(t - r, x) \, d^3x' = \int x'^i x'^j \nabla_k \nabla_l T^{kl}(t - r, x) \, d^3x'$$

Inserting this into the above equation, we arrive at

$$\bar{h}^{ij}(t, x) \approx -\frac{4}{r} \int x'^i x'^j \nabla_k \nabla_l T^{kl}(t - r, x) \, d^3x'$$

Finally, because we have chosen to work in coordinates for which $\nabla_\beta h^{\alpha\beta} = 0$, we know that $\nabla_\beta T^{\alpha\beta} = 0$. With a few simple manipulations, we can use this to prove that

$$\nabla_0 \nabla_0 T^{00} = \nabla_j \nabla_k T^{jk}$$

With this relation, the expression for the radiated field is

$$\bar{h}^{ij}(t, \vec{x}) \approx -\frac{4}{r}\frac{d^2}{dt^2} \int x'^i x'^j T^{00}(t - r, \vec{x}')\, d^3x'$$

In the linear case, $T^{00} = \rho$, the density of mass-energy.

To a very good approximation, the density of a simple binary can be described by a pair of delta-functions, which eliminates the integral. Explicitly, if the masses of the two objects are M_1 and M_2, and the positions are $\vec{x}_1$ and $\vec{x}_2$, then

$$\rho(t - r, \vec{x}') = M_1 \delta(\vec{x}' - \vec{x}_1(t - r)) + M_2 \delta(\vec{x}' - \vec{x}_2(t - r))$$

We can use this expression to do the integral above:

$$\dots \qquad \frac{4}{r}\frac{d^2}{dt^2} \left\{ \qquad \qquad \qquad \qquad \qquad \right\}$$

Using mass-centered coordinates, and assuming a circular binary, this is

$$\bar{h}^{ij}(t, \vec{x}) \approx -\frac{4}{r}\frac{M_1 M_2}{\qquad} \dots$$

where $\vec{n} = \vec{x}_1 / |\vec{x}_1|$. Plugging in the known values of $\vec{x}_1(t - r)$, we obtain the expressions given above for the radiation from a simple binary.

24.7 See also

- Gravitational wave background
- Cosmic gravitational wave background
- Big Bang Observer (BBO), proposed successor to LISA
- DECIGO "Deci-hertz Interferometer Gravitational wave Observatory", the planned laser interferometric detector in space
- Gravitational field
- Gravitomagnetism
- Graviton
- Gravitational wave astronomy
- Hawking radiation, for gravitationally induced electromagnetic radiation from black holes
- HM Cancri
- LIGO, VIRGO, GEO 600, and TAMA 300 — Gravitational wave detectors
- Linearised Einstein field equations
- LISA the proposed Laser Interferometer Space Antenna
- Peres metric

- pp-wave spacetime, for an important class of exact solutions modelling gravitational radiation

- Spin-flip, a consequence of gravitational wave emission from binary supermassive black holes

- Sticky bead argument, for a physical way to see that gravitational radiation should carry energy

- Tidal force

24.8 References

[1] Einstein, A (June 1916). "Näherungsweise Integration der Feldgleichungen der Gravitation". *Sitzungsberichte der Königlich Preussischen Akademie der Wissenschaften Berlin*. part 1: 688–696.

[2] Einstein, A (1918). "Über Gravitationswellen". *Sitzungsberichte der Königlich Preussischen Akademie der Wissenschaften Berlin*. part 1: 154–167.

[3] Finley, Dave. "Einstein's gravity theory passes toughest test yet: Bizarre binary star system pushes study of relativity to new limits.". Phys.Org.

[4] The Detection of Gravitational Waves using LIGO, B. Barish

[5] "First Second of the Big Bang". *How The Universe Works 3*. 2014. Discovery Science.

[6] Staff (17 March 2014). "BICEP2 2014 Results Release". *National Science Foundation*. Retrieved 18 March 2014.

[7] "First Direct Evidence of Cosmic Inflation". *http://www.cfa.harvard.edu". Harvard-Smithsonian Center for Astrophysics. 17 March 2014. Retrieved 17 March 2014*.

[8] Clavin, Whitney (17 March 2014). "NASA Technology Views Birth of the Universe". *NASA*. Retrieved 17 March 2014.

[9] Overbye, Dennis (17 March 2014). "Detection of Waves in Space Buttresses Landmark Theory of Big Bang". *New York Times*. Retrieved 17 March 2014.

[10] Cosmic inflation: 'Spectacular' discovery hailedAstronomers discover echoes from expansion after Big BangGravitational Waves: The Big Bang's Smoking GunGravitational Waves from Big Bang Detected

[11] Overbye, Dennis (24 March 2014). "Ripples From the Big Bang". *New York Times*. Retrieved 24 March 2014.

[12] Overbye, Dennis (19 June 2014). "Astronomers Hedge on Big Bang Detection Claim". *New York Times*. Retrieved 20 June 2014.

[13] Amos, Jonathan (19 June 2014). "Cosmic inflation: Confidence lowered for Big Bang signal". *BBC News*. Retrieved 20 June 2014.

[14] Ade, P.A.R. et al. (BICEP2 Collaboration) (19 June 2014). "Detection of B-Mode Polarization at Degree Angular Scales by BICEP2" (PDF). *Physical Review Letters*112: 241101. arXiv:1403.3985. Bibcode:2014PhRvL.112x1101A. doi:10.1103/PhysRevLett.112.241101. PMID 24996078. Retrieved 20 June 2014.

[15] Planck Collaboration Team (19 September 2014). "Planck intermediate results. XXX. The angular power spectrum of polarized dust emission at intermediate and high Galactic latitudes". *ArXiv*. arXiv:1409.5738. Bibcode:2014arXiv1409.5738P. Retrieved 22 September 2014.

[16] Overbye, Dennis (22 September 2014). "Study Confirms Criticism of Big Bang Finding". *New York Times*. Retrieved 22 September 2014.

[17] Clavin, Whitney (30 January 2015). "Gravitational Waves from Early Universe Remain Elusive". *NASA*. Retrieved 30 January 2015.

[18] Overbye, Dennis (30 January 2015). "Speck of Interstellar Dust Obscures Glimpse of Big Bang". *New York Times*. Retrieved 31 January 2015.

[19] LIGO Scientific Collaboration; Virgo Collaboration (2012). "Search for Gravitational Waves from Low Mass Compact Binary Coalescence in LIGO's Sixth Science Run and Virgo's Science Runs 2 and 3". *Physical Review D*85: 082002. arXiv:1111.7314. Bibcode:2012PhRvD..85h2002A. doi:10.1103/PhysRevD.85.082002.

[20] LIGO Scientific Collaboration; Virgo Collaboration (2012). "All-sky search for gravitational-wave bursts in the second joint LIGO-Virgo run". *Physical Review D* 85: 122007. arXiv:1202.2788. Bibcode:2012PhRvD..85l2007A. doi:10.1103/PhysRevD.85.122007.

[21] LIGO Scientific Collaboration; Virgo Collaboration (2013). "Search for gravitational waves from binary black hole inspiral, merger, and ringdown in LIGO-Virgo data from 2009-2010". *Physical Review D* 87: 022002. arXiv:1209.6533. Bibcode:2013PhRvD..87b2002A. doi:10.1103/PhysRevD.87.022002.

[22] Krauss, LM; Dodelson, S; Meyer, S (2010). "Primordial Gravitational Waves and Cosmology". *Science* 328 (5981): 989–992. arXiv:1004.2504. Bibcode:2010Sci...328..989K. doi:10.1126/science.1179541. PMID 20489015.

[23] Hawking, S. W. and Israel, W., *General Relativity: An Einstein Centenary Survey*, Cambridge University Press, Cambridge, 1979, 98.

[24] Landau, L. D. and Lifshitz, E. M., *The Classical Theory of Fields*. Fourth Revised English Edition, Pergamon Press., 1975, 356–357.

[25] Einstein, A (1918). "Über Gravitationswellen". *Sitzungsberichte, Preussische Akademie der Wissenschaften* 154.

[26] Gravitational Radiation

[27] LIGO Scientific Collaboration; Virgo Collaboration (2010). "Predictions for the rates of compact binary coalescences observable by ground-based gravitational-wave detectors". *Classical and Quantum Gravity* 27: 17300. arXiv:1003.2480. Bibcode: doi:10.1088/0264-9381/27/17/173001.

[28] Relativistic Binary Pulsar B1913+16: Thirty Years of Observations and Analysis

[29] The discovery of the first binary pulsar

[30] Crashing Black Holes

[31] Binary and Millisecond Pulsars

[32] L. P. Grishchuk (1976), "Primordial Gravitons and the Possibility of Their Observation", Sov. Phys. JETP Lett. 23, p. 293.

[33] Braginsky, V. B., Rudenko and Valentin, N. Section 7: "Generation of gravitational waves in the laboratory", *Physics Report* (Review section of *Physics Letters*), 46, No. 5. 165–200, (1978).

[34] Li, Fangyu, Baker, R. M L, Jr., and Woods, R. C., "Piezoelectric-Crystal-Resonator High-Frequency Gravitational Wave Generation and Synchro-Resonance Detection", in the proceedings of *Space Technology and Applications International Forum (STAIF-2006)*, edited by M.S. El-Genk, American Institute of Physics Conference Proceedings, Melville NY 813: 2006.

[35] Merritt, D. et al. (May 2004). "Consequences of Gravitational Wave Recoil". *The Astrophysical Journal Letters* 607 (1): L9–L12. arXiv:astro-ph/0402057. Bibcode:2004ApJ...607L...9M. doi:10.1086/421551.

[36] Gualandris, A.; Merritt, D. et al. (May 2008). "Ejection of Supermassive Black Holes from Galaxy Cores". *The Astrophysical Journal* 678 (2): 780–797. arXiv:0708.0771. Bibcode:2008ApJ...678..780G. doi:10.1086/586877.

[37] Merritt, D.; Schnittman, J. D.; Komossa, S. (2009). "Hypercompact Stellar Systems Around Recoiling Supermassive Black Holes". *The Astrophysical Journal* 699 (2): 1690–1710. arXiv:0809.5046. Bibcode:2009ApJ...699.1690M. doi:10.1088/0004-637X/699/2/1690.

[38] Komossa, S.; Zhou, H.; Lu, H. (May 2008). "A Recoiling Supermassive Black Hole in the Quasar SDSS J092712.65+294344.0?". *The Astrophysical Journal* 678 (2): L81–L84. arXiv:0804.4585. Bibcode:2008ApJ...678L..81K. doi:10.1086/588656

[39] Thorne, Kip S. (1995). "Gravitational Waves". *Cornell University Library*.

[40] David G. Blair (Ed.) (1991). *The detection of gravitational waves*. Cambridge University Press.

[41] For a review of early experiments using Weber bars, see Levine, J. (April 2004). "Early Gravity-Wave Detection Experiments, 1960–1975". *Physics in Perspective (Birkhäuser Basel)* 6 (1): 42–75. Bibcode:2004PhP.....6...42L. doi:10.1007/s00016-003-0179-6.

[42] Gravitational Radiation Antenna In Leiden

[43] de Waard, Arlette; Luciano Gottardi; Giorgio Frossati (July 2000). *Spherical Gravitational Wave Detectors: cooling and quality factor of a small CuAl6% sphere* (PDF). Marcel Grossmann meeting on General Relativity. Rome, Italy: World Scientific Publishing Co. Pte. Ltd. (published December 2002). pp. 1899–1901. Bibcode:2002nmgm.meet.1899D. doi: ISBN 9789812777386.

[44] The idea of using laser interferometry for gravitational wave detection was first mentioned by Gerstenstein and Pustovoit 1963 Sov. Phys.–JETP 16 433. Weber mentioned it in an unpublished laboratory notebook. Rainer Weiss first described in detail a practical solution with an analysis of realistic limitations to the technique in R. Weiss (1972). "Electromagetically Coupled Broadband Gravitational Antenna". Quarterly Progress Report, Research Laboratory of Electronics, MIT 105: 54.

[45] Einstein@Home

[46] Thorne, Kip (April 1980). "Multipole expansions of gravitational radiation". *Reviews of Modern Physics* **52** (2): 299–339. Bibcode:1980RvMP...52..299T. doi:10.1103/RevModPhys.52.299.

[47] C. W. Misner, K. S. Thorne, and J. A. Wheeler (1973). *Gravitation*. W. H. Freeman and Co.

24.9 Further reading

- Chakrabarty, Indrajit, "Gravitational Waves: An Introduction". arXiv:physics/9908041 v1, Aug 21, 1999.

- Landau, L. D. and Lifshitz, E. M., The Classical Theory of Fields (Pergamon Press),(1987).

- Will, Clifford M., *The Confrontation between General Relativity and Experiment*. Living Rev. Relativity 9 (2006) 3.

- Peter Saulson, "Fundamentals of Interferometric Gravitational Wave Detectors", World Scientific, 1994.

- J. Bicak, W.N. Rudienko, "Gravitacionnyje wolny w OTO i probliema ich obnarużenija", Izdatielstwo Moskovskovo Universitieta, 1987.

- A. Kułak, "Electromagnetic Detectors of Gravitational Radiation", PhD Thesis, Cracow 1980 (In Polish).

- P. Tatrocki, "On intuitive description of graviton detector", www.philica.com .

- P. Tatrocki, "Can the LIGO, VIRGO, GEO600, AIGO, TAMA, LISA detectors really detect?", www.philica.com .

24.10 Bibliography

- Berry, Michael, *Principles of cosmology and gravitation* (Adam Hilger, Philadelphia, 1989). ISBN 0-85274-037-9

- Collins, Harry, *Gravity's Shadow: the search for gravitational waves*, University of Chicago Press, 2004.

- P. J. E. Peebles, *Principles of Physical Cosmology* (Princeton University Press, Princeton, 1993). ISBN 0-691-01933-9.

- Wheeler, John Archibald and Ciufolini, Ignazio, *Gravitation and Inertia* (Princeton University Press, Princeton, 1995). ISBN 0-691-03323-4.

- Woolf, Harry, ed., *Some Strangeness in the Proportion* (Addison–Wesley, Reading, Massachusetts, 1980). ISBN 0-201-09924-1.

24.11 External links

- Gravitational waves at *Encyclopædia Britannica*

-

- Gravitational Waves on *In Our Time* at the BBC. (listen now)

- The LISA Brownbag – Selection of the most significant e-prints related to LISA science

- Astroparticle.org. To know everything about astroparticle physics, including gravitational waves

- Caltech's Physics 237-2002 Gravitational Waves by Kip Thorne **Video plus notes:** Graduate level but does not assume knowledge of General Relativity, Tensor Analysis, or Differential Geometry; Part 1: Theory (10 lectures), Part 2: Detection (9 lectures)

- www.astronomycast.com January 14, 2008 Episode 71: Gravitational Waves

- Laser Interferometer Gravitational Wave Observatory. LIGO Laboratory, operated by the California Institute of Technology and the Massachusetts Institute of Technology

- The LIGO Scientific Collaboration

- Einstein's Messengers – The LIGO Movie by NSF

- Home page for Einstein@Home project, a distributed computing project processing raw data from LIGO Laboratory, searching for gravitational waves

- The National Center for Supercomputing Applications – a numerical relativity group

- Caltech Relativity Tutorial – A basic introduction to gravitational waves, and astrophysical systems giving off gravitational waves

- Resource Letter GrW-1: Gravitational waves – a list of books, journals and web resources compiled by Joan Centrella for research into gravitational waves

- Mathematical and Physical Perspectives on Gravitational Radiation – written by B F Schutz of the Max Planck Institute explaining the significance and background of some key concepts in gravitational radiation

- Binary BH Merger – estimating the radiated power and merger time of a BH binary using dimensional analysis.